1st International Conference of Education on Sciences, Technology, Engineering, and Mathematics (ICE-STEM)

Journal of Physics: Conference Series Volume 948

Jakarta, Indonesia
17 - 19 October 2017

Editors:

Yoppy Wahyu Purnomo
Liszulfa Roza
Tian Abdul Aziz
Krisna Satria Perbowo
Syafika Ulfah

ISBN: 978-1-5108-5848-0
ISSN: 1742-6588

Printed from e-media with permission by:

Curran Associates, Inc.
57 Morehouse Lane
Red Hook, NY 12571

Some format issues inherent in the e-media version may also appear in this print version.

Printed by Curran Associates, Inc. (2018)

For permission requests, please contact the Institute of Physics at the address below.

Institute of Physics
Dirac House, Temple Back
Bristol BS1 6BE UK

Phone: 44 1 17 929 7481
Fax: 44 1 17 920 0979

techtracking@iop.org

Additional copies of this publication are available from:

Curran Associates, Inc.
57 Morehouse Lane
Red Hook, NY 12571 USA
Phone: 845-758-0400
Fax: 845-758-2633
Email: curran@proceedings.com
Web: www.proceedings.com

Table of contents

Volume 948

1st International Conference of Education on Sciences, Technology, Engineering, and Mathematics (ICE-STEM)

17–19 October 2017, Jakarta, Indonesia

Accepted papers received: 18 December 2017
Published online: 19 February 2018

Preface

Papers

Proceeding of 1st International Conference of Education on Science, Technology, Engineering, and Mathematics (1st ICE-STEM) 2017

Preface

Science, Technology, Engineering, and Mathematics (STEM) has become a remarkable headline in todays' world. It is an educational program developed to prepare students in improving competitiveness in the fields of science, technology, engineering, and mathematics. It aims to foster inquiring minds, logical reasoning, and collaboration skills. Mathematics as the queen of science can support the blooming STEM for better education in the 21st century. Mathematics pervades every part of our dynamic lives. To succeed in this new information-based and highly technological society, developing mathematics proficiency is the inevitable consequence. Hence, the role of mathematics in STEM development is essential. It is needed a forum to discuss the role of mathematics in STEM development. Therefore, the First International Conference of Education on Science, Technology, Engineering, and Mathematics (ICE-STEM) held and took place in Jakarta on October 17th-19th, 2017. The conference was organized by University of Muhammmadiyah Prof. DR. Hamka (UHAMKA) in collaboration with Indonesian Mathematical Society (Indo-MS).

There were 186 participants from countries all over the world attended the conference. The scientific program consisted of in total 136 talks, a big part of them presented in 15 mini-symposia. Four talks were invited plenary lecturers given by Budi Nurani Ruchjana (Indonesia), Somporn Chuai-Aree (Thailand), Kamisah Osman (Malaysia), and Imam Robandi (Indonesia). Topics included pure and applied mathematics, science, technology engineering, and science, technology, engineering, and mathematics education.

We would like to express our appreciation to many people who contributed to the success of the conference: the plenary and keynote speakers, the authors, the participants, the session chairs, and the members of the Committees who nominated plenary and keynote speakers. The editors are especially grateful to those who reviewed the manuscripts included in this special issue.

The Editors

Yoppy Wahyu Purnomo
Liszulfa Roza
Tian Abdul Aziz
Krisna Satria Perbowo
Syafika Ulfah

on Science,
Technology, Engineering and Mathematics (ICE-STEM)
OCTOBER 17th. 2017 UHAMKA JAKARTA
Theme :
“ Mathematics for all : The Role of Mathematics in STEM Developme

IOP Conf. Series: Journal of Physics: Conf. Series **948** (2018) 011001 doi:10.1088/1742-6596/948/1/011001

IOP Conf. Series: Journal of Physics: Conf. Series **948** (2018) 011001 doi:10.1088/1742-6596/948/1/011001

IOP Conf. Series: Journal of Physics: Conf. Series **948** (2018) 011001 doi:10.1088/1742-6596/948/1/011001

Proceeding of 1st International Conference of Education on Science, Technology, Engineering, and Mathematics (1st ICE-STEM) 2017

General Chair
Drs. Slamet Soro, M.Pd (Department of Mathematics Education, Universitas Muhammadiyah Prof. DR. HAMKA, Indonesia)

Advisory Committee
Assoc. Prof. Dr. Intan Muchtadi-Alamsyah. Institut Teknologi Bandung, Indonesia
Prof. Dr. Hans-Stefan Siller, Universitat Koblenz-Landau, Germany
Assoc. Prof. Tole Sutikno, Universitas Ahmad Dahlan, Indonesia
Prof. Anton Satria Prabuwono, Ph.D., King Abdul Aziz University, Arab Saudi
Prof. Dr. Khairurrijal, Institut Teknologi Bandung, Indonesia
Dr. Wanty Widjaja, Deakin University, Australia

Scientific Committee / Reviewers
Dr. Rattikan Saelim (*Prince of Songkla University, Thailand*)
Dr. Ridwan Maulana (*University of Groningen, Netherlands*)
Dr. Somporn Chuai-Aree (*Prince of Songkla University, Thailand*)
Dr. M. Shahrill (*Universiti Brunei Darussalam, Brunei*)
Dr. Liszulfah Roza, M.I.S. (*Universitas Muhammadiyah Prof. DR. HAMKA, Indonesia*)
Prof. Dr. Agus Suryanto (*Universitas Brawijaya, Indonesia*)
Tian Abdul Aziz, Ph.D. (*Universitas Muhammadiyah Prof. DR. HAMKA, Indonesia*)
Prof. Dr. (Eng.) Imam Robandy MT. (*Institut Teknologi Surabaya, Indonesia*)
Dr. Yoppy Wahyu Purnomo (*Universitas Muhammadiyah Prof. DR. HAMKA, Indonesia*)
Dr. Ervin Azhar, S.Si. M.Si. (*Universitas Muhammadiyah Prof. DR. HAMKA, Indonesia*)
Assoc. Prof. Dr. Kiki A. Sugeng (*Universitas Indonesia, Indonesia*)
Dr. Budi Akbar, M.Si. (*Universitas Muhammadiyah Prof. DR. HAMKA, Indonesia*)
Dr. Sigid Edy Purwanto, M.Pd. (*Universitas Muhammadiyah Prof. DR. HAMKA, Indonesia*)
Harry Ramza, MT., Ph.D. (*Universitas Muhammadiyah Prof. DR. HAMKA, Indonesia*)
Yuli Rahmawati, Ph.D. (*Universitas Negeri Jakarta, Indonesia*)
Dr. Dan Mugisidi, MT. (*Universitas Muhammadiyah Prof. DR. HAMKA, Indonesia*)

Public Relation & Secretarial Staffs
Supiat, M.Pd.
Luthfi Sapahi, M.Pd.
Gufron Amirullah, M.Pd.
Sulaiman, S.Pd.
Ambar Nugraheni, M.Pd.
Nurlina Rahman, S.Pd., M.Si.

Ceremony
Dr. Ishaq Nuriadin, M.Pd.
Isnaini Handayani, M.Pd.
Puri Pramudiani, S.Pd., M.Sc.
Drs. M. Soenarto, M.Si.
Abdurrahman Jupri, M.Pd.
Syarif Hidayatullah, M.Pd.
Bitha Dwi Rahmani, M.Pd.

Fundrisers & Sponsorship
Yunda Kurniawan, M.Pd.
Muntazhimah Nasution, M.Pd.
Drs. H. Hartono, M.Pd

Wira Adi Nugroho, S.Pd, M.Si
Hendrik, S.Pd
Faris Ning Hasan, S.Pd

Logistic
Ayu Tsurayya, S.Pd., M.Si.
Leni Marlena, S.Stat., M.Si.
Musriana Retnaningsih, M.Pd.
Hikmatul Husna, M.Pd.
Asih Miatun, M.Pd.
Nok Yeni Heryaningsih
Nurafni, M. Pd., M.Pd.
Windia Hadi, M.Pd.

Papers & Publications
Dr. Yoppy Wahyu Purnomo
Dr. Liszulfah Roza, M.I.S.
Tian Abdul Aziz, Ph.D.
Subhan Azis Awalludin, S.Pd, M.Sc
Syafika Ulfah, S.Pd, M. Sc
Krisna Satrio Perbowo, M.Pd.
Trisna Roy Pradipta, S.Pd., M.PMat.

Website Development
Dr. Sugema, M.Kom
Benny Hendriana, M.Pd.
Risqi Rahman, M.Pd.
Khoerul Umam, M.Pd.
Mirzanur Hidayat, M.Si.

Presentation & Program
Dr. Budi Akbar, M.Si.
Dr. Acep Kusdiwelirawan, M.Pd.
Dra. Imas Ratna, M.Sc
Dra. Maryanti Setianingsih, M.Pd
Kowiyah, M.Pd
Felicianda, S.Pd., M.Si.

Accomodation & Supporting Team
Samsul Maarif, M.Pd.
Sukirno, S.Ag.
Edi Supriadi, S.Pd.
Devi Anugrah, M.Pd.
Musodik, M.Pd.

Peer review statement

All papers published in this volume of *Journal of Physics: Conference Series* have been peer reviewed through processes administered by the proceedings Editors. Reviews were conducted by expert referees to the professional and scientific standards expected of a proceedings journal published by IOP Publishing.

The effect of creative problem solving on students' mathematical adaptive reasoning

A Muin[1], S H Hanifah[1] and F Diwidian[1]

[1]Universitas Islam Negeri Syarif Hidayatullah, Jakarta, Indonesia

E-mail: abdul.muin@uinjkt.ac.id

Abstract. This research was conducted to analyse the effect of creative problem solving (CPS) learning model on the students' mathematical adaptive reasoning. The method used in this study was a quasi-experimental with randomized post-test only control group design. Samples were taken as many as two classes by cluster random sampling technique consisting of experimental class (CPS) as many as 40 students and control class (conventional) as many as 40 students. Based on the result of hypothesis testing with the t-test at the significance level of 5%, it was obtained that significance level of 0.0000 is less than $\alpha = 0.05$. This shows that the students' mathematical adaptive reasoning skills who were taught by CPS model were higher than the students' mathematical adaptive reasoning skills of those who were taught by conventional model. The result of this research showed that the most prominent aspect of adaptive reasoning that could be developed through a CPS was inductive intuitive. Two aspects of adaptive reasoning, which were inductive intuitive and deductive intuitive, were mostly balanced. The different between inductive intuitive and deductive intuitive aspect was not too big. CPS model can develop student mathematical adaptive reasoning skills. CPS model can facilitate development of mathematical adaptive reasoning skills thoroughly.

1. Introduction

The mathematical adaptive reasoning is a very important reasoning in learning. The mathematical adaptive reasoning is one of the mathematical skills that must be possessed by students to demonstrate their learning ability. This mathematical skill is not the innate skill of the students, but the combination of knowledge and the abilities and beliefs that students gain with the help of teachers and other learning environments. Hayat and Yusuf state that reasoning ability is one of the most important competencies required today and in the future in mathematics learning [1]. Kilpatrick, Swafford, and Findell argue that adaptive reasoning involves not only deductive reasoning that only concludes based on formal proofing deductively, but adaptive reasoning also includes intuition and inductive reasoning with conclusions based on patterns, analogies, and metaphors [2].

Adaptive reasoning has a wider scope than general reasoning that only includes inductive and deductive reasoning only, because in the process of adaptive reasoning also involves intuition. The adaptive reasoning is a necessary part to support success in the mathematics learning process. Adaptive reasoning needs to be continuously trained and developed so that math learning becomes more meaningful and achieves the expected learning objectives. Through adaptive reasoning, students are expected to be able to solve the problems quickly and precisely. Students will build their minds to master the concept of mathematics intact both for now, later and become the basis of students in acting logically in math activities, or in other daily activities.

Some aspects of adaptive reasoning are inductive and deductive reasoning associated with intuition. Inductive reasoning is a general conclusion based on observed data, whereas deductive reasoning is a conclusion based on agreed rules [3]. According to Copi, deductive reasoning is the process of inferences that are derived in absolute terms according to statements that are not influenced by other factors [4]. Intuitive knowledge is a kind of knowledge that is not based on sufficient empirical evidence or a strictly logical argument and although like that, it is nevertheless accepted with confidence and clearness [5]. While intuitive understanding occurs if someone can predict or suspect the truth of something without hesitation and without first analyzing analytically [6].

To train mathematical adaptive reasoning, students need to be given problems that facilitate them to develop the ability to adapt the situation given. Students are given non-routine problems that train their creativity in finding solutions. It's Quite logical if they are guided by creative problem-solving learning. Problem-solving in mathematics mostly refers to Polya, which includes understanding the problem, devising a plan, carrying out the plan, and looking back [7]. Creative problem solving comes from the words creative, problem, and solving. Creative means many new and unique ideas in creating solutions and having value and relevance, the problem means a situation that provides challenges, opportunities, interrelated, while solving means planning a way to answer from a problem [8].

Creative Problem Solving learning model is the development of problem-solving learning. Stages in this learning model represent systematic procedures in identifying challenges, creating ideas, and implementing innovative solutions. Through the practice and implementation of the process on an ongoing basis, students can strengthen their creative techniques and adaptive reasoning and learn to apply them in new situations. Creative techniques need an intuitive ability. This learning model is not like the problem-solving method in general, this model prefers to ideas given and then there is no immediate final decision or there is still a delay of the solution [9].

The role of teachers in learning that apply creative problem-solving learning model puts more of them as facilitator (to facilitate students in learning process), motivator (giving motivation to students in learning activities), and dynamicator (teacher try to give stimulation in searching, collecting, and determining information for troubleshooting) [10]. During the learning process using the CPS learning model, students are given wide opportunity to solve the given problems. It is a prerequisite for students who want to learn independently. The goal to be achieved is that students are directed actively in building their knowledge through direct experience so that the expected objectives of mathematics learning can be achieved well.

This research was conducted to analyze the effects of creative problem-solving learning model on the students' mathematical adaptive reasoning. The student's mathematical adaptive reasoning skills in this study are the ability to think logically about the relationship between concepts and situations through intuitive-inductive and intuitive-deductive reasoning. Intuitive capabilities and either inductive or deductive reasoning are not a sequence, so intuitive ability can be in any part of the inductive or deductive reasoning process. The Intuitive-inductive reasoning is the conclusion from the special to the general that involves the process of intuition. The Intuitive-deductive reasoning is the conclusion from a general to a specific involving the intuition process. The process of intuition is a process or activity for guessing, establishing something with or without the use of representational assistance but without first proving or explaining formally. However, in this study focused on two indicators namely, the ability to guess and make general conclusions based on the amount of data observed (inductive-intuitive) and the ability to guess and make logical conclusions (deductive-intuitive). The method used in this study was a quasi-experimental with randomized posttest-only control group design. Cluster random sampling technique consisting of experimental class (CPS) and control class (conventional).

2. Methods and material

The method used in this study was a quasi-experimental with randomized posttest-only control group design. Cluster random sampling technique consisting of experimental class (CPS) as many as 40 students and control class (conventional) as many as 40 students. In the experimental group given treatment in the form of learning by using Creative Problem-Solving learning model, while the control group is given treatment by using conventional learning that is expository learning. After a certain period of time, both groups are observed, then given a posttest to know the difference of ability between the two groups. If there is a difference between the experimental group and the control group, it is can be concluded that the difference is due to the treatment was given. The design of experiment if R means random, X means treatment (CPS), and O means posttest, is as follows [11]:

R	X	O
R		O

Based on the explanation before, it can be formulated the learning with creative problem solving in this research, that is a model of problem-solving learning that emphasizes the discovery of various ideas or ideas that involve divergent and convergent thinking processes to find solutions in the form of the most efficient solution of a problem. Divergent thinking is generating ideas based on intuition in solving problems. Convergent thinking is a decision-making activity on the idea. The creative problem-solving stages used in this study were adapted from Mitchell and Kowalik [12], namely:

- Fact-finding,
  at this stage, students enroll all known facts related to the problem to be solved and look for other data required by involving divergent and convergent thinking processes.
- Problem finding,
  at this stage, the student identifies the problem statement and determines the important points underlying the problem by involving divergent and convergent thinking processes.
- Idea finding,
  at this stage, students are encouraged to develop a number of ideas that might be used to solve problems by involving divergent thinking processes.
- Solution finding,
  at this stage, students select the ideas acquired to find the most appropriate ideas in solving problems in a systematically way by involving divergent and convergent thinking processes.
- Acceptance finding,
  at this stage, students attempt to obtain the acceptance of problems solutions that have been found or to ensure successful solutions and to implement them by involving divergent and convergent thinking processes.

The instrument used in this research is a test to measure the students' mathematical adaptive reasoning, that is, the questions of 6 essays given in the form of posttest. This test instrument is given to the experimental class and control class on the rectangular and triangular subjects, where the tests given to the two classes are the same. The data needed in this research is the scores of students' mathematical adaptive reasoning in learning mathematics. Data collection is conducted by using the test technique, that is, the test of mathematical adaptive reasoning ability. The test is divided into two parts namely, a 10-minute for intuitive ability test and a 70-minute for reasoning ability test. Both parts of the test are conducted on the same day in each class. The test of mathematical adaptive reasoning is given to the experimental class, which in the learning process is given treatment with the CPS learning model, and control class is given conventional treatment that is by expository learning. The mathematical adaptive reasoning tests provided consist of six questions of the essay.

Instruments first tested before use so get a good instrument. This trial is intended to obtain the validity of items, the level of difficulty, the discriminant index, and the reliability of the instrument. The instrument was made of six questions about the ability of mathematical adaptive reasoning. After analyzing with statistical method, all items are valid. The level of difficulty analysis gives the result that the six items tested are at the medium level of difficulty. The analysis of Discriminant index gives the result that three of six items tested, between upper and lower class have the discriminant index in the medium category, two items are in a bad category and the others are in the good category. The reliability analysis gives the results that the tried instrument has a sufficient of reliability coefficient of 0.573, which means that this instrument is enough to measure the ability of mathematical adaptive reasoning.

The collected data from both the control class and the experimental class were processed and analyzed to explain the problem under study. The overall of data processing from the normality test to the test of equality of two means was performed using SPSS software. The data analysis used was the test of two means equality with a technique of Independent Samples T-Test. The two-mean equality test is performed on the overall mathematical adaptive reasoning test results. Before testing the hypothesis first tested the normality and homogeneity as the assumed prerequisite for the t-test. After the analysis of prerequisite test was performed, the distribution of the average score of mathematical adaptive reasoning ability in the experimental class as well as the controls was the normal distribution and had homogeneous variance. Therefore, to test the equality of two means is used Independent-Samples T Test analysis.

3. Results and discussion

In particular, the students' mathematical adaptive reasoning scores in both the experimental and control classes based on their indicators are presented in Table 1.

Table 1. The Comparison of Students Mathematical Adaptive Reasoning

Indicator	Ideal Score	Experiment (CPS)			Control (Expository)		
		Student Score	$\overline{x}$	%	Student score	$\overline{x}$	%
Intuitive-Inductive	12	401	10.03	83.54	335	8.38	69.79
Intuitive-Deductive	12	375	9.38	78.13	325	8.13	67.71
Overall	24	776	19.40	80.83	660	16.50	68.75

The percentage of mathematical adaptive reasoning scores obtained by the experimental class (CPS) is higher than the percentage of mathematical adaptive reasoning scores of control class for each indicator, both of intuitive-inductive or intuitive-deductive capabilities. The difference in intuitive-inductive capability in both classes is 13.75%, whereas in the intuitive-deductive ability it has a difference of 10.42% in both classes. The percentage of the overall score of mathematical adaptive reasoning of the experimental class is 12.08% higher than the control class.

Based on the result of hypothesis testing with the t-test at significance level of 5% as shown in Table 2, it was obtained that significance level of 0.0000 is less than $\alpha = 0.05$.

Table 2. The equality test of two means of mathematical adaptive reasoning

	Levene's Test for Equality of Variances		t-test for Equality of Means		
	F	Sig.	T	df	Sig. (2-tailed)
Equal variances assumed	3.246	0.075	4.940	78	.000

The result of homogeneity test using Levene's test, at the significant level of $\alpha = 0.05$ shows that the data of posttest score of mathematical adaptive reasoning ability of experiment class and control class is homogeneous, this is obtained by comparing the value of significance of calculation result with α which has been determined. The significance value listed on the homogeneity test results (sig. level = 0.075) is greater than the price of $\alpha = 0.05$.

Testing of normality and homogeneity has shown that posttest scores of mathematical adaptive reasoning abilities in both groups are normally distributed and the variants of both groups are also equal or homogeneous, therefore testing of two means equality can be performed using the Independent Samples T-Test analysis. The result of the equality test of the two means of the experimental class and the control for the mathematical adaptive reasoning ability indicates to receive the alternative hypotheses and reject null hypotheses. The alternative hypotheses stated that the mean of students mathematical adaptive reasoning ability whose learning using creative problem-solving learning model is higher than the mean of students mathematical adaptive reasoning ability using expository learning at 95% confidence level. It can be identified from the significance value (sig. level = 0.000) in Table 2 which less than the value of $\alpha = 0.05$.

The results as in Table 2 show that the students' mathematical adaptive reasoning skills who were taught by creative problem-solving model were higher than the students' mathematical adaptive reasoning skills of those who were taught by conventional model. The result of this research showed that the most prominent aspect of adaptive reasoning that could be developed through a CPS was inductive intuitive. Two aspects of adaptive reasoning, which were inductive intuitive and deductive intuitive, were mostly balanced. The different between inductive intuitive and deductive intuitive aspect was not too big. CPS model can develop student mathematical adaptive reasoning skills. CPS model can facilitate the development of mathematical adaptive reasoning skills thoroughly.

The students' mathematical adaptive reasoning skills who applied to creative problem-solving learning models are higher than students' mathematical adaptive reasoning skills who applied with conventional learning using expository model. This is in line with Pehkonen's opinion that creative thinking is a combination of divergent thinking based on intuition and logical thinking [13]. These creative aspects are needed in creative problem-solving learning model so that this model trains students to think divergently based on intuition and logical or convergent thinking. Divergent and convergent thinking based on intuition is both contained in students 'mathematical adaptive reasoning skills, so it can be concluded that creative problem-solving learning models can train and develop the students' mathematical adaptive reasoning skills.

The results of this study are also in line with the results of Desriyanti's research related to the influence of Thinking Aloud Pair Problem Solving (TAPPS) learning model to students' mathematical adaptive reasoning skills with an average value of 63.80, giving conclusion that TAPPS learning model is effective to improve student's mathematical adaptive reasoning ability [14]. However, student's mathematical adaptive reasoning skills applied by creative problem-solving model has higher mean value than student's mathematical adaptive reasoning skills applied by TAPPS learning model that is 80.83, so it can be concluded that the creative problem-solving learning model is more effective to improve the students' mathematical adaptive reasoning skills.

The results of Mahmudah's research about the influence of creative problem-solving model of the students' critical thinking skills with the average of 59, gives the conclusion that the creative problem-solving learning model is effective to improve students' critical thinking skills [15]. However, when it compared with the students' adaptive mathematical reasoning skills that also applied the creative problem-solving learning model has a higher average value of 80.83. This shows the creative problem-solving learning model is effective for developing students' mathematical adaptive reasoning skills and critical thinking skills, but the creative problem-solving learning model is more effective for improving students' mathematical adaptive reasoning skills.

4. Conclusion

In The student's mathematical adaptive reasoning skills whose applied creative problem-solving learning model has the best level of indicator on the ability to guess and make general conclusions based on the data observed (intuitive-inductive). However, the ability to guess and make logical conclusions (intuitive-deductive) is only slightly different with intuitive-inductive skills, even almost balanced. It can be said that the creative problem-solving learning model is an effective learning to develop students' mathematical adaptive reasoning skills on the aspects of intuitive-inductive and intuitive-deductive.

The students' mathematical adaptive reasoning skills who applied with conventional learning have the best level of indicator on the intuitive-inductive aspect rather than the intuitive-deductive aspect. In both aspects also have a difference that is not far away. Students who are learning with the conventional model have intuitive-deductive skills that are almost balanced with intuitive-inductive skills, despite the intuitive-inductive skills are greater. Overall, students taught with conventional learning models have mathematical adaptive reasoning skills that are balanced in both aspects, but low enough. It can be said that the conventional learning model has not been effective enough to develop students' mathematical adaptive reasoning skills.

The students' mathematical adaptive reasoning skills who applied to creative problem-solving learning model is higher than the students' mathematical adaptive reasoning skills who applied to the conventional learning model. The students taught by creative problem solving have an edge both aspects of intuitive-inductive and intuitive-deductive then the students taught by conventional learning models. The differences in each indicator in both classes are very visible. Thus, creative problem-solving learning model is better than conventional learning model in developing students' mathematical adaptive reasoning skills.

Based on the conclusions obtained, the creative problem-solving learning model can be used to improve students' mathematical adaptive reasoning abilities, so it can be used as an alternative learning model that can be used by teachers in mathematics learning. This study was conducted at the junior high school level, therefore to get comparison of the results of the research conducted it is necessary to conduct further research for other subjects, other aspects of ability or school level different.

References

[1] Hayat B and Yusuf S 2010 *Benchmark Internasional Mutu Pendidikan* (Jakarta: PT Bumi Aksara) pp 40–43
[2] Kilpatrick J Swafford J and Findell B 2001 *Adding It Up: Helping Children Learn Mathematics* (Washington DC: National Academy Press) pp 129–131
[3] Sumarmo U 2010 *Berpikir dan Disposisi Matematika: Apa, Mengapa, dan Bagaimana Dikembangkan Pada Peserta Didik* (Bandung: UPI Press) pp 5– 6
[4] Copi I M and Gould J A 1964 *Readings on logic* (New York: Macmillan)
[5] Fischbein E 1987 *Intuition Science and Mathematics: An Educational Approach* (Tel Aviv: School of Education Tel Aviv University) pp 26–55
[6] Kurniawati L 2006 *ALGORITMA: Jurnal Pendidikan Matematika* **1** pp 77–101
[7] Polya G 1973 *How To Solve It* (America: Pricenton University Press) pp xvi
[8] Mitchell W E and Kowalik T E 1999 *Creative Problem Solving*, (NUCEA: Genigraphics Inc.) pp 1–14
[9] Huda M 2013 *Model-Model Pengajaran dan Pembelajaran* (Yogyakarta: Pustaka Pelajar) pp 147–150
[10] Suryosubroto 2009 *Proses Belajar Mengajar di Sekolah* (Jakarta: PT Rineka Cipta) p 201
[11] Sugiyono 2008 *Metode Penelitian Pendidikan Pendekatan Kuantitatif, Kualitatif, dan R&D*, (Bandung: Alfabeta) p 114
[12] VanGundy A B 1987 *Creative Problem Solving: A Guide For Trainers and Management* (New York: Quorum Books) p 63
[13] Pehkonen E 1997 *ZDM* **29**(3) available at http://www.fiz.karlsruhe.de/fiz/publications/zdm
[14] Desriyanti Y 2014 Pengaruh Metode Pembelajaran Thinking Aloud Pair Problem Solving (TAPPS) Terhadap Kemampuan Penalaran Adaptif Matematis Siswa Graduate thesis Universitas Islam Negeri Syarif Hidayatullah
[15] Mahmudah R 2013 Pengaruh Model Pembelajaran Creative Problem Solving (CPS) Terhadap Kemampuan Berpikir Kritis Matematis Siswa Graduate thesis Universitas Islam Negeri Syarif Hidayatullah

IOP Conf. Series: Journal of Physics: Conf. Series **948** (2018) 012002 doi:10.1088/1742-6596/948/1/012002

Development of geometry materials based on scientific approach for junior high school students

Nurafni[1], R D Siswanto[1] and E Azhar[1]

[1]Universitas Muhammadiyah Prof. DR. HAMKA, Jakarta, Indonesia

E-mail: avnie_cute20@yahoo.com

Abstract. A scientific approach is a learning process designed so that learners can actively construct concepts, encourage learners to find out from various sources through observation, and not just be told. Therefore, learning by scientific approach offers a solution, because the goals, principles, and stages of the scientific approach allow for a good understanding of the students. Because of the absence of teaching materials "polyhedron geometry based on scientific approach" which is widely published in Indonesia, then we need to develop the teaching materials. The results obtained in this study are the tasks presented on teaching materials with a scientific approach both in defining the cube and the beam, identify and solve problems related to the properties and elements of cubes and beams, making cube and beam nets, solving problems related to cube and beam nets, solving problems related to cube and beam surface area. Beginning with the difficulties students face. Then, based on the results of interviews with teachers and analysis of student difficulties on each indicator, researchers revise the teaching materials as needed. Teaching materials that have not found any more student difficulties then the teaching materials are considered valid and ready for use by teachers and students.

1. Introduction

Curriculum in Indonesia today is sufficient to achieve mathematical ability, but still, the teacher plays a key role in the achievement of this matter. As Suryadi [1] said to achieve the competencies set in the curriculum, teachers are required to describe teaching and learning activities in the form of teaching planning. Because teaching materials are important tools that play a role in learning and thinking activities, the lack of meaningful teaching materials used by teachers is a problem in learning mathematics in the classroom.

According to Junior High School Mathematics Teachers who attended the PLPG in 2009 Rayon 037 and Rayon 035 from Jakarta and several Level II Regions in West Java (Bogor Regency, Bogor City, Sukabumi District and Sukabumi City), the most difficult material for students They are the matter of building a flat side room. Learning difficulties experienced by students is the lack of imagination and spatial ability of students to components wake flat so that students are less able to understand the concept given. In other words, students find it difficult when drawing or painting and making a mathematical model of the problem of waking up space. Meanwhile, painting plays an important role in the learning of geometry in schools because geometry paintings connect between physical space and theory. If in the area of Jakarta and West Java which in terms of educational facilities are more advanced than other regions, certainly not closed the possibility of difficulties are also felt in other areas of Indonesia.

To counter this, the scientific approach offers a solution. Learning with this approach has succeeded in improving learning outcomes in some primary and junior secondary schools in Indonesia, this can be seen in the research [2,3,4]. Based on some of the studies already mentioned, perhaps a scientific approach can also be used to improve students' learning outcomes in junior high schools especially for building a flat side room.

In order to implement a flat space based on a scientific approach, it is necessary to prepare a scientifically based teaching material that is currently not widely publicized. Therefore, the authors feel the need for an innovation of learning mathematics through the preparation of teaching materials to help students in understanding the material of building a flat side space. According to the 2013 curriculum that uses a scientific approach, the teaching materials are prepared according to the scientific approach. This research is very instrumental in the development of mathematics teaching materials, especially teaching materials based on a scientific approach. Because the scientific approach is a learning process designed in such a way that learners actively construct concepts, encourage learners to find out from various sources through observation, rather than just being told, so we need to develop their own scientifically based teaching-based teaching materials Junior high school students. This research is very important for the innovation of Indonesian mathematics learning so it is expected to become the basis for the development of interesting teaching materials for other materials in junior high school.

Based on the background of the problem and the scope of the research, the research formulation of this issue is "Development of Instructional Materials to Build a Flat Space Based Scientific Approach for Junior High Schools".

2. Methods

2.1 Place and Time of Research

This research was conducted in MTs. Islamiyah Ciputat Jl. Kihajar Dewantara No. 23 Ciputat, South Tangerang in the first semester of the academic year 2016 / 2017dated October 4, 2016, to November 3, 2016.

2.2 Research Procedure

This research is a development research whose product is a worksheet for three-dimensional material based on the scientific approach which will be used in MTs. Islamiyah Ciputat. Student worksheet developed the student textbook in the form of a collection of Student Worksheet. The development of this student worksheet refers to the model of development of the general teaching materials from Tjeerd Plomp [5] consisting of several phases namely (1) early investigation phase (identification of learning obstacle), (2) design phase, (3) realization, and (4)) Test phase, evaluation, and revision.

2.3 Instrument Development

There are 2 instruments developed in this research, namely: validation sheet and student worksheet.

3. Results and Discussion

3.1 Instrument Development Results

Prior to use, the instrument was validated first by one lecturer in Mathematics Education from University Semaragdan, one Mathematics teacher from MTs. Islamiyah Ciputat Based on the results of validation that has been done, the instrument that the researchers designed to use without proper revision.

3.2 Description of Interview Results with Teacher

Interviews with grade 9 teachers are conducted to find out the difficulties students experience in studying three-dimensional space materials based on their experiences in teaching. Based on the teacher's narrative, the weaknesses of the students are the first lack of imagination to imagine the three-dimensional castings component, so that the students are less able to understand the concept given. In other words, students find it difficult when drawing or painting and making a mathematical model of the problem of waking up space. And second, students find it difficult to work on a problem with a different context.

3.3 Hypothetical Learning Trajectory (HLT)

Hypothetical Learning Trajectory (HLT) is a series of lesson planning based on anticipation of the possibility of student mindset in learning to develop the purpose of spreading. HLT consists of three components, namely learning objectives, learning activities, and learning process hypotheses are predictions about the mindset and understanding of students organized in the context of learning activities. Based on the results of interviews with grade 9 teachers, interviews with an analysis of answers on the students' initial ability tests, HLT was prepared as provided in the appendix.

3.4 Experiment

The teaching materials that have been compiled are then tested to students in the 9th grade. Students work in groups, one group consists of 2-3 people. Each group gets an LKS. In this LKS there is the main task and the task of strengthening. All tasks are related to three-dimensional spaces. Students are given guidance in advance before working on teaching materials in the hope that students do not have difficulty in working on this resource. Student experiences and responses occurring during this phase will be collected as a reference in the improvement or modification of HLT for further learning.

3.5 Analysis of Student Difficulties in Initial Capability Test

The written test is done to find out the difficulties experienced by students in working on problems related to waking up the flat side space by paying attention to the indicator of material space dimension three. Here is an analysis of the difficulties experienced by students in answering each item given in accordance with the indicator.

3.5.1 Define cubes and beams

The difficulties students face on this point are: (1) to the Student does not understand the definition of a flat side room, as evidenced by the 12 groups working on the LAS, 7 groups cannot answer; and (2) build a flat side room where students are focused only on cubes.

3.5.2 Identify and resolve problems related to the properties as well as elements of cubes and beams

The difficulties students face on this item are as follows: (1) Three groups cannot answer at all the problem; (2) Students do not understand the concept of calculation so easy to forget and be fooled when solving the problem. Students are only capable of sketching sketches but not counting them.

3.6 Review Analysis

In the defining activity of defining cubes and beams in these three-dimensional spaces, there are several tasks. In this LKS, students do not have difficulty in doing all the tasks available. Based on the data obtained, the LKS or this task need not be revised, but teacher intervention is needed to help students determine the names of the dotted spots on the students' webs.

In determining activities identifying and solving problems related to the properties and elements of cubes and beams in this three-dimensional space there are several tasks. In this LKS, students do not have difficulty in doing all the tasks available. Based on the data obtained, the LKS or this task need not be revised, but teacher intervention is needed to help students determine the names of the dotted spots on the students' webs. Here's an example of student answers that indicate that students have no difficulty: Students not only sketch but also count them.

In determining activities identifying and solving problems related to the properties and elements of cubes and beams in this three-dimensional space there are several tasks. In this LKS, students do not have difficulty in doing all the tasks available. Based on the data obtained, the LKS or this task need not be revised, but teacher intervention is needed to help students determine the names of the dotted spots on the students' webs.

4. Conclusion

Based on the results of research and discussion in this study, got the conclusion that the tasks presented on teaching materials (LKS) with scientific approach both in defining cubes and beams, identify and solve problems related to the properties and elements of cubes and beams. Making cube and beam nets, solving problems related to cube and beam nets, solving problems related to the surface area of cubes and beams. Beginning with the difficulties students face. Then based on the results of interviews with teachers and analysis of student difficulties in each indicator above, researchers revise the teaching materials as needed. When it is no longer found students' difficulties then the teaching materials are considered valid and ready for use by teachers and students.

References

[1] Suryadi D 2010 Creating Active Learning Process: A Study from the Viewpoint of Learning Theory and The Didactic Theory [On line] retrieved from http://didi-suryadi.staf.upi.edu./files/2011/06/MENCIPTAKAN-PROSES-BELAJAR-AKTIF.pdf
[2] Bajdeber R 2014 *Development of Teaching Materials with a Scientific Approach to Class VIII Classroom Circle Materials* (Bandung: UPI)
[3] Khusna H 2014 *Development of Instructional Material for Fractional Counting Operation on Grade VII Junior High School Students* (Bandung: UPI)
[4] Hadi W 2015 *The Enhancement of Mathematical Representation Ability of Junior High School Students Through Discovery Learning by The Scientific Approach* In International Seminar on Mathematics Science And Computer Science Education pp 84-89
[5] Plomp T 1997 *Educational & Training System Design* (Enschede, Netherlands: Faculty of Education Science and Technology. University of Twente)

Non-linear learning in online tutorial to enhance students' knowledge on normal distribution application topic

Kartono[1, 2], D Suryadi[3] and T Herman[3]

[1] Faculty of Education and Teacher Training, Indonesia Open University
[2] Student of Doctoral Program, Indonesia University of Education
[3] Faculty of Mathematics and Natural Science Education, Indonesia University of Education

E-mail: kartono@ecampus.ut.ac.id

Abstract. This study aimed to analyze the enhancement of non-linear learning (NLL) in the online tutorial (OT) content to students' knowledge of normal distribution application (KONDA). KONDA is a competence expected to be achieved after students studied the topic of normal distribution application in the course named Education Statistics. The analysis was performed by quasi-experiment study design. The subject of the study was divided into an experimental class that was given OT content in NLL model and a control class which was given OT content in conventional learning (CL) model. Data used in this study were the results of online objective tests to measure students' statistical prior knowledge (SPK) and students' pre- and post-test of KONDA. The statistical analysis test of a gain score of KONDA of students who had low and moderate SPK's scores showed students' KONDA who learn OT content with NLL model was better than students' KONDA who learn OT content with CL model. Meanwhile, for students who had high SPK's scores, the gain score of students who learn OT content with NLL model had relatively similar with the gain score of students who learn OT content with CL model. Based on those findings it could be concluded that the NLL model applied to OT content could enhance KONDA of students in low and moderate SPK's levels. Extra and more challenging didactical situation was needed for students in high SPK's level to achieve the significant gain score.

1. Introduction

Education Statistics is one course that must be taken by students in several study programs in Faculty of Teacher Training and Education *Universitas Terbuka* (UT). The course is to teach the concepts, procedures, and the use of statistical theory that the students must use in completion of their final program reports. One most important topic is normal distribution application. The normal distribution is a statistical distribution that relates to other statistical topics, especially in inference statistic. Kokoska [1] stated that "*the bell shape density curve can be used to model many natural phenomena and the normal distribution is used extensively in statistical reference*." On the other side, Mulholland & Jones [2] stated: "*the most sets of random errors follow the normal ... it has been shown that the results obtained by assuming a no normal population to be normally distributed are reasonably accurate*."

Preliminary study shows that there are 36% students per semester in who do not pass final exam of Education Statistics, students' participation in online tutorial (OT) activities do not have significant influence on the achievement of the final exam, and the learning materials provided by tutors in OT activities are mostly in the pointers form packaged in PowerPoint presentations with learning process,

IOP Conf. Series: Journal of Physics: Conf. Series **948** (2018) 012003 doi:10.1088/1742-6596/948/1/012003

and which, therefore, are called "e-presentation" or "e-lectures" or "e-reading" [3]. In this study, the latter -shown diagrammatically below in Figure 1.a- is suspected as the main cause of the students' participation in OT do not have many influences to the readiness of students in working out the final exam. Without further explanation in multi-media forms (more detail explanation, oral presentation, video presentation, etc.), students have difficulties in understanding such materials. It is this consideration that this study is designed i.e. each material presentation need further explanation, especially for those who inadequate statistical prior knowledge (SPK). Figure 1.b shows the design of learning material called Non-Linear Learning (NLL) Model (while that of figure 1.a. -in this study- is called Conventional Learning or CL Model).

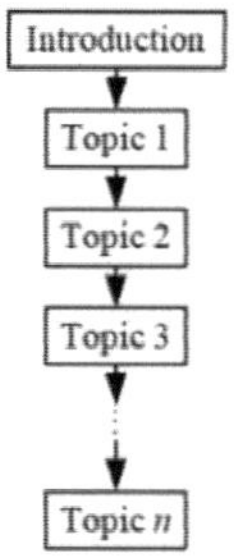

Figure 1.a. Flowchart of current design OT content in Education Statistics course

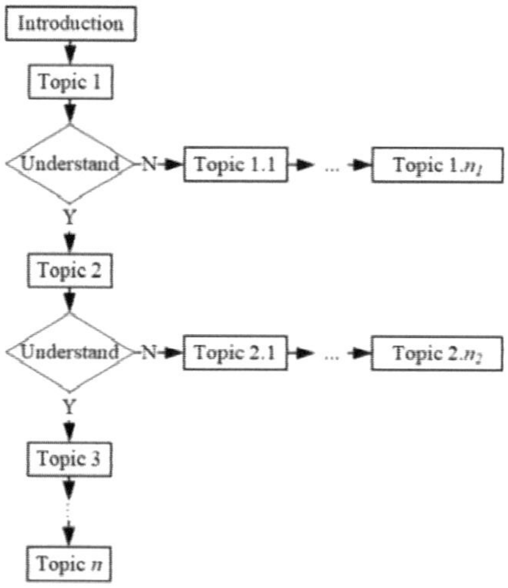

Figure 1.b. Flowchart of proposed design of OT content in Education Statistics course

Based on such reasoning, it can be formulated the problems of this research as follow:

- Is there gain score differences of knowledge on normal distribution application (KONDA) between students that learn OT content in NLL model and students that learn OT content in CL model?
- Is there gain score difference of KONDA between students who had low, moderate, and high SPK' scores after they learn OT content both in NLL and CL model?
- Is the student's KONDA who learn OT content in NLL model better than those who learn OT content in CL model?

2. Literature Review

According to Meyers [4], non-linear learning is a form of learning that is organized in complex ways by involving the ongoing systems and the underlying norms and values in substantial manners. Overall approach and prerequisites needed to produce and interpret learning are expected to alter the creative process of self-organization. Chen [5] states that non-linear learning applied in a hypermedia program can provide learners with many opportunities to explore and discover learning material that appropriate to their own individual needs. Some learners who are uncertain of the importance of a series of explanations in non-linear learning programs will be disoriented so that their learning cannot achieve better results. Therefore, to develop OT content with NLL model that is oriented to ameliorating students learning obstacle, a tutor can use Didactical Design Research (DDR) developed by Suryadi [6].

The development of learning material using DDR may be performed through three stages. The first stage of DDR is developing a hypothetical didactical design which is oriented to solve students learning obstacle. This stage can be performed before OT content presented to students. The second stage of DDR is by performing *metapedadidactical* analyses that is performed in the learning process. Lastly, the third stage in DDR is retrospective analyses that is performed after learning process finished. In developing the hypothetical didactical design, the tutor needs to perform didactical phenomenon observation in the previous semester. This observation is useful to know which topics that need to be emphasized in their OT content. It also helps the tutor to identify students' learning obstacle. The next step in development hypothetical didactical design is to develop a didactical situation that can identify students' comprehension level after they studied the printed learning material independently. The final step in the development of hypothetical didactical design is to develop several didactical-pedagogical anticipations which are expected to give students a better understanding of the given didactical situation.

To develop a didactical situation, an OT tutor can present a challenging problem for the students that are related to learning material which is presented by OT tutors. In the learning topic of normal distribution application, OT tutors need to stress concept, procedures, problem solution and statistical rules of normal distribution application. The problems of a didactical situation can be recognized from data in previous semester in the form of the students' answer in their final exam, their response in OT discussion forum, and the result of students' performance in working out of OT tasks. These three sources can be found from the database, the result of the final exam and OT log activity of the students. Besides, OT tutor can perform interview to the students who took the course in the previous semester to get further information about what the real problems are. The didactical situation can be differently responded by students. So, from an initial didactical situation can emerge new situation as a result of students' varied response to the initial didactical situation. The new situations might be as many as -or even more than- that varied response. The more various learning activity setting, the more various new didactical situation will be. As a result, the learning process can be very complex.

Related to the complexity of the new didactical situation, didactical-pedagogical anticipation can be developed by OT tutor as scaffolding that will direct the students to understand and to solve the problems that may emerge in the didactical situation. The didactical-pedagogical anticipation can be related to other concepts related to normal distribution application topic. As a result, related didactical-pedagogical anticipation will create NLL. By using NLL in OT content, the tutor must realize the individual difference and they need to pay attention to NLL as a way to accommodate different learning preference in each student. One facility in UT's OT application that can be used as NLL is "Lesson". Lesson is one feature in UT's OT application that offers the flexibility of website, interactive, quizzes while enabling the development of branched learning material [7]. By using Lesson, a student can have an opportunity to get the needed materials. In Lesson, a student can as well explore the consequences of their response. As an example of a hypothetical didactical design that covers the relationship between didactical situation and its anticipation is described in Figure 2 and Table 1 as follow.

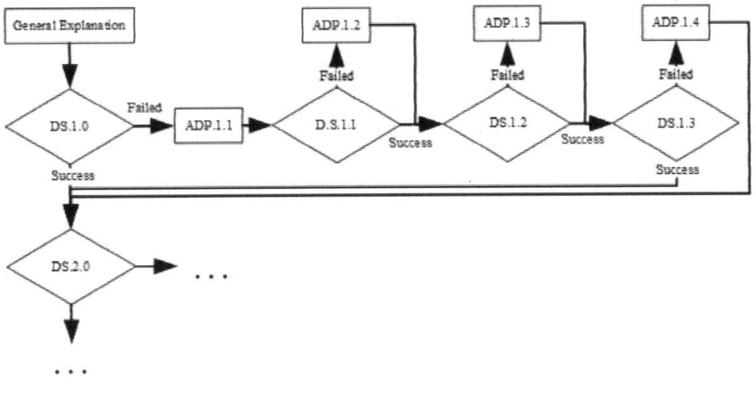

Figure 2. Flowchart of an example of hypothetical didactical design in OT content

Table 1. Flowchart' explanation on Figure 2

Frame Code	Activities of Learning	Navigation
DS.1.0	Given a case study containing mean and standard deviations values for large samples (N > 100) and normally distributed, students were asked to determine the area under the standard normal curve between the values $x_1 = \bar{x} = \mu$ and $x_2 > \bar{x}$	If the student can determine the area properly then the student can access DS.2.0 However, if the student cannot determine the area properly then the student must access ADP.1.1
ADP.1.1	Provides an explanation of the procedure of determining area under the standard normal curve between $x_1 = \bar{x} = \mu$ and $x_2 > \bar{x}$ of the case studies which contained mean and standard deviation values for large samples (N > 100) and normally distributed	If the student has understood the explanation of ADP.1.1 the student can continue accessing DS.1.1
DS.1.1	Given a case study that contains the mean and standard deviation values for large sample (N > 100) and normally distributed, the student is required to determine the values of z_1 for $x_1 = \bar{x} = \mu$ and z_2 for $x_2 > \bar{x}$	If the student can determine the z_1 and z_2 values correctly then the student can access DS.1.2. However, if the student cannot determine the z_1 and z_2 values correctly then the student must access ADP.1.2
ADP.1.2	Provides an explanation of the procedure of determining the values of z_1 for $x_1 = \bar{x} = \mu$ and z_2 for $x_2 > \bar{x}$ for the case study which contains the mean and standard deviation values for large samples (N > 100) and normally distributed	If the student has understood the explanation of ADP.1.2 the student can continue accessing DS.1.2
DS.1.2	Given a case study contains the mean and standard deviation values for large samples (N > 100) and normally distributed, students were asked to draw the area under the standard normal curve between values of z_1 for $x_1 = \bar{x} = \mu$ and z_2 for $x_2 > \bar{x}$	If the student can draw the area under the standard normal curve between values of z_1 for $x_1 = \bar{x} = \mu$ and z_2 for $x_2 > \bar{x}$ correctly then the student can access DS.1.2. However, if the student cannot draw correctly then the student must access ADP.1.3
ADP.1.3	Provides an explanation of the procedure of drawing the area under the standard normal curve between values of z_1 for $x_1 = \bar{x} = \mu$ and z_2 for $x_2 > \bar{x}$	If the student has understood the explanation of ADP.1.3 the student can continue accessing DS.1.3
DS.1.3	Given a case study containing mean, standard deviations, and the figure of area under the standard normal curve between the values $x_1 = \bar{x} = \mu$ and $x_2 > \bar{x}$ for large samples (N > 100) and normally distributed, students were asked to determine the area under the standard normal curve between the values $x_1 = \bar{x} = \mu$ and $x_2 > \bar{x}$	If the student can determine the area properly then the student can access DS.2.0 However, if the student cannot determine the area properly then the student must access ADP.1.4
ADP.1.4	Provides an explanation of the procedure of determining area under the standard normal curve between $x_1 = \bar{x} = \mu$ and $x_2 > \bar{x}$ of the case studies which contained mean and	If the student has understood the explanation of ADP.1.4 the student can continue accessing S.2.0

Frame Code	Activities of Learning	Navigation
	standard deviation values for large samples (*N* > 100) and normally distributed	
DS.2.0	…	
⋮		

Based on Figure 1 and Table 1, initial didactical situation (DS.1.0) can be used as a tool to know the learning materials the students needed to understand the concept, procedures, problem-solving, and statistical rules related with normal distribution application topic. The initial didactical situation will mean a lot if the tutor provides instruction to learn the printed learning material independently before learning OT content. For students who already have adequate knowledge of the concept of area under the standard normal curve between negative z value till z = 0, they can continue for the next learning. While, for students that still have difficulties in understanding the concept, they need to learn the concept of area under standard normal curve through ADP.1.1 till ADP.1.4.

3. Methods

The design used in this study is quasi-experiment. The subject is divided into two groups. The first group is experimental class given OT content in the NLL model, while, the second group is the control class given OT content in the CL model. Population in this study is the students of Basic Education in Graduate program of UT. Purposive random sampling technique is used to choose the sample. According to Creswell [8], purposive random sampling is the main strategy in sampling technique that combines both quantitative and qualitative criteria. There are 85 students in the experimental group and another 85 students in control group. Instruments used are objective tests to measure the statistical prior knowledge (SPK) and the knowledge of normal distribution application (KONDA). Before using the whole instruments, all items of the objective test are field-tested first to ensure their content validity, face validity, construct validity, and item validity. Samples used in this field testing are 30 students of Basic Education in Graduate program of UT that registered in Education Statistics course in a previous semester. Based on the result of field tests, it can be concluded that all SPK and KONDA instruments are eligible to be used [9].

Data used in this study includes SPK score, pretest score of KONDA and posttest score of KONDA. Data are collected by the online objective test in UT's OT application. Students are given enough time to finish all the questions of SPK and KONDA instruments. Data on the SPK and pretest of KONDAT are collected before the students learn different OT content. Meanwhile, data of posttest about KONDA are collected after the students learn different OT content. The students are grouped in according to the score level of SPK before learning OT content. For the students whose SPK score is less than $\bar{x} - s$, they will be grouped as low SPK, those whose SPK score between $\bar{x} - s$ and $\bar{x} + s$, are grouped in moderate SPK, and those whose SPK score more than $\bar{x} + s$, they are grouped in high SPK. Meanwhile, for testing the enhancement, the normalized gain score (G) of students' KONDA are used in statistical analysis. G is obtained by the following formula [10].

$$G = \frac{posttest\ scores(\%) - pretest\ scores(\%)}{100 - pretest\ score(\%)}$$

The preliminary statistical analysis is firstly conducted before data analyses. The analysis is on the normality of SPK score both for students in experimental class and control class to determine which statistical test of homogeneity that appropriate. The analysis also on normality test of a normalized gain score of KONDA both for students in experimental class and control class to determine which statistical test that is appropriate to show OT content influence on KONDA enhancement.

4. Findings and Discussion

It can be shown that students' SPK average score before they learn different OT content is 54.15 with standard deviation as 18.3. Based on the results, the students are categorized into three levels. Those who have SPK score less than 36.20 are in low SPK level, those who have SPK score between 36.20

and 72.82 are in moderate SPK level, while those who have SPK score more than 72.82 are in high SPK level. It can be also shown that average SPK score of the experimental class is 56.31 with standard deviation as 18.69, while those from control class is 52.71 with standard deviation as 17.75. Meanwhile, the Kolmogorov-Smirnov test statistic in Table 2 can be used to determine the homogeneity statistical test that appropriate.

Table 2. The summary of normality test for students' SPK score

	Experimental	**Control**
N	85	85
Test Statistic	0.116	0.116
Asymp. Sig. (2-tailed)	0.007	0.007

Table 2 shows that the students' SPK score in both classes is not normally distributed. Therefore, homogeneity statistical test for the students' SPK score of both classes is performed by using Levene test statistic. The result of Levene test statistic shows the value 0.224 with Sig value as 0.637. This means that students' SPK score in experimental and in control class are equal. Next, the result of normality test statistic using Kolmogorov-Smirnov testing of students' KONDA gain score for students who learn OT content in NLL and CL models are presented in Table 3.

Table 3. The summary of normality test for students' KONDA score

	OT Content Model		**SPK Level**		
	NLL	**CL**	**Low**	**Moderate**	**High**
N	85	85	25	109	36
Test Statistic	0.110	0.272	0.179	0.270	0.126
Asymp. Sig. (2-tailed)	0.013	0.000	0.037	0.000	0.163

Table 3 shows that almost all statistical tests of the students' KONDA gain score in each group are not normally distributed (only high SPK level is normally distributed). Meanwhile, the result of normality test of students' KONDA gains score based on OT content models in each SPK level are presented in Table 4.

Table 4. The summary of normality test for students' KONDA gain score based on OT content models in each SPK level

OT Content Model	**NLL**			**CL**		
SPK Level	**High**	**Moderate**	**Low**	**High**	**Moderate**	**Low**
N	23	52	10	13	57	15
Test Statistic	0.170	0.117	0.125	0.151	0.307	0.247
Asymp. Sig. (2-tailed)	0.084	0.074	0.200	0.200	0.000	0.014

Table 4 shows that almost all of students' KONDA gain score are normally distributed. Only two groups (moderate and low SPK level of students that learn OT content in CL models) are not normally distributed. Based on the normality test in Table 3 and 4 it can be concluded that the appropriate statistic to test the influence of different OT content models to students' KONDA gain score in a different level of SPK is non-parametric test statistic. The appropriate non-parametric statistic to test the influence of two different models of OT content is the Mann-Whitney test statistics.

Table 5. The summary of influence test for difference OT content models to students' KONDA score

OT Content Model	***N***	***Median***	***W***	***Z***	***p-value***
NLL	85	0.2778	9193.0	0.3000	0.0000
CL	85	0.0000			

Table 5 shows that there is a significantly different of KONDA gain score between the students that learn OT content in NLL model and in CL model. Next, appropriate non-parametric statistic to test the influence of three different SPK levels to students' KONDA gain score is Kruskal-Wallis test statistic.

Table 6. The summary of influence test for difference SPK levels to students' KONDAT score

SPK Level	*N*	*Median*	*Z*	*H*	*p-value*
Low	23	0.000	-2.01		
Moderate	108	0.125	-0.51	8.05	0.018
High	32	0.250	2.36		

Table 6 shows that there is significantly different KONDA gain score among students in low, moderate and high SPK levels. Based on the results of Tables 5 and 6 the statistical testing can be continued to test the influence of differences OT content models in each SPK levels to gain score of students' KONDA. The non-parametric statistic to test the influence of interaction between OT content models and SPK levels to students' KONDAT gain score is Kruskal-Wallis test statistic.

Table 7. The summary of influence test for interaction between OT content models on each level of SPK to student's KONDAT improvement

OT Content Model	SPK Level	*N*	Median	*Z*	*H*	*p-value*
NLL	Low	10	0.381	1.77		
	Moderate	52	0.232	2.94		
	High	23	0.500	3.59	42.60	0.000
CL	Low	15	-0.143	-3.32		
	Moderate	57	0.000	-4.32		
	High	13	0.167	-0.06		

Table 7 shows that there is a significant difference of KONDA gain score between students in deferent SPK levels that learn different models of OT content. Based on the results in Table 5, Table 6 and Table 7 show that statistical testing can be continued to compare KONDA gain score between students that learn OT content in NLL model and in CL model. That testing not only compares the two models in general but also compare those in each SPK levels. Furthermore, based on the normality test in Table 3, a statistical testing that can be used to compare KONDA gain score between students that learn OT content in NLL model and in CL model is the Mann-Whitney test statistics.

Table 8. The summary of compare means test for students' KONDA score between two models of OT content

OT Content Model	*N*	*Median*	*W*	*Z*	*p-value*	Conclusion
NLL	85	0.2778	9193.0	0.3000	0.0000	Rejected H_0
CL	85	0.0000				

Table 8 shows that the students' KONDA gain scores of the students who learn OT content in NLL model are significantly better than those who learn OT content in CL model. On the other hand, based on normality test in Table 4, the statistical testings that can be used to compare KONDA gain score between OT content models in each SPK levels are the *t*-test and Mann-Whitney tests statistic.

Table 9. The summary of *t*-test for compare KONDA score for students in high level of SPK between two models of OT content

SPK Level	OT Content Model	*N*	Average	Std. Dev	*df*	*t*	*p-value*	Conclusion
High	NLL	23	0.354	0.409	34	1.62	0.057	Rejected H_1
	CL	13	0.126	0.401				

Table 10. The summary of Mann-Whitney test statistic for compare KONDA score for students in low and moderate levels of SPK between two models of OT content

SPK Level	OT Content Model	*N*	*Median*	*W*	*Z*	*p-value*	Conclusion
Moderate	NLL	52	0.232	3639.5	0.2321	0.0000	Rejected H_0
	CL	57	0.000				
Low	NLL	10	0.381	179.5	0.5313	0.0033	Rejected H_0
	CL	15	-0.143				

Table 9 and Table 10 show that for students in the low or moderate level of SPK, the KONDA gain score between the students who learn OT content in NLL model is significantly better than those who learn OT content in CL model. Meanwhile, for students in the high level of SPK, the KONDA gain score between the students who learn OT content in NLL model is not significantly different with those who learn OT content in CL model. Based on those results shows that, in general, NLL model that implemented in OT content can enhance students' KONDA. Furthermore, the enhancement of KONDA of students who have high SPK's score, they have understood the explanation of normal distribution application from the printed material before they learn OT content in both models. Therefore, extra and a more challenging didactical situation is needed to make them achieve the significant gain score.

5. Conclusion and Implication

The results of data analysis in this study show that there is the difference in the gain score of knowledge on normal distribution application between students who learn online tutorial content in non-linear learning model and those who learn online tutorial content in conventional learning model. This results also states that there is the difference in the gain score of KONDA between students who have low, moderate and high statistical prior knowledge after they learn online tutorial content in two different ways. It can also be shown that there is the difference in the gain score of knowledge on normal distribution application between students who learn online tutorial content in non-linear learning model and those who learn online tutorial content in conventional learning model on each level of students' statistical prior knowledge. The study also gives empirical evidence that the gain score of knowledge on normal distribution application for students who learn online tutorial content in non-linear learning model better than those who learn online tutorial content in conventional learning model. Finally, for students with low and moderate statistical prior knowledge, this research finds that the gain score of knowledge on normal distribution application for students who learn online tutorial content in non-linear learning model is better than those who learn online tutorial content in conventional learning model. Therefore, in general, online tutorial content in non-linear learning model can help students who have learning difficulties in understanding printed material independently. Meanwhile, for students who have high SPK's score, they have understood the explanation of normal distribution application from the printed material before they learn OT content in both models. This can only mean that extra and a more challenging didactical situation is needed to make them achieve the significant gain score.

References

[1] Kokoska S 2011 *Introductory Statistics: A Problem-Solving Approach* **2nd** ed. (New York: W. H. Freeman and Company)
[2] Mulholland H and Jones C R 1968 *Fundamentals of statistics* (New York: Springer Science)
[3] Bozarth J 2008 *Better than bullet points: Creating engaging e-Learning with PowerPoint* (San Francisco: Pfeiffer)
[4] Meyers P W 1990 Non-linear learning in large technological firms: Period four implies chaos *Research Policy* **19** 2 pp 97-115
[5] Chen S 2002 *British Journal of Educational Technology* **33** 4 pp 449-460
[6] Suryadi D 2013 *Far East Journal of Mathematical Education* **10** 1 pp 91-107
[7] Rice W 2011 *Moodle 2.0 e-Learning course development: A complete guide to successful learning using Moodle* (Birmingham: Packt Publishing)

IOP Conf. Series: Journal of Physics: Conf. Series **948** (2018) 012003 doi:10.1088/1742-6596/948/1/012003

[8] Creswell J W 2009 *Research design: Qualitative, Quantitative, and Mixed Methods Approaches* **3rd** ed. (California: SAGE Publications, Inc)
[9] Guilford J P 1942 *Fundamental Statistics in Psychology and Education* (New York: McGraw-Hill Book Company, Inc)
[10] Coletta V P and Phillips J A 2005 *American Journal of Physics* **73** 12 pp 1172-1182

IOP Conf. Series: Journal of Physics: Conf. Series **948** (2018) 012004 doi:10.1088/1742-6596/948/1/012004

Improving mathematical problem solving skills through visual media

S A Widodo[1,2], Darhim[2] and T Ikhwanudin[2]

[1]Universitas Sarjanawiyata Tamansiswa, Yogyakarta
[2]Universitas Pendidikan Indonesia, Bandung

E-mail: sriadi@ustjogja.ac.id

Abstract: The purpose of this article was to find out the enhancement of students' mathematical problem solving by using visual learning media. The ability to solve mathematical problems is the ability possessed by students to solve problems encountered, one of the problem-solving model of Polya. This preliminary study was not to make a model, but it only took a conceptual approach by comparing the various literature of problem-solving skills by linking visual learning media. The results of the study indicated that the use of learning media had not been appropriated so that the ability to solve mathematical problems was not optimal. The inappropriateness of media use was due to the instructional media that was not adapted to the characteristics of the learners. Suggestions that can be given is the need to develop visual media to increase the ability to solve problems.

1. Introduction

The characteristic of mathematics is to have an abstract object [1]. This abstract nature causes many students to have difficulty in math. In fact, many mathematics teachers teach without regard to it. This causes the majority of learners to have negative perceptions of mathematics, such as lazy students and avoid mathematics because mathematics is difficult and complicated [2]. In addition, the low achievement of learning can be attributed to the ability of learners in solving math problems is not adequate because in solving math problems irregular and inconsistent [3]. Student irregularities in solving mathematical problems can be seen from how to solve problems that just write down the answer, while for other stages not done.

In mathematics education, problem-solving is also an important thing to be instilled in students and must be owned by students in learning mathematics [3]. There are several reasons that problem-solving is important and becomes one of the basic skills of a person in solving math problems. First, problem-solving cannot be separated in everyday life, with the ability to solve problems that can be used to provide solutions or answers to problems faced more analytic so that someone can be a problem solver. In other words, when students are trained to solve problems, learners will be able to make decisions because learners have become skilled about how to gather relevant information, analyze information, and be aware of the need to re-examine the results that have been obtained [4].

Secondly, in mathematics learning, problem-solving can be used to formulate concepts, develop ideas or ideas, and capital success for students in solving mathematical problems because a concept or principle will be meaningful if it can be applied in problem-solving [5]. Third, the mathematical standards in schools should include standard content and process standards. Process standards include problem-solving, reasoning and verification, interconnection, communication, and representation [6].

Fourth, one of the objectives of mathematics learning in Indonesia is that students are expected to develop critical, logical, systematic, accurate, effective, and efficient in solving problems [7].

The use of media is one of the many problems in school learning. SMP IPS teachers in Kota Banjarmasin tend to use package books and whiteboards in learning [8]. Even though the learning media is one of the learning tools that must be prepared to support the learning process [9], besides the use of learning media can facilitate communication patterns during the learning process [10].

Learning media can be used to channel messages from the sender (teacher) to the recipient (student) to stimulate the thoughts, feelings, attention or willingness of students so that will encourage the learning process. Because the learning media has a purpose to bring messages or information to the students, the message or information brought by the learning media can be a message that is prepared to meet the learning needs and the ability of students so that students can actively participate in the learning process.

This article aims to discuss the improvement of students' mathematical problem-solving skills by using visual learning media. The ability to solve mathematical problems using the steps of understand the problem, make a plan, carry out our plan, and look back at the completed solution.

2. Method

This study was a preliminary research and did not intend to make a teaching material or model of learning, but only did a conceptual approach prior to a research development. So, the main method carried out was to compare the various literature related to problem-solving ability and associate to learning media.

3. Discussion

3.1 Problem-Solving in Math

The gap that exists between the current situation with the goal to be achieved is one factor the occurrence of the problem. With these problems, we will be required to seek an answer or a solution to the gap. The process of finding such an answer is called problem-solving [11, 4, 12], problem-solving is an individual process or effort to respond or overcome obstacles or obstacles when an answer or answer method is not obvious [13]. Problem-solving involves the interaction between a student's (knowledge) scheme with an application process that uses cognitive and affective factors in solving problems [14]. Problem-solving is a high-level mental process and requires a more complex thought process [15]. This is because in solving mathematical problems, one is faced with the concept, skills, and mathematical process to solve mathematical problems [16].

The students 'ability to solve problems is closely related to the students' ability to read and understand the language of the problem, to present and design it in mathematical models, to plan the calculations of the mathematical model, to solve the calculations of the problems encountered, and to interpret the solutions that have been obtained [17, 18, 19]. Achieving the ability to solve mathematics requires the obedience (demands) of learners in using the steps to solve the problem. If students are not coherent in solving the problem can be ascertained that the ability of learners is not satisfactory, so that student achievement becomes low. As expressed by [17] problem-solving ability is an effort or way student in solving a problem by using systematic steps.

To solve the problem of mathematics required a method or steps that system so that the process of completion becomes easy and directed. One way to solve mathematical problems such as using a way of Polya. There are Understand the problem, Make a plan, Carry out our plan, and Look back at the completed solution [11].

The Understand the problem stage refers to an understanding of what is known, what is asked, or whether sufficient, insufficient, exaggerated or contradictory terms to seek out the question [20]. To believe or understand a problem can be done by questioning yourself about what is known or what is asked?, What data is given?, What is the condition of the problem?, Is it possible that the condition is expressed in the form of an equation or other relationship?, What is the condition given Enough to find the question?, Is the condition not enough or the condition is excessive or the condition is contradictory?, And whether it can be drawn and writing the appropriate notation to facilitate solving the problem? [17].

At the stage of understanding the problem of providing a foundation for learners to be able to step on the next stage [21], this is because students are not possible to solve problems correctly without any understanding of the problems faced [22].

The learner is said to have understood the problem if the learner expresses the question along with answers such as (1) what data or information is known of the problem, (2) what is the essence of the problem that requires solving, (3) is there any question about the formula, tables, or special signs, and (4) there are important conditions to note in the question [23], students can identify known, questioned elements and the adequacy of the necessary elements [24, 25, 18]. So the aspects that students should include at the stage of understanding the problem include what is known and what is asked [20].

In the make, a planning phase, referring to how the associated settlement strategy [20], making plans to solve problems is a mental activity that links between existing knowledge and problem-solving results [26]. A problem cannot be solved properly without good planning [27]. Planning for problem-solving depends largely on the creative experience of students in putting together a problem solving [22], the more varied their experiences are, the more creative the students tend to be in preparing a problem-solving plan [24].

At the stage of making plans, students can formulate mathematical problems or develop mathematical models, implement strategies to solve problems [25, 18, 24], looking for links between concepts or theories that are mutually Support and look for the formulas needed to solve the problem [23], besides that students need to know first some math problem solving strategies so that problem solving can be done more effectively [28].

Errors such as procedural errors can cause the student's plan to become unfavorable, so students are expected to make procedures to resolve the problem appropriately. Searching for links between known and unknown information, or performing calculations on unknown variables may prevent the student from error in the procedure for solving the problem. So students get the question of how the information that has been known will be interconnected to obtain the things that are not known or students do self-question like never had this problem before? Or has there ever been the same or similar problem in another form?, Did you know something similar to this? Which theory can be used in this problem?, Pay attention to the question! Think about a question once familiar with one or similar question!, If there is a similar problem to the problem that was ever solved, can the experience be used in the present problem? Can the results and methods used here?, Whether looking for other elements in order to take advantage of the original problem?, Can repeat the problem earlier?, Can declare in another form? Return to the definition!, If the new problem cannot be resolved, think about a similar problem and finish it! [17].

If a plan for solving a problem has been made (whether written or not), and has determined which strategy the student will use to continue the next stage [28,24]. The third stage is the implementation of the plan, which is the stage of implementation of the plan is the student has been prepared to perform calculations with all kinds of necessary data including the concepts and formulas or equations as appropriate [22].

At the stage of implementing the plan, the student must be able to form a more standard problematic system, in the sense that the formulas to be used are formulas that are ready to be used in accordance with what is used in the problem, then students begin to enter the data that have been obtained from the previous step, After which the students carry out the plan steps that have been made, so that the problems faced can be proved or resolved [22, 23]. In addition, students can question themselves about how to implement the settlement plan and check each step, check that each step is correct ?, and how to prove that the selected step is correct? [17]. Using that method, calculation errors or algorithm errors and procedural errors that can cause the process of solving the problem to be not maximally can be avoided.

At the last stage or look back at the completed solution stage the student will look back at the answer to make sure that the answer to the problem is correct, re-checking in problem-solving is a mental activity that links the existing knowledge to the problem-solving steps [26]. This step is important to check whether the results obtained are in accordance with the provisions and there is no contradiction with the question [23, 27].

Steps that can be used by the student to conduct a re-examination stage include matching the results obtained with the item in question, interpreting the answers obtained, identifying whether there is any other way to get the problem solved, and identifying whether there are answers or other results that meet [28]. Besides, it can also ask how to check the results of truth obtained?, Can be checked the rebuttal? Can you search for that result in another way?, Can you see it at a glance? Can the result or the way it is used for other questions? [17].

Although the back-end of the answer (check back at the completed solution) has the least weight among the other stages of Polya, but if this step is not done, the student cannot make corrections or correct mistakes such as conceptual errors, procedural errors and miscalculations has been done in the previous stages [23]. Students can minimize errors that may arise in the previous stage by doing this last step [24]. Students can check the systematics and stages of completion whether it is good and right or not, students can re-examine every step of the solution that does

3.2 The Use of Visual Media in Mathematics Problem Solving

Ability to solve problems is a prayer one ability that must be possessed by each learner in addition to the ability to think critically, logically, systematically. Even the problem-solving ability became the standard of mathematics learning process [26] and one of the objectives of mathematics learning in Indonesia [27]. Because of the importance of problem-solving skills, learners are expected not to avoid any mathematical problems encountered. However learners need to be given the motivation to solve math problems so that learners realize that math is an important lesson because it is implicitly in daily life.

For example, everyday events related to mathematics Among others are a student went to the grocery store to buy 5 kg of rice type A and 3 kg of rice type B. The price of each kg of type B rice more expensive Rp. 500, - of type A, if both the price of the rice is added to Rp. 6000, -. To find out what the price of rice A and the price of rice B do not pounds, the student can use the linear equations obtained when studying mathematics.

The success of learners in learning depends on the way of presentation of learning materials, instructional media and teaching methods used by teachers in teaching and learning process [29]. Learning media is a component that is interconnected with other components, so the learning media is expected to concretely abstract the material, increase student interest to learn, and reduce student misconception [30, 31, 32]. Unfortunately, the current learning of the students tends to be just the recipient of the message so that the communication that occurs in the learning is not going both ways, but one way. In fact, learning requires students not only act as the recipient of the message, but also acts as a communicator or messenger so that there are two-way communication and even a lot of direct communication [30].

In learning communication, learning media is needed to improve the effectiveness of learning achievement goals. That is, the learning process will occur if there is communication between the recipient of the message with the source/channel message through the media. In addition to good learning planning, the success of teaching and learning process is also influenced by the suitability between the subject matter and the level of students' thinking ability [30].

Planning a good learning, learning strategies that have been prepared very well, as well as learning materials that have been adapted to the level of students' cognitive understanding to be in vain in learning if the media used, is not appropriate. Because the success of learning is determined by two main components of teaching methods and learning media [32]. These two components are interrelated and inseparable, the use and selection of any particular teaching method have consequences on the appropriate use of the type of instructional media.

Planning a good learning, learning strategies that have been prepared very well, as well as learning materials that have been adapted to the level of students' cognitive understanding to be in vain in learning if the media used, is not appropriate. Because the success of learning is determined by two main components of teaching methods and learning media [32]. Both of these components are interrelated and Cannot be separated, use and choice of one particular teaching method have consequences on the use of appropriate types of learning media.

Visual-shaped media can be used as an alternative to mathematics learning in junior high, considering that junior high school students are in a transition phase from concrete to abstract. The results of the study show that visual media are used in learning [34, 35, 36] especially in mathematics learning [37, 38, 39].

Visual media is a media that only relies on the senses of vision and plays a very important role in teaching and learning process, can facilitate understanding and strengthen memories, foster student interest and can provide support for the content of subject matter to the real world [40]. Visual media are divided into two, namely silent visual media and visual media motion. Still visual media such as photographs, books, encyclopedias, magazines, newspapers, reference books, and other printed products, drawings, illustrations, clippings, film frames, coupling films, transparencies, microphones, projector overheads, graphs, charts, diagrams and sketches, Posters, cartoon images, maps and globes, while the motion visual media is like a soundless movie [41]. Sometimes to attract users, visual media combined with audio becomes an audio-visual media like movies.

In order for effective visual learning media to be used, it should be placed in a meaningful context and students must interact with the visual (image) itself to ensure the process of information. This is because the effectiveness of learning through learning media one of them can be seen from the level of student enjoyment when learning to use pictorial text [10].

Based on that, for effective instructional media used to improve students' ability in solving problems, it is necessary to develop instructional media that reflect problem-solving steps, as done by creating an algebra equation by applying Polya problem-solving steps in solving the problem [19]. So that students indirectly feel guided in solving problems encountered. With guided problem solving is expected students can reveal or write every step that is on Polya, in other words, students can solve math problems with a structured and systematic.

4. Conclusion

The ability to solve mathematical problems is the ability of students to overcome the problems that were not clear the answer. Problems that arise in solving problems are ways that students use in solving mathematical problems has not been systematic or sequential, so the ability of students in solving math problems has not been maximized. In addition, the media used in mathematics learning in junior high school has not accommodated the cognitive development of children in the transition phase. One alternative to improve the ability to solve math problems for junior high school students is to use visual learning media that reflect the troubleshooting steps. Based on these conclusions it is necessary to develop a visual media to increase the ability to solve problems.

References

[1] Zinurie 2009 *Pakar matematika Bicara tentang Prestasi Pendidikan Matematika Indonesia.* Available at http://zinurie.wordpress.com

[2] Widodo S A 2011 Efektifitas Model Pembelajaran Team Accelerated Instruction Pada Siswa Kelas X SMK Tunas Harapan Tahun Pelajaran 2008-2009 *Pros. Sem. Nas. MIPA: Pemantapan Keprofesionalan Peneliti, Pendidik, dan Praktisi MIPA Untuk Mendukung Pembangunan Karakter Bangsa*, edited by Kismiantini, et al, pp PM1 – PM6

[3] Pardimin and Widodo S A 2016 *Journal of Education and Learning* **10** 4 pp 390 – 395

[4] Widjayanti D B 2009 Kemampuan Pemecahan Masalah Matematis Mahasiswa Calon Guru Matematika: Apa dan Bagaimana Mengembangkannya *Pros. Sem. Nas. Matematika dan Pendidikan Matematika* pp 402 – 413.

[5] Widodo S A 2013 *Jurnal Pendidikan dan Pengajaran* **46** 2 pp 106 – 113

[6] NCTM 2000 *Principles and Standards for School Mathematics* (Virginia, USA: NCTM)

[7] BSNP 2006 *Lampiran Peraturan Menteri Pendidikan Nasional No 22 Tahun 2006* (Jakarta: Depdiknas)

[8] Rahmatullah M 2011 *Jurnal Penelitian Pendidikan* **12** 1 pp 178 – 186
[9] Prasetyo Z K, et al 2011 *Pengembangan Perangkat Pembelajaran Sains Terpadu Untuk Meningkatkan Kognitif, Keterampilan Proses, Kreativitas serta Menerapkan Konsep Ilmiah Peserta Didik SMP* (Yogyakarta: Pascasarjana UNY)
[10] Haryoko S 2009 *Jurnal Edukasi Elektro* **5** 1
[11] Polya G 2014 *How To Solve It: A New Aspect Of Mathematical Method* (USA: Princeton university press)
[12] Elvina A 2012 *Hubungan Antara Self Regulated Learning Dengan Kemampuan Memecahkan Masalah Pada Pembelajaran Matematika Pada Siswa SMUN 53 Di Jakarta Timur* Available at publication.gunadarma.ac.id
[13] Siswono T Y E 2008 *Model Pembelajaran Matematika Berbasis Pengajuan dan Pemecahan Masalah Untuk Meningkatkan Kemampuan Berpikir Kreatif* (Surabaya: Unesa University Press)
[14] Webb N L 1973 *Journal for Research in Mathematics Education* **10** 2 pp. 83-93
[15] Haryani D 2014 Pembelajaran Matematika Dengan Pemecahan Masalah Untuk Menumbuhkembangkan Kemampuan Berpikir Kritis Siswa *Pros. Sem. Nas. Penelitian, Pendidikan dan Penerapan MIPA* pp PM121 – PM126
[16] Meidawati Y 2014 *Jurnal Pendidikan dan Keguruan* **1** 2
[17] Anisa W N 2014 *Jurnal Pendidikan dan Keguruan* **1** 1 pp. 1-8
[18] Windari F, Dwina F and Suherman 2014 *Jurnal Pendidikan Matematika* **3** 2 pp 25 – 28
[19] Widodo S A 2017 *Infinity* **6** 1 pp 61 – 70
[20] Suci A A W and Rosyidi A H 2012 Kemampuan Pemecahan Masalah Matematika Siswa Pada Pembelajaran Problem Posing Berkelompok *MATHEdunesa* **1** 2
[21] Widodo S A 2015 *Kreano: Jurnal Matematika Kreatif – Inovatif* **6** 2 pp 142-151
[22] Indarwati D, Wahyudi and Ratu N 2014 *Satya Widya: Jurnal Penelitian Pengembangan Kependidikan* **30** 1 pp 17-27
[23] Utomo D U 2012 Pembelajaran Lingkaran Dengan Pendekatan Pemecahan Masalah Versi Polya Pada Kelas VIII di SMP PGRI 01 Dau *Widya Warta* XXXVI **1**
[24] Herman T 2000 Strategi Pemecahan Masalah (Problem Solving) Dalam Pembelajaran Matematika. *Paper unpublish*
[25] Suherman E, et. al 2003 *Strategi Pembelajaran Matematika Kontemporer* (Bandung: UPI dan IMSTEP JICA)
[26] Susanto H A 2011 Pemahaman Pemecahan Masalah Pembuktian Sebagai Sarana Berpikir Kreatif *Pros. Sem. Nas. Penelitian, Pendidikan, dan Penerapan MIPA* pp. 189 – 196
[27] Suryana 2015 *Jurnal Jurnal Cakrawala Pendas* **1** 2 pp 12 – 18
[28] Wahyudi and Budiono I 2011 *Pemecahan Masalah Matematika* (Salatiga: Widya Sari Press)
[29] Herawati E 2017 *JNPM (Jurnal Nasional Pendidikan Matematika)* **1** 1 pp 66 – 87
[30] Nuroso H and Siswanto J 2010 *Jurnal Penelitian Pembelajaran Fisika (JP2F)* **1** 1 pp 35 – 46
[31] Nurseto T 2011 *Jurnal Ekonomi Dan Pendidikan* **8** 1 pp 19 – 34
[32] Ali M 2009 *Jurnal Edukasi Elektro* **5** 1
[33] Sudrajat A 2008 *Media Pembelajaran.* Available at akhmadsudrajat.wordpress.com
[34] Hakim A A 2010 Perbanding Efektifitas Media Visual Running Text dan Media Visual Text Book Berbahasa Jawa dalam Pembelajaran Membaca pada Siswa Kelas XI IPA 1 dan XI IPA 2 SMA Negeri 1 Tahunan Jepara (Semarang: Unnes)
[35] Rohmatun A S 2008 *Efektifitas Media Visual Dalam Meningkatkan Kemampuan Memahami Isi Berita Pada Anak Tunarungu Di SMPLB Karya Mulia Surabaya* (Surabaya: UIN Sunan Ampel)
[36] Nussofa I 2015 *Efektifitas Media Visual Dalam Pembelajaran Terhadap Penevgkatan Hasil Belajar Siswa Mata Pelajaran Fiqih Kelas XIDI Madrasah Aliyah Al-Ikhlas Tanjung Bintang Lampung Selatan Tahun Pelajaran 2015/2016* (Lampung: Universitas Muhamadiyah Metro)
[37] Marti M, Syamswisna S and Panjaitan R G P 2016 *Jurnal Pendidikan Matematika dan IPA* **5** 2

[38] Suriwati L M, Nilakusmawati D P E and Sumarjaya I W 2014 Efektivitas Pembelajaran Dengan Media Visual Untuk Meningkatkan Hasil Belajar Siswa Tuna Rungu Pokok Bahasan Pecahan Senilai *Pros. Seminar Nasional Matematika*
[39] Susanti M A 2016 *Efektivitas Penggunaan Media Pembelajaran Visual Bangun Ruang Pada Mata Pelajaran Matematika Ditinjau Dari Prestasi Belajar Siswa Sekolah Dasar* (Yogyakarta: Universitas PGRI)
[40] Nugrahani R 2007 *Lembaran Ilmu Kependidikan* **36** 1 pp. 35 – 44
[41] Muhson A 2010 *Jurnal Pendidikan Akuntansi Indonesia* **8** 2 pp 1 – 10

Acknowledgment
This publication does not work well without the support of Lembaga Pengelolaan Dana Pendidikan (LPDP), for which we would like to thank the LPDP who has participated as a sponsorship of our publications.

Improving students' creative mathematical reasoning ability students through adversity quotient and argument driven inquiry learning

W Hidayat[1], Wahyudin[1] and S Prabawanto[1]

[1]Universitas Pendidikan Indonesia, Indonesia

E-mail: wahyu@stkipsiliwangi.ac.id

Abstract. This study aimed to investigate the role factors of Adversity Quotient (AQ) and Argument-Driven Inquiry (ADI) instruction in improving mathematical creative reasoning ability from students' who is a candidate for a math teacher. The study was designed in the form of experiments with a pretest-posttest control group design that aims to examine the role of Adversity Quotient (AQ) and Argument-Driven Inquiry (ADI) learning on improving students' mathematical creative reasoning abilities. The population in this research was the student of mathematics teacher candidate in Cimahi City, while the sample of this research was 90 students of the candidate of the teacher of mathematics specified purposively then determined randomly which belong to experiment class and control class. Based on the results and discussion, it was concluded that: (1) Improvement the ability of mathematical creative reasoning of students' who was a candidate for a math teacher who received Argument-Driven Inquiry (ADI) instruction is better than those who received direct instruction is reviewed based on the whole; (2) There was no different improvement the ability of mathematical creative reasoning of students' who is a candidate for a math teacher who received Argument-Driven Inquiry (ADI) instruction and direct instruction was reviewed based on the type of Adversity Quotient (Quitter / AQ Low, Champer / AQ Medium, and the Climber / AQ High); (3) Learning factors and type of Adversity Quotient (AQ) affected the improvement of students' mathematical creative reasoning ability. In addition, there was no interaction effect between learning and AQ together in developing of students' mathematical creative reasoning ability; (4) mathematical creative reasoning ability of students' who is a candidate for math teacher had not been achieved optimally on the indicators novelty.

1. Introduction

Mathematics learning teaches people to solve problems by paying attention to a procedure or process that prioritizes pedagogic aspects through a scientific approach so that learners can better understand meaningfully through the process of observing, asking, trying, reasoning, presenting and creating [1]. Development of learning mathematics should shape the mindset of learners that can be measured by the ability. One of the abilities that can form the mindset is the ability of mathematical reasoning. Mathematical reasoning is a process of concluding a problem-solving from a given problem [2,3]. The mathematical creative reasoning is one of the reasoning processes that a person undertakes in solving problems through non-routine procedures [4,5].

One of the success factors in learning mathematics is Adversity Quotient (AQ). AQ is a person's persistence in facing obstacles to achieve success. AQ is expected to make a strong motivation so that it can better support his desire in achieving success to solve the problems encountered. AQ can also be said to be a person's fighting power to solve the problems he is facing [6,7].

Currently, there are still learners who have difficulty in doing reasoning process because the problem solving is still based on the algorithm memorized or routinely done by the teacher. This is because there are still reliable mathematics teachers, but not accompanied by a good conceptual understanding [8,9,10,11].

Achievement and improvement of the ability of mathematical reasoning and AQ of a person cannot be separated from meaningful learning process [12,13]. The learning model is supposed to optimize the improvement of mathematical creative reasoning ability and AQ is Argument-Driven Inquiry ADI), because ADI learning emphasizes the provision of challenging mathematical tasks based on the level of difficulty so that the achievement and improvement of mathematical creative reasoning and AQ of prospective teachers can be better [14,15,16].

Based on that, problems and objectives in this research are to know and examine about: (a) Improving students' mathematical creative reasoning abilities who are candidates for mathematics teachers who get Argument-Driven Inquiry (ADI) learning with those who have direct instruction reviewed based on the whole; (b) Improving students' mathematical creative reasoning abilities who are candidates for mathematics teachers who get Argument-Driven Inquiry (ADI) learning with those who have direct instruction reviewed by Adversity Quotient (Quitter, Champer, and Climber); (c) Is there any interaction effect between learning factors and types of Adversity Quotient (Quitter, Champer, and Climber) in developing improvement of students' mathematical creative reasoning ability; (d) How is the achievement of students' in mathematical creative reasoning abilities.

2. Methods

The study was designed in the form of experiments with a pretest-posttest control group design that aims to examine the role of Adversity Quotient (AQ) and Argument-Driven Inquiry (ADI) learning on improving students' mathematical creative reasoning abilities. Population in this research is the student of mathematics teacher candidate in Kota Cimahi, while the sample of this research is 90 students of the candidate of the teacher of mathematics specified purposively then determined randomly which belong to experiment class and control class. The test of mathematical creative reasoning abilities in this study is based on the characteristics of mathematical creative reasoning ability and good test preparation guidelines. Data were analyzed by using Two-Way Annova statistic test to see the difference and interaction effect between learning and type of Adversity Quotient (Quitter, Champer, and Climber) in generating improved students' mathematical creative reasoning ability.

To find out how much improvement of students' creative reasoning ability before and after learning activity, a normalized gain score analysis was calculated using the following formula:

$$g = \frac{\text{postest score} - \text{pretest score}}{\text{maximum ideal score} - \text{pretest score}}$$

Normalized gain score levels are grouped into three categories:

0,70	<	(g)		: High
0,30	≤	(g)	≤ 0,70	: Medium
(g)	<	0,30		: Low

Finding out grouping of AQ types is based on the following categories:

144	<	(AQ)		: Climber
95	≤	(AQ)	≤ 144	: Champer
(AQ)	<	95		: Quitter

3. Results and Discussion

3.1. Results

Findings on students' mathematical creative reasoning abilities based on learning and Adversity Quotient types are presented in Table 1.

Table 1. Gain of mathematical creative reasoning ability based on learning factors and types of AQ

Type of Adversity Quotient	The Increase of Mathematical creative reasoning ability			
	ADI Class (n = 45)		Direct Learning Class (n = 45)	
	Mean	SD	Mean	SD
Climber	0,78	0,04	0,63	0,15
Champer	0,72	0,09	0,59	0,14
Quitter	0,62	0.10	0,57	0,10
Total	0,71	0,10	0,59	0,13

Notes: Ideal Score 1,00

Based on the above description, it can be seen that the mathematical creative reasoning ability of students learning through ADI learning is better than those learned through direct learning, whether reviewed as a whole or by type AQ (Climber, Champer, and Quitter). However, improvements for ADI class AQ (Climber and Champer) types are classified into a high category, while ADI class on type AQ (Quitter) and direct learning class on AQ type (Climber, Champer, and Quitter) are classified. In addition, if reviewed based on factors that affect the increase of the ability of mathematical creative reasoning, then based on the description in Table 1 shows that the learning factors and the AQ type affect the development of creative reasoning ability of mathematics students.

To support the description of the students 'mathematical creative reasoning abilities that have been described above, then test data analysis about students' mathematical creative reasoning ability through the statistical test of average difference. After the normality test of the data distribution of students' mathematical creative reasoning ability, it is obtained that the data is normally distributed. Based on these findings, the above-average difference test was performed using Two-Way ANOVA test presented in Table 2.

Table 2. Summary of Two-Way ANOVA test the increase of students' mathematical creative reasoning ability based on learning factors and type of AQ

Sources	JK	Dk	RJK	F_{count}	Sig
Learning Approach (A)	0,119	1	0,119	8,937	0,004
Type of AQ (B)	0,065	2	0,032	2,435	0,094
A x B	0,055	2	0,028	2,079	0,131
Inter	1,116	84	0.013		

(output SPSS. 23)

In relation to the learning factors presented in Table 2, the sig = 0.004; It can be concluded that there is a significant difference between improving students' creative reasoning abilities that learn to use Argument-Driven Inquiry learning with those who learn to use direct learning at 5% significance level. Based on Table 2 also obtained the value of sig for the factor of AQ to creative reasoning ability of 0.094; It can be concluded that there is no difference in improving the students' creative reasoning abilities that learn to use Argument-Driven Inquiry learning with those who learn to use direct learning at 5% significance level.

In addition, Table 2 also found that there is no significant interaction effect between the learning approach (ADI and Direct) with Type AQ (Climber, Champer, and Quitter) in generating improved students' mathematical creative reasoning at 5% significance level (sig = 0.131).

IOP Conf. Series: Journal of Physics: Conf. Series **948** (2018) 012005 doi:10.1088/1742-6596/948/1/012005

3.2. Discussion

Based on the results of research can be seen that basically the student candidate math teacher has been able to perform a good reasoning process based on completion through a process that contains in it a logical statement and can be maintained truth. This is because mathematical reasoning is a process of concluding problem solving based on logical statements derived from a given proposition and testing the assumptions of a particular problem [3,17].

After analyzed in depth about the results of research, the students' reasoning process is the result of heterogonous thinking. It can be seen that the problem-solving process done by students still tends to imitative reasoning [4,5,11].

In addition, although most students solve the problem with the ability of imitative reasoning, but not a few students who have good creative reasoning ability. This is because students can solve problems based on novelty aspect, plausible and mathematical foundation.

Student's mathematical creative reasoning skill based on Adversity Quotient (AQ) grouping is seen that students who have AQ Quitter have difficulty and often give up easily in solving problems. But for students who have AQ Camper and Climber, the students tend to be careful and try to re-examine the given solution, so the solution is considered to be a true job. It can be interpreted that AQ is one of the factors supporting the success of students in facing difficulties encountered in the problem-solving process [18,19,20].

In relation to the learning factor, Argument-Driven Inquiry (ADI) has a positive influence on the achievement of mathematical creative reasoning ability because ADI learning contains four important aspects that can improve students' mathematical creative reasoning ability: 1) construct and produce argumentation ability in support of mathematical explanation ; 2) the ability to use mathematical explanations in problem solving; 3) participate in the discussion process; 4) write the conclusions based on his own opinions as well as input from the opinions of others. This is also because the process of effective interaction between educators and learners in constructing knowledge is a critical determinant of the success of the learning process [16].

Based on the findings in the field, the achievement of students' mathematical creative reasoning abilities is presented in Table 3.

Table 3. The achievement of mathematical creative reasoning ability

Indicators of Mathematical Creative Reasoning	Type of AQ	*ADI* Class	Direct Learning Class
Novelty	Climber	66%	61%
	Champer	53%	51%
	Quitter	51%	52%
	Total	58%	55%
Plausible	Climber	82%	80%
	Champer	78%	76%
	Quitter	73%	75%
	Total	79%	77%
Mathematical Fondation	Climber	88%	86%
	Champer	85%	84%
	Quitter	84%	81%
	Total	86%	85%

The findings in Table 3 can be interpreted that the achievement of mastering students' mathematical creative reasoning ability has not been optimally achieved on the novelty indicator. Based on the observation, the problem is that students are not used to solve mathematical problems by making new solutions, but they are used to the usual completion steps in their daily problem-solving.

4. Conclusion

Based on the results and discussion, then obtained the following conclusions: (1) Improvement the ability of mathematical creative reasoning of students' who is a candidate for math teacher who received Argument-Driven Inquiry (ADI) instruction is better than those who received direct instruction is reviewed based on the whole; (2) There is no different improvement the ability of mathematical creative reasoning of students' who is a candidate for math teacher who received Argument-Driven Inquiry (ADI) instruction and direct instruction is reviewed based on the type of Adversity Quotient (Quitter / AQ Low, Champer / AQ Medium, and the Climber / AQ High); (3) Learning factors and type of Adversity Quotient (AQ) affect the improvement of students' mathematical creative reasoning ability students'. In addition, there is no interaction effect between learning and AQ together in developing of students' mathematical creative reasoning ability; (4) mathematical creative reasoning ability of students' who is a candidate for math teacher has not been achieved optimally on the indicators novelty.

References

[1] Quigley M 2011 The Centrality of Metaphor in The Teaching of Mathematics *PPG LINGUÍSTICA/UFJF – JUIZ DE FORA* **1** 2 pp. 57-69
[2] Killpatrick J, Swafford J and Findell B 2001 *Adding it up: Helping Children Learn Mathematics* (Washington, DC: National Academy Press)
[3] Ponte J P, Pereira M J and Henriques A 2012 O raciocínio matemático nos alunos do ensino básico e do ensino superior *J. Praxis Educativa* **7** 2 pp 355-377
[4] Bergqvist E 2007 *The Journal of Mathematical Behavior* **26** 4 pp 348–370
[5] Lithner J 2008 *Educational Studies in Mathematics* **67** 3 pp 255-276
[6] Stoltz P G 2004 *Adversity Quotient: Mengubah Hambatan Menjadi Peluang Edisi Kelima* (Jakarta: Grasindo)
[7] Syah M 2010 *Psikologi Belajar* (Jakarta: Rajagrafindo Persada)
[8] Bardini C, Pierce R, Vincent J and King D 2014 *Journal on Mathematics Education* **5** 2 pp 85-107
[9] Hendriana H, Rohaeti E E and Hidayat W 2016 *Journal on Mathematics Education* **8** 1 pp 55-64
[10] Hendriana H 2017 *J. Infinity* **6** 1 pp 51-58
[11] Hidayat W 2017 *KALAMATIKA Jurnal Pendidikan Matematika* **2** 1 pp 15-28
[12] Muijs D and Reynolds D 2005 *Effective Teaching-Introduction & Conclusion 2nd edition* (London: Sage Publications)
[13] Oppendekker M C and Van Damme J 2006 *Teaching and Teacher Education* **22** pp 1-21
[14] Fraivillig J 2001 *Teaching Children Mathematics* **7** 8 pp 454-459
[15] Samuelsson J 2010 *International Electronic Journal of Mathematics Education* **5** 2 pp 61-78
[16] Alfieri L 2011 *Journal of Educational Psychology American Psychological Association* **103** 1 pp 1-18
[17] Velez I and Ponte J P D A 2013 Representations and reasoning strategies of Grade 3 students in problem solving *Proc. of the Eight Congress of the European Society for Research in Mathematics Education* 383-391 (Ankara: Middle East Technical University)
[18] Phoolka S K 2012 *Research Journal of Social Science and Management* **2** 7
[19] Parvathy U and Praseeda M 2014 *Journal of Humanities and Social Sicence* **19** 11 pp 23-26
[20] MZ Z A, Risnawati, Kurniati A and Prahmana R C I 2017 *International Journal on Emerging Mathematics Education* **1** 2 pp 169-176

IOP Conf. Series: Journal of Physics: Conf. Series **948** (2018) 012006 doi:10.1088/1742-6596/948/1/012006

Pre-service mathematics teachers' reasoning ability in solving mathematical non-routine problem according to cognitive style

A Faradillah[1], W Hadi[1] and A Tsurayya[1]

[1]Universitas Muhammadiyah Prof. DR. HAMKA, Jakarta, Indonesia

E-mail: ayufaradillah@uhamka.ac.id

Abstract. This research described the level of mathematical reasoning ability of preservice teachers in solving the mathematical non-routine problem based on cognitive style. The cognitive style used was fast-accurate, reflective, impulsive and slow-less accurate cognitive style. Researchers used Matching Figure Familiar Test (MFFT) to find each cognitive styles. The total subject of this research was four mathematics education's preservice teachers. In addition, the category of research method used was explorative research with a qualitative approach where researchers tried to describe the level of preservice teachers' mathematical reasoning ability in solving the mathematical non-routine problem according to their cognitive style. Moreover, based on the result of this research, there were three of four mathematics education's preservice teachers who have mathematical reasoning ability in level 2. Meanwhile, there was one preservice teacher who got level 3 of mathematical reasoning ability. Furthermore, it related to their cognitive style. Since the preservice teacher who got level 3 of mathematical reasoning ability, she had fast-accurate cognitive style. Whereas, they who got level 2 had reflective, impulsive and slow-less accurate cognitive styles.

1. Introduction

Mathematics appears from the results of human thought related to ideas, processes, and reasoning. As mathematical material and mathematical reasoning cannot be separated, this is because mathematics is understood through reasoning and reasoning understood and trained through mathematical material [1]. So that the ability of mathematical reasoning is very important and necessary in mathematical learning. Moreover, mathematical reasoning as a part of the mathematical thinking that involves forming generalizations and drawing valid conclusions about ideas and how they are related. This identifies that a person who has mathematical reasoning ability will be able to generalize a problem to form a valid conclusion [2].

Pre-service teachers said to be capable of reasoning when they are able to use reasoning in patterns and traits, perform mathematical manipulations in generalizing, compiling evidence, or explaining mathematical ideas and statements [3]. Therefore, pre-service teachers must have good mathematical reasoning ability so that they are able to apply to the learning process in the school. This is reinforced by Baroody [3] reveals that a good reasoner should be introduced to situations of problems related to reasoning as early as possible including in mathematical learning in the school.

The mathematical reasoning ability of pre-service teachers can help them see mathematics as logical and plausible, thus it will help them to develop their beliefs that mathematics is something that they can understand, think, justify and evaluate. Based on some of the above opinions show that the ability of mathematical reasoning is an important aspect that should get the attention of teachers.

Accordingly, the teacher must also have the ability of mathematical reasoning inhabiting to be able to assist students in improving their reasoning. In fact, teachers often only invite students to learn routine questions, so students have not honed their mathematical reasoning ability in mathematical materials. Whereas, non-routine problems that can help students in improving the ability of mathematical reasoning. Non-routine problems lead to process problems, it is requiring more than simply translating problems into mathematical sentences and using known procedures. This is the focus in learning mathematics in Education Unit Level Curriculum Non-routine questions can be used to improve reasoning and non-routine problems focus on high level of interpretation and organizing problems. Non-routine problems in mathematics can encourage logical thinking, add conceptual understanding, develop mathematical reasoning, develop abstractive thinking skills and transfer math skills to unfamiliar situations.

Based on the statement, a teacher must also have the high mathematical reasoning ability to solve non-routine problems. This indicates that before becoming a teacher, it must first be analyzed to what extent reasoning ability possessed by pre-service teachers. They are difficult in guessing the strategies that will be used to solve the problems given and also less precise in generalizing the problem. This is because they are unable to use their reasoning in solving problems [4]. Though the objects studied in mathematics are abstract so that students need to use their rational power to solve the given problem.

Besides, there are other aspects that need to be considered that is cognitive style. The cognitive style is characterized as a possible cognitive trait to express the difference in the individual's ability in the quality of his thinking. [5], cognitive style is a typical way of studying students, both related to the way of receiving and processing information, attitudes toward information, and habits related to the learning environment. This is reinforced one group of student-made decisions after briefly looking at the figures, so they were cognitively impulsive, while the other was cognitively reflective [6]. One group of children who in making decisions briefly after viewing the image, so they are impulsive cognitive style, while other groups carefully choose before making a decision, they are reflective cognitive style. Moreover, there is a cognitive style in which students are fast and correct in answering and cognitive style where students are slow an incorrect in answering.

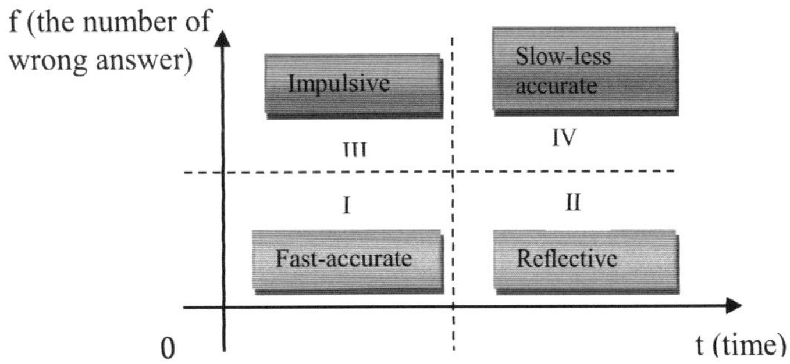

Figure 1. Fast-accurate, Reflective, Impulsive and Slow-less accurate Cognitive Style

According to figure 1, some definitions of cognitive style put forward by the experts, hence the cognitive style in this research is a person's way of processing, thinking and solving that is consistent where the type of cognitive style is a cognitive style fast-accurate, reflective, impulsive and slow-less accurate. Cognitive style associated with the use of time used by students to answer the problem and the number of errors made by students in solving problems given.

2. Methods

This research aims to describe the level of mathematical reasoning ability of pre-service teachers in solving non-routine mathematical problems according to their cognitive style. This research is categorized as explorative research with a qualitative approach. Before testing the ability of mathematical reasoning, the researcher will be selected subject of research. The subjects of research were selected by using MFFT to categorize subjects based on the four cognitive styles that were focused on this study. After obtaining the subjects, the subject was asked to solve a non-routine mathematical problem in writing first. Furthermore, the written results are stored by researchers and researchers conducted interviews related to non-routine mathematical problems that have been completed in writing. After obtaining the required data the researcher analyzed the data obtained in accordance with the purpose of this study. The level of mathematical reasoning ability that is measured in this research is level 4 (Superior), level 3 (satisfying with a few shortcomings): Level 2 (Quite Satisfying), level 1 (Satisfactory).

The analysis of qualitative data is done in a process, means that data analysis can be done since the data collection in the field and ended at the time of preparation of research report [7]. Suggests the analysis during data collection provides an opportunity for researchers to re-examine existing data and develop strategies for obtaining quality data [8]. Furthermore suggested there are three stages of analyzing qualitative data that is data reduction, data presentation, and conclusion [9]. Data analysis techniques in this study refers to the stages of data analysis proposed [9].

2.1. Data Reduction
Data reduction is intended to select, refocus, abstract, and transform raw data. In this research, the raw data obtained from field research is reduced to get the data that is really needed in analyzing the mathematical reasoning ability of the prospective teacher in solving the non-routine mathematics problem based on the cognitive style.

2.2. Presentation of data
The presentation of research data includes classification and identification. This is intended to make it easier to draw conclusions from the data. Thus, the written data has been well organized and categorized.

2.3. Conclusions
The conclusion is to give meaning and explanation to result in data presentation. Furthermore, the conclusion of this study is aimed to reveal the mathematical reasoning ability of prospective teachers in solving non-routine mathematical problems based on cognitive style.

3. Findings and Discussion
The selection of research subjects started with the Matching Familiar Figure Test (MFFT) which was held on June 2, 2017. Researchers conducted MFFT in the 2F class of pre-service teachers of Mathematics Education Study Program of University of Muhammadiyah Prof. DR. HAMKA (UHAMKA), where the number of pre-service teachers in the class is 26 students. The result of MFFT class 2F can be seen in Table 1.

Table 1. The Result of MFFT

Cognitive style	Total Pre-service Teachers
Fast-accurate	10
Reflective	5
Impulsive	10
Slow-less accurate	1
Total	**26**

According to the results of MFFT class 2F students of UHAMKA mathematics education program in Table 1. above, it is known that from 26 pre-service teachers attending MFFT there are 10 pre-service teachers who have fast-accurate cognitive style, 5 pre-service teachers who have reflective cognitive style, 10 pre-service teachers cognitive impulsive style and 1 pre-service teachers who have slow-less accurate cognitive style. The selected research subjects can be seen in Table 2. as follows.

Table 2. The List of Research Subjects According to MFFT

Subject	M/F	Cognitive style	*f*	*t*
S1	F	Fast-accurate	11	15.55
S2	M	Reflective	9	26.57
S3	F	Impulsive	3	14.03
S4	F	Slow-less accurate	3	16,14

Note:
f : Number of incorrect answers
t : Time students use to answer the given questions

After getting the research subject, the researcher did the validation of the problem-solving problem of non-routine mathematics. Researchers asked one lecturer of mathematics education at the State University of Surabaya, East Java, Indonesia to validate the question. The result of the validation stated that the test question to analyze the mathematical reasoning ability of the pre-service teacher is feasible to be used with the improvement in the sentence used by the problem.

Mathematical reasoning test is used to measure the level of reasoning ability possessed by students of mathematics teacher candidates. There are four levels of mathematical reasoning abilities, namely level 4 (superior), level 3 (satisfactory with a few shortcomings), level 2 (quite satisfactory) and level 1 (satisfactory). The following is the presentation of the results of the analysis of the level of mathematical reasoning ability of prospective teachers based on their cognitive style.

3.1. Analysis of The Level of Mathematical Reasoning Ability of Pre-Service Teachers' Who has Fast-Accurate Cognitive Style (S1)

S1 who has a cognitive style quickly-able to solve the problem non-routine given by researchers for approximately 70 minutes. Based on the result of the completion of S1 on the problem of mathematical reasoning test, the level of mathematical reasoning ability possessed by S1 is at level 3. This is because S1 is able to show understanding of mathematical concepts, using appropriate solutions in analyzing most mathematical situations, Indirectly most of it is true by using the volume formula and using the ribs that have been obtained, using the appropriate strategy that is correct in searching the ribs for the cube with the expanded cubical surface area, able to make the image most appropriate, calculate (computation) mostly true And meet the desired demand.

3.2. Analysis of The Level of Mathematical Reasoning Ability of Pre-Service Teachers' Who has Reflective Cognitive Style (S2)

S2 who has a reflective cognitive style to solve non-routine problems provided by researchers for approximately 60 minutes. Based on the results of completion of S2 in solving non-routine mathematical problems given, the level of mathematical reasoning capabilities possessed by S2 is at level 2. This is because S2 is satisfactory even though the results are less precise. S2 demonstrates an understanding of some mathematical concepts, using improper solutions that result in imprecise results, making indirect proofs largely wrong and using less appropriate, less systematic and organized strategies to solve the problem.

3.3. Analysis of The Level of Mathematical Reasoning Ability of Pre-Service Teachers' Who has Impulsive Cognitive Style (S3)

S3 who has a reflective cognitive style solve non-routine problems provided by researchers for approximately 60 minutes. Based on the result of the completion of S3 in solving the non-routine mathematical problem given, the level of mathematical reasoning ability possessed by S3 is at level 2. This is because S3 is satisfactory even though the result is less precise. S3 demonstrates an understanding of some mathematical concepts, using improper solutions that result in imprecise results, making indirect proofs largely wrong and using less appropriate, less systematic and organized strategies to solve the problem.

3.4. Analysis of The Level of Mathematical Reasoning Ability of Pre-Service Teachers' Who has Slow-Less Accurate Cognitive Style (S4)

S4 who has a reflective cognitive style resolves non-routine problems given by researchers for approximately 60 minutes. Based on the result of the completion of S4 in solving the nonroutine mathematical problem given, hence level of mathematical reasoning ability possessed by S4 is in level 2. This is because S4 is satisfactory even though the result is less precise. S4 demonstrates an understanding of some mathematical concepts, using improper solutions that result in imprecise results, making indirect proofs largely wrong and using less appropriate, less systematic and organized strategies to solve the problem.

In this part will describe the discussion of research results based on data obtained related to the ability of reasoning of pre-service teachers in solving non-routine mathematical problems based on cognitive style fast-accurate, reflective, impulsive and slow-less accurate. S1 who has a fast-accurate cognitive style and level 3 mathematical reasoning ability in accordance with the definition of cognitive style stated [10]. A pre-service teacher with fast-accurate cognitive style has fast characteristics in answering the problem and accurate so the answer is always right [10]. This is seen from the level of reasoning that S1 is the highest among other subjects. Although the timing of the test S1 does not match the characteristics it should have based on its cognitive style.

While the S2 who has reflective cognitive style and level 2 of mathematical reasoning ability. At the beginning of the test, S2 had difficulty in determining the way to solve the given problem. In line with the opinion subjects with reflective cognitive style experience confusion when determining the strategy that will be used to solve the problem because the subject needs to remember and match some problem solving previously used to find a way to solve the given problem [11]. S2 spends his time thinking at the beginning of work but is not able to solve the problem in the end. Furthermore, S3 has an impulsive cognitive style and level 2 of mathematical reasoning ability. At the beginning of the test, S3 looks confident in working on the given problem. She was able to solve problems earlier than other subjects although in the end waiting for other subjects to collect the results of the first answer to the new S3 participate collect. In accordance with that expressed by impulsive subject is more concerned with the speed of answering than the fear of answers [12]. And the last S4 who has a slow-less accurate cognitive style and level 2 of mathematical reasoning ability. At the beginning of the test, the S4 looks tense when the problem is given. She also did not directly do the question but instead draw other things that are not related to the problem. As expressed in the definition of cognitive style slow-less careful that students who have slow characteristics in answering problems and less accurate so often wrong answers.

4. Conclusion

According to the results of the research that has been described above, the conclusions that researchers can reveal in this study is the exposure level analysis of mathematical reasoning ability of students prospective teachers based on cognitive style, there is one pre-service teacher who obtains level of mathematical reasoning ability at level 3 (satisfactory with a few shortcomings) where pre-service teacher's cognitive style is fast-accurate. While 3 other pre-service teachers get the level of mathematical reasoning ability at level 2 (quite satisfactory) where each cognitive style it has is reflective, impulsive and slow-less accurate.

References

[1] Sa'adah W N 2010 *Peningkatan Kemampuan Penalaran Matematis Siswa Kelas VIII SMP Negeri 3 Banguntapan dalam Pembelajaran Matematika Melalui Pendekatan Pendidikan Matematika Realistik Indonesia (PMRI)* (Yogyakarta: Thesis Universitas Negeri Yogyakarta)

[2] Castro 2004 *Journal of Asia Pacific Education* **5** 2
[3] Widyasari N 2013 *Meningkatkan Kemampuan Penalaran dan Disposisi Matematis Siswa SMP Melalui Pendekatan Metaphorical Thinking* (Bandung: Tesis Sekolah Pascasarjana, Universitas Pendidikan Indonesia)
[4] Faradillah, Ayu 2014 *Profil Penalaran Matematis Siswa SMP dalam Memecahkan Masalah Matematika Berdasarkan Gaya Kognitif Reflektif-Impulsif* (Surabaya: Tesis Universitas Negeri Surabaya)
[5] Uno H B 2006 *Orientasi Baru dalam Psikologi Pembelajaran*s (Jakarta: PT. Bumi Aksara)
[6] Froehlich 2003 *Cognitive Styles: A Review of the Major Theories and Their Application to Information Seeking in Virtual Environments* (Online) http://www.personal.kent.edu/~plucasst/Cognitive%20Styles.pdf (8 Maret 2017)
[7] Moleong, Lexy J 2010 *Metodologi Penelitian Kualitatif (edisi revisi)* (Bandung: Remaja Rosdakarya)
[8] Rohendi, Tjetjep 1993 *Analisis Data Kualitatif* (Jakarta: Universitas Indonesia Press)
[9] Miles B M and Huberman 1992 *Analisis Data Kualitatif (terjemahan)* (Jakarta: Universitas Indonesia Press)
[10] Warly 2010 *Profil Kreativitas Siwa yang Bergaya Kognitif Reflektif dan Siswa Begaya Kognitif Impulsive dalam Memecahkan Masalah Geometri* (Surabaya: Dissertation Universitas Negeri Surabaya)
[11] Sulisawati 2013 *Pengaturan Diri Siswa SMP dengan Gaya Kognitif Reflektif-Impulsif dalam Menyelesaikan Masalah Aljabar* (Surabaya: Thesis Universitas Negeri Surabaya)
[12] Kenny, Robert F 2007 *International Jurnal of Social Sciences* **2** 3

IOP Conf. Series: Journal of Physics: Conf. Series **948** (2018) 012007 doi:10.1088/1742-6596/948/1/012007

Characteristics of students in comparative problem solving

M Irfan[1], Sudirman[2] and R Rahardi[2]

[1]Universitas Sarjanawiyata Tamansiswa, Yogyakarta, Indonesia
[2]Universitas Negeri Malang, Indonesia

E-mail: muhammad.irfan@ustjogja.ac.id

Abstract. Often teachers provided examples and exercised to students with regard to comparative problems consisting of one quantity. In this study, the researchers gave the problem of comparison with the two quantities mixed. It was necessary to have a good understanding to solve this problem. This study aimed to determine whether students understand the comparison in depth and be able to solve the problem of non-routine comparison. This study used qualitative explorative methods, with researchers conducting in-depth interviews on subjects to explore the thinking process when solving comparative problems. The subject of this study was three students selected by purposive sampling of 120 students. From this research, researchers found there were three subjects with different characteristics, namely: subject 1, he did the first and second questions with methods of elimination and substitution (non-comparison); subject 2, he did the first question with the concept of comparison although the answer was wrong, and did the second question with the method of elimination and substitution (non-comparison); and subject 3, he did both questions with the concept of comparison. In the first question, he did wrong because he was unable to understand the problem, while on the second he did correctly. From the characteristics of the answers, the researchers divided into 3 groups based on thinking process, namely: blind-proportion, partial-proportion, and proportion thinking.

1. Introduction

In order for students to more easily understand mathematics, teachers need to present real situations during the lesson. Real situations are usually presented when learning in the classroom, but only in words. For example, on a broad topic area. The teacher gives an example of the problem of determining the area of the pool, frame, or other objects that are often encountered. Teachers seldom invite students to take advantage of the school environment for learning. For example, students are invited to the school garden, then given the task to determine the area of various types of leaves, tile fragments, or other objects. Teachers are more frequently present a problem with the word (phrase) rather than inviting students into the real world that exists around the school [1]. They argue that by presenting a contextual problem, students can visualize the problem in real situations. Presenting math in everyday life increasingly necessary as technological developments, environmental, social and cultural constantly [2]. Students need to be exposed to real situations to apply their mathematical skills [3]. Mathematical problems that are often encountered in everyday life is a comparison.

Comparisons are always taught at every level of education from elementary to high school. In SD, the concept of comparison is taught about the scale and ratio. Then the material is taught again in junior high school and followed by a comparison of value and turning value. In general, the problem presented in the comparison material is only a comparison of one quantity. For example, "Rafi requires 10 workers to complete home construction in a month. He wants the house construction to finish faster and add 3 workers. How many days the house was completed? ". Or "Aaron needs 1 liter of oil to fry 50 chickens, how many liters does Aaron need to fry 150 chickens?". In reality, we often encounter comparative problems with more than one quantity. Suppose that at the time of building a house, not only the workers and the duration of the course are compared. Let's say the comparison of cement, water, and sand; comparison of workers, length of work, and money spent; iron, wire, and cement; and so forth. In this study, the researchers tried to present the problem of comparison with two objects: coffee and sugar on problem A and meatballs and noodles on problem B. Researchers got the idea after looking at research conducted by Van Dooren [4]. In the study, he discovered the phenomenon that students used comparative methods to solve arithmetic problems [5] and to solve the problem of finding unknown value [4].

Other studies discussing the problem of comparison [6,7,8,9]. Research conducted by Fernandez [9] shows that there is an increase in the use of addition method in elementary students, and there is a decrease in the use of the method of addition to junior high school students. The use of proportional methods in the summing situation increases in elementary and junior high. In addition, the presence or absence of integer ratios strongly influences this behavior but the nature of the quantity has little effect on the use of the comparison method. Comparative problems that allow for multiple representations make it difficult for students to understand and solve comparative problems [7]. While research Van Dooren [6] found that in the early years of elementary school, there was an improvement from the application of the sum method wherever the situation. In the following year, they use proportional methods wherever the situation is. Between these two stages, many students experience a transitional stage in which they simultaneously use the sum method for proportional problems and proportional methods of addition problems, switching between them based on the numbers given in the problem. This means that the type of number used in the problem also determines the selection of the settlement method. Another case is that students use the concept of comparative comparison to solve the problem of value-turning ratio and the students use the concept of reversing the value of the comparison to solve the problem of comparative worth. This shows that students' understanding of the problem of comparison is still not good.

The level of student understanding determines the truth of the settlement. In order for students to understand, teachers need to relate problems with everyday situations. Many teachers teach comparison only to solve the problem, i.e. by multiplying cross to find the unknown value. To understand the problem of comparison, one thing we have to understand is the sum and its relation to multiplication [10,11]. When students are wrong in solving comparative problems, there are two possibilities. First, the student has not understood the concept of complete comparison, so that he is wrong in choosing the method or wrong in the calculation. Second, teachers are unable to communicate the knowledge they have for students to understand.

Research with student subjects with comparative problems has been widely studied [6,9,12]. While research with subject teachers has been investigated by [1,2]. Lemonidis [13] found that prospective students experienced pseudo-proportion when making samples of comparative problems. In his research, there are students who make examples of non-comparison questions that are considered comparisons. Based on these problems, researchers are interested to describe the characteristics of students in solving the problem of comparison with two quantities mixed.

2. Methods and Material

This research was conducted on students of 4th semester academic year 2016/2017 in Yogyakarta. Of the 120 students divided into 3 classes, it is given the problem of comparison problem, as in Table 1.

Table 1. Comparison Problem

Problem	
Rafi every day always makes coffee with the same dosage. For 30 days, Rafi spends 600 grams of coffee and 1 kg of sugar. If today he bought 500 grams of coffee and ¾ kg of sugar, then how many days would Rafi be able to make coffee with existing supplies?	A
Bu Tumini requires 90 grains of meatballs and 3 kg of noodles to make 30 servings of meatballs. How many grains of meatballs and noodles are needed to make 100 servings of meatballs?	B

The matter is a matter of comparison with two objects compared in one question. Researchers assume that there are students who solve the problem by using two linear equations. The reason, on the matter, each of the two things. In question A, there is sugar and coffee; on problem B, there are meatballs and noodles. This is not very familiar to students, however, the problem can be found in everyday life. From the students' work, the researchers grouped into two groups, the group of students who completed correctly each of the questions and groups of students who completed with the wrong answers. The researcher chose the research subject of the group of students who answered wrongly by the purposive method. Researchers consider student completion and communication skills. Finally, selected three students who became the subject of research. Selected students are then interviewed for data. After the data collected, then performed the analysis, then interpreted. The process of data analysis in this study contains the following steps: (1) transcoding the collected interview data, (2) understanding all existing data from various sources (interviews, photos, videos, field notes, student answers), (3)) (4) categorize / organize the collected data by making the coding, (5) describe the student's thinking structure, (6) analyze the students' thinking process, (7) analyze the unique things (if any), and (8)) Make a conclusion.

3. Results and Discussion

3.1. Results

After examining the students' work, the researcher found that 15% of students were able to answer correctly about A and 45% for problem B. From that fact, it showed that students had difficulty in solving the problem of comparison of two objects in one problem. Uniquely, there are 5 students who use the method of elimination and substitution to solve the problem. We know that both methods are found on the subject of a two-variable linear equation system (SPLDV). It is possible that students use SPLDV because they have studied SPLDV before. In SPLDV learning, students will encounter problems with two variables and three conditions (including those asked). For example: *"Rafi bought 5 packs of coffee and 1kg of sugar, he had to pay Rp 25.000,00. At another time, Rafi bought 15 packs of coffee and 1.5 kg of sugar and had to pay Rp 50.000,00. What is the price of 1 pack of coffee and 2 kg of sugar?"*. From the example of the problem, there are two variables, namely coffee, and sugar. The first condition, 5 packs of coffee and 1kg of sugar, he must pay Rp 25.000,00, the second condition, 15 packs of coffee and 1.5 kg of sugar and must pay Rp 50.000,00, the third condition, the price of 1 pack of coffee and 2 kg of sugar. Unlike the A question: For 30 days, Rafi spends 600 grams of coffee and 1 kg of sugar. If today he bought 500 grams of coffee and ¾ kg of sugar, then how many days would Rafi be able to make coffee with existing supplies? This problem is not a problem that can be solved by the method of elimination or substitution. This problem is a matter of determining a value that does not yet exist, i.e. the number of days. Facts in the field, the researchers found there are three subjects with different characteristics different, namely: subject 1, he answered the first and second questions with methods of elimination and substitution (non-comparison); Subject 2, he answered the first question with the concept of comparison although the answer was wrong, and answered the second problem with the method of elimination and substitution (non-comparison); And subject 3, he answered both questions with the concept of comparison. On the first question, he answered wrong because he was unable to understand the problem, while on the second question he answered correctly. From the characteristics of the answer, the researchers divide into 3 groups based on thinking process, namely: blind-proportion, partial-proportion, and proportion thinking.

3.2. Discussion

Comparative problem is a complex problem. This makes students have to think deeply to solve it. Typically, the comparative problems that teachers present in schools only compare one object only. Suppose the mother needs 1 kg of sugar to make 40 cups of tea. How much tea is made if the supply of sugar as much as 750 grams? Of course, such problems will easily be solved. In this study, researchers present the problem of comparison with two objects at once. On problem A, the objects that are compared are coffee and sugar, while the problem B is meatballs and noodles.

Of all students, there are students (subject 1) who answered similarly to the answers written researcher. Actually, the subject has found that the dose of Rafi makes coffee per day is 20 grams of coffee and 33.3 grams of sugar. He also found that 500 grams of coffee would run out for 25 days, and 750 grams of sugar would run out for 22.5 days. However, he did not understand the sentence: Rafi every day always makes coffee with the same dosage. The same dosage means that the coffee and sugar ratio is always the same. Therefore, he wrote, "So, for 500 g coffee can be used for 25 days and for 750 grams sugar can be used for 22.5 days".

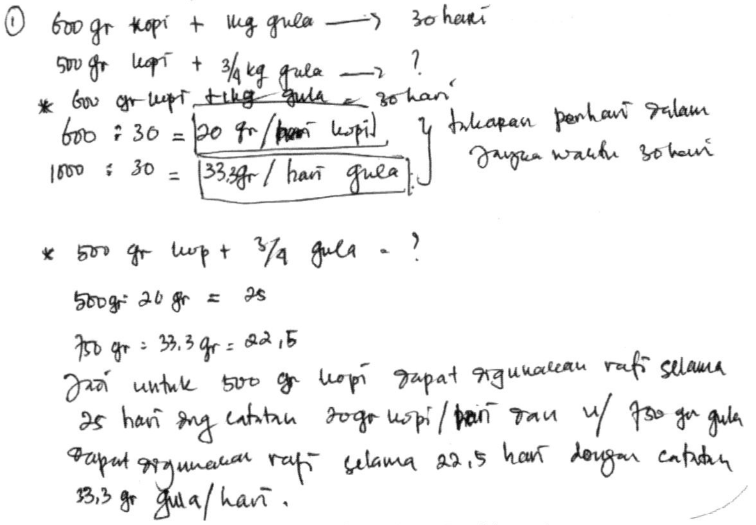

Figure 1. Subject 1 on Problem A

To explore further information, researchers conducted interviews on the subject of answers to the problem given. The interview snippet as presented in Table 2.

Table 2. Interview Researcher with Subject

Researcher	: From the answers, you have obtained, how many days Rafi can make coffee?
Subject	: If the coffee will run out for 25 days if the sugar will run out in 22.5 days.
Researcher	: So, there are two answers? Which one for coffee, the other for sugar?
Subject	: Yes, because there are two objects, then there are two answers.

From the interview transcript in Table 2, it shows that the Subject does not understand the situation in everyday life that if making coffee requires coffee and sugar. However, he considers coffee and sugar to be something separate. Thus, he found two answers, for coffee and sugar. If the job is seen, the Subject

IOP Conf. Series: Journal of Physics: Conf. Series **948** (2018) 012007 doi:10.1088/1742-6596/948/1/012007

has actually used the comparison method correctly. He also found the time coffee and sugar will run out. But he was not able to interpret the phrase "Rafi every day always make coffee with the same dose".

In contrast to problem A, on problem B, the subject does not use the comparison method to solve the problem. Instead, he uses the methods of elimination and substitution to solve comparative problems. This is a unique phenomenon. From the Subject's work, he gets the answer x = -3.1 and y = 103 (see Figure 2). At the beginning of the work, the subject did the example, namely by using variables x for meatballs and y variables for noodles. So, 90 grains of meatballs and 3 kg of noodles to make 30 servings of meatballs were changed to 90x + 3y = 30, and the meatballs and noodles needed to make 100 servings of meatballs were changed to x + y = 100. Then, from the equation, two linear equations (SPLDV). After using the method of elimination and substitution found x = -3.1 and y = 103. According to him, the interpretation of x = -3.1 and y = 103 is to make 100 servings of meatballs require -3.1 meatballs and 103 noodles. From the interpretation of the answers obtained, it appears that the subject just answers the problem by performing imitation procedures. He does not reflect on what he has written. Could meatballs amount to -3.1?

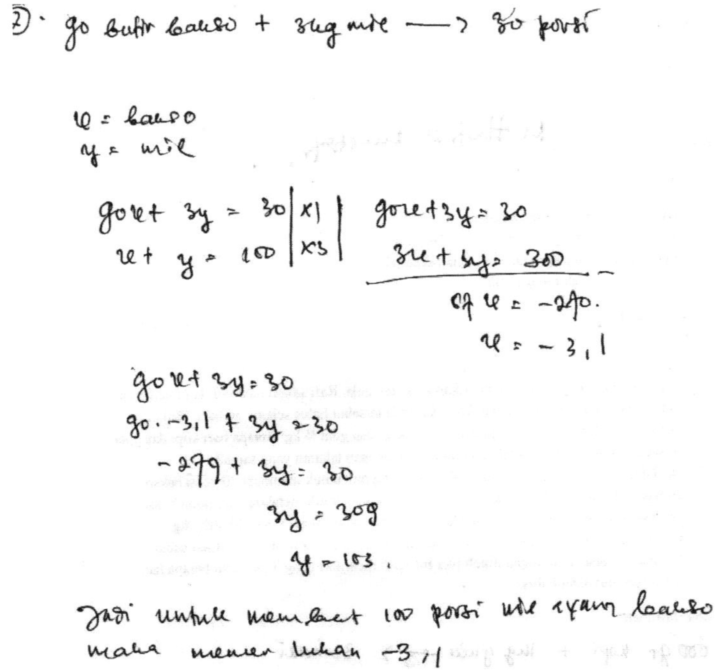

Figure 2. Subject 1 on problem B

To explore further information, researchers conducted interviews on the subject of answers to the problem given. The interview snippet as presented in Table 3.

Table 3. Interview Researcher with Subject

Researcher	: This second question, included in what subject?
Subject	: System of linear equations of two variables, sir.
Researcher	: Why?
Subject	: Because there are two variables, meatballs, and noodles.
Researcher	: Because there are two variables, then use SPLDV?
Subject	: Yes, as far as I know, sir.
Researcher	: Why is the first problem not using SPLDV?
Subject	: Mmm, which is the first question of comparison, sir.
.	
.	

Researcher	: The intent of x = -3.1 and y = 103 what is it?
Subject	: X the number of meatballs, and the number of noodles needed for 100 servings.
Researcher	: The number of meatballs is negative?
Subject	: Mmm, yes sir.
Researcher	: Can?
Subject	: No sir
Researcher	: then?
Subject	: Confused sir.

From the interview in Table 3, the subject understood the problem of comparison as a two-variable linear equation system problem. It uses the method of elimination and substitution to find the values of x and y. Having obtained the values x = -3.1 and y = 103, the subject does not reflect his work. Is the value of x = -3.1 and y = 103 satisfying the existing equation? Could x be negative? When the researcher asks why x is negative, it cannot answer correctly.

In contrast to subject 1 work, the subject's work produces the wrong answer to both questions. On the first question, he understood the problem as a matter of comparison. He writes the comparison relationship of coffee, sugar, and the day by writing $\frac{600}{500} = \frac{1}{3/4} = \frac{30}{x}$. From the equation, it should be split into two forms of the equation, i.e. $\frac{600}{500} = \frac{30}{x}$ to find the number of days to spend coffee and $\frac{1}{3/4} = \frac{30}{x}$ to find the number of days to spend sugar. Subject 2 uses cross product at once from three comparisons (see Figure 3).

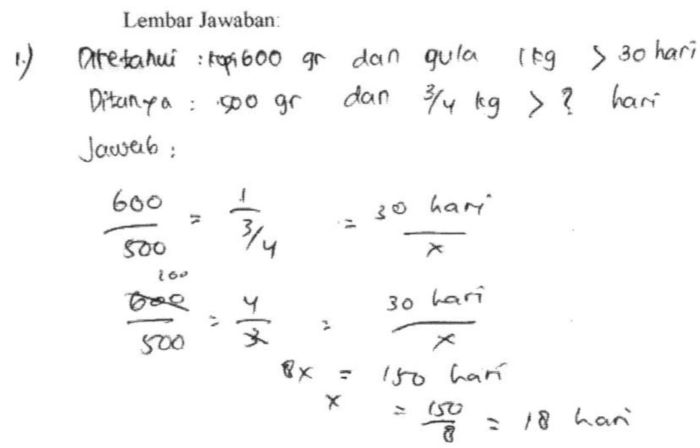

Figure 3. Subjek 2 on problem A

To explore further information, researchers conducted interviews on the subject of answers to the problem given. Interview fragments as presented in Table 4.

Table 4. Interview Researcher with Subject

Researcher	: What does the equation $\frac{600}{500} = \frac{1}{3/4} = \frac{30}{x}$ mean?
Subject	: A comparison of 600 grams of coffee with 500 grams of coffee equals 1 kg of sugar compared to ¾ kg of sugar equal to 30 days compared with x days.
Researcher	: Why use 'equal to' sign?
Subject	: Because the same proportion, sir.
Researcher	: Then, the next step, why are you streaking 600 and 3?
Subject	: Let it be simpler sir.
Researcher	: obtained 8x = 150 from where?
Subject	: 200 equals 500 simplified, then 2 multiplied by 4 and 5 multiplied by 30.
Researcher	: Why is that?

Subject	: This is a comparison, sir, so it is way with cross product.

From the conversation quote in Table 4, it can be observed that the Subject at the beginning of the job he wrote down the formula correctly. However, he did not understand the meaning of the comparison in depth. Thus, he only runs the procedure of completing the comparison by means of cross product. The cross-product he's working on is not right. If it is correct, it certainly will not multiply 2 by 4 and 5 by 30 simultaneously.

In the second question, Subject 2 is working like subject 1 in the same question. It uses a two-variable linear equation system to answer questions. What distinguishes Subject 1's answer from subject 2 is on the value of x. If subject 1 gets the value of x = -3.1 in subject 2 the value of x is 3.1 ≈ 3. In fact, as in Figure 4, the subject writes 3x + 3y = 300 - (90x + 3y = 30). Supposedly, it gets -87x = 270, and x = -3.1. This means, in addition to the notion of comparisons that have not been good, subject 2 also has an understanding of the operation of algebraic forms that have not been good.

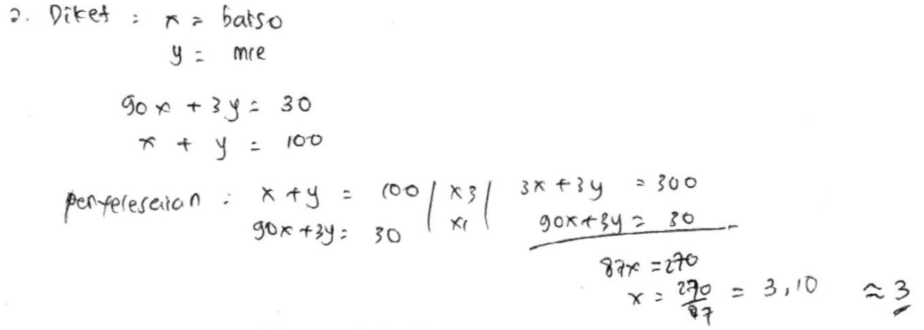

Figure 4. Subjek 2 on problem B

The researcher conducted an interview with Subject 2 concerning the work on the second problem. As for the interview quote as in Table 5.

Table 5. Interview Researcher with Subject

Researcher	: Why are you using SPLDV?
Subject	: Because there are two objects, sir, meatballs and noodles.
Researcher	: 87x = 270 obtained from where?
Subject	: From 3x + 3y = 300 minus 90x + 3y = 30.
Researcher	: Already correct?
Subject	: Mmm, it should be -87x, sir.
Researcher	: Why are you writing 87x?
Subject	: So x is not negative, sir.

Subject 2 is aware that he made a mistake while writing 87x = 270. He wrote 87x = 270 to avoid negative x values. He realizes that the value of x is unlikely to be negative. However, he is like subject 1, using SPLDV to solve a comparison problem with two objects.

In subject 3, he worked on A and B with the concept of comparison. On problem A, he answers wrongly, and on problem B he answers correctly. In problem A, he initially makes two equations: 600x + y = 30 and 500x + 0.75y = a. It exemplifies coffee as x, and sugar as y, as in Figure 5.

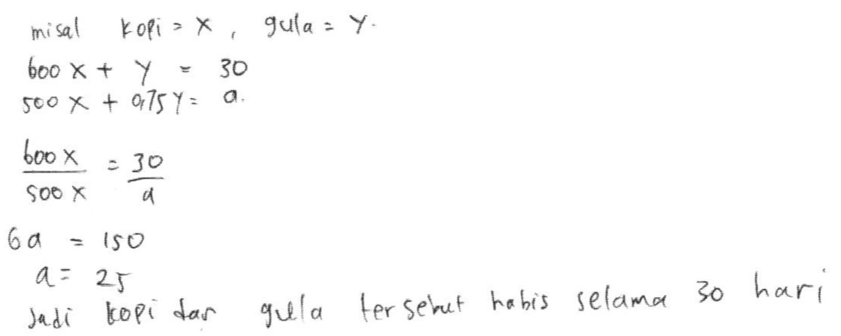

Figure 5. Subjek 3 on problem A

To get more information, the researcher conducted an interview with Subject 3 on the job on the first problem. As for the interview quote as in Table 6.

Table 6. Interview Researcher with Subject

Researcher	: What is the meaning of 600x + y = 30 and 500x + 0.75y = a?
Subject	: 600 grams of coffee and 1 kg of sugar can make coffee for 30 days. The second, 500 grams of coffee and ¾ kg of sugar to make coffee during 'a' day.
Researcher	: Why did you make those two equations?
Subject	: Because there are two variables.
Researcher	: Then, you use what way to solve it?
Subject	: Comparison.
Researcher	: Your purpose to write the two equations for what?
Subject	: To make it easy.
Researcher	: What does $\frac{600}{500} = \frac{30}{a}$ mean?
Subject	: Comparison of coffee and day.
Researcher	: So, that's it?
Subject	: Yeah, because when the coffee is gone, can not make more coffee.

From the conversation, it can be seen that the subject understands if the given problem is a matter of comparison, although at first, he makes two equations. He makes two similarities in order to facilitate workmanship. In contrast to subject 1 which makes two equations to be used as a system of two linear equations. Subject 3 means that if the coffee has run out, it can not make more coffee. He does not pay attention to the sugar and the dosage. In fact, supplies of sugar and coffee are not balanced, 500 grams of coffee and ¾ kg of sugar. Of course one will be exhausted first.

On the question of B, it starts as in A. It tells meatballs as x and noodles as y. After that, he makes two equations 90x + 3y = 30 and ax + by = 100 which is used to facilitate the workmanship. In contrast to A's work, he makes a comparison relationship $\frac{100}{30} = \frac{ax}{90x} = \frac{by}{3y}$ as in Figure 6. After that, it separates into $\frac{100}{30} = \frac{ax}{90x}$ and $\frac{100}{30} = \frac{by}{3y}$. Then, he got the answer 300 grains of meatballs and 10 kg of noodles to make 100 servings of meatballs.

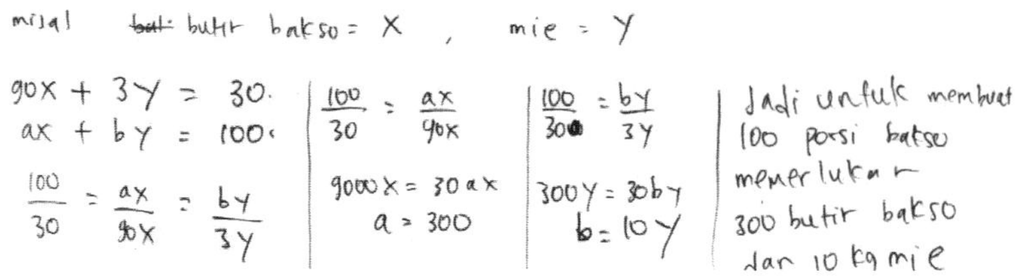

Figure 6. Subjek 2 on problem B

To get more information, the researcher conducted an interview with Subject 3 on the job on the first problem. As for the interview quote as in Table 7.

Table 7. Interview Researcher with Subject

Researcher	: What is the meaning of 90x + 3y = 30 and ax + by = 100?
Subject	: 90but meatballs and 3 kg noodles to make 30 servings of meatballs. And how many meatballs and noodles for 100 servings.
Researcher	: Why did you make those two equations?
Subject	: Because there are two variables.
Researcher	: Then, you use what way to solve it?
Subject	: Comparison.
Researcher	: Your purpose to write the two equations for what?
Subject	: To make it easy.
Researcher	: What the purpose of $\frac{100}{30}=\frac{ax}{90x}=\frac{by}{3y}$?
Subject	: Comparison of 100 portions versus 30 servings and ‘a’ grain of meatballs with 90 grains of meatballs, and ‘b’ kg of noodles and 3 kg of noodles.
Researcher	: After that, why do you split up into $\frac{100}{30}=\frac{ax}{90x}$ with $\frac{100}{30}=\frac{by}{3y}$.
Subject	: Because of the different, one for meatballs and the other for noodles.

From the conversation, it can be seen that the subject understands if the given problem is a matter of comparison, although at first he makes two equations. He makes two similarities in order to facilitate workmanship. Unlike the work on the first problem, that he interpreted that if the coffee is up, it can not make more coffee. In the second work, the subject separates the comparisons into two parts, namely the ratio of meatballs to the portion and the ratio of noodles to the serving. He realized that there were two types of things that needed to be determined.

4. Conclusion

The problem of comparison is closely related to the problems in everyday life such as those expressed by Silvestre and da Ponte, and Boyer and Levine [7,12]. To understand the problem of comparison is not something easy. Comparative problem is a complex and basic problem for understanding algebra. The complexity of the problem in comparison because in comparison can be attributed to other material. The findings of this study may be one of the reasons why students have difficulty in solving comparative problems [6,8,9]. Student prospective teachers are the next generation of teachers who will be teaching students at the school. If teachers have an understanding of the concept of superficial comparison, of course the students also experience the same thing. Therefore, prospective teachers need to review lessons in elementary, junior and senior high schools.

References

[1] Verschaffel L, Greer B and De Corte E 2000 *Making Sense of Word Problems* (Netherlands: Swets & Zeitlinger) ***6*** 2
[2] Rosa M and Orey D C 2015 *A trivium curriculum for mathematics based on literacy, matheracy, and technoracy: an ethnomathematics perspective* **4** 587-598
[3] Rosa M and Orey D C 2010 *Revista Latinoamericana de Etnomatemática* **3** 2
[4] Van Dooren W, De Bock D, Evers M and Verschaffel L 2009 *Journal for Research in Mathematics Education* 187-211.
[5] Van Dooren W, De Bock D, Hessels A, Janssens D and Verschaffel L 2004 Students' Overreliance on Proportionality: Evidence from Primary School Pupils Solving Arithmetic Word Problems *International Group for the Psychology of Mathematics Education*
[6] Dooren W V, Bock D D and Verschaffel L 2010 *Cognition and Instruction* **3** 360-381
[7] Silvestre A I and da Ponte J P 2012 Missing value and comparison problems: What pupils know before the teaching of proportion *PNA* **3** 73-83 retrieved from http://hdl.handle.net/10481/19500
[8] Irfan M 2017 *Jurnal Penelitian Pendidikan dan Pengajaran Matematika (JP3M)* **1** 4
[9] Fernández C, Llinares S, Van Dooren W, De Bock D and Verschaffel L 2012 *European journal of psychology of education* **3** 421-438
[10] Son J W 2013 *Educational studies in mathematics* **1** 49-70
[11] Vera F and Lloyd P 2016 Proportional reasoning ability of school leavers aspiring to higher education in South Africa *Pythagoras* **1** 1-10
[12] Boyer T W and Levine S C 2012 *Journal of Experimental Child Psychology* **111** 516-533 doi:10.1016/j.jecp.2011.11.001
[13] Lemonidis C 2008 Prospective teachers' application of the mathematical concept of proportion in real life situations *Research in Mathematics Education* p 163-172

Acknowledgments

Thanks to all those who assist in the implementation of this research, especially LPDP as the main sponsor.

Correlation among science process skill, concept comprehension, and scientific attitude on regulation system materials

P M Sari[1], F Sudargo[2] and D Priyandoko[2]

[1]Universitas Muhammadiyah Prof. Dr. HAMKA, Jakarta, Indonesia
[2]Biology Education, Department of School of Postgraduate Studies, Indonesia University of Education, Bandung, West Java, Indonesia

E-mail: primamutia@gmail.com

Abstract. The aim of this research was to analyze the correlation between science process skill and concept comprehension as well as scientific attitude. The research method was a descriptive correlational. The research population consisted of all students in class XI in one of senior high school in West Sumatra with a total population of 100 students. The sample consisted of 26 students which were taken by a cluster random sampling. The instrument consisted of a science process skills test, concept comprehension test, and scientific attitude scale. The data were analyzed using linear regression test. The results showed that there was a significant correlation between science process skill and concept comprehension. In addition, there was a significant correlation between concept comprehension and scientific attitude. Nevertheless, there was not a significant correlation between science process skill and scientific attitude.

1. Introduction

Science is basically a process of discovery. This is consistent with the background of the importance of science in the Ministry of Education [1] which stated that the Natural Sciences deals with how to find out (inquiry) about the systematic nature. Therefore, science is not just a mastery of knowledge in the form of a collection of facts, concepts or principles, but also a process of discovery.

Learning science should be more emphasis on the process, where students are active during learning to build knowledge through a series of learning activities that are meaningful to students. Biology as one of the areas of science provides a variety of learning experiences to understand the concepts and processes of science.

Science process skill is basic abilities which are needed for use and understanding of science [2]. Furthermore, according to German [3] there are two skills that are used in the world of science. Firstly, basic science process skills that related to the ability to conduct empirical - inductive reasoning as well as providing basic skills in conducting scientific investigations. Secondly, integrated science process skills that associated with ability to perform the hypothetic-deductive reasoning. These skills are used to solve problems or conducting experiments. Science process skills cannot be separated in practice understanding of the concepts involved in a study and application of science especially biology.

Moreover, according to Rooser [2] concept is an abstract that represents objects, events, activities, and relationships which have similar attributes. A concept is a mental abstraction that represents level of stimulus. Concept comprehension could be measured by multiple choice test of concept comprehension. A set of test was made suitable to new Bloom Taxonomy. There are six levels of cognitive domain that appropriate with new Bloom taxonomy. They are remembering, understanding, applying, analyzing, evaluating, and creating.

In addition, science learning is also expected to build students' scientific attitude. According to Baharudin [4] scientific attitude is basically the attitude shown by scientists as they perform activities as a scientist. If a good scientific attitude has been embedded in a student then it is expected this attitude will also remain inherent in everyday life. Scientific attitude is needed in order to build the character of the nation as one of the efforts in overcoming various problems facing the nation, such as student brawl, corruption and so on.

All this time, the research that aimed to reveal the relationship between science process skills, concept comprehensions and scientific attitudes of students were still rare. Therefore, the researcher wanted to analyse these aspects. Whether the student who has a high science process skill showed a high concept comprehension and scientific attitude and vice versa. Furthermore, whether student who has a high concept comprehension showed a high scientific attitude and vice versa.

2. Methods

The research method was a descriptive correlational to see the correlation between science process skill and concept comprehension as well as scientific attitude of regulation system materials.

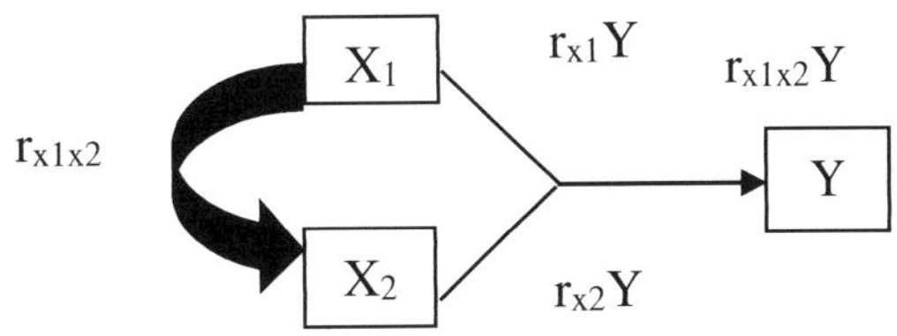

Explanation:

X_1 : science process skill
X_2 : scientific attitude
Y : concept comprehension

The research population consisted of all students in class XI in one of senior high school in West Sumatra with a total population of 100 students. The sample consisted of 26 students which were taken by a cluster random sampling.

3. Results and Discussion

The data was analysed using linear regression test. The result of the test can be seen in Table 1.

Table 1. The Correlation Between Score of Science Process Skill, Score of Concept Comprehension and Score of Scientific Attitude

Variable	*Sig-1 tailed*		Criteria of significance level	Conclusion
	Pearson Correlation	*Spearman* Correlation		
Science process skill and concept comprehension	0,007	0,03	0,05	There is a significant correlation
Scientific attitude and concept comprehension	0,012	0,014	0,05	There is a significant correlation
Science process skill and scientific attitude	0,052	0,1	0,05	There is no significant correlation

For each variable that has significant correlation, the regression equation could be searched as shown in the table below.

Table 2. The Correlation Coefficient, Determination Coefficient and Regression Equation of Science Process Skill, Scientific Attitude and Concept Comprehension

Variable	Correlation Coefficient (r)	Determination Coefficient (R^2)	Regression Equation
Science process skill (X_1) and Concept comprehension (Y)	0,475	0,226	$Y = 0{,}567X_1$
Scientific attitude (X_2) and Concept comprehension (Y)	0,442	0,196	$Y = 0{,}771X_2$

Based on analyzing of correlation between science process skill (X1) and scientific attitude (X2) and concept comprehension (Y), we could see there was a significant correlation between science process skill and concept comprehension, moreover, there was a significant correlation between concept comprehension and scientific attitude. Meanwhile, there was no significant correlation between science process skill and scientific attitude.

The science process skill affected concept comprehension. It meant that if the science process skill was high, then the concept comprehension also increased. This was suitable for the opinion of Rustaman, et al. [5] which stated that student could learn a concept as a science learning goal learning through a scientific process approach. Thus, science process skill also supported concept comprehension on regulation system materials.

Moreover, it can be seen that correlation coefficient (r) was 0.475 and determination coefficient (R2) was 0.226. It meant that 22.6% of concept comprehension was affected by science process skill. Meanwhile, the remaining 77.4% was influenced by other factors such as motivation, interest in student learning and the ability of teacher for explaining a concept.

Moreover, scientific attitude influenced concept comprehension. It meant that if the scientific attitude was high, then the concept comprehension also increased. This was in line with Sukaesih's opinion [6] which stated that scientific attitude was affected by the level of student information mastery.

According to Slameto [7], other factors that affected student learning outcomes was the attitude. Attitude is something that is learned, and attitude determines how individuals react to situations and determine what individuals search for their life.

Lack of student positive attitude in learning could lead to low student learning outcome. Scientific attitude strongly supported student learning activities in a positive direction. Therefore, it could be concluded that student scientific attitude influenced student learning outcome which in this case was the concept comprehension of regulation system.

Furthermore, it can be seen that correlation coefficient (r) was 0.442 and determination coefficient (R2) was 0.196. It meant that 19.6% of concept comprehension was affected by scientific attitude and 80.4% was affected by other factors such as student age, student background, and student learning motivation.

According to the data, there was no significant between science process skill and scientific attitude. This was not suitable with the result of Zeidan [8] that revealed where there was a significant positive correlation between science process skills and attitudes toward science among Palestinian secondary school students has been found. The difference of this result might be caused by the number of samples that used was low. Therefore, this result could not be generalized.

IOP Conf. Series: Journal of Physics: Conf. Series **948** (2018) 012008 doi:10.1088/1742-6596/948/1/012008

4. Conclusion

Based on the research that has been done can be concluded that there was a significant correlation between science process skill and concept comprehension. In addition, there was a significant correlation between concept comprehension and scientific attitude. Nevertheless, there was not a significant correlation between science process skill and scientific attitude. Based on the conclusions that have been drawn, the researchers gave the following recommendations. First, teachers should develop learning that can improve the science process skill, because science process skill has a correlation with the students' conceptual understanding. Second, the number of participants for correlation research should be larger.

References

[1] Depdiknas 2006 *Badan Nasional Standar Pendidikan* (Jakarta: Depdiknas)
[2] Dahar R W 1996 *Teori-Teori Belajar* (Jakarta: Erlangga)
[3] Karamustafaoğlu and Sevilay 2011 *Eurasian Journal Physics Chemistry. Education* **3** 1 pp 26-38
[4] Baharudin 1982 *Peranan Kemampuan Dasar Intelektual, Sikap dan Pemahaman dalam Fisika terhadap Kemampuan Siswa di Sulawesi Selatan Membangun Model Analog dan Model Mental* (Disertasi IKIP Bandung)
[5] Rustaman, Nuryani Y, Soendjojo D, Suroso A Y, Yusnani A, Ruchji S, Diana R and Mimin N K 2003 *Strategi Belajar Mengajar Biologi* (Bandung: Jurusan Pendidikan Biologi FPMIPA UPI)
[6] Sukaesih and Sri 2010 *Pembelajaran Berbasis Praktikum Dengan Menerapkan Asesmen Tes Lisan Pada Topik Keanekaragaman Hayati Untuk Mengembangkan Kemampuan Berpikir Kritis dan Sikap Ilmiah Mahasiswa* (Tesis pada SPS UPI Bandung)
[7] Slameto 1995 *Belajar Dan Faktor-Faktor Yang Mempengaruhi* (Jakarta: Bina Aksara)
[8] Zeidan Afif H and Majdi R J 2015 *World Journal of Education* **5** 1

The influence of mathematics learning using SAVI approach on junior high school students' mathematical modelling ability

H Khusna[1] **and N Y Heryaningsih**[1]

[1]Universitas Muhammadiyah Prof. Dr. HAMKA, Jakarta, Indonesia

E-mail: hikmatulhusna@uhamka.ac.id

Abstract. The aim of this research was to examine mathematical modeling ability who learn mathematics by using SAVI approach. This research was a quasi-experimental research with non-equivalent control group designed by using purposive sampling technique. The population of this research was the state junior high school students in Lembang while the sample consisted of two class at 8th grade. The instrument used in this research was mathematical modeling ability. Data analysis of this research was conducted by using SPSS 20 by Windows. The result showed that students' ability of mathematical modeling who learn mathematics by using SAVI approach was better than students' ability of mathematical modeling who learn mathematics using conventional learning.

1. Introduction

School is an educational institution which gives a conducive space for the individual to acquire education. Producing good human resources, an educational institution must facilitate students to explore their skill and support them in having life skill. Syaodih (Turmudi & Yakub) stated that the benefits of education are to help students in developing all of potential, competent, and improve their personal characteristics in a positive direction, both within themselves and their surroundings [1]. So in the process of learning needs to be considered various aspects so that students' development can be achieved well.

Tata argues that improving students' mathematical modeling abilities is an effort to improve students' ability to problem-solving through enclose modeling as a part of mathematics learning process [2]. Mathematical modeling is used in math word problems. The students are asked to formulate real problems into mathematical modeling which students then look for such solutions using mathematical modeling. However, not all students can easily understand mathematical modeling [3]. Crouch & Haines stated that the difficulty aspect of students in modeling is the interface between real-world problems and mathematical models, that is the transition from the real world problems to the mathematical model and the transition from mathematical model to the real world.[3] While the results of the application do not indicate a particular difficulty for students to stage the modeling process.

The preliminary study revealed that students have difficulty in formulating problems in everyday life into mathematical models, such as interpreting the real problem context into mathematical modeling, understanding the structure of mathematics (including regularity, connection, and pattern) in problem [4]. Parlaungan stated that the students' failure in modelling can be caused by the failure of students to understand (identify) the problem, inability to transform the real problem to mathematical modelling, ignorance of the underlying mathematical concepts toward modelling, inability to relate data with mathematical principles and inability to solve the given mathematical model [3].

Many causal factors make students' mathematical modeling abilities are low. One other thing is the habit of learning students in school. In learning of mathematics, many students know mathematics but cannot understand mathematics concept, students prefer work mathematics in mathematical modeling with calculation mathematics only, and student not trying to understand the means of that mathematical modeling. Furthermore, finding the final solution is the main purpose of students in learning mathematics and ignore the process to find the mathematical modeling.

Ebutt & Strakker argue that to make the potential of students are develop and optimize in mathematics, assumptions about the characteristics of the students are given include: (1) students can learn mathematics if they have motivation; 2) students learn with their style; 3) Students learn mathematics with themselves or teamwork with their friends [5]. One of learning which makes develop the potential of students and learns mathematics optimally with involving learning style is learning mathematics with SAVI approach.

Learning with SAVI approach is learning which involve learning style who have students. Acronym of SAVI is somatic, auditory, visual, intellectual. The principal of SAVI approach is students active in physical activity or intellectual activity in the learning process. Furthermore, SAVI approach is followed modern cognitive science which argues that the best of learning is involve emotion, all of the body, all of the sense, grace other learning styles with consciously that everybody learns with their ways [6].

Meier stated that students can improve the ability to express their idea (Intellectual) if they are move something (Somatic) to produce pictorial, diagram, graphics, and others (Visual) while discussing what they doing (Auditory) [7]. The activity of SAVI approach in this study is: somatic learning is learning with involving move their body, active learning can facility student somatic cause in active learning the student acquitted to move their body in learning. Auditory learning is learning with involve the sense of hearing, the students discuss with their group and listen to the result of discussion other groups. Visual learning is learning with involve the sense of vision, the student makes mathematical modeling by pictures and make mind map in the end of learning to make light of students construct their knowledge. intellectual learning is learning to solve problems of mathematical modeling.

Based on the study has been described previously, SAVI approach can influence students' ability to mathematical modeling. Study on the improvement of students' ability of mathematical modeling has done. The result of study was then written in the form of report title the article "The Influence of Mathematics Learning Using SAVI Approach to Students' Ability of Mathematical Modelling at Junior High School"

2. Methods

This study is designed in the form of experiment with the design of nonequivalent control group which part of the quasi experiment. The sample in this study was not randomly selected but classes by using existing classes. The experimental class was treated SAVI approach (SA) and the class is given control of conventional learning (CL). Before the study was held, the researcher did documentation observation in the class were treated SAVI approach and class were tread conventional learning to find out the similarity of students' ability both. After the study is completed the final test is held (O), which test the students' ability of mathematical modeling (AMM).

The instrument used in this study is the ability of mathematical modeling test. The mathematical modeling test is a test that aims to measure the students' ability to mathematical modeling. Mathematical modeling is abilities that include the ability to perform the structuring, mathematization, solving, interpreting, validating.

This study is performed in junior high school students in Lembang, West Bandung, West Java Province, Indonesia. Population in this study is eighth-grade students in junior high school in Lembang.

The sample is taken in two class which chosen with purposive technique. Sampling with purposive technique is technique sampling which based on certain consideration [8]. The Consideration sampling in this study is characteristics and capabilities of class equivalent. To decided that class has characteristics and capabilities equivalent, the researcher get information based on teacher who teaches in school of research.

3. Result and Discussion

Based on the data analysis, the students' ability of mathematical modeling which given treated SAVI approach is better than the students' ability of mathematical modeling which given treated conventional learning. It seen from statistic test, it indicates that students' ability to mathematical modeling in SA and CL is different significant.

Documentation studies done to determine the quality of mathematical prior knowledge students have prior to this study and to see equivalent samples. Documentation studies data source by data middle semester test (MST). The average value of MST shown in Table 1.

Table 1. Average Value of MST for Class SA and Class CL

Type of Ability	Group	
	Experiments (n = 38)	Control (n = 40)
The average MST value ($\bar{x}$)	47.05	52.83
Standard deviation (s)	16.86	16.57
Ideal maximum Score	100	100

Based on Table 1 average value of MST for class SA and class CL, SA score average is 47.05 and CL score average 52.83. The standard deviation of SA is 16.86 and 16.57. Despite average in SA and CL is different, but then given the similarity of the two test results mean MST between the two groups shows no difference in average values of MST between the two groups at the 0.05 significance level.

Table 2. Average Value of AMM for Class SA and Class CL

Type of Ability	Group	
	SA (n = 38)	CL (n = 40)
The average AMM value ($\bar{x}$)	69.37	63.78
Standard deviation (s)	9.67	11.18
Ideal maximum Score	100	100

Based on Table 2 average value of AMM for class SA and class CL shows that scoring average SA is 69.37 and score average CL is 63.78. The standard deviation of SA is 9.76 and standard deviation of CL is 11,18. The fact shows that data CL more spread that data SA. The average between two groups shows that average value of AMM in class SA higher than the average value of AMM in class CL. Based on average value which obtains in description of statistic, it shows that AMM in class SA is better than AMM in class CL.

Table. 3 Result of Treatment Value Average of AMM

		t-test for Equality of Means						
							95% Confidence Interval of the Difference	
		t	df	Sig. (2-tailed)	Mean Difference	Std. Error Difference	Lower	Upper
Data	Equal variances assumed	2,358	76	,021	5,593	2,372	,868	10,319

Thus, the result of similarity of the two test results means AMM shows that there is a different average values AMM between two groups at the 0.05 significance level. It showed that result of significant value in the similarity of the two test results mean AMM is 0.021, it less than the significance level, 0.05. It means that AMM in class SA is better than AMM in class CL.

After the study about students' ability of mathematical modeling has done, the researcher found some particular difficulty for students to learn the mathematical modeling. Students difficulty to determine the concept of mathematics that used to solve the problem in real-world context. Students are still difficult to perform counting operations, there are students who are still confused do the multiplication of decimal numbers, this resulted in fewer students mastering the stage in doing mathematical modeling is the stage of solving. Another difficulty is changing the solution of a mathematical problem into a real-world problem-solving. This can be seen from the exercise of the problem and the result of the test post that is done, in the solving stage the students do not experience difficulties, but in the interpreting stage, many students experience errors. Students' habits do math problems solely on the solving stage, solving mathematical problems with math, making it difficult for students to interpret the numbers they can. What's more, the math problems that students normally use do not use real-world contexts. As a result mathematics and the real world as separate, whereas the emergence of mathematics because of a problem in the real world which is then solved by using mathematics and translated back as a solution in the real world.

4. Conclusion

Based on the result and discussion, it can be concluded that students' ability of mathematical modeling who acquire mathematics using SAVI approach is better than using conventional class.

References

[1] Turmudi and Yakob H 2009 *Panduan Teknis Peningkatan Kemampuan Siswa melalui Proses Pembelajaran Berbasis Motivasi* (Jakarta: Direktorat Jendral Sekolah Menengah Atas, Departemen Pendidikan Nasional)

[2] Tata 2013 Mathematical Modeling dalam Pendidikan Matematika *Prosiding Seminar Nasional Matematika dan Pendidikan Matematika* **1** pp 408-414

[3] Parlaungan 2008 *Pemodelan Matematika untuk Peningkatan Bermatematika Siswa Sekolah Menengah Atas (SMA)* (Medan: Thesis Sekolah Pascasarjana, Universitas Sumatra Utara)

[4] Edo S I, *et.al* 2013 *IndoMS. J.M.E* **4** 1 pp 41-58

[5] Depdiknas 2006 *Kurikulum Tingkat Satuan Pendidikan (KTSP)* (Jakarta: Pusat Kurikulum Balitbang Depdiknas)

[6] Meier D 2000 *The Accelerated Learning Handbook* (New York: McGraw-Hill)

[7] Fitrianingsih I 2009 *Pembelajaran Matematika dengan Pendekatan SAVI Ditinjau dari Motivasi Belajar Siswa* (Surakarta: Thesis Universitas Muhammadiyah Surakarta)

[8] Sugiyono 2013 *Statistika untuk Penelitian* (Bandung: Alfabeta)

Acknowledgments

I would like to thank Prof. Darhim M.Si, the Indonesia University of Education for his guidance in listening to me and revising this manuscript. And I would also like thank Khaerunissa, M.Pd for her help in translate this manuscript.

Fulfilling the law of a single independent variable and improving the result of mathematical educational research

H Pardimin[1] and N Arcana[1]

[1]Pendidikan Matematika, Universitas Sarjanawiayata Tamansiswa

E-mail: nyoman.arcana@ustjogja.ac.id

Abstract. Many types of research in the field of mathematics education apply the Quasi-Experimental method and statistical analysis use t-test. Quasi-experiment has a weakness that is difficult to fulfil "the law of a single independent variable". T-test also has a weakness that is a generalization of the conclusions obtained is less powerful. This research aimed to find ways to reduce the weaknesses of the Quasi-experimental method and improved the generalization of the research results. The method applied in the research was a non-interactive qualitative method, and the type was concept analysis. Concepts analysed are the concept of statistics, research methods of education, and research reports. The result represented a way to overcome the weaknesses of quasi-Experiments and T-test. In addition, the way was to apply a combination of Factorial Design and Balanced Design, which the authors refer to as Factorial-Balanced Design. The advantages of this design are: (1) almost fulfilling "the low of single independent variable" so no need to test the similarity of the academic ability, (2) the sample size of the experimental group and the control group became larger and equal; so it becomes robust to deal with violations of the assumptions of the ANOVA test.

1. Introduction

The success of teachers in teaching is influenced by many factors. One factor is the method of teaching. Therefore, many students of the Faculty of Education who are writing their thesis conduct researches on a comparison of two or more methods of teaching. In the field of Mathematics education, for example, students do a comparison between the effectiveness of Example and Non-Example method with the conventional method in learning geometry. The research method used is usually a Quasi-experimental method. Some examples of researchers using quasi-experimental research methods are studies conducted by [1,2]; this research aims to know the difference of effectiveness between two teaching methods in learning Mathematics. The quasi-experimental method is used because it is not possible to do sampling using simple random sampling directly on students to form experimental group and control group. This is because students have been formed in the classroom (intact group).

The quasi-experimental method is a research method that uses all students who have formed in one intact group as a sample, an example is a group of students in one class [3]. Therefore, the samples are taken in the class, not the students [4]. The quasi-experimental method according to its definition, is weak in its randomization. The word "quasi" in Latin means almost or near, so a quasi-experiment can be interpreted as "near-experiment" [5].

This method requires the researcher to identify the control group as similar or as close as possible to the experimental group in the underlying characteristics [6], such as academic ability and student motivation. Since it is difficult to find two "similar" classes, the main problem faced by the quasi-experimental method is that it is difficult to meet a "law of the single independent variable". The law of a single independent variable is a law essentially stating that all conditions in the experimental class must be the same as the control class except the teaching method (Mill, 1986 in [7]).

The data in the research [1,2] were analyzed using a t-test of two mean differences. The significance of the mean differences between the two sets of data can be tested using the t-test of two mean differences [8]. However, the t-test applied to investigate the significance of the difference in effectiveness between the two teaching methods has weaknesses. The disadvantage is that the conclusions obtained through the t-test are incapable of seeing whether differences of effectiveness occur in groups of students with high, moderate or low academic ability [9]. So the conclusions of the research obtained through the t-test are less extensive in its generalization.

From the above description, it appears that the quasi-experimental method and t-test has a weakness. The weakness of the use of quasi-experimental research method is difficult to fulfill "the law of a single independent variable," a law stating that the only one difference is allowed to exist between the two classes, that is the teaching method. The weakness of using the t-test is only able to see if there are differences in the effectiveness of the classical but is unable to see how the differences in each group of students' academic ability (the ability of high, medium and low). In fact, the difference in academic ability is very necessary to be considered in the learning of Mathematics.

This research aims to find ways to reduce the weaknesses of the quasi-experimental method and improves the generalization of the research results.

2. Method

The research applied non-iterative qualitative method, also called analysis research. The type of analysis used was the concept analysis. The concepts analyzed were the concept of statistics, educational research methodology, and Mathematics education research report.

The data were collected with a literary technique. This technique collects data by reviewing documents and archives on matters related to the research. The data sources were the documents.

The data analysis was carried out by identifying, analyzing, and synthesizing concepts from the collected data. Then, the interpretation of the results of the synthesis was carried out to discover a new concept.

3. Results and Discussion

The results of this research are the concept of the relationship between quasi-experimental method and the law of single independent variable; the concept of the relationship between quasi-experiment and Balanced Design; the concept of relationship among t-test, Factorial ANOVA, and the generalization of research result; concept of the robust of the F-test, concept of Balanced-Factorial Design, and the procedure of applying Balanced-Factorial Design.

There are two general categories of experimental research methods, namely: true experimental and quasi-experimental [5]. Quasi-experimental research may be regarded as the best effort of experimental research when it is impossible or unreasonable to meet all the criteria of a true experimental. In the true experimental method, the study subjects (students) are taken randomly, while on quasi-experiment randomization is done to the class.

As a result of randomization performed in the class, there are many problems in interpreting the observed difference between groups [10]. In Mathematics for example, perhaps most of the experimental class backgrounds are better at Math than the control class. It shows that quasi-experiments are weak in the fulfillment of the fundamental similarity of characteristics between the experimental class and control class.

Basically, the experimental method is based on two assumptions about variables, [7] one of which is: if the two situations are the same in all respects except for the variables are added or removed from one situation, then the differences later emerged between the two situations can be associated with these variables. This statement is called "the law of a single independent variable".

The core of the experimental method is control. One of the variables that must be controlled is the variable of students' academic ability. The academic ability of the students in the class whose learning using X_1 method should be the same as the students' academic ability in the class that got X_2 method.

One way to achieve this similarity is by rotating classes in intervals during the experiment. Suppose the experimental period is divided into two equal intervals of time, the materials will be taught are also divided into two units, namely unit 1 and unit 2. The complexity of the material in unit 1 has to be balanced with that of unit 2. During the first interval material unit 1 is taught, one class (class A) applies X_1 method and the second class (class B) applies X_2 method. During the second interval unit, 2 material is taught, but class A applies method X_2 and class B applies X_1 method. If such rotation is done then it is said that research applies counterbalanced or Balanced Design [7]. Table 1 is a Balanced Design that is used to compare the two methods of teaching, namely X1 and X2.

Table 1. The Balanced Design with Two Teaching Methods

Period interval & the subject matter	X_1 teaching method	X_2 teaching method
Period interval I, Unit 1	Class A	Class B
Period interval II, Unit 2	Class B	Class A
Combination	Score $X1_A + X1_B$	Score $X2_B + X2_A$
Mean	Mean of score X_1	Mean of score X_2

Based on Table 1, X_{1A} score is combined with X_{1B} score. Likewise, the X_{2A} score is combined with the X_{2B} score.

Testing assumptions of normality and homogeneity of variance are required by the t-test and the ANOVA (Analysis of variance) test since both of these tests include the parametric test. There are 4 assumptions for parametric tests [9,11], those are: (1) normally distributed data, (2) the variance between homogeneous groups, (3) the data interval, and (4) the independent data. If these assumptions are not met, then, the conclusions of ANOVA will deviate from what it should be. However, fortunately, in general, the consequences caused by not fulfilling assumptions with respect to ANOVA is not too bad [12]. In other words, a moderate deviation from the terms outlined in assumptions is not too dangerous. For example, if there is a slight deviation from normal assumption or assumption of homogeneity of variance it would only affect very small on the test results and the conclusions.

ANOVA (also called F test) includes a robust test, that is, it will be not too problematic if the assumption is not met [9]. If the sample size is large, then the sampling distribution will have a normal distribution (based on Central Limit Theorem). If the sample size is the same, F test is very robust against violations of normality; thereby also, F test remains accurate even though there is a slight violation of the homogeneity assumption. However, if the sample size is not the same and it results in a violation of the homogeneity assumption, the conclusions derived from the F test can be biased.

The above statement is supported also by [11], for samples of the same size or almost the same, ANOVA is less sensitive to non-fulfillment of normality and homogeneity of variance. This is a beneficial use of Balanced Design. Even many modern statistical experts feel that the testing of assumption is only necessary for small samples [13].

One-Factor ANOVA and t-test have proved to be very useful in some studies of the exact sciences, such as Agriculture and Pharmaceuticals, but poor when they are used in the field of science concerning behavior, for example in the field of Mathematics teaching [14]. Because One-Factor ANOVA will not be able to reveal the interactive effect between teaching methods and student academic ability.

The analysis which can reveal this interaction is Factorial ANOVA. The term "way" or "factor" is often used to describe the number of independent variables measured by an ANOVA test. The design is called a factorial design. This design was first developed by R.A. Fisher in Agriculture [14]. Another advantage of using a factorial design is that it increases the significance of the differential effectiveness of two methods of teaching [8]. Besides, it can also increase the generalizability of research results.

Supposedly, there is a question on whether the effectiveness of a method of teaching also depends on other factors, such as students' academic ability factor. Such question can be answered by using ANOVA Factorial. The effectiveness of a teaching method can be seen from the students' learning achievement (achievement of test scores). If the learning achievement is seen from two factors, for example, factors of teaching methods and level of academic ability, then Two Factors ANOVA is applied to analyze the data. If the teaching method consists of 2 (two) levels (X1 and X2 method), while the academic ability of students consists of 3 (three) levels (High, Medium, and Low), then the design is called a 2 x 3 Factorial Design. Table 2 below is 2 x 3 Factorial Design.

Table 2. The 2 x 3 Factorial Design

Academic Ability	X_1 Teaching method	X_1 Teaching method	
High	AT score[a)]	AT score	Average row
Moderate	AT score	AT score	Average row
Low	AT score	AT score	Average row
	Average column	Average column	

Note [a)] AT score is Achievement Test score

Based on Table 2, it appears that researchers can compare test scores for each level of academic ability.

Based on the advantages of Balanced Design and Factorial Design, the authors propose to apply a Factorial-Balanced Design to counteract the weaknesses of the quasi-experimental method and the t-test. Factorial-Balanced Design is a combination of Factorial Design with Balanced Design. This design name is the authors' own creation. The procedure for carrying out the research on the use of the Factorial-Balanced Design is presented in Figure 1. This procedure has been practiced [15], in fact, the application of this procedure is not simple, but the results provide information that is unbiased and it strengthens generalization.

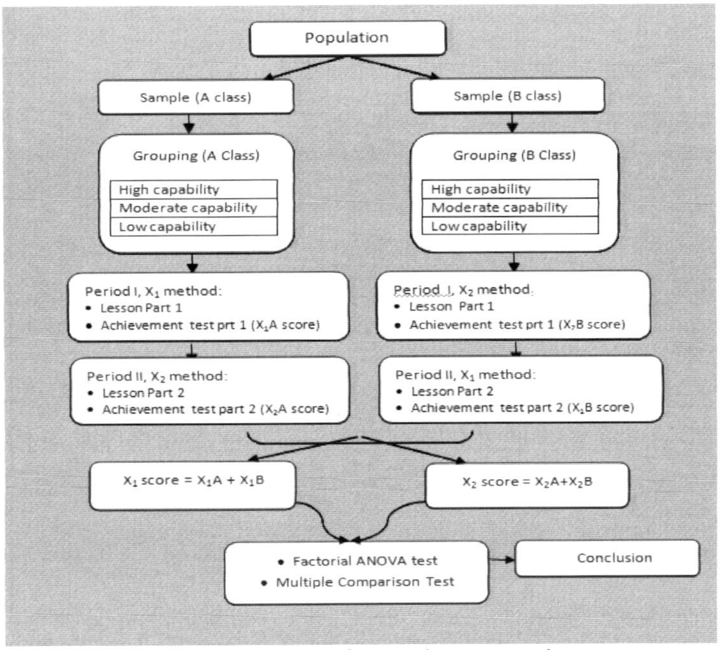

Figure 1. Cart of procedure research

IOP Conf. Series: Journal of Physics: Conf. Series **948** (2018) 012010 doi:10.1088/1742-6596/948/1/012010

Figure 1 shows that two classes were taken as samples. The sampling method used cluster random sampling, and therefore it obtained the class A and class B. Class A was divided into three ability groups, those were groups of high, moderate and low. The basic grouping was based on a summative test score and it has been obtained before the research was done. Grouping was also performed in class B.

The subject matter was divided into two units, namely unit 1 and unit 2. The experimentation time was divided into two, namely the first period and the second period. The first period was taught unit 1 material: class A got X1 method, while class B got X2 method.

The first period ended with achievement test for unit 1. The second period was taught unit 2 material: class A applied X2 method, while class B applied X1 method. The second period ended with achievement test for unit 2. Achievement test scores as a result of applying the X1 method in class A were merged with those in class B. As well, achievement test scores as a result of applying the X2 method in class A were merged with those in class B. The form of a merger is shown in Table 2.

The data in Table 2 were analyzed using Factorial ANOVA test followed by multiple comparison tests. The multiple comparison tests used was the Scheffe test. Scheffe test is one of the multiple tests most frequently used [11]. It is even guaranteed as the best test [8]. This test has several advantages, one of which is that a pair can be selected in accordance with the desired pair.

Conclusion

The law of single independent variable is difficult to be fulfilled by the quasi-experiment method. The approach of the fulfillment of this law can apply Balanced Design. Through this design, the class is rotated in intervals during the experiment so that each student is paired with him/herself. The advantages of this design are: first, the sample size in both groups becomes larger and equal; this condition increases the robustness of ANOVA against the assumptions of normality and homogeneity of variance; second, it does not need to test the similarity of average academic ability between groups because each student is paired with him/herself.

The use of t-test cannot conclude whether the effectiveness of the teaching method is the same at every level of academic ability of students, so the generalization is less extensive. A Factorial Design is applied to determine whether the effectiveness of teaching methods depends on the level of academic ability. Based on the previous explanation, a way to fulfill the law of a single independent variable and increase generalization of mathematics education research is the application of a Factorial-Balanced Design. The design is a combination of Factorial Design and Balanced Design. The procedure of applying the Factorial-Balanced Design is not simple, but the results provide information that is unbiased and it strengthens generalization.

4. Conclusions

The law of single independent variable is difficult to be fulfilled by the quasi-experiment method. The approach of the fulfillment of this law can apply Balanced Design. Through this design, the class is rotated in intervals during the experiment so that each student is paired with him/herself. The advantages of this design are first, the sample size in both groups becomes larger and equal; this condition increases the robustness of ANOVA against the assumptions of normality and homogeneity of variance; second, it does not need to test the similarity of average academic ability between groups because each student is paired with him/herself.

The use of t-test cannot conclude whether the effectiveness of the teaching method is the same at every level of academic ability of students, so the generalization is less extensive. A Factorial Design is applied to determine whether the effectiveness of teaching methods depends on the level of academic ability. Based on the previous explanation, a way to fulfill the law of a single independent variable and increase generalization of mathematics education research is the application of a Factorial-Balanced Design. The design is a combination of Factorial Design and Balanced Design. The procedure of applying the Factorial-Balanced Design is not simple, but the results provide information that is unbiased and it strengthens generalization.

References

[1] Rochimawati N and Kusmanto B 2016 *Union* **4** 1 pp 79-88
[2] Alfiah Y and Arigiyati T A 2015 *Union* **3** 2 pp 169-175
[3] Aznam N, Sumarno and Rahmat A 2005 *Metodologi Penelitian untuk Peningkatan Kualitas Pembelajaran, Penelitian Kuasi Eksperimen dan PPK* .Jakarta: Depdiknas pp 4-5
[4] McGowan H M 2011 *Journal of Statistics Education* **19** 2 pp 4-5
[5] Bradley K 2009 *Quasi Experimental Research, A Comparison to Experimental Research* .The Gale Group, Inc. All rights reserved
[6] White H and Sabarwal S 2014 *Quasi-experimental Design and Methods, Methodological Briefs: Impact Evaluation* ***8*** Florence UNICEF Office of Research 2
[7] Ary D, Jacobs L C, Sorensen C and Razavieh A 2010 *Introduction to Research in Education., 8th ed* USA: Wadsworth, Cengage Learning pp 267-319
[8] Moortgat L R 1980. *Analysis of Variance from t-Test to Analysis of Covariance* (Manila: De la Salle University) pp 28
[9] Field A 2009 *Discovering Statistics Using SPSS, 3th ed* (London: SAGE Publications Ltd) p 133
[10] Fife-Schaw C 2006 *Research Methods in Psychology* Quasi- experimental Designs-Semantic Scholarhttps://pdfs.semanticscholar.org pp 79-80
[11] Deauna C M 1982 *Applied Statistics for Education Psychology and the Social Sciences, 1st ed* (Manila: De la Salle University)
[12] Sudjana 1985 *Desain dan Analisis Eksperimen* (Bandung: Tarsito) pp 48-49
[13] Hayes W L 1973 *Statistics for the Social Sciences, 2nd* (New York: Holt, Rinehart & Winston)
[14] Best J W 1981 *Research in Education* (USA: Prentice-Hall, Inc) pp 58
[15] Ngubaedilah A, Arcana I N and Suyadi A 2016 *Efektivitas Model Example Non Example dengan Game Puzzle Terhadap Prestasi Belajar Matematika Pokok Bahasan Geometri pada Siswa Kelas VII SMP Negeri 2 Mlati* (Yogyakarta: Universitas SarjanawiyataTamansiswa)

Acknowledgments

The authors would like to extend their gratitude to Ahmad Ngubaedilah who has been practicing Balanced-Factorial Design (design created by the author) in his research for his thesis in Mathematics Education of Sarjanawiyata Tamansiswa University, Yogyakarta, also to Th. Laksmi Widyarini, M.Hum., for having proofread this article.

IOP Conf. Series: Journal of Physics: Conf. Series **948** (2018) 012011 doi:10.1088/1742-6596/948/1/012011

The effect of ice-breaking using stand-up comedy on students' mathematical belief system

M D Kurniasih[1], D H Lenaldi[1] and Wahidin[1]

[1]Universitas Muhammadiyah Prof DR HAMKA, Jakarta, Indonesia

E-mail: meyta.dkurniasih@uhamka.ac.id

Abstract. The purpose of this study was to determine the effect of ice breaking by doing stand-up comedy on the result of mathematical belief. This research was conducted at SMPN 37 Jakarta on 2nd grade in the 2nd-semester academic year 2016-2017. The research design used was the treatment by level 2x2 with learning method as a dependent variable which is the ability to understand the mathematics concept as a moderator variable and mathematical belief as an independent variable. The sampling technique used cluster random sampling. The result showed that the learning through icebreaking by doing stand-up comedy is able to increase the mathematics' belief of students. In addition, the learning method has the effect of mathematics' belief which depends on the ability to understand the mathematics concept of the student. Scholars who study by ice breaking with stand-up comedy teaching have the mathematics' belief higher than scholars who study by the conventional method.

1. Introduction

Mathematics is something that is not foreign to the community because of its usefulness in every activity of life, whether consciously or unconsciously. Learning mathematics is implemented at every level of formal education, but until now mathematics has not become an idolized lesson for students. Meaning idolized here relates to the sense of affective aspects, aspects that are often forgotten by teachers in teaching, so in learning that put forward only the results of learning from the cognitive aspect only. Affective domain includes feelings, emotions, mental or psychological belief of mathematics in mathematics lessons [1]. Confidence can arise from the desire to manifest itself to act and succeed in students. The belief of mathematics has a big role in welcoming the lesson. Abstract mathematics and many formulas, plus a rigid way of presenting material by the teacher make it even more difficult to understand. It can also be a cause for belief in low student mathematics. As Lazim pointed out, et al that beliefs of students towards mathematics relate to the role of teachers in learning [2]. Belief in mathematics is not innate but evolves through the socialization and activeness of each individual. There are three things that influence the beliefs of math such as self, object (mathematical education) and context (class) [3]. These three aspects are interrelated in shaping the beliefs of student mathematics.

Learning that only puts the students to solve routine problems and as the object of the listener only, so indirectly the students are only trained in numeracy skills in routine workmanship, without being directed to understand this concept is increasingly making belief to low student math. Ice breaking with standup comedy mathematics can be one of the learning solutions that can create a fun and comfortable learning environment so that the learning process seems meaningful. Ice breaking is one of the learning techniques included in the Quantum Learning model. According to DePorter, Quantum Learning is rooted in Dr. Georgi Lozanoc, a Bulgarian educator who experimented with what he called "suggestology" or "suggestopedia".

The principle is that suggestion can and does affect learning outcomes, and any detail gives positive or negative suggestions. Some of the techniques he uses to give positive suggestions are to seat students comfortably and provide teachers who are well-trained in teaching suggestive arts. [4]

The principle of Quantum Learning is in line with Ice Breaking which puts students at ease by learning in an atmosphere without tension. Ice Breaking did if the conditions in the class already feel bored and not conducive. One type of Ice Breaking is a form of funny stories or light humor that provokes the spirit to learn. By using Ice Breaking Humor Math can refresh the learning atmosphere, eliminate boredom, and relieve drowsiness [5]. By inserting Ice Breaking in every learning, it is hoped that the students' capability can be maximized and the learning atmosphere in the classroom becomes fresh again [6]. In addition, by implementing Ice Breaking with stand-up comedy mathematics can reduce the impression of rigid and spooky to the process of learning mathematics that goes on, because the stand-up itself basically dig up the anxiety or anxiety experienced by the ways more relaxed and fun so that students who feel The same anxiety about mathematics becomes mutually reinforcing and mutually expressed in order to get a pleasant solution. In the process of learning mathematics using Ice Breaking with standup comedy mathematics, students feel anxiety about mathematics that had been hit, will be increasingly lost and belief students in math are increasing.

In addition to learning methods, other things that need to get attention to the success of learning mathematics students is the individual difference is about the ability to understand the concept of mathematics. This difference will certainly affect the acceptance of each individual's material, so belief in mathematics will also influence. Understanding concepts have an important role for students to solve a mathematical problem [7]. By applying an understanding of mathematical concepts associated with other mathematical concepts to obtain a broader understanding of concepts. Understanding concepts is a skill that deals with comprehending mathematical ideas that are thorough and functional [8]. This causes the ability to comprehend the concept of mathematical role in the success of student learning. Students can be said to have the ability to understand the concept of mathematics if it can restate a concept that has been studied previously and can apply to various problem solving, so students are not just required to memorize the existing mathematical formulas.

2. Methods

This study was experimental in which the dependent variable was mathematical belief and independent variables were treatment and moderator. Treatment variables consisted of learning with ice breaking standup comedy mathematics (A_1) and conventional learning (A_2). Moderator variable was the students' understanding of concept mathematics ability consisting of high understanding of concept ability (B_1) and low understanding of concept ability (B_2). The study design used treatment by level.

Table 1. Study Design with Treatment by Level 2x2

Understanding Ability (B)	Learning Methods (A)	
	Ice Breaking by doing Stand-up comedy (A_1)	Conventional (A_2)
High (B1)	A_1B_1	A_2B_1
Low (B2)	A_1B_2	A_2B_2

Remark:

A_1B_1 = Mathematical belief score for group of students receiving learning with Ice Breaking by doing Stand-up comedy with high understand the mathematics concept ability.
A_1B_2 = Mathematical belief score for group of students receiving learning with Ice Breaking by doing Stand-up comedy with low understand the mathematics concept ability.
A_2B_1 = Mathematical belief score for group of students receiving conventional learning with high understand the mathematics concept ability.
A_2B_2 = Mathematical belief score for group of students receiving conventional learning with low understand the mathematics concept ability.

The population in this study is the target population of all students of SMP Negeri 37 Jakarta in the even semester of the academic year 2016/2017, the population reaches all students of class VIII. Sampling technique is done by Cluster Randomized Sampling. Randomization is done by drawing 4 groups of students of class VIII. Of the 5 classes, two classes were randomly assigned, namely class VIII-A and class VIII-B where for class VIII-A consisted of 36 students and class VIII-B consisted of 36 students. The two classes were randomly assigned to the experimental class and control class, namely class VIII-B as an experimental class that received ice breaking learning with standup comedy mathematics and class VIII-A as a control class with conventional learning. Then after performing the students' ability to master the mathematical concept, the determination of the group's ability to develop high mathematical concepts and the ability of the low learning mathematical concepts is done by arranging the sequence of respondents based on the scores of the mathematical concepts of mathematical concepts obtained, i.e. the highest scores to the lowest scores. To get a high group and low group of S. Naga [9] suggests, the size to determine the high and low group is $33\frac{1}{3}\%$ ($M_T = M_R = 33\frac{1}{3}\%$) Where this number is quite a contrast and reliable. Thus, the sample of this study consisted of two classes, each class divided into two groups, taking $33\frac{1}{3}\%$ high initial skill students and $33\frac{1}{3}\%$ low initial ability students. The sample distribution can be seen in Table 2 below.

Table 2. Sample Distribution in Each Class

Understanding Ability(B)	Learning Methods(A)		Total
	Ice Breaking by doing Stand-up comedy (A_1)	Conventional (A_2)	
High (B_1)	12 persons	12 persons	24 persons
Low (B_2)	12 persons	12 persons	24 persons
Total	24 persons	24 persons	48 persons

Belief instrument developed was Likert scale with 5 options consisting of 47 valid items. The instruments consisted of five indicators, namely the certainty of knowledge, the role of lecturer, systematic process, innate ability, quick learning. The respondents' responses were analyzed to test the construct validity using factor analysis with the results. Value of $KMO = 0{,}814 > 0{,}7$ means the mathematical belief instrument is acceptable. Extraction (PCA), Eigenvalue is 5 factors. The result of *Rotated Component Matrix*, with loading factor criteria < 0.30 and > -0, 30, means that point 9, 12, 40 is deciduous (not valid) so that only 47 items that are valid and can be used as measuring tool.

While the instrument ability to understand the mathematical concepts of students by using essay test. From the validity test with Product Moment Person formula [10]. There are 10 questions got 8 valid problems with reliability test with *Alpha Cronbach* [11] $r_{hitung} = 0{,}484 > 0{,}339 = r_{tabel}$. So the instrument of the understanding ability of students' mathematical concept is reliable.

3. Results and Discussion

Table 3, shows that the students' average mathematical belief score for groups A_1 and A_2 are 152,88 and 148,42, group A_1B_1, A_1B_2, A_2B_1 and A_2B_2 are 170,58; 135,17; 151,17 and 145,67. This score is much different descriptively can be inferred group A1 and A_2. So that group A_1B_1, A_1B_2, A_2B_1.

Table 3. Average Mathematical belief Score

Understanding Ability(B)	Statistics	Learning Methods(A) Ice Breaking by doing Stand-up comedy (A_1)	Conventional (A_2)
High (B1)	N	12	12
	Mean	170,58	151,17
	St. Deviation	16,65	15,23
Low (B2)	N	12	12
	Mean	135,17	145,67
	St. Deviation	11,56	15,71
Total	N	24	24
	Mean	152,88	148,42
	St. Deviation	22.88	15,39

The result of analysis of two-way ANOVA, compare belief of mathematical score on group A1 and A2 obtained sig value. = 0.005 <α = 0.05 and the average score of group A1 mathematical beliefs is higher than group A2. This means that there is a mathematical belief score of the student group subjected to Ice Breaking by doing Standup comedy (A1) higher than the conventional belief mathematical score (A2). Mathematical learning with stand-alone math has four benefits: building relationships and improving communication, stress reduction tools, making learning interesting, and strengthening memory [12]. So that learning more fun, not tense and student love with math will be better [13]. Between beliefs of mathematics and mathematics learning are interrelated forming a circular process [14]. Thus, the more varied and effective the learning of make belief towards the higher the mathematics.

The result of interaction test between the learning method and the comprehension ability of mathematical concept (A * B) obtained sign = = 0.007 > α = 0,05, this means there is the significant influence of learning method interaction and the ability to understand mathematical concept of the student to believe mathematics score. Learning with Ice Breaker is energizer before giving the material and creation of learning atmosphere from passive to active, from rigid to motion (familiar), from saturated to cheerful (fresh), and able to arouse passion learn [15]. While the ability to comprehend the mathematical concepts of students is a conceptual understanding refers to the ability of students to connect new ideas in mathematics with ideas they know [16]. Students who have the ability to understand high mathematical concepts mean it can be easier to follow the learning and connect ideas so that math lessons become more fun.

4. Conclusion

Belief in mathematics is one of the effective aspects that influence students' success in learning mathematics. Learning by using icebreaking with mathematical standup comedy is one of the learning methods that will help students improve belief in mathematics compared to conventional learning.

Individual differences such as the ability to comprehend mathematical concepts have a significant effect on belief in student math. Furthermore, for students with the high understanding of mathematical concepts if given learning with ice breaking standup comedy mathematics average score belief to mathematics higher than that given conventional learning. While for students with low comprehension of mathematical concepts, statistic test results showed no differences between groups of students using learning with ice breaking standup comedy mathematics and conventional learning. Therefore, further study is needed to determine the appropriate learning for the group of students with the ability to comprehend low mathematical concepts.

From this research is expected teachers in providing education services to students, not only pay attention to the results of learning but also pay attention to the affective aspects. Teachers are also expected to apply learning to ice breaking standup comedy mathematics in the learning process, to believe in mathematics of students, which will also have an impact on improving student learning achievement.

References

[1] W Rahayu and M D Kurniasih 2014 The influence of react strategy towards mathematical belief *Proc. International Seminar on Innovation in Mathematics and Mathematics Education 1st* (Yogjakarta)

[2] e. Lazim 2004 *The Statistical Evidence In Describing The Students' Beliefs About Mathematics* (Kuala Lumpur: University College of Science and Technology)

[3] P Eynde E Corte and V L 2002 Framing Student's Mathematics-Related Beliefs: A Quest for Conceptual Clarity and a Comprehensive Categorization in *Beliefs; A Hidden Variable in Mathematics Education?*, G. Leder, P. and G. Torner, Eds., London, Kluwer Academics Publisher

[4] B DePorter and M Hernacki 2000 *Quantum Learning* (Bandung: Kaifa)

[5] Darmansyah 2009 *Pembelajaran Menggunakan Sisipan Humor dalam Mata Pelajaran Matematika* (Padang: Universitas Negeri Padang)

[6] Sunarto 2012 *Ice Breaker dalam pembelajaran aktif* (Surakarta: Cakrawala Media)

[7] S Wardhani 2008 *Analisis SI dan SKL Mata Pelajaran Matematika SMP/MTs untuk Optimalisasi Tujuan Mata Pelajaran Matematika* (Yogyakarta: P4TK Matematika)

[8] K E Lestari and M R Yudhanega 2015 Penelitian Pendidikan Matematika (Bandung: Refika Aditama)

[9] D Naga Teori Skor pada Pengukuran Mental (Jakarta: PT Nagarani Citrayasa)

[10] S Arikunto 2001 *Dasar-Dasar Evaluasi Pendidikan* (Jakarta: Bumi Aksara)

[11] A Jihad and A Haris 2013 *Evaluasi Pembelajaran* (Yogjakarta: Multi Pressindo)

[12] Darmansyah 2010 *Strategi Pembelajaran Menyenangkan dengan Humor* (Jakarta: Bumi Aksara)

[13] M Said 2010 *80+ Ice Breaker Games Kumpulan Permainan Penggugah Semangat* (Yogjakarta: ANDI)

[14] E Pehkonen and A Pietila On Relationship Between Beliefs and Knowledge in Mathematics Education *European Research in Mathematics Education III*

[15] A N Cahyani 2014 *Pengaruh Reward dan Ice Breaker terhadap minat belajar Tematik Siswa Kelas IV SDN Ngadirejo 01. Kec. Kartasura, Kab. Sukoarjo. Tahun Pelajaran 2013/2014 Universitas Muhammadiyah Surakarta, Surakarta*

[16] F E Purnamasari 2014 *Peningkatan Kemampuan Pemahaan Konsep Matematika melalui pendekatan Open Ended bagi siswa kelas VIII Semester Genap SMP Muhammadiyah 10 Surakarta Tahun 2013/2014 Universitas Muhammadiyah Surakarta*

IOP Conf. Series: Journal of Physics: Conf. Series **948** (2018) 012012 doi:10.1088/1742-6596/948/1/012012

Task design for improving students' engagement in mathematics learning

Khairunnisa[1]

[1]Indonesia University of Education, Jl. Dr Setiabudhi No.229, Bandung 40154

E-mail: khairunnisa.58@student.upi.edu

Abstract. This article analysed the importance of task design as one of the instruments in the learning and its application in several studies. Through task design, students engage in learning caused them enthusiastically in expressing ideas, opinion or knowledge of them. Thus, the teacher was able to gain an idea of knowledge belonging to students. By using this information, teachers are able to develop the thinking ability of students.

1. Introduction

Learning mathematics currently is directed at the dominance of the activity of the students. They explore, collaborate, rediscover the formula, and build an understanding of the concept with their own language. These activities are carried out to meet the learning objectives of mathematics that is making mathematics as a means of critical thinking, survival tools, cooperative learning and capable of expressing his opinion of his own.

Teachers have an important role in learning mathematics, but not as the sole source of student learning. They are tasked to facilitate and manage the students to keep learning in accordance with the learning objectives to be accomplished. Therefore, teachers are no longer just delivering learning with oral, but also capable of designing tasks that can involve students in learning.

Tasks which used in learning mathematics are not just a form of exercise problem for the coaching numeracy skill of students. Instead, those are able to develop the thinking ability of students. The ability to think is expected like problem-solving, reasoning, representation, communication or connection. With the task of learning that accommodates such things, then through those tasks can improve and hone students ' thinking ability. To be able to devise such a task, there are things that still need to be noticed by the teacher. When the students struggling with the task later, do not let them lose the element of education or pedagogy in learning mathematics.

This article will explain things that can be accomplished through task design. There are 1) for closing gap between the goal of teacher and student activity, 2) a tool of decentering, 3) generalizing pattern, 4) honing a concept and 5) means of collaboration

2. Methods

2.1. Task Design

The task of learning mathematics is meant as anything that teachers use to display mathematics so that students are actively engaged in learning or to ask students to do something. Tasks can also be meant any students decide to do by him in certain situations [1]. So that through the assignment can result in activities that give the opportunity for students to learn math concepts, ideas, strategies, and also to use and develop mathematical thinking and doing inventions [1]. Thus, the task is an intermediate used in the mathematics teaching-learning activities so that students gain experience and understanding of mathematics.

As a learning tool, the task is the instrument of cognitive processes to gain knowledge. Adherents of Vygotsky even assess a tool of learning, e.g., student worksheets; can become the tools of Psychology within the context of a growing culture of social interaction through the zone of proximal development and the process of internalization. The involvement of students in learning mathematics that enable the majority of sensory will provide a strong link between the activities of the students with learning activities itself.

In a class initially using lecture, question and answer happen later and ends with awarding the assignment to the students. But the task is not just accidental events and only one time in learning. The task is part of a sequence of learning activities that can help students to understand the material. Thus a task should have a clear arrangement. The arrangement can be either a similar Problem but the use of numbers ranging from simple to complex, questions or problems that developed with more complex and increased variable also resulting solution which requires a lot of stages, or Starts from a simple concept into a more sophisticated and complex. The importance of the order in the task that shows the order of difficulty, this becomes the focus of attention on the lesson study as well as Realistic Mathematics Education.

Thus, learning includes selecting, modify, designing, composing, arranging, observing and evaluating tasks. This role is often given in textbooks, student worksheets, module or other sources that are designed by someone else, not the teachers themselves. As a result, sometimes between the tasks of the resource and referrals from teachers not aligned. However, the use of the task from another source does not become the issue. It's just that when emerging questions from students relating to assignments from other sources, the teacher must be responded even can anticipate things that pertain to the task.

The tasks are arranged must notice the ability of the students. Better yet, if the task can accommodate differences in learning styles of students. Because the task or task order is indeed designed and administered so that students can acquire mathematical knowledge and honed their mathematical thinking ability. However, teachers should be aware of the student's responses regarding the task. It is possible that students complete tasks are given due regard the important thing for them is the task finish. If it happens then the student not getting mathematical knowledge, even less honed their mathematical thinking ability. This can occur, particularly for the younger students also underachieving because teacher helping them complete the tasks so that those students concerned are not left behind by his friends. Thus, the teachers strived to give different tasks to different students to facilitate learning.

2.2. Task Desaign for Closing the Gap Between the Goal of Teacher and Student Activity

The fundamental principles that are used in the design of the task for closing the gap between the goal of teacher and student activity, using the principles expressed by Alf Coles and Laurinda Brown [2]. The first three principles are used to cover the gap between the goals of a teacher with student activity. While the next four is the special way which is applied by Coles and Brown to get the desired math activities in their research. The principles are as follows:

- Present at least two conflicting examples, if possible in the form of pictures, and then collect responses
- Ask students to give an opinion on the similarities and differences between the conflicting examples or pose questions.
- Introduced the term and the granting of notation based on statements of students
- Start the covered activity can be accompanied by learning new skills
- Have challenge prepared for students if there were no questions coming
- Opportunities for the teacher to teach further new skills and for students to train their skills in different contexts.
- Gives students the opportunity to notice the pattern, make a conjecture and work on proof them

The principle of first to third is linked to the inactive views in knowing and learning with their emphasis on the difference in the example given. By using at least two examples (principle 1) and ask for the opinion of the students about the similarities and differences (principle 2), will support students in sharing the distinctions and through sharing this will create a new distinguishing as well as learning from the perspective of the inactive. Terminology and notation were introduced to distinguish name that made students (principle 3) supports a new way of looking. Usage examples (principle 1) intended to capture a variety of possibilities, including images, animations, and procedures.

Whereas the principle of four to seven relates to the work of Caleb Gattegno (Coles & Brown, 2016). Closing activities may involve something visible or affordable and give students something to do or discussed (principle 4). Having challenges (principle 5) and the opportunity to teach skills in different contexts (principle 6) associated with extraction capabilities of teachers. When teachers use something with different objectives and contexts are different, the skills can be honed and potential to survive without tradition. As delivered by Tall that the Crystal structure of mathematical concepts make the students make the necessary connections if they are able to make distinguishing in mathematical context [2]. Principle 7 comes from the view that mathematics is essentially about relationships, so it certainly refers to the pattern and a possible conjecture no matter what material is covered [2].

It can be concluded that to achieve the learning objectives of mathematics, teachers choose or design tasks and sorts the task, then select the appropriate media for the presentation of the task. Therefore the students can communicate the result, design pedagogy and relate the opportunities to make it happen in the task, determine the difficulty level of the task for students including how adjustment for them and to anticipate the process of assessing student learning. Each of this ruling influenced the understanding of math teachers. It is preceded by the relevant assessment of teachers over the readiness of students, experience or creativity or access the teacher against teacher, hopes the top source of student involvement, the commitment of digitally learning associate teachers to student life and awareness as well as the willingness of teachers to apply the appropriate pedagogy.

Task design uses the 7 principles previously mentioned. Through this design task, the teacher trying to involve students in the learning activities, and it is managed. Students who have been involved in the study took out what they think about a given task. When the students pulled out their thoughts about a given task, the teacher knew what was in the minds of his disciples, making it easier for the teacher to direct the students to the concept that was intended. Nevertheless, the involvement of students preferred. So the selection of the material presented in the task more on that make students interested and active in the learning, though it is not included in the curriculum is essential. However, with the active involvement of students and time of learning, then it is hoped they can enthusiastically direct at the intended concept.

In practice, there are some students who developed his mind based on the given task, for example, initially the teacher presents the number of integers in the form of a concept, but students can develop it in the form of fractions, decimal, or without regular referrals from teachers. With the involvement of students in learning mathematical reasoning, then triggers the students, so as to develop the student's way of thinking about the mathematical material studied. Students also make distinctions and questions without the need to be asked. Then the students have done inquiry or discovery through his involvement, not just because it is a task that has been arranged. The purpose of this research on task design in support of mathematical thinking and teach skills in a meaningful context for students that allows immediate applied these new skills [2].

2.3. Task Design as a Tool of Decentering

Decentering introduced Piaget (1955) is an activity to characterize the actions of the observers who tried to understand the perspective of different individuals. Similar to the activity of the Coles & Brown, in this case else teacher trying to steer students in the intended thing after finding out what is in the mind of the student. What is in the minds of students considered a teacher as the first model of the concept that will be studied. Then the teacher by asking questions that dig will be directed to the second model or a formal model. Teacher argued that is not the student's answer is right or wrong, but rather the meaning of a concept that can be understood. There is no misconception, misconceptions is a conception that for some reason, the teacher does not want students to use it. In other words, the misconception is the initial data for teachers to develop a model of the next level.

To be able to do so, the teacher should be able to ask questions that can bring the students understand what they're doing. The giving task continued by questioning students about thinking for the display of the task. It should be noted, that the teacher's role was to develop the thinking ability of students, don't get stuck to direct the procedures or steps solving problems students. So the first level of the briefing, informal, to the second level, formal models, depending on the perspective of the teacher to think their students [3]

2.4. Task Design in Generalizing Pattern

A study was done by Rivera and Becker in using task design for generating activity of generalized patterns. This is based on previous studies suggesting that children and adults have a natural disposition towards generalizing patterns. Then Rivera and Becker use this research to uncover the tendency of students in making the pattern.

This research uses a task with the following arrangement. Originally a teacher submits an image sequence to 4 pieces (4 stage) then students are asked to specify the pattern. The next task, students are given a picture to phase one; the students further forward it myself until the fourth stage. Students use the figures to determine the pattern of each.

The results of the study show that in making generalizing patterns students tend to use a linear formula involving addition and multiplication. A small portion of students who used the formula not in generating a linear pattern, and there's still that makes the images on the next stage without having a specific pattern [4]. So to develop the ability to generalize patterns, granting make students actively create images and determining the pattern of its own. It is more engaging students compared to merely provide a picture and ask students to determine the pattern.

2.5. Task Design for Honing a Concept

One of the mathematical content that needs to be controlled by the student according to the National Council of Teachers of the Mathematics (NCTM) is the geometry. Research using task design aimed to observe the student's work in making a ceramic pattern by using the reflection of a given format.

Students are expected to be able to complete the task given by the use of the concept of a perpendicular bisector. It turns out that only a small portion of students who use those concepts to solve problems. Most of the students do a reflection by using the process of folding and flipping [5]. After getting the data through a task that has been given, the students got the direction of the teacher to be able to use the concept is meant to complete the pattern ceramics by using reflection. Through task design, teachers can document and identify various sources of knowledge of students is reflected from his mathematical work. Thus, teachers can find out opportunities to be able to develop the thinking ability of students.

2.6. Task Design as a Means collaboration

Task design is not the dominance of teachers only. Researchers, mathematicians can also contribute to task design. Learning design team consisting of 3 teachers and author of designing learning with an early form of problem-solving in the context of the outbreak. Students complete the task independently and submit the results. The team checked the answers students and creates a new task by modifying the first task. The team made a list of predictions of the difficulty students in completing new tasks and the anticipation questions advice and referrals. Once ready, teachers give new tasks to be done in a group. During the activity, the teacher has twice conducted discussions with the students. The first to review the results of one of the groups, the second session of the panel, such as by asking two groups. After the study is completed, the team returned, having discussions to study the learning process in the classroom.

The task provides a starting point for teachers and students in the attachment class activities [6]. This design task relating not only to the activity of the students but also for people like teachers who have a variety of different education also researchers or experts, as teachers brought the task as an instrument that can be used in the practice of learning.

3. Conclusions

Task design has a role in the learning of mathematics; one of them is the bridge between students and teachers to be able to achieve learning objectives. The task may be released through the workings of mathematics students, which varies, so teachers have preliminary data to direct students to the learning objectives in accordance with the initial perception of the students. Some task design expressed include discussions between students and teachers to get confirmation of what the students think about the mathematical tasks related to their working on. The task of making the design not only by the teacher can be from other experts, textbooks or materials that can be downloaded. Even collaboration in any task design is possible to get a variety of views in anticipation of learning.

References

[1] A Watson and M Ohtani Eds 2015 *Task Design In Mathematics Education* Springer International Publishing Switzerland
[2] A Coles and L Brown 2016 *J. Math. Teach. Educ* **19** 2 pp 149–168
[3] D Teuscher, K C Moore and M P Carlson 2015 *J. Math. Teach. Educ.*
[4] F D Rivera and J R Becker 2016 *J. Math. Behav* **43** pp 53–69
[5] A F Dejarnette, G González, J T Deal and S L Rosado 2016 *J. Math. Behav* **43** 35–52
[6] G Wake, M Swan and C Foster 2016 *J. Math. Teach. Educ* **19** 2 pp 243–260

IOP Conf. Series: Journal of Physics: Conf. Series **948** (2018) 012013 doi:10.1088/1742-6596/948/1/012013

Analysis of students' self-determination in learning mathematics

H Wilujeng[1]

[1]Mathematics Education Study Program, Universitas Muhammadiyah Tangerang, Banten, Indonesia

E-mail: hestuwilujeng@gmail.com

Abstract. Self-determination (SDT) is the ability in identifying and achieving the purpose based on knowledge and the assessment of the individual against himself. Three aspects in the SDT includes autonomy, competence and relationships become an important part for students to be able to understand the capabilities of themselves, having a positive competitiveness to other students and can interact well between friends. Therefore, teachers need to know the ability of students SDT after making the learning process. This research was conducted to improve the process of learning mathematics by knowing the ability of students SDT. The researcher gave the question form to 38 students and analyzed the ability of SDT. The Results of the study showed that the student SDT ability is still poor. Students were lack of confidence to solve math problems. In addition, the competitiveness of students was low that have made them looked lazy. This can be resolved by making learning more interesting for students so that it can increase the student SDT ability.

1. Introduction

Self-determination (SDT) includes three psychological needs, i.e. need competence, autonomy, and relationship. The third of these requirements, when successful, can increase self-motivation and mental health, on the other hand, if they fail, they will reduce motivation and well-being. So it could be interpreted that the three important psychological needs are important in the field of healthcare, education, employment, sports, religion, and psychotherapy [1].

In education, students and teachers need to have those three needs in the learning process. The needs for autonomy, competence and relationship can increase the student's motivation in learning so as to alleviate the difficulties of the students in the learning process. Teachers must be able to facilitate students to be able to improve these three needs in the learning process.

Field research has shown that teachers who support autonomy (in contrast with control) catalyze intrinsic motivation, curiosity, and their desire for a challenge. Students who are taught with an approach that is more in control are not only losing their initiatives but also learning effectiveness, especially when learning requires a conceptual and creative process [2].

The learning process is demanding students to always think conceptually, systematically and creatively. Students who have the ability to think creatively will love the challenges in resolving problems that are given by the teacher. Students will use ideas freely, yet remain conceptual and systematic, so it will be a reliable problem solver. Habits of students in solving a problem, it will affect the attitude and way of thinking in the face of problems in daily life.

In addition, research shows that not only tangible rewards but also threats, deadlines, leads, depressed evaluations, and forced goals to diminish intrinsic motivation because, as a real appreciation, they are leaning to the perceived external causality locus. In contrast, choice, recognition of feelings, and opportunities for selfdirection are found to increase intrinsic motivation because it allows people to feel greater autonomy [3].

Teachers who do not reprimand the students when they do not do the job or do not pay attention to the teacher whenever he explains the lesson had resulted in no motivation in learning. The impact of this is the result of student learning is low and they do not like math lessons. Appropriate appreciation can increase students' motivation in learning, so that students will love the math lesson, would seek to resolve the matter and the results of learning towards mathematics will increase.

Additionally, Niemic [4] stated the evidence suggests that teacher support for students' basic psychological needs for autonomy, competence, and relevance facilitates student self-regulation for learning, academic achievement, and well-being. Thus, SDT has strong implications for classroom practice and educational reform.

SDT is useful in the learning process because the three needs of autonomy, competence and relationships are needed in the learning process in the classroom. Learning that facilitates these three needs is owned by students will make students become human beings who have the character and can compete. This can increase the quality of creative and superior human resources.

2. Methods

The method used in this research is descriptive. This research aimed to analyze the self-determination of students. Students who serve as samples are as many as 60 students with 24 male students and 36 female students. The selection of the number of students is not determined by previous researchers. Researchers gave the question form SDT to 60 SMP students 16 in Tangerang. Question form given by the researchers was the result of the adaptation of BNSP developed by Decy and Ryan [1]. The adaptation was done by researcher by aiming at adjusting the character material junior and math students.

3. Results and Discussion

3.1. Aspects of the SDT

Overall aspects of SDT and all charge indicators will be spelled out in the table below;

Table 1. The Overall Aspects of The SDT

Aspects of SDT	Indicator	Answers			
		SA	A	D	SD
Autonomy	Recognition of the feeling of acting of their own accord	1	14	37	8
		11	38	10	1
		4	37	16	0
	Total	16	**89**	63	9
	The percentage	8.9%	49.4%	35%	5%
	Choice over their own experience	3	37	19	1
		4	30	23	1
	Total	7	**67**	42	2
	The percentage	5.8%	55.8%	35%	1.7%
	The opportunity for the promoted direction	2	24	33	0
		0	12	34	14
	Total	2	36	**67**	14
	The percentage	1.7%	30%	55.8%	11.7%
Competence	Fighting power	15	33	12	0
		5	37	17	1
	Total	20	**70**	29	1
	The percentage	16.7%	58.3%	24.2%	0.8%
	Feedback	3	36	20	1
		2	42	14	1
		3	24	29	3
	Total	8	**102**	63	5
	The percentage	4.4%	56.7%	35%	2.8%
	Awards	6	35	18	1

	Total	6	**35**	18	1
	The percentage	10%	58.3%	30%	1.7%
Relation	The development of self confidence	9	37	10	1
		18	34	7	1
		10	32	10	5
		2	10	35	10
		15	38	7	0
		19	35	3	2
	Total	73	186	72	19
	The percentage	24.3%	62%	24%	6.3%
	Fulfilling relationships with others	27	28	6	1
		6	30	20	4
	Total	33	58	26	5
	The percentage of	27.5%	48.3%	21.7%	4.2%

Table 1 describes that the elaboration is overall of the three aspects of SDT, most students responded in agreement. Many students responded disagree on the indicator of opportunity for the promoted direction that is equal to 55.8%. Although many students answered in every SDT indicator agree but the percentage of students' SDT capability is still low under 65%. It can be concluded that in the autonomous ability of students still lack the sense of confidence and recognition that in themselves, there is the ability to decide the problem or have great ability. While in the competence, students still lack the fighting power and good feedback, so that during the learning in the class seemed resigned and did not want to try to ask if experiencing problems in learning mathematics. In the aspect of relations, students lack the ability to develop confidence so that the interaction with others are low. The explanation of each aspect of SDT can be explained in the following table;

Table 2. Percentage of SDT

Aspect of SDT	Answers			
	SA	A	D	SD
Autonomy	5.5%	45.3%	41.9%	6.3%
Competence	10.6%	57.8%	29.8%	1.8%
Release	22.9%	52.3%	19.8%	5.3%

Table 2 can be described as follows: (1) On the aspect of autonomy, 45.3% respondents answered agree and disagree as much as 41.9%. Students answered strongly disagree by 6.3% and strongly agree 5.5%; (2) On the aspect of competence, most students answered agreed at 57.8% while the answer did not agree equal to 29.8%. Students strongly agree as much as 10.6% and strongly disagree as much as 1.8%; (3) On the aspect of Relation, students agree most at 52.3% and strongly agree as much as 22.9%. Students answered disagree as much as 19.8% and strongly disagree as much as 5.3%; (4) On the calculation of SDT aspect, it is seen that for the most dominant aspects of the students answered agree; and (5) The relation aspect is superior to the aspect of autonomy and competence.

3.2. The elaboration of every aspect of the SDT in learning mathematics
The elaboration of every aspect of the SDT in learning math can be outlined as follows:

3.2.1. Aspects of autonomy
In Table 3, the indicator on the recognition of the feeling of acting on their own accord, the students who responded agree as much as 49.4% this looks like some students are not depressed in the face of a math problem, students are free to express their ideas in solving math problems.

Table 3. The percentage of indicators on

the recognition of the feeling of acting on their own accord

Indicator	Answers			
	SA	**A**	**D**	**SD**
Recognition of the feeling of acting on their own accord	1	14	37	8
	11	38	10	1
	4	37	16	0
Total	16	89	63	9
The percentage	8.9%	49.4%	35%	5%

Table 4. Percentage on the experience over the choice of indicators

Indicator	Answers			
	SA	**A**	**D**	**SD**
Choice over their own experience	3	37	19	1
	4	30	23	1
Total	7	67	42	2
The percentage	5.8%	55.8%	35%	1.7%

In Table 4, the indicator of choice on their own experience, most students answered agree as much as 55.8%. It is seen that some students in learning mathematics or solving math problems have used his experience of prerequisite materials in accordance with the matter. this shows the students' material understanding is due to they still remember the previous mathematical material. So the teacher can improve this ability by giving the problem that can be represented in many ways.

Table 5. Percentage on the indicator of chance for the promoted direction

Indicator	Answers			
	SA	**A**	**D**	**SD**
The opportunity for the promoted direction	2	24	33	0
	0	12	34	14
Total	2	36	67	14
The percentage	1.7%	30%	55.8%	11.7%

In Table 5, the indicator of opportunity for the promoted direction, most students responded not agree that is equal to 55.8%. It is seen in the learning process that students are less confident that they have potential in math so that teachers and friends recognize that talent in mathematics is more dominant. As a result, when told to explain to his friends, they are still shy and seemingly refused. This can be overcome by teachers motivate students more to dare to be wrong and convince them that there is great potential that can benefit themselves and others.

3.2.2. Aspects of Competence

Table 6. The percentage indicator of fighting power

Indicator	Answers			
	SA	**A**	**D**	**SD**
Fighting power	15	33	12	0
	5	37	17	1
Total	20	70	29	1
The percentage	16.7%	58.3%	24.2%	0.8%

In Table 6, the indicator of fighting the power, most students agreed are as much as 58.3%. This is seen though not all of the students, there are some students who have low fighting power in solving math problems. Most students who have the fighting power are students who have the ability to go higher. students always ask a friend who is smarter or ask the teacher.

Table 7. The percentage indicator of feedback

Indicators	Answers			
	SA	A	D	SD
Feedback	3	36	20	1
	2	42	14	1
	3	24	29	3
Total	8	102	63	5
The percentage	4.4%	56.7%	35%	2.8%

On the Table 7, feedback indicator, most students answered agree as much as 56.7%. This is apparent in the learning process that most students who have the higher capability often do feedback to friends or teachers when facing difficulties in understanding the material or completing a math problem.

Table 8. Indicator percentage of awards

Indicator	Answers			
	SA	A	D	SD
Awards	6	35	18	1
Total	6	35	18	1
The percentage	10%	58,3%	30%	1.7%

In Table 8, the award indicator, students who answered agreed as much as 58.3%. This is seen in the learning process that students always do the cooperation in groups well, the students always solve the problem in accordance with the ability and receive feedback from teachers and friends well.

3.2.3. Relation Aspect

Table 9. Indicator Percentage of Confident Development

Indicator	Answers			
	SA	A	D	SD
Developing confidence	9	37	10	1
	18	34	7	1
	10	32	10	5
	2	10	35	10
	15	38	7	0
	19	35	3	2
Total	73	186	72	19
The percentage	24.3%	62%	24%	6.3%

In Table 9, confident development indicators, most students answered agreed as much as 62%. This is seen in the learning process that students interact well, have mutual care and friendliness among friends.

Table 10. Indicator Percentage of Relationship Fulfillment with Others

Indicator	Answers			
	SA	A	D	SD
Relationship fulfillment with others	27	28	6	1
	6	30	20	4
Total	33	58	26	5
The percentage	27.5%	48.3%	21.7%	4.2%

In Table 10, the indicator of the relationship fulfillment with others, most students answered agreed are amounted to 48.3%. This is seen in the learning that students are befriended with friends who are considered in accordance with their wishes. Although in general, they are all friends, for closer interaction students are grouped together

4. Conclusion

The learning process requires students to always think conceptually, systematically and creatively. Students who have creative thinking skills will love the challenge of solving problems given by teachers. Students will use ideas freely but conceptually and systematically, so that will be a reliable problem solver. Habits of students in solving problems will affect the attitude and way of thinking in dealing with problems in everyday life. SDT is useful in the learning process because the three needs of autonomy, competence and relationships are needed in the learning process in the classroom. Learning that facilitates these three needs that are owned by students will turn them into human beings who have the character and can compete. This can increase the quality of creative and superior human resources.

References

[1] Deci E L and Ryan R M 2000 *Basic Psychological Needs Scales*
[2] Deci and Ryan 2000 *The American Psychological Association* **55** 1 pp 68-78
[3] Ryan, Deci, and Baard 2004 *Journal of Applied Social Psychology, 2004* **34** 10 pp 2045-2068
[4] Niemic and Ryan 2009 *Theory and Research in Education* **7** 2 pp 133 -144

Developing self-concept instrument for pre-service mathematics teachers

M W Afgani[1, 2], **D Suryadi**[1] **and J A Dahlan**[1]

[1]Department of Mathematics Education, Indonesia University of Education, Indonesia
[2]Department of Mathematics Education, State Islamic University of Raden Fatah, Indonesia

E-mail: afgani@student.upi.edu

Abstract. This study aimed to develop self-concept instrument for undergraduate students of mathematics education in Palembang, Indonesia. Type of this study was development research of non-test instrument in questionnaire form. A Validity test of the instrument was performed with construct validity test by using Pearson product moment and factor analysis, while reliability test used Cronbach's alpha. The instrument was tested by 65 undergraduate students of mathematics education in one of the universities at Palembang, Indonesia. The instrument consisted of 43 items with 7 aspects of self-concept, that were the individual concern, social identity, individual personality, view of the future, the influence of others who become role models, the influence of the environment inside or outside the classroom, and view of the mathematics. The result of validity test showed there was one invalid item because the value of Pearson's r was 0.107 less than the critical value (0.244; $\alpha = 0.05$). The item was included in social identity aspect. After the invalid item was removed, Construct validity test with factor analysis generated only one factor. The Kaiser-Meyer-Olkin (KMO) coefficient was 0.846 and reliability coefficient was 0.91. From that result, we concluded that the self-concept instrument for undergraduate students of mathematics education in Palembang, Indonesia was valid and reliable with 42 items.

1. Introduction

Self-concept is an individual view of him/her self that can influence his/her life. That view can be the individual appraisal of him/her self. That appraisal is yielded from a mental appraisal of the individual of his/her characteristic that can be positive or negative. Positive self-concept indicate that individual accept him/her self in an individual position that has excess and inadequacy, and this leads to individual confidence in socialization, exhibit behaviour as an independent individual, responsible, not easy disconsolate, enthusiasm to face a challenging duty, give a positive influence to others, whereas negative self-concept reflect a feeling paltry, incapable or incapacity, avoid a situation that can make worried, like to blame others, impressionable by others, and often frustration [1, 2].

The self-concept consists of the way of someone see him/her self as an individual, individual feeling about him/her self, and the point of view of the individual as a human that he/she expects. Leonard and Supardi [3] also stated that self-concept is an individual responsibility which is positive about his/her life and self. Individual self-concept relates to his/her past, present, and future life [4, 5, 6, 7]. Individual future is determined by his/her perception what he/she may become later, what he/she will like to become, or what he/she is afraid of becoming. Those things correspond with hopes, fears, objectives, goals, and threats from individual condition today. The condition is an incentive for future behaviour. So, individual perception about his/her past and future life associates to his/her perception today [4].

Self-concept is a factor that is formed through individual experience by interaction with others. Gunawan implied that self-concept is notched because an influence from peoples that is deemed important by the individual which is a role model for him/her [3]. The peoples are the parent, family member, friend, teacher, or others [8]. Besides that, Jalaludin phrased self-concept is determinant in interpersonal communication which every individual behavior describe his/her self-concept [3]. According to Takaria [9], self-concept is individual perception to him/her self, a point of view from another person to excess, inadequacy, a capacity that inside the individual, and individual attitude. A positive self-concept inside individual can be seen from his/her activity, such as activity in learning mathematics, individual perception is showed through high enthusiasm, hardworking, confident, and high motivation to solve varying mathematics problem.

Every individual has a unique self-concept that can be identified his/her characteristic. First, physique characteristic is an important component of self-concept because other people use it to appraise individual. People often mention his/her physique characteristic to show that he/she different positively or negatively from the others because he/she understands that his/her characteristics prominent and probably is used by others to appraisal him/her. Ethnic individuality in self-concept is an exclusive function of the individual in a social environment. The second characteristic of self-concept reflects individual involve in social group concerning his/her existence and caring. In this matter, the general response is such as "I am an artist", "my ethnic is Melayu", and "I am a college student at State Islamic University". Individual participation in social group notches an important part of self-concept because it gives his/her social identity. Social identity is an individual appraisal that involves the social group. The last characteristic of the self-concept is personality. Personality is an individual characteristic that is permanent and special which describe his/her self, for example, "I am a friendly person", "I mousy", and "I am persistent". This individual difference is important to determine individual behavior, and this aspect of self-concept reflects people diversity [10].

Hurlock stated there are three principal components that are as follows [9]:

a. Perception is an individual point of view that concern to his/her physique characteristic and others point of view to him/her too.
b. Conceptual is an individual conception that concern to his/her capability an incapability, background, and cogitation about future.
c. Attitude relate to feeling, belief which is inside individual about his/her self, and others perception of that individual.

From the components of self-concept, aspects that can be used to measure an individual self-concept, that is 1) individual point of view about his/her capability, 2) Attitude and individual belief, 3) individual perception about him/her self, 4) individual cogitation about his/her future.

From several affirmations above, there are many aspects that form self-concept of the individual. It can affect directly or indirectly. One of them is the influence of surrounding environment because it has an important role in the development of individual self-concept [11]. It means that self-concept of individual who is in X environment cannot be measured by using self-concept instrument that is valid in Y environment. An instrument which is used directly can make the result may risk oversimplification [12]. Hence, the instrument ought to be modified to correspond with the environment. According to Villegas and Tomasini, an instrument which is used as a measurement tool and evaluation depends on the specific condition of the environment [13]. Consequently, self-concept instrument should be developed to measure individual self-concept in specific environment and condition. In this study, we developed a self-concept instrument that focuses on the undergraduate student of mathematics education in Palembang, Indonesia. So, self-concept can be concluded as a point of view, idea, and individual appraisal to his/her self that is formed from academic and non-academic experience which is felt when he/she interact with his/her surrounding environment so that its influence positively or negatively to his/her behavior. The aspects of self-concept that we analyzed, that are the individual concern, social identity, individual personality, view of the future, the influence of others who become role models, the influence of the environment inside or outside the classroom, and view of the mathematics. We did not analyze physique characteristic because we assume that aspect is not appropriate with a questionnaire as the instrument, but the observation is more enable.

2. Methods

The type of this study was development research of non-test instrument. The procedure was as follows, that is (1) determine all possible aspects that form self-concept, (2) write the statements for each aspect, (3) discuss the instrument with mathematics education experts, (4) field test, (5) instrument analysis, (6) determine the final instrument. The respondents of this study were 65 undergraduate students of mathematics education at one of the universities in Palembang, Indonesia. The form of the instrument was questionnaire in Likert scale with four options. For positive statement, the score is 4 = Strongly Agree, 3 = Agree, 2 = Disagree, and 1 = Strongly Disagree. A negative statement is vice versa. The respondents must choose only one of the options. The technique of data analysis used (1) expert judgment for construct validity test theoretically, (2) Pearson product moment and factor analysis for construct validity test empirically, and (3) Cronbach's alpha for the reliability test. To help data processing, SPSS 16.0 was used in this study.

3. Results and Discussion

The aim of this study was to develop self-concept instrument for undergraduate students of mathematics education in Palembang, Indonesia. The focus in this study was self-concept of undergraduate students of mathematics education. The instrument was developed from previous studies [9, 13, 14, 15, 16]. After literature review, We designed 43 items with 7 aspects of self-concept, that are individual concern with 4 items (example: If my friend has a difficulty in doing a calculus task, then I will be happy help him/her), social identity with 6 items (example: I am ashamed to introduce myself to others as an undergraduate students of mathematics education), individual personality with 10 items (example: If I do not understand about calculus, then I will strive to find out for myself first), view of the future with 5 items (example: Being a mathematics teacher is my dream), the influence of others who become role models with 5 items (example: I feel Ki Hadjar Dewantara's thinking in education is feasible to be an example), the influence of the environment inside or outside the classroom with 8 items (example: I am delighted with the method of teaching that applied by calculus lecture), and view of the mathematics with 5 items (example: I find that mathematics is beautiful, because the rules are logical). For construct validity test theoretically, the instrument was reviewed by the experts in mathematics education. From discussion with four doctors and one professor in mathematics education as the experts, the instrument was revised. They suggested the number of positive and negative statements for every aspect should equal, the statement like "If I cannot do mathematics task, then I will not shame to ask assistance to a friend who is can" must avoid, because there are two negative sentences, and the influence of others who become role models as lecture or instructor must explicit, is the lecture mathematics education or else?. After revising, the detail of 7 aspects of academic self-concept, that are individual concern with 2 positive and 2 negative statements, social identity with 3 positive and 3 negative statements, individual personality with 5 positive and 5 negative statements, view of the future with 3 positive and 2 negative statements, the influence of others who become role models with 3 positive and 2 negative statements, the influence of the environment inside or outside the classroom with 4 positive and 4 negative statements, and view of the mathematics with 3 positive and 2 negative statements.

For construct validity test empirically, the instrument was tested by 65 undergraduate students of mathematics education at one of the universities in Palembang, Indonesia. The items validity test used Pearson product moment. The result showed there is one item is invalid, because the value of Pearson's r is 0.107 less than its critical value (0.244; $\alpha = 0.05$). The item is "I will angry, if someone disparage mathematics education department". That item is included in social identity aspect. After invalid item was removed, we tested the normality of the data. We found that Mean of the data is 141.35 and standard deviation is 12.65. The self-concept of students, according Arikunto [17], can be categorized by using that data. The result was 16.92% students are low, 67.69% students are moderate, and the rest of them are high. It showed that there are still students need to improve their self-concept. The result from Kolmogorov-Smirnov test also showed significant value is 0.200. It means that the distribution of the data is normal, because its significant value is more than 0.05. Next analysis was factor analysis to 7 variables of self-concept. Construct validity test empirically with factor analysis showed Kaiser-Meyer-Olkin (KMO) of *Measure of sampling Adequacy* (MSA) is 0.846. That value is more than 0,5. It show adequacy of sample, while reliability test used Cronbach's alpha. The value of reliability coefficient is

0,91. Besides that, the value of Bartlet's test was 232.806 and significance was 0.000 which showed that there are correlations among variables and proper to process furthermore. The values also indicated that the instrument has satisfied valid requisite.

Hereafter, to know which variables that can be processed further or removed can be observed from the MSA values in anti-image correlation (Table 1). All MSA values for each variable were more than 0.5. It shows all variables can be processed further and valid as formers variable of academic self-concept.

Table 1. Anti-image Correlation

Variable	A	B	C	D	E	F	G
A	**0.852**[a]						
B	-0.221	**0.882**[a]					
C	-0.280	0.014	**0.795**[a]				
D	-0.020	-0.143	-0.389	**0.846**[a]			
E	-0.024	-0.362	0.072	0.017	**0.871**[a]		
F	0.110	-0.147	-0.541	0.134	-0.111	**0.819**[a]	
G	0.036	-0.179	-0.045	-0.378	-0.240	-0.263	**0.876**[a]

a. Measures of Sampling Adequacy(MSA)

A: individual concern
B: social identity
C: individual personality
D: view of the future
E: the influence of others who become role models
F: the influence of the environment inside or outside the classroom
G: view of the mathematics

The next analysis was to know how much the factors are formed can explain all variables. It can be observed from communalities value. The result showed that the factor can explain 30.60% of individual concern, 65.50% of social identity, 69.50% of individual personality, 63.60% of view of the future, 49.60% of the influence of others who become role models, and 66.20% of the influence of the environment inside or outside the classroom, and 72.80% of view of the mathematics. Because an average of explanation is more than 50%, then the fixed factor will be determined.

Table 2. Communalities

Variable	Initial	Extraction
Individual Concern	1.000	0.306
Social Identity	1.000	0.655
Individual Personality	1.000	0.695
View of The Future	1.000	0.636
The Influence of Others Who Become Role Models	1.000	0.496
The influence of The Environment Inside or Outside The Classroom	1.000	0.662
View of The Mathematics	1.000	0.728

Extraction Method: Principal Component Analysis.

To know how many possible factors are formed, it can be observed from Total Variance Explained. The variance could be explained by one factor, that was 59.686% and there was no other factor. It means that seven independent variables only formed one factor of self-concept. The variables could not be simplified to become more than one factor. All that variables were needed to explain self-concept for undergraduate students of mathematics education. 59.686% could be explained by them, and the rest by other factors were not observed. While that, the correlation between independent variables with the factor can be seen in the component matrix.

Table 3. Component Matrix[a]

Variable	Component
	1
Individual Concern	0.553
Social Identity	0.809
Individual Personality	**0.834**
View of The Future	0.798
The Influence of Others Who Become Role Models	0.704
The influence of The Environment Inside or Outside the Classroom	0.814
View of The Mathematics	**0.853**

Extraction Method: Principal Component Analysis.
a. 1 components extracted.

It shows that 'view of the mathematics' and 'individual personality' have a strong correlation with the factor of undergraduate students self-concept of mathematics education. That finding is different with Yahaya, et al [18]. They claimed that the correlation between students' personality and his/her self-concept is very weak. They observed secondary school students. So, their claim cannot be generalized to all students' level. This because of individual personality will develop from time to time. From the result, we could imply that the factor can be described as formers factor of self-concept for undergraduate students of mathematics education. Besides that, the correlation between independent variables can be seen from this table below.

Table 4. The Correlation between Independent Variables

Variable	A	B	C	D	E	F	G
A	1						
B	0.425	1					
C	0.461	0.536	1				
D	0.366	0.557	0.675	1			
E	0.284	0.622	0.408	0.428	1		
F	0.315	0.570	**0.738**	0.535	0.491	1	
G	0.329	0.638	0.627	0.682	0.582	0.659	1

A: individual concern
B: social identity
C: individual personality
D: view of the future
E: the influence of others who become role models
F: the influence of the environment inside or outside the classroom
G: view of the mathematics

Based on Arikunto [19], Table 4 shows that view of the mathematics has strong correlation with other variables except for individual concern and the influence of others who become role models, individual personality has strong correlation with the influence of the environment inside or outside the classroom and view of the future, and social identity has strong correlation with the influence of others who become role models. Each of variables also showed sig. < 0.05. It means that each of variables has significant effect difference on self-concept for undergraduate students of mathematics education in Palembang, Indonesia.

4. Conclusion

From this study, we have succeeded to develop self-concept instrument for undergraduate students of mathematics education in Palembang, Indonesia. The instrument consisted of 7 aspects, that are the individual concern, social identity, individual personality, view of the future, the influence of others who become role models, the influence of the environment inside or outside the classroom, and view of the mathematics. All that variables were needed to explain academic and non-academic self-concept for undergraduate students of mathematics education. 59.686% could be explained by them, and the rest by other factors were not observed. The result from this study, we also could conclude that the instrument is valid and reliable with 42 items and could be used to measure the self-concept of undergraduate students of mathematics education in Palembang, Indonesia.

References

[1] Alipour Z M and Nasab H M 2012 Effectiveness of Life Skills Training in Enhancing Academic Self-Concept of Pre-University Girl Students in Tehran *An International Peer-Reviewed Journal* **4** 1
[2] Sumartini T S 2015 Mengembangkan Self Concept Siswa melalui Model Pembelajaran Concept Attainment *Jurnal Pendidikan Matematika* **5** 2
[3] Andinny Y 2015 Pengaruh Konsep Diri dan Berpikir Positif Terhadap Prestasi Belajar Matematika Siswa *Jurnal Formatif* **3** 2 pp 126 – 135
[4] Barongo S and Nyamwange C 2013 Contribution of Self-Concept in Guidance and Counselling Among Students *Research on Humanities and Social Sciences* **3** 13 pp 7 – 12
[5] Deshmukh K P M 2015 Comparison of Self-Concept of School Children Belonging to High and Low Fitness Groups *International Journal of Physical Education, Sports and Health* **1** 3 pp 61 – 64
[6] Lone P A and Lone T A 2016 A Study on Relation between Self Concept and Academic Achievement among Secondary School Students of Jammu District *Journal of Education and Practice* **7** 31 pp 19 – 23
[7] Rajeswari V and Kalaivani M 2016 The Role of Academic Motivation and Academic Self Concept in Students Academic Achievement *International Journal of Research Granthaalayah* **4** 9 pp 37 – 49
[8] Othman N and Leng K B 2011 The Relationship between Self-Concept, Intrinsic Motivation, Self-Determination and Academic Achievement among Chinese Primary School Students *International Journal of Psychological Studies* **3** 1 pp 90 – 98
[9] Takaria J 2015 Peningkatan Literasi Statistis, Representasi Matematis, dan Self Concept Mahasiswa Calon Guru Sekolah Dasar melalui Model Collaborative Problem Solving *Dissertation* Indonesia University of Education
[10] Stangor C 2012 Social Psychology Principles Available at http://lardbucket.org/
[11] Baran M and Maskan A K A Study of Relationships Between Academic Self Concepts, Some Selected Variables and Physics Course Achievement *International Journal of Education* **3** 1 pp 1-12
[12] Hsu Y W and Lu F J H 2013 The Development and Validation of the Physical Self-Concept Scale for Older Adults *Educational Gerontology* 39 pp 501 – 513
[13] Villegas G O and Tomasini G A 2014 Development of An Academic Self Concept for Adolescent (ASCA) Scale *Journal of Behavior, Health, & Social Issues* **5** 2 pp 117 – 130
[14] Ferla J, Valcke M, and Cai Y 2009 Academic Self-Efficacy and Academic Self-Concept: Reconsidering Structural Relationships *Learning and Individual Differences* 19 pp 499 - 505
[15] Matovu M 2012 Academic Self-Concept and Academic Achievement among University Students *International Online Journal of Educational Sciences* **4** 1 pp 107 - 116
[16] Flowers L O, Raynor J E, and White E N 2013 Investigation of Academic Self-Concept of Undergraduates in STEM Courses *Journal of Studies in Social Sciences* **5** 1 pp 1 – 11
[17] Arikunto S 2012 *Dasar-Dasar Evaluasi Pendidikan* (Jakarta: Bumi Aksara)

[18] Yahaya A, Ramli J, and Boon Y 2009 Relationship between Self Concepts and Personality and Students' Academic Performance in Selected Secondary Schools *European Journal of Social Sciences* **11** 2 pp 302 - 316
[19] Arikunto S 2005 *Prosedur Penelitian: Suatu Pendekatan Praktek* (Jakarta: Rineka Cipta)

Acknowledgments
The author extends gratitude to Prof. Dr. Didi Suryadi, M.Ed. and Dr. Jarnawi Afgani Dahlan, M.Kes. as co-author for their guidance, so that this paper as a part of my dissertation could be finished. Especially thanks to Prof. Dr. Kasinyo Harto, M.Ag. and Prof. Dr. Aflatun Muchtar, MA. which have given opportunity and scholarship, so that author could participate as a student in Indonesia University of Education. Last but not least, the author renders thanks and gives respect very grateful to all other individuals who have supported author to writing this paper.

Pre-service mathematics teachers' ability in solving well-structured problem

R Paradesa[1]

[1]Universitas PGRI Palembang, Palembang, Indonesia

E-mail: retni.paradesa@yahoo.com

Abstract. This study aimed to describe the mathematical problem-solving ability of undergraduate students of mathematics education in solving the well-structured problem. The type of this study was qualitative descriptive. The subjects in this study were 100 undergraduate students of Mathematics Education at one of the private universities in Palembang city. The data in this study was collected through two test items with essay form. The results of this study showed that, from the first problem, only 8% students can solve it, but do not check back again to validate the process. Based on a scoring rubric that follows Polya strategy, their answer satisfied 2 4 2 0 patterns. But, from the second problem, 45% students satisfied it. This is because the second problem imitated from the example that was given in learning process. The average score of undergraduate students mathematical problem-solving ability in solving well-structured problems showed 56.00 with standard deviation was 13.22. It means that, from 0 – 100 scale, undergraduate students mathematical problem-solving ability can be categorized low. From this result, the conclusion was undergraduate students of mathematics education in Palembang still have a problem in solving mathematics well-structured problem.

1. Introduction

On mathematics curriculum, the problem-solving ability is a very important. In the learning process, Students gain an experience by using their previous mathematics knowledge to solve the routine or non-routine problem. Mathematics problem solving is a process that uses the concepts of mathematics in solves a mathematics problem. In mathematics education, problem-solving is one of important ability to develop students' mathematics thinking. With mathematical problem solving, students will understand mathematics meaningful because the concepts or principles can be applied to solving a problem. The importance of mathematical problems solving ability, according to Wilson [1], is the main goal of mathematics teaching and learning.

Based on the author experience, most students had difficulty when solving a mathematics problem. This was indicated by the results of a test that the author gives the lecture, while the problem is basic. This was a reason that encourages the authors to conduct this study. Preliminary observation to undergraduate students of mathematics education at one University in Palembang who take applied mathematics class showed that they are not able enough to solve mathematical problems relate to the real world and do not get used to delivering a mathematics idea in oral and written form. They difficult determined the exact problem, found a strategy, and determined a usable pattern. The students could solve mathematics problem if they were given mathematics symbols and figures directly so that they knew what to solve immediately without interpreting the problem first. A question of "why" is hardest to answer them.

NCTM [2] stated that mathematics problem solving is a tool and an objective in learning mathematics. It is very recommended to be given from elementary school until university. Through problem-solving, the way of student's thinking will be used to face nonprocedural problems. According to Kirkley [3], there are three types of problem, which is well-structured problems, moderately-structured problems, and ill-structured problems. Well-structured problems have characteristic, that is the strategy of problem-solving is usually more than one, and all information available in order to solve the problem directly. The characteristic of moderately-structured problems, that is the strategy of problem-solving is usually more than one, has one solution, and needs additional information in order to solve the problem. The last, ill-structured problems have characteristic, that is difficult to allegedly, the strategy of problem-solving has many perspectives, any solutions, and needs additional information in order to solve the problem.

Mathematics problem solving has two significances in learning. As a learning approach, mathematics problem solving is an approach that provides a contextual problem as starting point and then reinvention gradually and understands mathematics concept or principal. As an ability, It must be reached after learning process, problem-solving is an activity which the solution to a problem doesn't know yet or no immediately found. In solving a problem, the student must integrate his/her knowledge by developing his/her new understanding. Generally, problem-solving is nonroutine. Hence, this ability is classified as high order mathematics thinking ability [4]. Nonroutine problem is interpreted by Arslan and Altun [5] as a problem that different from usual and cannot be solved with general strategy. This problem can relate to the real-life situation or not, but it's always be conceived in real life model situation.

The mathematical problem solving that was used in this study refers to Polya [6]. The four steps to solve the mathematics problem are (1) understanding the problem, (2) making a plan, (3) executing the plan, and (4) re-examining the answer. At the stage of understanding the problem, the student must be convinced that a problem can be solved in several ways, read repeatedly, determine what is known, what is unknown, and seek the purpose of the mathematical problem. At the stage of creating a plan, students can observe the relationship between known and unknown information. At this stage, the calculation is performed to an unknown variable. It will obtain the answer to the question about how the known information interconnects to unknown. At the stage of executing the plan, the students check each step that outlines in the plan and writes it in detail to ensure that each step is correct. At the re-examining stage, students look back at the answer to make sure that the answer to the problem is correct. Based on that background, the purpose of this study was to describe mathematical problem-solving ability of undergraduate students of mathematics education in solving well-structured problem.

2. Methods

The method of this study was descriptive qualitative. The instrument in this study tested. The subject of this study was 100 students of Mathematics Education in one of the private Universities at Palembang who take Applied Mathematics class of academic year 2016/2017. Mathematical problem-solving ability was measured using an essay test. The indicators of mathematical problem-solving ability in this study were: (1) understanding the problem; (2) plan for problem solving; (3) implementing a problem-solving plan; (4) re-examination. The following is a scoring rubric of mathematics problem-solving ability that pursues Polya strategy:

Table 1. The Scoring Rubric of Mathematics Problem Solving Ability Following to Polya Strategy

Score	Understanding problem	Constructing plan	Performing calculation	Checking back the result
0	Misinterpretation or incorrect at all	No plan, irrelevance of constructing plan	Not performing calculation	Not checking back
1	Misinterpretation partially, disregard of problem condition	Constructing a plan, but it cannot workable	Performing the right procedure and probably produce a correct answer but miscalculate	Checking back but incomplete
2	Understanding problem clearly	Constructing the right plan but incorrect in the result or no result	Performing the right procedure and getting a correct answer	Checking back to see the validity of process
3		Constructing the right plan but incomplete		
4		Constructing a plan that appropriates with procedure and set in a correct answer		
	Max score is 2	Max score is 4	Max score is 2	Max score is 2

Source: Sumarmo [4]

3. Results and Discussion

The general description of this study was mathematical problem-solving ability of undergraduate students of mathematics education in solving well-structured problem. The test was conducted after the lecture about applied mathematics is finish. Learning method that is applied in learning process was expository. The first test item is "a cup of milk with a temperature of 60° C is placed indoors with a temperature of 10 C, in 30 minutes the temperature of milk decreases to 40° C. How long does it take for the temperature to cool from 60° C to 30° C, then how long does it take to cool from 50° C to 30° C". The second test item is "a bacterial colony increases at a rate proportional to the amount of material present. If the number of bacteria tripled within 5 hours, how long will it take for the number of bacteria to multiply 5". Students' answer analysis on both problems used the rubric of Sumarmo [4].

Table 2. The percentage of the pattern of students' answer based on the first problem

Score of students answer analysis				Number of Students Percentage
Understanding Problem	Constructing Plan	Performing Calculation	Checking Back	
2	4	2	0	8%
2	4	1	0	4%
2	3	1	0	22%
2	2	1	0	15%
1	3	2	0	10%
1	3	1	0	5%
1	2	1	0	27%
1	2	0	0	2%
1	0	1	0	1%
0	0	0	0	6%

Table 2 shows that 8 of 100 students can understand the problem clearly, can construct the plan correctly and completely, can perform the procedure correctly so that produce the correct answer, but do not check back again to validate the process. 4 of 100 students can understand the problem clearly, can construct the plan correctly and completely, can perform the procedure correctly but the final answer is not correct, and they do not check back. 22 of 100 students can understand the problem clearly, can construct the plan correctly but incomplete, can perform the procedure correctly but the answer is not correct, and do not check back. 15 of 100 students can understand the problem clearly, construct the plan inappropriately and incompletely, can perform the procedure correctly but the answer is wrong, and they do not check back. 10 of 100 students do some errors. One of the errors is to ignore the conditions in determining the temperature decrease from 50 to 30. This can be seen from incomplete plan, but the process for the initial conditions is correct. 5 of 100 students make mistakes in understanding the problem, can construct the plan correctly and incomplete, can perform the procedure but produce the wrong answer, and do not check back. 27 of 100 students make mistakes in understanding the problem, construct the plan inappropriately and incompletely, can perform the procedure correctly but the answer is wrong, and they do not check back. Errors that occur are the students ignore the conditions in determining the temperature dropped from 50 to 30, make the right plan but wrong in the result because they implement procedures that probably far from correct answer. 2 of 100 students make mistakes in understanding the problem, not constructing the plan, can perform the procedure with the answer is not correct, and do not check back. One of them wrote, $s(0)\ e^{pt} = 3\ s(0)$. It should be written $s(0)\ e^{pt} = 5\ s(0)$. It seems the plan that he/she makes indicate the calculation will correct, but the answer is irrelevant so that it leads to impossibility to produce the correct answer. The rest of them were unable to solve the first problem ranging from not being able to interpret the problem until not checking again.

The pattern of students' answer based on the first problem forming 10 patterns. It is different with the patterns that occur from students' answer based on the second problem. 12 patterns were formed by them. The percentage of the pattern is shown detail in table 3.

Table 3. The percentage of the pattern of students' answer based on the second problem

Score of students answer analysis				Number of Students Percentage
Understanding Problem	Constructing Plan	Performing Calculation	Checking Back	
2	4	2	0	45%
2	4	1	0	1%
2	3	1	0	21%
2	2	1	0	9%
2	1	1	0	1%
1	3	2	0	3%
1	3	1	0	3%
1	2	1	0	5%
1	2	0	0	3%
1	0	1	0	6%
1	0	0	0	1%
0	0	0	0	2%

Table 3 shows that 45 of 100 students can understand the problem clearly, can construct the plan correctly and completely, can perform the procedure correctly so that produce the correct answer, but do not check back again to validate the process. It was different with the result from the first problem. This is because the second problem imitated from the example that was given in learning process by lecture. One of 100 students can understand the problem clearly, can construct the plan correctly and completely, can perform the procedure correctly but the final answer is not correct, and do not check back. 21 of 100 students can understand the problem clearly, can construct the plan correctly but incomplete, can perform the procedure correctly but the answer is not correct, and do not check back. 9 of 100 students can understand the problem clearly, construct the plan inappropriately and incompletely, can perform the procedure correctly but the answer is wrong, and they do not check back. One of 100 students can understand the problem clearly, constructing a plan, but it cannot workable, can perform the procedure correctly but the answer is wrong, and they do not check back. 3 of 100 students do some errors. This can be seen from incomplete plan, but the process for the initial conditions is correct. 3 of 100 students make mistakes in understanding the problem, can construct the plan correctly and incomplete, can perform the procedure but produce the wrong answer, and do not check back. 5 of 100 students make mistakes in understanding the problem, construct the plan inappropriately and incompletely, can perform the procedure correctly but the answer is wrong, and they do not check back. 3 of 100 students make mistakes in understanding the problem, do not construct the plan, can perform the procedure with the answer is not correct, and do not check back. 6 of 100 students make mistakes in understanding the problem, do not construct the plan, can perform the procedure with the answer is not correct, and do not check back.

Description from the pattern of students' answer leads the conclusion that more than 50% undergraduate students of mathematics education still difficult to solve well-structured problem as long as the problem does not imitate the example that is given in learning process. The average score from both problems showed that undergraduate students' mathematical problem-solving ability in solving well-structured problems was 56.00 with standard deviation was 13.22. It means that, from 0 – 100 scale, undergraduate students' mathematical problem-solving ability can be categorized low. The result indicated that they have difficulties in mathematics problem-solving. This result espouses the studies by Tambychik and Meerah [7], Klymchuk and Zverkova [8], Julita [9], and Rusyda, et al. [10]. Student difficulties were revealed through their mistake in answer the problems. Students' mistakes occur because (1) the students are unable to convey what is the question about, (2) students are unable to state what is known and asked, and (3) students have errors in interpreting the language [11, 12]. This is because they make mistakes in understanding the meaning of the problem. It also expressed by Sumargiyani [13] and Widodo [14]. They revealed that there are three types of errors that are often done by students. The three errors are a concept, language interpretation, and computing. Concept errors can be seen to students' misconceptions about what is known and what is asked in the question. The error of language interpretation occurs when the students make mistake in understanding the problem and change mathematics problem in the form of everyday language into the form of a mathematics sentence. Error in computing is the error when students do calculation to find the correct answer. Students who make misconceptions will unable to solve the problem at a later stage. However, students who make misconstruction to both problems will get difficulty to solve a problem at the next stage.

4. Conclusions

In this study, I have succeeded describe mathematical problem-solving ability of undergraduate students of mathematics education in solving well-structured problem at one of private universities in Palembang. The result showed that more than 50% of them still difficult to solve well-structured problem. The average score was 56.00 with standard deviation was 13.22. It means that their ability can be categorized low. From this result, I could conclude that undergraduate students of mathematics education in Palembang still have a problem in solving mathematics well-structured problem.

References

[1] Wilson 1993 *Mathematical Problem Solving* (New York: Macmilan Publishing Company)
[2] National Council of Teachers of Mathematics 2000 *Principles and Standards for School*

Mathematics

[3] Kirkley J 2003 *Principles for Teaching Problem Solving* (Indiana University: Plato Learning, Inc)

[4] Ibrahim 2012 Peningkatan Kemampuan Komunikasi, Penalaran, dan Pemecahan Masalah Matematis serta Kecerdasan Emosional melalui Pembelajaran Masalah pada Siswa Sekolah Menengah Atas *Disertasi Sekolah Pascasarjana Universitas Pendidikan Indonesia* (Bandung: Unpublished)

[5] Arslan C and Altun M 2007 Learning to Solve Non-Routine Mathematical Problems *Elementary Education Online* **6** 1 pp 50 – 61

[6] Polya G 1973 *How To Solve it: A New Aspect of Mathematical Method* (USA: Pricenton University Press, New Jersey)

[7] Tambychik T and Meerah T S M 2010 *Procedia-Social and Behavioral Sciences* **8** pp 142 – 151

[8] Klymchuk S and Zverkova T 2010 *Mathematics Education Research Journal* **22** 1 pp 81 – 91

[9] Julita 2017 *Journal of Mathematics Education* **6** 1 pp 37 – 50

[10] Rusyda N A, Kusnandi K and Suhendra S 2017 *International Conference on Mathematics and Science Education ICMScE IOP Conference Series: Journal of Physics: Conference Series* **895** (012081)

[11] Subaidah 2006 *Mathedu* **1** 2 pp 171 – 178

[12] Sahriah S, Muksar M and Lestari T E 2012 *Jurnal Pendidikan Matematika* **1** 1 pp 1 – 10

[13] Sumargiyani 2013 *Jurnal Karya Pendidikan Matematika* pp 1 – 7

[14] Widodo S A 2013 *Jurnal pendidikan dan pengajaran* **46** 2 pp 106-113

Acknowledgments

Author expressed her gratitude to the Dean of the Faculty of Teacher Training and Education of Universitas PGRI Palembang, Dra. Andinasari, M.M., M.Pd., who has given support to this study.

An analysis of primary school students' representational ability in mathematics based on gender perspective

Kowiyah[1] and I Mulyawati[1]

[1]Universitas Muhammadiyah Prof DR. HAMKA, Jakarta, Indonesia

E-mail: kowi_agil@yahoo.co.id

Abstract. Mathematic representation is one of the basic mathematic skills that allows students to communicate their mathematic ideas through visual realities such as pictures, tables, mathematic expressions and mathematic equities. The present research aims at: 1) analysing students' mathematic representation ability in solving mathematic problems and 2) examining the difference of students' mathematic ability based on their gender. A total of sixty primary school students participated in this study comprising of thirty males and thirty females. Data required in this study were collected through mathematic representation tests, interviews and test evaluation rubric. Findings of this study showed that students' mathematic representation of visual realities (image and tables) was reported higher at 62.3% than at in the form of description (or statement) at 8.6%. From gender perspective, male students performed better than the females at action planning stage. The percentage of males was reported at 68% (the highest), 33% (medium) and 21.3% (the lowest) while the females were at 36% (the highest), 37.7% (medium) and 32.6% (the lowest).

1. Introduction

In the process of doing mathematical problems, students can change an object into a verbal, symbolic, tabular, or graphical form in order for the process to contain information that can be used to solve the problems. Students rarely produce their own representations [1]. Students more often follow the steps their teacher use to solve problems. For example, to solve a mathematical problem, students need to write down what information is provided and what needs to be solved. This process requires mathematical representation.

Representation is defined differently by several experts. Fuad [2], for example, saw representation as a problem solver for students where a precise mathematical representation will ease students in solving problems. Hwang [3] viewed that students' ability in problem solving will be exceptional if they are able to manipulate verbal representations, picture representations (figures, graphs), and formal representations (sentences, equations, phrases). Meanwhile, Kartini [1] viewed representation as a combination of three components, which are symbols (writing), real objects, and mental pictures. According to Mustangin [4], the variety of representations that are most often used to convey mathematics are (1) visuals, such as tables, figures, graphs; (2) mathematical equations or notations and; (3) written text. According to Dahlan and Juandi [5], mathematical representations can be formulated from problems, such as visual mathematical representations in the form of figures and tables, mathematical expressions or equations, descriptions, or statements.

Fuad, Hwang, Mustangin, Dahlan and Juandi, and Kartini [2,3,4,5,1] pointed out that mathematical representations are related to problem solving. Those five studies interpreted representations as tools to present mathematical ideas. In this research, representation is defined as a method to communicate mathematical ideas through visual means in the form of figures or tables, mathematical expressions, and equations.

NCTM [6] stated that representations formulated by students while solving problems and delving into mathematical ideas play an important role in helping students understand, solve problems, and provide a meaningful way to outline other methods. For students, mathematical representations are important not only for studying conventional representations, but also for building and using their own representations for studying mathematics [7]. However, in reality, the ability of students in presenting mathematics is still considered low. A study conducted by Purnomo, Kowiyah, Aliyani, and Assiti [8] shows that students' mathematical representation skill towards the value of 1.25% is still low. This shows that students experience misconceptions with the presentation of numbers.

Mathematical representation abilities between male and female students in answering problems are different to each other. Male students are more dominant in spatial abilities, while females are more dominant in logical reasoning [9]. This is in line with the research by Mairing, Budayasa, and Juniati [10] who expressed the difference in problem solving schemes used by male and female students. There are several studies (for example, Fryer & Levitt, [11]) that showed that male and female students are not different when entering school at kindergarten level where they have equal abilities in mathematics and reading. However, by the end of fifth grade, it is shown that females have a standard deviation of 0.2 higher than males. Other than that, females also have a way of thinking and a way of receiving information that is different from males.

Several studies had been done regarding the difference in problem solving between males and females. For example, studies were done by Fuad, Asis et al, and Mairing et al [2,9,10]. These studies tend to describe differences in mathematical achievement between male and female students. Fuad [2] described in detail the mathematical representation used by high school students in solving problems viewed by differences in gender. However, there are still only a few studies that describe primary school students' mathematical representations in problem solving based on gender. Other than that, this research also discusses mathematical representations towards problem solving among high-, middle-, and low- ability male and female students.

This research was done by primary school students in number sense. Number sense is a part of primary school curriculum that is tied to everyday life. Several studies that had been done showed that student mathematical representation skills are low. In a research conducted by Fuad [2] where a problem was given to several individuals, several responses with different ways of problem solving were observed. This is caused by the difference in students' representation abilities while solving the problem. Other than that, student mastery of numbers is also considered low [12]. This is similar to a study conducted by Japa [13] that indicates the lack of problem solving skills among primary school students.

2. Methods

This research used mixed methods by using quantitative and qualitative methods to determine whether a single understanding would emerge from the problem being researched.

2.1. The Participants

The subjects of this research were fifth graders at an Integrated Islamic Elementary School in East Jakarta consisting 30 male and 30 female students. The school has their classes divided, separating the male and female students.

2.2. Data Collection

2.2.1. Instruments. The data collection technique was done using problem solving ability tests, assessment rubric, interviews, and documentation. During the initial stage, a problem solving ability test was given to the male and female students of the school. Mathematical representations generally are made up of three representations which are visuals in the form of figures or tables, mathematical expressions or equations, and descriptions or statements. Indicator representation visual in form of figure and table is present data or information from a representation in the form of a table, diagram, graph, etc.; a variety of visual representations; elaborate a geometrical pattern figure; draw from a geometrical figure. Indicator mathematical expression or equations is find the relation in a mathematical pattern; formulate an equation from a different representation; formulate a conjecture from the pattern found; solve a problem with a mathematical equation. Indicator descriptions or statements is state everyday problems into words or written text; write the solution to a problem through written sentences.

2.2.2. Procedure. The main procedure of this research was to administer a problem solving ability test to fifth-grade students at the Integrated Islamic Elementary School, East Jakarta in order to measure their problem solving abilities. The steps undergone in this research were as follow: (1) Provided a problem solving ability representation test that was adapted from Polya's approach; (2) Analyzed and converted the result of problem solving ability tests into three categories which are 27% for the high-ability group, 27% for the low-ability group, and the other percentage goes to the middle-ability group; (3) Determined the subject for the interview by grouping the students into groups with high, middle, and low- problem solving abilities; (4) Conducted thorough interviews to generalize the problem solving abilities of the students. The interview involved 8 male and 8 female students that represented 3 students from the high-ability group, 3 from the middle-ability group, and 2 from the low-ability group. The subjects were interviewed regarding the results of their work; (5) Analyzed students' answers and interview results to draw a conclusion.

3. Results and Discussion

The result of individual analysis of the mathematical representation indicator on problem solving based on gender is shown in Table 1 below.

Table 1. Percentage of the Respondents' Correct Answer on Mathematical Representation

Representation	Percentage	Males	Females
Visuals in the form of - Figures - Tables	62.3%	69.17%	55.35%
Mathematical expressions or equations	22.3%	22.57%	22.08%
Descriptions or statements	8.6%	10.56%	6.67%

Table 1 above shows that the highest representation aspect is visual in the form of figures and tables (62.3%). The lowest representation aspect is descriptions or statements (8.6%). Regarding representations that are constructed by the students, the result of the analysis of students' answers shows that the ability to visually represent figures and tables is higher than other representations. This is in line with the research by Van Ganderen [14] that visual representations are more often used by students for solving problems.

Table 2. Percentage of Respondents' Correct Answer on Mathematical Representation Towards Solving Mathematical Problems

IOP Conf. Series: Journal of Physics: Conf. Series **948** (2018) 012016 doi:10.1088/1742-6596/948/1/012016

Mathematical Representation	**Problem Solving Aspect**	**Males**			**Females**		
		High	**Middle**	**Low**	**High**	**Middle**	**Low**
Visuals in the form of figures and tables	Understanding problems	84%	70%	58%	78%	72%	55%
	Problem solving planning	96%	78%	61%	78%	60%	46%
	Solving the problem	86%	67%	57%	76%	50%	40%
	Checking answers	74%	46%	33%	64%	28%	13%
Mathematical exploration/ mathematical equation	Understanding problems	56%	24%	5%	30%	30%	21%
	Problem solving planning	66%	15%	3%	38%	25%	10%
	Solving the problem	56%	13%	1%	35%	23%	10%
	Checking answers	44%	7%	0%	24%	10%	0%
Description or Statements	Understanding problems	28%	7%	0%	0%	11%	22%
	Problem solving planning	42%	7%	0%	0%	10%	9%
	Solving the problem	24%	3%	0%	0%	9%	9%
	Checking answers	28%	4%	0%	0%	0%	0%

From the table above we can observe that visual representations in the form of figures and tables are higher among males than females. The highest mathematical representation on problem solving for males is the aspect of problem solving planning, which is 68% in the high-ability group, 33% in the middle-ability group, and 21.3% in the low-ability group. Meanwhile, the highest mathematical representation on problem solving for females is the aspect of understanding problems, which is 36% in the high-ability group, 37.7% in the middle-ability group, and 32,6% in the low ability group. The findings of this study confirm the study by Fuad (2016) that males are able to do mathematical representations well and are able to correctly answer the questions. The visual representation aspect in the form of figures and tables for the indicator of representing data or information into a table, diagram, or graphical form for male and female students at the high-ability groups are shown in Figure 1 and Figure 2 below.

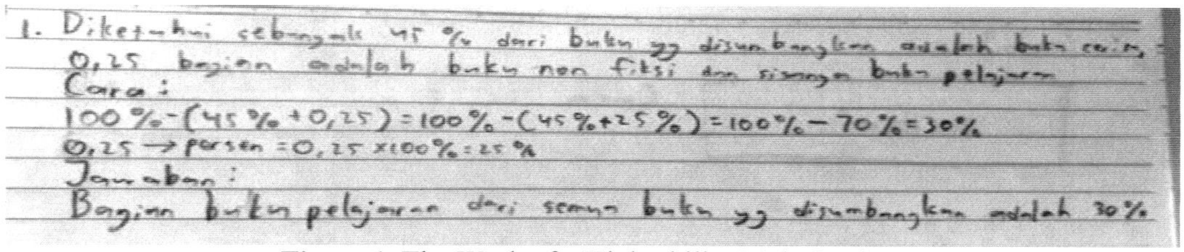

Figure 1. The Work of a High-ability Male Student

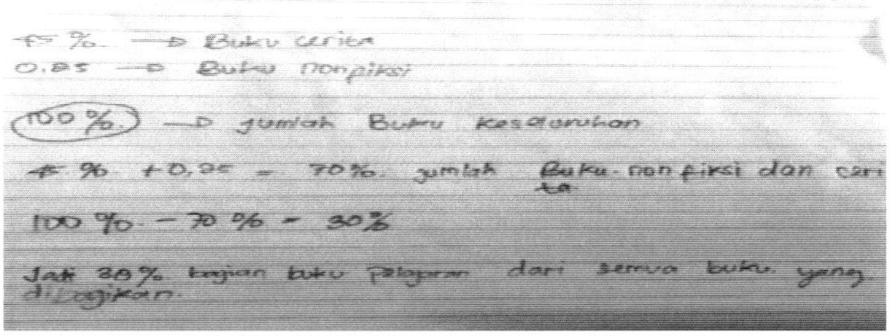

Figure 2. The Work of a High-ability Female Student

During the process of understanding the question, the male student of the high-ability group wrote down what was known, which was that 45% of books donated were story books, 0.25 of that were non-fiction books, and the rest were school books. Meanwhile, the female student also wrote down what was known, but did not write down what was being asked the question. During the interview, the students expressed what was known and were being asked the question. The representation formulated by both male and female students were in the form of words, or written text.

While planning to solve the problem, the male student changed the decimal form into a percentage form, in this case, the non-fiction books, which was written as 0.25, was changed to 25%. Other than that, during the interview, the male student stated that the total number of books donated was 100%, while the female student wrote down that the total number of books donated was 100%. This meant that the male student used a verbal representation, while the female student used a written representation.

When solving the problem, the male student directly counted that the number of school books was 100% subtracted by 45% and 0.25. The male student interpreted the result as "the number of school books out of the total number of books donated is 30%". The female student counted that the number of non-fiction and school books was 70% and used subtraction to count the school book as 100% - 70% equalling 30%. The representation formulated by both the male and female student was by using written words in the form of symbol manipulation. This is in line with the research by Fuad (2016) where during the problem solving stage, male students used verbal words, while female students used written words as representations.

For the checking the answers stage, it was observed that the male and female student did not re-check their answers and therefore no representation was used. This is different from the research by Fuad (2016). In Fuad's (2016) research, it was found that while the male students did not check their answers, the female students checked their answers through verbal words.

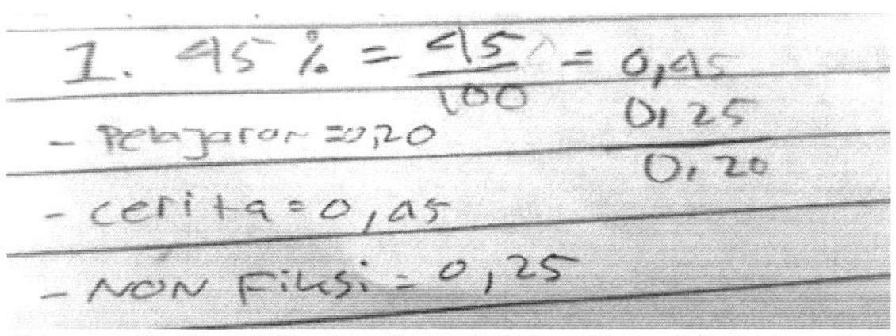

Figure 3. The Work of a Middle Class Male Student

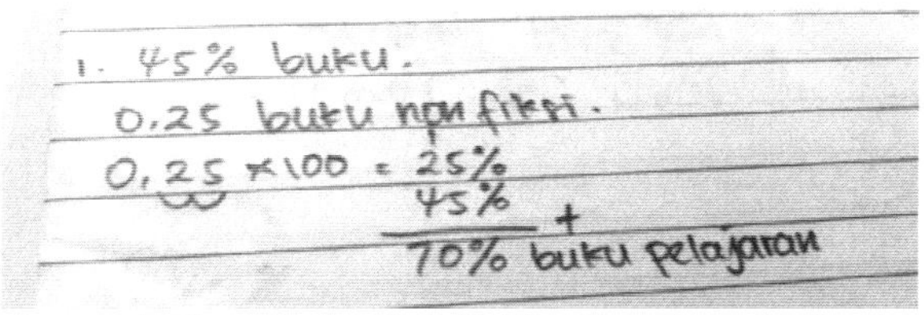

Figure 4. The Work of a Middle Class Female Student

While understanding the problem, both the male and female student did not completely write down what was known and what was asked the question. During the interview process, the students described what they know from the question. The male student wrote down 0.45 story books and 0.25 non-fiction books without writing down what was being asked the question. Meanwhile, the female student wrote down 45% books and 0.25 non-fiction books without writing down what was being asked the question. The representation used by both the male and female student was in the form of words and the representation was not perfectly written down.

In the problem solving planning stage, the male student changed the percentage form into a decimal form, or from 45% books to 0.25. Other than that, during the interview, the student stated that the total number of school books is 0.45 subtracted by 0.25. At the same time, the female student changed the decimal form into a percentage, or from 0.25 to 25%. During the interview, the students stated that to count the number of school books, they need to add up the number of story books and non-fiction books. This mistake in planning will have an influence on the outcome of the problem. This mistake on the hand of the students was caused by the student not knowing that the total number of books donated was 100%. This showed that both the male and female student used a verbal representation.

During the problem solving stage, the male student right away counted the number of leftover school books as 0.45 subtracted by 0.25. The male student did not interpret the result of the problem by writing down "the acquired result". The female student calculated that the number of school books was 70%, acquired from adding the number of non-fiction books (25%) and story books (45%). This mistake in answering was caused by the students not understanding the problem well. The representation formulated by the male and female students were in the form of text without writing down the acquired result so that the representation is not complete.

During the checking the answer stage, both male and female students did not check their answers so that no representation was done.

Figure 5. The Work of a Low Class Male Student

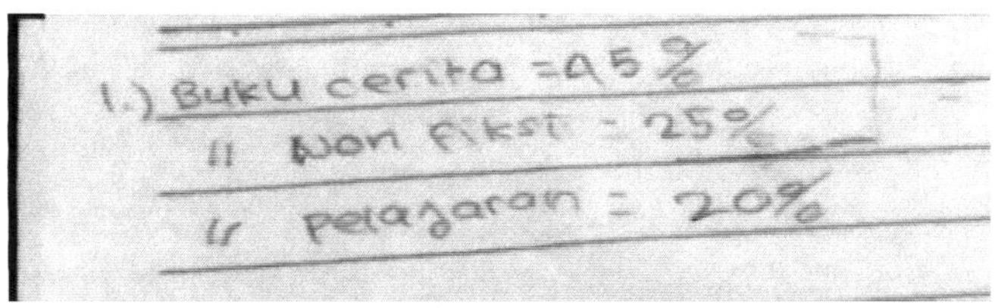

Fligure 6. The Work of a Low Class female Student

During the process of understanding the problem, the male and female student of the low-ability group only wrote down what was known from the problem without stating what was asked. The representation that was formulated by both the male and female student was in the form of incomplete writing.

While planning to solve the problem, both the male and female student did not formulate a strategy to complete the problem. During this stage, the students tend to directly attempt to solve the problem. Neither the male or female student did a representation.

During the problem solving stage, neither the male or female student was able to finish the problem as there were some students that did not realize that the total number of books donated by the donator was 100%. As a result, there were still students who counted wrongly by counting the number of school books as the number of story books subtracted by the number of non-fiction books. Other mistakes were when the students counted the number of school books by adding the total of story books and non-fiction books. They did not know that the correct way of calculating the number of school books is by calculating 100% - (the number of story books added by the number of non-fiction books). The main cause of all this was that the students were not able to understand the problem well.

During the stage of checking the answer, neither the male or female student checked their answer. Therefore, there was no representation done during this stage.

4. Conclusion

The highest representation aspect was done visually in the form of figures and tables as much as 62.3%. The lowest representation aspect was through descriptions or statements as little as 8.6%. The mathematical representation of problem solving towards male students is the highest during problem solving planning, which is, consecutively, 68%, 33%, and 21.3% at the high-, middle-, and low-ability group. Meanwhile, mathematical representation on problem solving towards female students is the highest at the understanding the problem stage, which is, consecutively, 36%, 37.7%, and 32.6% at the high-, middle-, and low-ability group.

Representations of problem solving during the (1) understanding the problem solving stage show that both male and female students in the high-ability group described what they knew and what was being asked by the question, and the representation formulated was in the form of words, or written text. Meanwhile, male and female students in the middle-ability group did not fully write what was known and what was being asked from the question and formulated their representation in the form of words. On the other hand, male and female students in the low-ability group only wrote down what was known from the problem, but not what was being asked. The representation formulated by the male and female students was in the form of incomplete written texts; (2) during the problem solving planning stage, the representation formulated by high-ability male students was in the form of verbal words, while the representation of female students was in the form of written words. In the middle-ability group, the representation of both male and female students was in the form of verbal words. In the low-ability group, both male and female students did not formulate a representation; (3) During the problem solving stage, the representation made by male and female students of the high-ability group was by words or written text in the form of manipulated symbols. In the middle-ability group, the representation formulated by the male and female students were written texts that did not include the final result and therefore the representation process was incomplete. In the low-ability group, many errors occurred due to the lack of understanding of the question; (4) for the stage of checking the answer, none of the high, middle, or low-ability students checked their final answer.

References

[1] Kartini 2009 The Role of Representation in Mathematics Learning. Proceedings of the National Seminar on Mathematics and Mathematics Education UNY on December 5, 2009
[2] Fuad M N 2016 *Kreano Jurnal Matematika Kreatif-Inovatif* **7** 2 pp 145 – 152
[3] Hwang, et al 2007 *Educational Technology & Society* **10** 2 pp 191-212
[4] Mustangin 2015 *Jurnal Pendidikan Matematika* **1** 1 pp 15 – 21
[5] Dahlan J W and Junandi D 2011 *Jurnal Pengajaran MIPA* **16** 1 pp 128 – 138
[6] NCTM 2000 Principle and Standards for School Mathematics Reston V A: The National Councils of Teachers of Mathematics, Inc.
[7] Fadillah S 2009 *Mathematical Problem Solving Ability in Mathematics Class* Proc. Seminar Nasional Penelitian, Pendidikan dan Penerapan MIPA Fakultas MIPA, Universitas Negeri Yogyakarta
[8] Purnomo Y W, Kowiyah, Aliyani F and Assiti S S 2014 *International Education Studies* **7** 8 pp 74 – 84
[9] Asis M, Arsyad N and Alimuddin 2015 *Jurnal Daya Matematis* **3** 1 pp 78 – 87
[10] Mairing J P, Budayasa K and Juniati D 2012 *Jurnal Ilmu Pendidikan* **18** 2 pp 125-134

[11] Fryer R G and Levitt S D 2009 *An Empirical Analysis of the Gender Gap in Mathematics.* http://www.nber.org/papers/w15430.
[12] Herman 2001 *Mental Strategy Used by Primary school Students for Counting* Paper Presented in National Seminar on Mathematics Education at Yogyakarta State University dated 21 April 2001
[13] Japa I G N 2008 *Jurnal Penelitian dan Pengembangan Pendidikan* **2** 1 pp 60 – 73
[14] Van Gardene D 2006 *Journal of Learning Disabilities* **39** 6 pp 496 – 506

Analysis of genetic diversity and genome relationships of four eggplant species (*Solanum melongena* L) using RAPD markers

Susilo[1] and M Setyaningsih[1]
[1]Departement of Biology Education, Universitas Muhammadiyah Prof. DR. HAMKA. Jalan Tanah Merdeka, Kampung Rambutan, Ciracas, East Jakarta, Indonesia, 13830

E-mail: susilo@uhamka.ac.id

Abstract. *Solanum melongena* (eggplant) is one of the diversity of the Solanum family which is grown and widely spread in Indonesia and widely used by the community. This research explored the genetic diversity of four local Indonesian eggplant species namely leuca, tekokak, gelatik and kopek by using RAPD (*Random Amplified Polymorphic DNA*). The samples were obtained from Agricultural Technology Assessment Institute (BPTP) Bogor, Indonesia. The result of data observation was in the form of *Solanum melongena* plant's DNA profile analyzed descriptively and quantitatively. 30 DNA bands (28 polymorphic and 2 monomorphic) were successfully scored by using four primers (OPF-01, OPF-02, OPF-03, and OPF-04). The Primers were used able to amplify all of the four eggplant samples. The result of PCR-RAPD visualization produces bands of 300-1500 bp. The result of cluster analysis showed the existence of three clusters (A, B, and C). Cluster A (coefficient of equal to 49%) consisted of a gelatik, cluster B (coefficient of 65% equilibrium) consisted of TPU (Kopek) and TK (Tekokak), and cluster C (55% equilibrium coefficient) consisted of LC (Leunca). These results indicated that the closest proximity is found in samples of TK (Tekokak) and TPU (Kopek).

1. Introduction

One of the national problems faced and must be solved the way out was energy and food tenacity. Therefore, global and fundamental thinking were needed in improving agricultural productivity in order to boost food tenacity in Indonesia. One such effort was to increase the production of vegetable or olericulture crops. Eggplant (*Solanum melongena*) is one type of vegetable that needs to be improved because it has many nutrients including protein, carbohydrate, fiber, iron, vitamins and calcium [22, 12]. These plant also contains phytonutrient anthocyanin or nasunin on purple eggplant skin that highly efficacious for nutrition one of the most important organs of our body like the brain [1, 24]. In addition to nourishing the brain, nasunin is also useful for protecting brain cell membranes from fatal harmful [2, 14, 3].

In Indonesia, many eggplant plants are used by the community as vegetable crops such as kopek, gelatik, tekokak, and leunca. These plants do have morphological character differences, but for molecular data between cultivars have not been studied. The data is very important because it is one of the genetic resources of Indonesia which can be used to improve the germplasm of the nation [19, 23, 9]

Given the countless benefits and the abundant number of eggplant cultivars, it appears to be a must to review and catalog the plant's genetic resources [8]. Molecular data collection of Solanum crops is of paramount importance to support further research [16]. It can also be used as a nation-wide database of germplasm collections in Indonesia as well as further studying the benefits of each plant [2]. The large diversity of Solanum crops in Indonesia requires further research to examine its benefits. Based on the above description, the researcher is interested to examine the DNA character of some *Solanum melongena* or eggplants by using RAPD markers

Sequencing-based on random amplified polymorphic DNA (RAPD) provides better resolution at intra-genus and levels above. RAPD can provide a tool for classifying individuals into the nominal genotypic category and are largely suitable for intra-species genotypic variation study [7, 5]. RAPD is based on the amplification of genomic DNA with single primers of arbitrary nucleotide sequence [10, 13]. These primers detect polymorphisms in the absence of specific nucleotide sequence information and the polymorphisms function as genetic markers and can be used to construct genetic maps [14]. The objective of this paper was to use clearly amplified RAPD fragments to examine the genetic variation within and between populations of four local *Solanum melongena* in Indonesia.

2. Methods

2.1. Plant Collection

A total of fours *Solanum melongena* namely leuca (LC), tekokak (TK), gelatik (TBU) and kopek (TPU) were included in the present investigation. Young leaf tissue was taken on each species and directly inserted into the plastic and stored into the coolbox for further DNA testing in the laboratory [6]. DNA test by RAPD technique was done at Balai Besar and Development of Biotechnology and Genetic Resources of Bogor Agriculture, Indonesia.

2.2. DNA Extraction

The eggplant leaf tissue was grinded to powder in liquid nitrogen using a mortar and pestle and then inserted into 2.0 ml microtube. 200 µl CTAB buffer was added that has been heated at 650C for 5 minutes. The sample was incubated in a water bath with a temperature 650C for 1 hour, flipped back and forth every 10 minutes after cooling at normal room's temperature for approximately 10 minutes. The next step was added 60 µl of NaOac, 700 µl Chisam (mixed chloroform: isoamic alcohol with ratio 24:1). Then centrifuged for 10 minutes with a speed of 12,000 rpm at 200C. The supernate obtained was taken 400 µl and transferred to 1.5 ml microtube. 900 µl of Chisam was added and centrifuged for 10 minutes with a speed of 12,000 rpm at 200C. Then Separated 400 µl supernatant and added 1000 µl of absolute ethanol and then centrifuged again for 5 min with speed 12.000 rpm at 200C. The pellet was washed with 70% ethanol to get rid of salts, air dried and dissolved in sterile water and then settled for over a night [20].

2.3. The Quality and Quantity of DNA

Quality testing and concentration on DNA isolation are important for the manipulation of all substances. The DNA concentration and it is purity can be determined by spectrometers using UV light at wavelengths of 260 and 280 nm. The comparison in the test using UV light with a length of 260 / 280nm measures a proportion of proteins contamination in a DNA, and the use of UV with a length of 260 nm aims to calculate the concentration of a DNA in the sample. However, the UV absorption process cannot separate RNA contamination and contaminants such as proteins, phenols, oligo or polysaccharides contained in DNA.

2.4. RAPD Process

Four arbitrary oligonucleotide primers obtained from Genetic Science Inc. (Singapore). The primers used in this study were OPF-01 (ATGTCACCAC), OPF-02 (TCAGGACTCC), OPF-03 (TCGAGGACCT), OPF-04 (TAGCTTCACG). Pure genomic DNA was amplified using a single primer dekanucleotide. PCR cocktail was used 1 µl DNA template, 2 µl of ultrapure water, 5 µl mix (kappa 2g fast reading mix) and 1 µl primer. PCR was performed under conditions of pre-denaturation at 960C, denaturation at 940C, annealing at 370C, elongation at 720C and termination at 720C. Then the dye loading solution was added to increase the weight of the DNA molecule. The amplification results were visualized using horizontal electrophoresis with 1% agarose gel (w/v) in a 1x TAE buffer. The agarose gel was immersed in EtBr solution so the ribbon pattern could be seen under the ultraviolet Geldoc.

2.5. Data Analysis

The results of DNA bands emerging from the RAPD process showed the molecular weight of DNA. The banding pattern profile from DNA analysis was detected using 254 nm UV rays and sprayed with general spray reagent (Cerium (IV) Sulfate). The parameters observed were the presence or absence of the band on the agarose gel after UV irradiation. The emulated RAPD ribbon is assigned a value of "1" and that does not appear given a value of "0". Only clear, consistent bands and polymorphic bands were used to create binary matrices for statistical analysis. The amplification products obtained for all samples were compared to each other. In order to assess the genetic structure within and among eggplant, all 4 diagnostic RAPD markers were analyzed using NTSYS-pc. Genetic similarities were calculated between populations for producing a UPGMA cluster (Unweighted Pair Group Method Averaging).

3. Results and Discussion

3.1. DNA Profile

DNA quality and quantity test has been done with Spectrophotometry and Agarose Gel Electrophoresis. The amplification results of the total quality and quantity test of eggplant genome with agarose and spectrophotometry resulted in the purity and concentration of DNA seen in Table 1.

Table 1. Total number of purity and genomic DNA concentrations

Sample code	Purity(µg/µl)	Concentrations (nm)
Gelatik (TBU)	1,1272	1780,24
Kopek (TPU)	1,2219	1196,19
Tekokak (TK)	1,2316	1830,97
Leunca (LC)	1,5127	1266,21

Based on Table 1, the result of quantity test using spectrophotometer showed the highest purity result was in LC sample with purity of 1, 5127 M and concentration 1266, 21 nm, while the lowest purity was in TBU sample with purity 1, 1272 M and concentration 1780, 24 nm. Four primers which showed intra-and interspecific banding pattern for these species were used for further analysis of four *Solanum melongena* sample. All samples yield a readable and scored banding pattern, so the results can be used for analysis. All the primers showed interspecific RAPD polymorphism while some primers showed intraspecific variation (Table 2).

Table 2. The number of primers used for DNA amplification of genomic DNA and level of polymorphism

Primers	Sequence	Polymorphic	Monomorphic	Total bands	Polymorphic Percentage
OPF-01	ATGTCACCAC	5	0	5	100
OPF-02	TCAGGACTCC	2	1	3	150
OPF-03	TCGAGGACCT	11	1	12	100
OPF-04	TAGCTTCACG	10	0	10	100
	Total	28	2	30	

3.2. Species Relationships

PCR-RAPD amplification of total genomic DNA from four accessions of *Solanum melongena* by using four RAPD primers (OPF-04, OPF-03, OPF-02, OPF-02, and OPF-01) produced 30 DNA bands that could be scored. The number of bands produced ranges from 3 (OPF-02) to 12 (OPF-03) with the resulting band sizes ranging from 300 bp to 2000 bp.

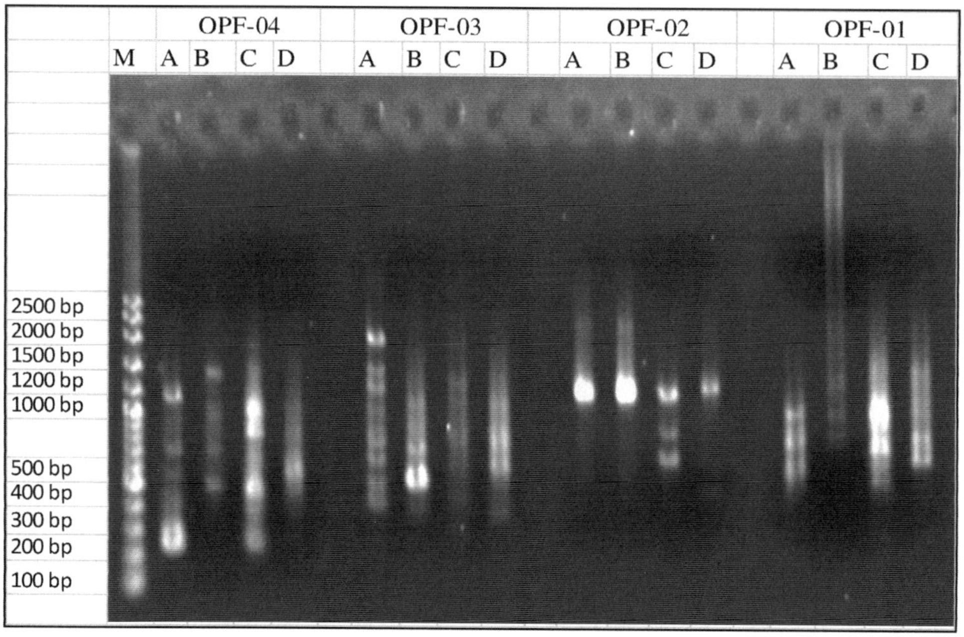

Figure 1. RAPD profiles of total genomic DNA from four accessions of *Solanum melongena* amplified by using random primer OPF-04, OPF-03, OPF-02, OPF-01. (M: Marka /Ledder, A: Gelatik, B: Kopek, C: Tekokak and D: Leunca).

Cluster analysis showed the separation of eggplant into clusters clustered by type Figure 3. The coefficient of the genetic similarity of the four accessions of eggplant ranged from 25% to 100%. The result of cluster analysis showed the existence of three clusters (A, B, and C). Cluster A (coefficient of equality of 49%) consisted of a gelatik. Cluster B (coefficient of 65% equilibrium) was consisting of kopek and tekokak. Cluster C (55% equation coefficient) consisted of Leunca (Figure 2). The dendrogram was constructed based on the percentage similarity calculated after pooling the data from 30 loci RAPD of four accessions of *Solanum melongena.*

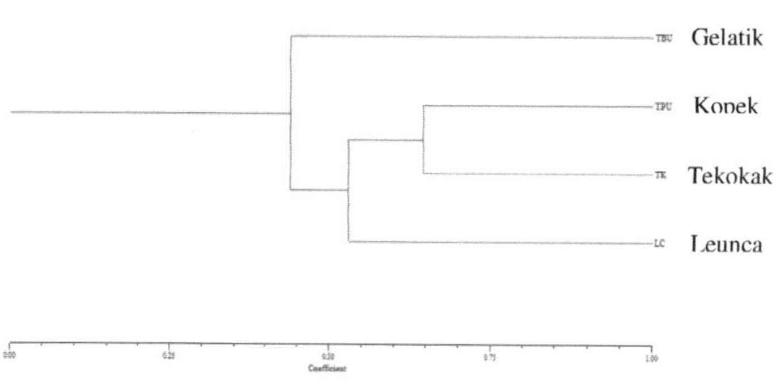

Figure 2. Dendrogram depicting the genomic relationship among four accessions of *Solanum melongena*.

Four *Solanum melongena* cultivars local from Indonesia namely leuca (LC), tekokak (TK), gelatik (TBU) and kopek (TPU) has been successfully justified with four primers. In early testing, Cetyl Trimethyl Ammonium Bromide (CTAB) was used to extract genomic DNA. CTAB is a common method of extraction of plant genomic DNA containing polysaccharides and polyphenolic compounds [18]. Basically, genomic DNA isolation consists of three major steps, namely the destruction of cell walls (lysis), separation of DNA from solid materials such as cellulose and proteins, and DNA purification [4]. In this study, our early stages we tested the quality and concentration of DNA. Quality testing and concentration on DNA isolation are important for the manipulation of all substances. DNA concentration and its purity can be determined by spectrophotometer using UV light with 260 and 280 nm [224]. The comparison in the test using UV light with a length of 260/280 nm gives a measure of the contamination of proteins in a DNA, while the use of UV light with a length of 260 nm aims to calculate the concentration of a sample DNA [17, 3]. UV absorption process could not be separate RNA contaminants, and contaminants such as proteins, phenols, oligo or polysaccharides contained in DNA are difficult to estimate [16, 4].

In general, the total amplification of the DNA genomes of the Solanaceae sample using selected primers yields a series of bands, among which there are bands commonly encountered throughout the sample and those of specific bands found only by specific species/cultivars. The overall band profile resulting from this RAPD primer amplification is the DNA of each type/cultivar. RAPD techniques are evaluated to distinguish individuals with close family ties [16, 13]. Based on phylogenetic tree Figure 2 showed that kopek (TPU) was closer to kinship with tekokak (TK).

The result of quantity test with spectrophotometer showed good result. Four accessions obtained DNA genomes with high purity and concentration. Primers used in this study were primers OPF-04, OPF-03, OPF-02, and OPF-01. Of the four primers used, all successfully amplify genomic DNA. At the time of testing the quality and quantity of DNA kopek (TPU) to get results that were not significant, electrophoresis results showed not too clear the DNA band cultivars. Measurements by using the speculator produce a pretty good number, there was DNA but the amount obtained was less. Inadequate quality and quantity test results do not guarantee that genomic isolation actually fails [9]. However, if there is too little DNA, then the test results may show negative results due to the minimum limitations of the tool [5]. Very few genomic DNAs so difficult to detect by means of equipment can still be used as templates in the PCR process [11]. PCR has the ability to multiply very powerful DNA fragments. PCR requires only a few templates to be able to reproduce certain DNA fragments [14, 24].

4. Conclusion

The dendrogram results showed that TPU (kopek eggplant) was closer to its hazard with TK (tekokak) with a similarity of 65%, whereas in LC (lenca) had 55% similarity, and in TBU (eggplant) had the furthest kinship with 49% equality.

References

[1] Alarcón-Flores, M. I., Romero-González, R., Vidal, J. L. M. dan Frenich, A.G 2015 *Antioxidants* **4** p 345-358

[2] Arivalagan, M., Gangopadhyay, K. K., Kumar, G., Bhardwaj, R., Prasad, T.V., Sarkar, S. K. and Roy, A 2011 *Journal of Food Composition and Analysis* **26** p 173–176

[3] Braga, C. P., Scalzo, R. L., Sasso, M. D., Lattuada, N., Greco, V. and Fibiani, M 2016 *Journal of Functional Foods* **20** p 411-421

[4] Caguiat, X. G. I. dan Hautea, D. M. 2014 *SABRAO Journal of Breeding and Genetics* **46** (2) p 183-201

[5] Cericola, F., Portis, E., Toppino, L., Barchi, L., Acciarri, N., Ciriaci, T., Sala, T., Rotino, G.L., dan Lanteri, S 2013 *PLoS ONE* **8** (9)

[6] Dharmayanti, Indi 2011 *Wartazoa* **21** (1) p 1-10

[7] Demir, K., Bakır, M., Sarıkamis, G., dan Acunalp, S 2010 *Genet. Mol. Res* **9** (3) 1568-1576

[8] Fatchiyah, Estri Laras, Sri Widyarti, Sri Rahayu 2011 *Biologi Molekuler Prinsip Dasar Analisis*. (Jakarta: Erlangga)

[9] Ge, H., Liua, Y., Jianga, M., Zhangb, J., Hana, H. dan Chen, H 2013 *Scientia Horticulturae* **162** 71–75
[10] Ibrahim, A. A., Mohammad, A. B., Haseeb, A. K., Ahmad, H. A., Ali, A. A., Ali, H. B., Mohammad, A. S. and Mohammad S 2010 *Int. J. Mol. Sci.* **11** 2079-2096
[11] Jose´, R.S., Plazas, M., Sa´nchez-Mata, M.C., Camara, M., and Prohens, J 2016 *Journal of Food Composition and Analysis*. **45** 130–140
[12] Kaur, C., Nagal, S., Nishada, J., Kumar, R., and Sarika 2014 *Food Research International* **60** 205–211
[13] Kelly, Y. C. et al 2015 *Molecules* **20** 22454–22462
[14] Kumchai, J., Wei, Y.-C. Lee, C.-Y., Chen, F.-C. and Chin, S.-W 2013 *Genet. Mol. Res* **12**(1) 755-764
[15] Mondini, L., Noorani A. and Mario A. P 2009 *Diversity* **1** 19-35
[16] Nevena, N., Ksenija, T., Goran, B., Aleksandar, B., Ivana S., Dragan M. and Slobodan K 2011 *Int. J. Mol. Sci* **12** 5449-5460
[17] Néjib, H., Messaoud, M., Jemni, C and Mokhtar, T 2010 *Diversity* **2** 107-114
[18] Noroozi, M., Hishamuddin, O., Soon, G. T. and Suhaimi, N 2011 *Molecules* **16** 2598-2608
[19] Nunome, T., Negoro, S., Kono I., et al 2009 *Theor. Appl. Genet* **119** 1143-1153
[20] Olga, S., José A. P. and Paula, B 2011 *Int. J. Mol. Sci* **12** 4156-4164
[21] Pamela, E. A., Michael, A. G., Uterdzua, O., Dorcas, O. I. and Oyeronke, A. O 2016 *Resources* **5**(6) 1-15
[22] Ravella, R., Reddy, M., Taylor, K. and Elobeid, A 2013 *Agronomy* **3** 248-255
[23] Suparman 2012 *J. Bioedukasi* **1** 59-68

Acknowledgments
This research was financially supported by the Research and Development Institutions (LEMLITBANG) Universitas Muhammadiyah Prof. Dr. HAMKA, Jakarta.

IOP Conf. Series: Journal of Physics: Conf. Series **948** (2018) 012018 doi:10.1088/1742-6596/948/1/012018

Development of syntax of intuition-based learning model in solving mathematics problems

Nok Yeni Heryaningsih[1] and Hikmatul Khusna[1]
[1]Mathematics Education - Universitas Muhammadiyah Prof. Dr. Hamka, Jakarta

Email: yeni.herison@yahoo.com

Abstract. The aim of the research was to produce syntax of Intuition Based Learning (IBL) model in solving mathematics problem for improving mathematics students' achievement that valid, practical and effective. The subject of the research were 2 classes in grade XI students of SMAN 2 Sragen, Central Java. The type of the research was a Research and Development (R&D). Development process adopted Plomp and Borg & Gall development model, they were preliminary investigation step, design step, realization step, evaluation and revision step. Development steps were as follow: (1) Collected the information and studied of theories in Preliminary Investigation step, studied about intuition, learning model development, students condition, and topic analysis, (2) Designed syntax that could bring up intuition in solving mathematics problem and then designed research instruments. They were several phases that could bring up intuition, Preparation phase, Incubation phase, Illumination phase and Verification phase, (3) Realized syntax of Intuition Based Learning model that has been designed to be the first draft, (4) Did validation of the first draft to the validator, (5) Tested the syntax of Intuition Based Learning model in the classrooms to know the effectiveness of the syntax, (6) Conducted Focus Group Discussion (FGD) to evaluate the result of syntax model testing in the classrooms, and then did the revision on syntax IBL model. The results of the research were produced syntax of IBL model in solving mathematics problems that valid, practical and effective. The syntax of IBL model in the classroom were, (1) Opening with apperception, motivations and build students' positive perceptions, (2) Teacher explains the material generally, (3) Group discussion about the material, (4) Teacher gives students mathematics problems, (5) Doing exercises individually to solve mathematics problems with steps that could bring up students' intuition: Preparations, Incubation, Illumination, and Verification, (6) Closure with the review of students have learned or giving homework.

1. Introduction

Mathematics learning in Indonesia usually still used conventional method, where a teacher more dominant learning in the classroom. In the other word, teachers who more active in solving mathematics problems. In this case, teachers often did not allow their students to solve mathematics problems individually by trying to bring up the ideas that students had. These ideas were usually used by students to solve mathematics problems. To bring up these ideas need to develop an "intuition" from within students.

According to Fischbein [5] an intuition is a cognition that appears subjectively as self-evident, directly acceptable, holistic, coercive and extrapolative. When students have to face a notion of advanced mathematics that turns intuitively unacceptable for them, they produce, deliberately or unconsciously, more acceptable intuitive substitutes: the intuitive models, that may be understood, represented and manipulated like other concrete realities [5]. Intuition can be used as a "Bridge of cognition" that connected students understanding to help them to relate mathematics problems with the several alternative solutions. Thus, intuition from within students was very useful to determine what steps should be done to solve the problems. This is linear with the result of Fichbein and Grossman research that Intuitions may be thought of as being based on some structuring rules expressing both levels of maturation and levels of training. If the objective is to influence intuition favourably – and this is an important task of the learning process – one has to identify the structuring schemata and the specific ones which underlie the respective intuitions. Intuition can be an idea based on some structuring rules that expressing levels of maturation and levels of beginning [4]. Therefore, it is necessary to develop a students' intuition in solving mathematics problems.

Wallis said that intuition can be developed by developing work process of creative thinking that consisted of four phases. They were preparation phase, incubation phase, illumination phase and verification phase. Furthermore, the phases can be explained as follows: (1) in the preparation phase, students define the problem, need, or desire and gather any related information of the solution or response need to account for, and set up criteria for verifying the solution's acceptability. In this case, students prepare to solve the problem by thinking, seeking answers, and being able to ask others. (2) in the incubation phase, students step back from the problems to break away for a while and let our mind to work behind the scene. Like preparation, incubation can last minutes, weeks, even years. (3) in the Illumination phase, ideas arise from our mind to provide the basic of a creative response [15]. These ideas can be pieces of the whole or the whole itself, i.e. seeing the entire concept or entity all at once. Different from another phase, illumination is often very brief, involving a tremendous rush of insight within a few minutes or hours. This stage is commonly called as "Aha! Moment", where at the moment is the ideas appear. Then that ideas we called intuition. (4) in verification, the final phase, students carries our activities to demonstrate whether or not what emerged in illumination satisfies the need and the criteria defined in the preparation phase. Through these phases, then the intuition that arises from within student can be raised. Therefore, learning model with developing creative thinking process with four phases in the classroom is needed to bring up students' intuition.

According to Alison "The model is structured of teaching problems, which can illuminate processes teachers engage in during their daily planning, thus providing a useful lens to understand the nature of teachers' planning routines and reasons underlying their decisions during this phase of teaching" [1]. In the other hand, Joyce and Weil said that there are five important elements which describe a learning model, including syntax, social systems, reaction principles, supporting systems, and the impact of instructional and companion impact [6]. The term 'syntax' according to Matthew is quoted by Van Vallin and LaPolla is from the Ancient Greek syntaxes, a verbal noun which literally means "arrangement" or "setting out together". Traditionally, it refers to the branch of grammar dealing with the ways in which words, with or without appropriate inflections, are arranged to show connections of meaning within the sentence [14]. Furthermore, the syntax is the study of the principles and processes by which sentences are constructed in particular languages. Syntactic investigation of a given language has as its goal the construction of a grammar that can be viewed as a device of some sort for producing the sentences of the language under analysis [2]. Therefore, the syntax of learning model in this research was defined as a framework or planning that describes an arrangement and processes the learning procedure systematically to achieve certain learning objectives which contain objectives, syntax, learning environment, and management system.

Implementation of appropriate learning model in the classroom will make students learn effectively. Students are able to solve mathematics problems and make students as good problem solver. One criterion of the problem is called a mathematical problem if students have never solved that problem and it cannot be solved by the routine solution. According to Polya, problem-solving is defined as an attempt to find a solution of a goal to achieve an objective [10]. The process of the solving problem requires a person to coordinate his experience, knowledge, understanding, and intuition to solve the problem. Therefore, intuition becomes an important role in solving mathematics problems.

Some relevant research results that used intuition-based learning models such as the results of the research were conducted by Usodo [13], the results of the study indicate that learning by using stages that can bring up the intuition who initiated by Wallis was able to improve students' ability in solving mathematics problems. In addition, other research conducted by Syahputra et al. in SMP 6 Medan mentioned that the relationship between students' intuitive thinking with students' mathematics achievement in SMP 6 Medan was positive and real [12]. This means that students with intuitive thinking can improve students learning achievement of mathematics in the classroom. Therefore, it was developed a syntax of Intuition Based Learning (IBL) model in solving mathematics problems to improve students achievement.

The syntax of Intuition Based Learning (IBL) model is learning steps in the classroom which contains the phases of Wallis model that can exercise students to bring up students' intuition in solving mathematical problems. The preparation phase, incubation phase, illumination phase and verification phase.

Based on the descriptions above, then it was developed a syntax of Intuition Based Learning (IBL) model in solving mathematics problems. The purpose of the research was to produce syntax of Intuition Based Learning (IBL) model in solving mathematics problem that valid, practical and effective.

2. Methods

The type of this research was a Research and Development (R&D). Research and Development (R & D) is a research method that uses to produce a certain product, and then do test the effectiveness of the product [11]. The development process was adopted by Plomp and Borg & Gall (in Khabibah) development model, i.e, Preliminary investigation step, Design step, Realization step and Test, Evaluation and Revision step [7,9]. The process of development can be explained as follows, (1) Preliminary investigation step. In this step conducted all related information, context analysis, and conducted several related studies, (2) The design step. By examining the results of the preliminary investigation, then designed the draft of syntax Intuition Based Learning (IBL) model. The syntax of learning models was designed will include steps that exercising students to bring up students' intuition in solving mathematics problems. In this case were used open-ended questions, including the preparation phase, incubation phase, illumination phase and verification phase. (3) Realization step. In this step, all the designs that have been designed at the previous step will be realized more clarity and detail. The results of this step were named as the first draft. (4) The testing, evaluation, and revision step. In the test validation, was did validation of the first draft to the expert validator, and then after being declared that first draft was valid, it was evaluated by tested the syntax of IBL model. The testing was conducted to determine the practicality and effectiveness of the model syntax. In this test was observed by two observers. After testing, it will be analyzed whether the syntax of IBL model had practical and effective criteria or not. If not yet, then the syntax of IBL model will be revised and re-tested. At this step, activities continued until the syntax of IBL model was valid, practical and effective, (5) Focus Group Discussions (FGD) were conducted to evaluate the result of syntax model testing in the classroom that has been conducted, and then to get some suggestions for the next testing. This FGD involved some practitioners of mathematics education and involved an expert who understood about the learning model.

In the validation and the testing of the syntax of IBL model, there were several supporting research instruments required. These instruments were also developed and adapted with the implementation of the syntax of IBL model. Some of the instruments were developed in this study included, (1) student response questionnaire, (2) observation sheets of model implementation, and (3) validation sheets.

2.1. Data Collection Methods

Data collection methods that used to collect the data in this study were, (1) check list method, this method was used to ask validator in assessing syntax of IBL model by giving a sign (Ö) on column which has been provided, (2) observation method, this method was used to collect the model implementation data by observing the learning activities in the classroom using the syntax of IBL model by two observers, (3) the questionnaire method, the questionnaire data collection method was used by distributed a response questionnaire to the students whose learning implemented syntax of IBL model, and (4) test method, this test method was used by distributed STS (Student Test Sheet) to be solved by students individually. This test was used to measure the effectiveness of syntax of IBL model syntax.

2.2. *Data Analysis*
The data that had collected then analyzed quantitatively and directed to answer whether the syntax of IBL model had valid, practical and effective or not. Analysis of the validity data by: did recapitulation of all results from the validator, calculated the average value of each validator, calculated the total mean of the three validators, categorized the value of the validity by the predetermined category, and then decided the value of the validity of syntax of IBL model. References that used by researchers to determine the category interval in this study were adopted from Darwis [3]. Practical data analysis was done by recapitulated all result that given by the observer about the implementation of the syntax of IBL model in the classroom, calculated the percentage of the value of each observer, calculated the total average of the two observers, determined the category of model syntax implementation based on predetermined category, and then decided the practicality of syntax of IBL model. Effectiveness data analysis was performed by recapitulated all students' scores that obtained from Student Test Sheet (STS) and student response questionnaires, calculated the mean of all students' scores from each STS, calculated the increasing of the first test result to another test, calculated the percentage of the number of students who giving positive responses to each category, determined the category of effectiveness, and then the last was to decide the value of the effectiveness of syntax of IBL model.

3. Results and Discussion
The development of the syntax of IBL model began with a preliminary investigation step, the researcher did some studying and collected all related information. The results of the preliminary investigation step were: (1) based on pre-survey at SMA 2 Sragen, students were gave less problem solving training, which gave students opportunity to express their ideas, (2) Reference of intuition that used in this study refers to Fischbein, (3) learning theory that supported the syntax of IBL model was social constructivism Vygotsky learning theory, (4) the topic that used for this IBL research tested was the Slice of Two-Circle.

The results of the studying from preliminary investigation step were used as a reference for designing the syntax of IBL model and research instruments. The syntax of IBL model that had designed then realized and had been produced was referred to as the first draft. The first draft of the syntax of IBL learning was as follows: (1) Introduction, (2) Grouping, (3) Group discussions, (4) Students solving mathematics problem with the step that can bring up students intuition, (5) Closing. In addition, in this step also produced the design of research instruments, they were students response questionnaires, model implementation sheet and model validation sheet and instrument validation sheet.

The first draft was validated to 3 validators, they are (1) Prof. Drs. Tri Atmojo Kusmayadi, M.Sc., Ph.D., as a pure mathematician, (2) Drs. Paidi, M.Pd., as an education and model development expert and (3) Partono, M.Pd., a high school mathematics teacher.

The result of validation about the syntax of IBL model from three validators was valid. The average assessment of each validator was 4.00, 4.08 and 3.75. Based on predetermined criteria, the result in intervals $3.5 \leq V < 4.5$ was the valid category. Therefore, the syntax of IBL model was valid because it meets all aspects of validity. The validity of this syntax of IBL model was determined by the assessment of the validator. According to Nieveen [8], models that meet the value of validity are models which based on strong theoretical rationale. The theoretical rationale that used in this study was the theory of Vygotsky's constructive learning. In the syntax of IBL model that developed had relation with constructivism theory. There was a syntax of IBL models which students were directed to build their own knowledge as there was group discussion among students in learning process.

After the syntax of IBL model was valid and usable, then did test 1 to see the practicality and effectiveness of the syntax of IBL model. Test 1 was conducted in class XI IPA 6 SMA N 2 Sragen with 32 students. Test activity 1 was observed by 2 observers to see the implementation of the syntax of IBL model in the classroom and to determine the practicality of the model. According to Nieveen [8], that a model is said to be practical if in reality shows that the model can be applied in the classroom. Therefore, the practicality of the syntax of IBL model in this study was determined by the level of the syntax of IBL model implementation in solving mathematics problem should be high. The result of the percentage of total model implementation on test 1 was 75.92%. Based on predetermined criteria, the syntax of IBL models had been implemented. This means that the syntax of IBL model had met the criteria of practicality.

To measure the effectiveness of the model, it can be observed from the results of student tests on student response questionnaires and STS (Student Test Sheet). According to Nieveen to declare effectiveness of a model that in the reality, the model has to give results in high expectations [8]. Therefore, in this study the effectiveness of the syntax of IBL model was determined by two things: first, by the students' test results on the STS that given at the end of each meeting and the second from the positive responses that given by the students to the syntax of IBL model.

The result of average STS 1, STS 2 and STS 3 on test 1 were 34.32, 65.94 and 66.09 respectively. While for the average value of STS 1, 2 and 3 is 55.67. Although there was an increase in student test results, based on predetermined criteria, the syntax of IBL model was not effective yet. Furthermore, the effective value when seen from the results of students responses questionnaire. The average percentage of students who responded positively to positive questions was 81.8%. While the average percentage of students who gave positive responses to negative questions was 52.08%. However, the syntax of IBL model is not effective yet based on predetermined criteria. From the result of student test and the result of students responses questionnaire on test 1, can be stated that model not effective yet.

The analysis result of test 1 it was founded that syntax of IBL model not effective yet, hence did some revision on the syntax of IBL model. This revision was based on suggestions from the FGD results that had been held, also by considered some aspects that ware still lacking and then considered to validator suggestions. After the syntax of IBL model was revised and then redesigned and realized to be the second draft. The second draft was then tested. Test 2 was conducted in class XI IPA 4 SMA N 2 Sragen with 32 students. The second test activity was observed by 2 observers, to see the implementation of the syntax of IBL model in the classroom. The result of the percentage of total model implementation on test 2 is 93.52%. Based on predetermined criteria, most of the syntax of IBL models had been implemented. This means that the syntax of IBL model had met the criteria of practicality.

The result of average STS 1, STS 2 and STS 3 on Test 2 were 76.76, 88.91, and 93.14 respectively. While the average value of STS 1, 2 and 3 on test 2 was 86.27. Furthermore, there is an increase in student test results, from STS 1 to STS 2 was 15.83%, STS 2 to STS 3 was 4.76% and from STS 1 to STS 3 was 21.34%. Based on predetermined criteria, the syntax of IBL model was declared effective when viewed from students test results. Furthermore, the average percentage of students who responded positively to positive questions was 96.00%, while the mean percentage of students who responded positively to negative questions was 75.56%. Because positive responses to positive and negative questions were more than 75%, then the syntax of IBL model had been effective if seen from the results of student responses.

From the results of the test and questionnaire results of student responses above, the syntax of IBL model was said to be effective in solving the mathematics problem. Related to the results of research conducted by Usodo shows that learning by using the step that can bring up the intuition who initiated by Wallis was able to improve students' ability in solving problems [13]. This means that IBL model effective to improve student's mathematics achievements in the classroom.

Based on the description above, the syntax of IBL model had met all aspects of validity, validity, and effectiveness. Therefore, had been obtained the final product of syntax of IBL models that valid, practical and effective.

4. Conclusion

The development process of the syntax of IBL model referred to development model by Plomp and modified by Borg & Gall development model. The process of developing the syntax of IBL model has been done, they were, (1) at the preliminary investigation step, conducted the study of intuition, the study of model development, reviewed the condition of the students, and topic analyzed at SMA 2 Sragen, (2) designed the syntax of the IBL model in which there were steps that can give bring up intuition. In addition, at this step also designed research instruments, (3) realized the syntax of IBL model that has been compiled into the first draft, (4) validated the first draft to the validator. And then, after valid, the first draft was tested to class XI IPA 6 SMA N 2 Sragen. Because test 1 has not been effective yet, then it was revised syntax of IBL model, (5) conducted FGD to evaluated the test, then revised based on the FGD result suggestion for next test, (6) Test 2 was implemented in class XI IPA 4 SMA N 2 Sragen, and obtained the final product of syntax of IBL model.

Model Intuition Based Learning (IBL) was a learning model that had met the valid, practical and effective criteria in solving mathematical problems to increase students' achievement, especially on the topic of Two-Circle Slice. Here, was the final product of syntax of IBL model:
(1) Opening with apperception, motivations and build students' positive perceptions,
(2) Teacher explains the material generally,
(3) Group discussion about the material,
(4) Teacher gives students mathematics problems,
(5) Doing exercises individually to solve mathematics problems with steps that could bring up students' intuition: Preparations, Incubation, Illumination, and Verification,
(6) Closure with the review of students has learned or giving homework.

References

[1] Alison C 2008 *The Mathematics Educator Journal* **18** (2) 11-22
[2] Chomsky Noam 1957 & 2002 *Syntactic Structures* (Berlin: Walter de Gruyter GmbH)
[3] Darwis, M 2007 *Model Pembelajaran Matematika yang Melibatkan Kecerdasan Emosional* (Surabaya: Dissertation Universitas Negeri Surabaya,)
[4] Fischbein E and Grossman 1997 *Educational Studies in Mathematics* **34** p 27–47
[5] Fischebin E 1999 *Educational Studies in Mathematics* **38** 11-50
[6] Joyce B R and Weil, M 2009 *Model of Teaching, 8th Edition* (US: Pearson)
[7] Khabibah S 2006 *Pengembangan Model Pembelajaran Matematika dengan Soal Terbuka untuk Meningkatkan Kreativitas Siswa Sekolah Dasar* (Surabaya: Dissertation Universitas Negeri Surabaya)
[8] Nieveen, N 1999 *Prototyping to Reach Product Quality* In J. van den Akker, R. Branch, K. Gustafson, N. Nieveen, & T. Plomp (Eds.) *Design approaches and tools in education and training* pp.125-136 (Dordrecht: Kluwer Academic Publishers)
[9] Plomp Tj 2013 *Educational Design: Introduction* (Eschende: SLO)
[10] Polya G 1985 *How to Solve It: A new aspect of mathematics method (2 ed)* (Princeton N J: Princetonn University Press)
[11] Sugiyono 2012 *Metodologi Penelitian Administrasi* (Bandung: Alfabeta)
[12] Syahputra H, Mansyur A, Muliyono, Indra Z and Iskandar S 2012 *Proses Berpikir Intuitif Siswa SMA 6 Medan* (Medan: Unimed Press)
[13] Usodo B 2012 *Penerapan Pembelajaran yang Berbasis Pada Pengembangan Intuisi untuk Meningkatkan Kemampuan Pemecahan Masalah Siswa SMAN 1 Sragen (RSBI)* (Surakarta: Research Report Universitas Sebelas Maret)
[14] Van Valin Jr Robert and LaPolla Randy J 1997 *Syntax: Structure, Meaning and Function (*UK: Cambridge University Press)
[15] Wallis G 2006 *Intuition Effect in Creativity* available at http://www.itpin.com/blog/category/mind-thinking/intuition/

Student teachers' mathematical questioning and courage in metaphorical thinking learning

H Hendriana[1], W Hidayat[1] and M G Ristiana[2]
[1]STKIP Siliwangi, Jalan Terusan Jenderal Sudirman, Cimahi, Indonesia
[2]Universitas Pendidikan Indonesia, Jalan Setiabudi No.229, Bandung 40154, Indonesia

E-mail: erishen@stkipsiliwangi.ac.id

Abstract. This study was designed in the form of experiments with control group design and post-test only which aimed to examine the role of metaphorical thinking learning in the mathematical questioning ability of student teachers based on the level of mathematical courage. The population of this study was student teachers of mathematics education study program in West Java Province, while the sample of this study was 152 student teachers which were set purposively and then randomly to be included in the experimental class and control class. Based on the results and discussion, it was concluded that: (a) the mathematical questioning ability of student teachers who received Metaphorical Thinking learning was better than those who received conventional learning seen from mathematical courage level; (b) learning and mathematical courage level factors affected the achievement of student teachers' mathematical questioning ability. In addition, there was no interaction effect between learning and mathematical courage level (high, medium, and low) simultaneously in developing student teachers' mathematical questioning ability; (c) achievement of mastering mathematical questioning ability of student teacher was still not well achieved on indicator of problem posing in the form of non-routine question and open question.

1. Introduction

Students' achievement of learning mathematics can be seen through their way of working out and understanding of students in solving problems given. Measuring the extent to which a person's level of understanding goes can be known through the questions raised in his resolving steps [1,2,3].

In raising the question, one must have courage first. Courage is an ability that must be possessed by someone so that it can control him or her in term of when to ask, when to answer questions, and when to respond to questions or answers given. Courage in mathematics includes one's ability to answer questions, ask questions, and provide ideas in responding answers or questions of a given problem.

But a question can be good or bad. Inadequate questions can make learning less conducive and create confusion. To raise good questions posed is a good question, then, the role of educators should be better in managing the teaching and learning process in class [4].

In order for learners to create an innovative solution that makes them understand the problem and solve it, it is necessary to have meaningful learning. In addition, the need for strategies that must be done by teachers in providing structured learning support so that students can be encouraged to learn independently, at least in addressing the problems faced both in the form of questions and statements [5,6].

Development of questioning ability and courage owned by students is also supported by the factor of questioning ability and courage owned by the teacher of mathematics. In fact, there are still teachers who are not already explored in mathematical questioning ability. Therefore, to motivate math student teachers in developing questioning ability on the subjects of mathematics is to teach them metaphorical thinking. This learning is able to connect the problem to the given question into a referred statement so as to produce accurate information [7,8,9,10].

Based on that, the problems and objectives in this research are to know and examine about: (a) is the mathematical questioning ability of math student teachers who received metaphorical thinking learning is better than those who received conventional learning seen from mathematical courage level. (b) Is there any interaction effect between mathematical courage level and learning in developing the mathematical questioning ability of math student teachers? (c) How is the achievement of the mathematical questioning ability of the math student teachers?

2. Methods

This study is designed in the form of experiments with control group design and posttest only that aims to examine the role of learning metaphorical thinking on the mathematical questioning ability of student teachers seen from mathematical courage level. The Population of this study is student teachers of mathematics education study program in West Java Province, while the sample of this study are 152 student teachers set purposively and then randomly to be included in the experimental class and control class.

Indicator of mathematical questioning ability in this research is the ability of student teachers which include: (a) Posing problems related to the context of the given statement; (b) Posing problems in the form of a non-routine question of the given statement; (c) Posing problems in the form of open questions of given statements.

The degree of mathematical courage is grouped into the three categories presented in Table 1 below:

Table 1. Categorization of mathematical courage level

Aspect of Mathematical Courage	Level
Students can answer questions; ask questions, and give ideas in responding answers or questions of given problems.	High
Students can answer questions and ask questions of given problems.	Moderate
Students can answer questions of given problems	Low

The data of the research are analyzed by using Two-Way ANOVA statistic test to see the difference and interaction effect between the learning of metaphorical thinking and the level of mathematical courage (high, moderate, and low) in producing mathematical questioning ability of student teachers.

3. Results and Discussion

Findings on the mathematical questioning ability of student teachers are presented in Table 2.

Table 2. Description of mathematical questioning ability of students prospective teachers based on metaphorical thinking (MT) learning and mathematical courage level

Mathematical Courage Level	Mathematical Questioning Ability			
	Metaphorical Thinking (n = 76)		Conventional Learning (n = 76)	
	Mean	SD	Mean	SD
High	7,684	1,003	7,227	1,020
Moderate	6,857	1,352	6,414	1,053
Low	5,833	0,811	5,480	0,770
Overall	6,579	1,278	6,342	1,172

Based on the above description, the following interpretations are obtained:

Overall, there are differences in the mathematical questioning ability of student teachers who received Metaphorical Thinking learning and conventional learning. The students' mathematical questioning ability in the classroom with learning Metaphorical Thinking and ordinary learning is high (76.84% > 72.27%). This is caused by existence of association between questioning ability and metaphorical thinking, namely: (i) Students are able to connect problem from statement; (ii) Students are able to find new concepts that they do not have; (iii) students are able to create creative ideas that come from the problem; (iv) the student is able to apply the results of his thinking in the form of question [6].

In addition, if reviewed based on the level of mathematical courage, the ability to ask mathematically of students who received Metaphorical Thinking learning also looks different and shows mathematical questioning ability with Metaphorical Thinking learning is better than those who learn with conventional learning. Mathematical questioning ability at the mathematical courage of high level for both learning is high (76,84% > 72,27%). Meanwhile mathematical ability at mathematical courage of moderate and low level for both learning is moderate (68,57% > 64,14% and 58,33% > 54,80%).

When seen from which factors affecting the development of mathematical questioning ability, it seems that the two factors (learning and mathematical courage level) affect each other on the development of mathematical questioning ability. In addition, there is no interaction effect between learning and mathematical courage level in developing the student teachers mathematical questioning ability.

To support the description of the mathematical questioning ability of the student teachers, then data analysis about the mathematical questioning ability of the teacher candidate students through the statistical test of average difference. After testing the normality of data distribution of students' mathematical questioning ability obtained that the data is normally distributed. Based on these findings, the test of means difference test is performed by Two-Way ANOVA test (Table 3).

Table 3. Summary of two-way ANOVA test of students' mathematical questioning ability based on the learning factor and level of mathematical courage

SOURCES	Sum of Squares	DF	Mean Square	F_{Count}	Sig
Learning Approach (A)	6,338	1	6,338	6,390	0,013
Mathematical Courage Level (B)	79,999	2	40,000	40,327	0,000
A x B	0,083	2	0,042	0,042	0,959
Inter	144,815	146	0,992		

Based on Table 3 we get sig = 0,013; So it can be concluded that there is a significant difference between student teachers' mathematical questioning ability whose learning using Metaphorical Thinking learning with the use of conventional learning.

If viewed on the basis of the level of mathematical courage (high, moderate, and low) seen from Table 3, it can be concluded that there is at least one group of certain levels of mathematical courage level whose mathematical inquiring abilities differ significantly with another mathematical courage level. To know which degree of math is different is a significantly Scheffe test. The results are presented in Table 4.

Table 4. Scheffe test of mathematical questioning ability development based on mathematical courage (MC) level

MC Type (I)	MC Type (J)	Mean Difference (I – J)	Sig	Interpretation
High	Moderate	0,8390*	0,001	Different
Moderate	Low	0,9115*	0,000	Different
High	Low	1,7505*	0,000	Different

Based on Table 4, it is concluded that mathematical questioning ability has developed at all mathematical courage level. In addition, it is also concluded that there is no interaction effect between learning and mathematical courage level (high, medium, and low) simultaneously in developing student teachers' mathematical questioning ability.

Other findings related to the achievement of student teachers' mathematical questioning ability it is also found that mathematical questioning ability is not achieved well as expected. The achievement result is attached to Table 5.

Table 5. Achievement of mathematical asking ability

Indicators of mathematical questioning ability	Mathematical Courage Level	Learning Class MT	Learning Class
Posing problems is connected to contexts of given statements	High	100%	100%
	Moderate	96%	89%
	Low	82%	79%
	Overall	92%	90%
Posing problems is in the form of non-routine of given statements	High	84%	82%
	Moderate	70%	67%
	Low	66%	65%
	Overall	78%	77%
Posing problems is in the form of open questions of given statements	High	73%	71%
	Moderate	70%	68%
	Low	69%	66%
	Overall	72%	69%

Based on the findings in Table 5, it can be concluded that the achievement of the mathematical questioning ability of the prospective teacher students is still not well achieved on the indicator of problem submission in the form of non-routine questions and open questions. Based on the observation, the obstacle faced by the students is still difficult to construct non-routine questions by expecting open an answer from other students such as:

a. Students' are not trained to think openly in their previous learning.
b. Student's thinking is more concerned with the results than the process that has become the learning culture.
c. The constraint can also be interpreted that learners often have difficulty solving problems caused by the lesson taught is still monotonous (not centered on the students) [8]. In addition, the learning culture experienced by learners in the class has not been accustomed to thinking abstract in solving the problem [6,11].

4. Conclusion

Based on the results and discussion, it can be concluded that: (a) Mathematical questioning ability of student teachers who received metaphorical thinking learning is better than those who received conventional learning seen from mathematical courage level (high, moderate, and low); (b) Learning and mathematical courage level factors influence students' achievement in mathematical questioning ability. In addition, there is no interaction effect between learning and mathematical courage level (high, medium, and low) simultaneously in developing student teachers' mathematical questioning ability; (c) Achievement of mastering mathematical questioning ability of student teacher is still not well achieved on indicator of problem posing in the form of non-routine question and open question.

References

[1] Rahman, A 2013 *Jurnal Ilmu Pendidikan* **19** (2) pp 244-251
[2] Tofade, T, Elsner, J and Haines, S T 2013. *American journal of pharmaceutical education* **77** (7) pp 155
[3] Hendriana, H 2017 *Infinity Journal* **6** (1) pp 51-58
[4] Syah, M 2010 *Psikologi Belajar* (Jakarta: Rajagrafindo Persada)
[5] Zhang, C, Wang, H and Li, W 2012 *Engineering Education and Management* 437-441
[6] Hendriana H, Rohaeti E E, and Hidayat, W 2016 *Journal on Mathematics Education* **8** (1) pp 55-64
[7] Widodo A 2006 *Jurnal Pendidikan dan Pembelajaran* **4** (2) pp 139-148
[8] Hendriana H 2012 *Infinity Journal* **1** (1) pp 90-103
[9] Japa I G 2014 *Jurnal Ilmu Pendidikan* **20** (1) pp 9-16
[10] Hidayat W 2017 *KALAMATIKA Jurnal Pendidikan Matematika* **2** (1) pp 15-28
[11] Piatek-Jimenez K 2010 *Mathematics Education Research Journal* **22** (3) pp 41-56

The effect of team accelerated instruction on students' mathematics achievement and learning motivation

Agustina Sri Purnami[1], Sri Adi Widodo[1], and Rully Charitas Indra Prahmana[2]
[1]Universitas Sarjanawiyata Tamansiswa, Jl. Batikan UH III/1045, Yogyakarta
[2]Universitas Ahmad Dahlan, Jl. Pramuka Kav. 5, Pandeyan, Umbulharjo, Yogyakarta

E-mail: sriadi@ustjogja.ac.id

Abstract. This study aimed to know the improvement of achievement and motivation of learning mathematics by using Team Accelerated Instruction. The research method used was the experiment with descriptive pre-test post-test experiment. The population in this study was all students of class VIII junior high school in Jogjakarta. The sample was taken using cluster random sampling technique. The instrument used in this research was questionnaire and test. Data analysis technique used was Wilcoxon test. It concluded that there was an increase in motivation and student achievement of class VII on linear equation system material by using the learning model of Team Accelerated Instruction. Based on the results of the learning model Team Accelerated Instruction can be used as a variation model in learning mathematics.

1. Introduction

Students have a perception that math lessons are a very difficult and complicated lesson, unpleasant, less varied teacher in presenting the material so that students are lazy to learn it [1]. This lazy attitude that became one of the causes of student achievement in mathematics cannot be optimal and mathematics has always been a scourge in learning in school. Mathematics is very concerned with abstract ideas and hierarchical concepts and deductive reasoning [2]. So, the material or ideas/concepts of mathematics at the previous level related to understanding the concept of mathematics in the next level [3].

Also, educators have an important role in the learning system for the students, especially to attract student motivation [4]. In general, it stated that one's motivation could affect the learning achievement of the person. In general, someone who has a high motivation will have a good learning achievement as well [5]. Vice versa, if someone has a low motivation, there is a tendency will have a low learning achievement as well. So that motivation can be used as a benchmark for someone to obtain the achievement expected.

Several ways can be used to foster motivation in learning activities among other are scoring; prizes; competition/competition; Ego-involvement; replicate; notify results; praise; punishment; the desire to learn; interests; recognized goals [6-9]. The learners will be keen to learn if they know there will be repetition or evaluation [7, 8]. Therefore, giving a test or evaluation is also a means to increase or foster motivation in the lesson.

Every human being does something because of motivation by a certain motivation [6]. The motivation is the impulse existing the human being. The motivation is also the desire that comes from the human being to do something. Attribution theory is now ending its third decade as a dominant conception in motivation [10]. Achievement motivation as a force associated with the achievement of some standard of excellence or intelligence, which is an impulse within a person so that he seeks in all activities as high-end.

Also, the low achievement of student learning caused by the ability of learners in solving math problems is less adequate. It shows that the learner in solving the problem of math is unstructured and not systematic, meaning that students only do the step answer only. While the step of understanding, planning and looking back answers sometimes not done [11].

Along with the growing education system and demands, not infrequently schools that still use direct instruction or conventional learning [12]. Where the conventional learning is done solitaire, meaning that learning starts from planning, implementation, to the evaluation of student learning done by one teacher. In fact, the curriculum of education has grown, teachers are required to be more innovative and creative in determining/choosing the learning method used which tailored to the subject matter that presented to the students.

Teachers also have an important role in the learning system, especially improving the quality of learning. Teachers should use varied learning models as well as tailored to the characteristics of the subject, and the learner taught to attract students' learning. In addition to the direct learning model, some learning models used in learning mathematics such as Jigsaw, Team Accelerated Instruction (TAI), and Student Team Achievement Division (STAD) [13].

Although direct learning has shown a pattern of interaction between teachers with learners or a group of students [11], this learning model causes students to be less interested in learning. Fewer students on learning by using direct learning model because the communication patterns that occur only go one way so that students feel not involved directly in learning.

Cooperative learning is one of the learning models that arise as a result of the notion of constructivism, where students will construct knowledge if students find and understand the concept in discussion and cooperate in mastering the material given by the teacher [13-16]. With cooperative learning, students learn to cooperate mutually beneficial not to compete with each other. At least there is Team Accelerated Instruction or Team Assisted Individualization that has the characteristics of the responsibility of learning to be in individual students [11]. Students should build their knowledge and not only accept the finished form of the teacher. On the other hands, the teacher-student communication pattern is negotiation and not the impositions, with TAI learners, can learn at their level. So learners can negotiate with friends in groups or negotiate with teachers regarding the material to be learned in groups [16].

In addition to students, teachers also have an important role in learning, especially improving the quality of learning [17]. To attract student learning, the teacher must use the learning model besides the learning model of language learning. So that learning using direct learning model can cause learners feel bored. It is because the direct learning model of communication patterns run in one direction so that learners feel not directly involved in learning.

Based on the background of the problem, the purpose of this research is to know the improvement of achievement and motivation to learn mathematics by using Team Accelerated Instruction.

2. Methods

Among the problems to be studied, then this type of research included in the experimental research. This research intends to give treatment to the sample; then the researcher wants to know the improvement of achievement and motivation to learn [18].

The population is the overall subject of research, based on that opinion, the population in this study is all students of class VIII junior high school SMP Negeri or private SMP in the city of Yogyakarta Lesson Year 2016 - 2017. The sample representative of the population cannot be proved, but can only be approached methodologically through known and recognized parameters whether or not they are theoretical or experimental. The parameters that can be used to determine the representative of a sample are population variability, sample size, sample determination technique, and accuracy includes the characteristics of the population in the sample. The larger the sample size, the greater the representative level. But in the case of a homogeneous population, the sample size does not affect the representative level of the sample. In this research the sampling technique using cluster random sampling. The sampling step by cluster random sampling technique is population divided by school. Since public and private schools are considered equal, the population has 57 schools. Furthermore, the school group was randomly using the lottery to obtain six schools used as research sites. Furthermore, the six schools taken by lottery method were randomized back by using cluster random sampling technique based on the classes held in each school. It aims to obtain two classes used as a research sample. From the cluster, random sampling technique used obtained six schools with the sample size of 134 students.

The instruments used in this study were questionnaires and tests. Questionnaires are used to obtain data about learning motivation. The questionnaire contained several statements with category answers. The answers chosen by the students describe the individual states of their students. The indicator of motor learning is the giving of prizes; Competition/competition; Ego-involvement; Give a repeat/task; Notify results; praise; punishment; Desire to learn; interest. Questionnaire Motivation gave to students before and after treatment. Of the 38 items prepared, 30 items have eligible alignment levels that have a correlation index of more than 0.300. The Cronbach alpha calculation results from 30 items indicate that the alpha index is 0.96, it indicates that the compiled motivation questionnaire can be used to obtain the data. The test instrument is used to collect preliminary (pre-test) and student achievement data on linear equation system (posttest) material. Both pre-test and posttest are multiple choices with five choices of answers. Of the 20 items prepared, all items have an eligible alignment level that has a correlation index of more than 0.300, KR calculation results from 20 items indicates that the alpha index is 0.819, it indicates that the compiled test can be used to obtain data

Descriptive data analysis technique used to test the proposed hypothesis is done using Wilcoxon test statistic. The hypothesis tested is TAI learning can improve the motivation and achievement of learning mathematics.

3. Results and Discussion

Based on the data of learning motivation before using Team Accelerated Instruction obtained that the highest score is 103, the lowest score is 52, the average of 75.49 and the standard deviation is 11,630. While the results of the questionnaire calculation of learning motivation after using Team Accelerated Instruction obtained that the highest score is 112, the lowest score is 66, the average is 93.63, and the standard deviation is 10.876.

Based on the test results before using Team Accelerated Instruction, it was found that the highest score was 20, the lowest score was 3, the mean of 14.34 and the standard deviation was 3,614. While the results of test calculations after using Team Accelerated Instruction obtained that the highest score is 20, the lowest score is 2, the average of 15.62 and the standard deviation is 3.869.

Results of non-parametric statistical calculations with Wilcoxon test on learning motivation obtained that the coefficient of Z equal to -9.705 with sign 0.000 while on learning achievement of -4.255 with sign 0.000. Based on this result, it concluded that there is the difference between the mean of motivation and learning achievement between before and after using a model of learning Team Accelerated Instruction. The average frequency difference results presented in Table 1.

Table 1: Frequency of pre and post differences

	Post < Pre	Post = pre	Post > pre
Motivation	6	2	126
Achievement	47	22	109

To find out whether the difference is a positive difference seen in Table 1, where most of the difference that occurs is a positive difference means that the average post is better when compared with the average pre-test. Based on these results it can be concluded that the learning model of Team Accelerated Instruction can improve the motivation and achievement of student teaching mathematics class VII on linear equation system material.

Motivation to learn is the impetus that it is for someone to do learning activities. Motivation learning is the very important role for students in the effort to achieve high learning achievement [19].

Students with high motivation to learn tend to show passion and excitement in learning, they usually look more seriously at learning and actively participate in learning activities, both in class and outside the classroom.

Motivation is one of the factors that determine learning achievement. The motivation to learn has a positive relationship to the preservation of learning so that learning motivation influences achieving one's goals [20, 21]. In the learning activities, intrinsic and extrinsic factors are not able to stand alone, but both factors must be together to guide one's behavior toward a target to be desired or intended. According to need, motivation differentiated into the need for achievement and the need for social relationships.

In essence, the motivation of each is different. Some have high motivation, and some have low motivation. Motivation differences are one of the causes of differences in achievement of learning achieved each. With a strong motivation then someone will be easy to achieve what desired, so to obtain a good learning achievement becomes easier to achieve and vice versa, if the motivation is weak then there is a possibility to reach something becomes more difficult so to obtain a good learning achievement to be difficult.

Students who have high motivation to learn will be more diligent, passionate, more resilient and have higher ambitions in achieving better learning achievement, compared with students who lack or have no motivation to learn [22]. Those who do not have the motivation to learn will seem less or less passionate about learning and learning in the classroom, not paying attention to lessons learned, apathetic and not actively participating in learning. The condition of students who lack the motivation to learn is certainly not able to produce satisfactory achievement.

Team Accelerated Instruction, can motivate students in learning seen from the discussion process that occurs in the group, where students who have better ability compared with his friends in the group provide the impetus to solve the problems of mathematics [11, 22]. It is what encourages students motivated in solving math problems.

Improving student achievement in learning mathematics is one of the tasks of a teacher. Factors that greatly affect student achievement are factors from outside students. The results of the previous research indicate that there is a relationship between learning and mathematics learning achievement [13-15] this result indicates that learning achievement influenced by learning motivation owned by learners. Also, the learning model used also affects the learning achievement [5, 6, 16]

Team Accelerated Instruction model requires students to work together in groups and is expected to be able to become a group tutor for students who cannot pre-pay as the next material supporters. Also, teachers can also help students become tutors in groups of students to present material in groups and solve mathematical problems. So that students controlled their ability to understand the material other than that, students with a variety of abilities are also served maximally by the teacher and the pattern of communications created and the direction.

In addition to learners with high initial skills, educators can also help students become tutors in groups of students to present the material to the group and solve mathematical problems [22, 23]. So that students controlled their ability to understand the material. Also, students with a variety of abilities are also served maximally by the teacher and the pattern of communications created and the direction. It is what causes student achievement increased between before and after treatment.

4. Conclusion

The results of research concluded that there is an increase in motivation and student achievement of class VII on linear equation system material by using learning model of Team Accelerated Instruction (TAI).

References

[1] Syamsuri, Purwanto, Subanji, and Irawati S 2017 *International Journal on Emerging Mathematics Education* **1** 135
[2] Widodo S A, Purnami A S, and Prahmana R C I 2017 *International Journal on Emerging Mathematics Education* **1** 193
[3] Sundayana R, Herman T, Dahlan J A, and Prahmana R C I 2017 *World Transactions on*

Engineering and Technology Education **15**
[4] Siemens G and Long P 2011 *Educause Review* **46** 30
[5] Knight P T and Yorke M 2003 *Teaching in Higher Education* **8** 3
[6] Dickey M D 2007 *Educational Technology Research and Development* **55** 253
[7] Deci E L, Vallerand R J, Pelletier L G, and Ryan R M 1991 *Educational Psychologist* **26** 325
[8] Linnenbrink E A and Pintrich P R 2002 *School Psychology Review* **31** 313
[9] Barron K E and Harackiewicz J M 2001 *Journal of Personality and Social Psychology* **80** 706
[10] Weiner B 2000 *Educational Psychology Review* **12** 1
[11] Widodo S A 2017 *Infinity* **6** 61
[12] Tanujaya B, Prahmana R C I, and Mumu J 2017 *World Transactions on Engineering and Technology Education* **15**
[13] Slavin R E 1995 *Cooperative Learning: Theory, Research, and Practice* (New Jersey: Prentice Hall)
[14] Eggen P D and Kauchak D P 1993 *Learning and Teaching* (Massachusetts: Allyn and Bacon)
[15] Eggen P D and Kauchak D P 1996 *Strategies for Teacher Teaching Content and Thinking Skills* (Boston: Allyn and Bacon)
[16] Artzt A F and Newman C M 1990 *Mathematics Teacher* **83** 448
[17] Prahmana R C I and Kusumah Y S 2016 *Pedagogika* **42**
[18] Prahmana R C I 2017 *Design Research. Teori dan Implementasinya: Suatu Pengantar* (Jakarta: Rajawali Pers)
[19] Makur A P and Prahmana R C I 2015 *Jurnal Elemen* **1** 1
[20] Aritonang K T 2008 *Jurnal Pendidikan Penabur* **7** 11
[21] Siswanto I 2017 *Jurnal Mathematics Paedagogic* **1** 180
[22] Alsa A 2011 *Jurnal Psikologi* **38** 82

The effect of discovery learning and problem-based learning on middle school students' self-regulated learning

A Miatun[1] and Muntazhimah[1]

[1]Faculty of Teacher and Education Universitas Muhammadiyah Prof. DR. HAMKA, Jl. Tanah Merdeka, Kp Rambutan, Pasar Rebo, Jakarta Timur 13838, Jakarta, Indonesia

E-mail: asihmiatun@gmail.com

Abstract. The aim of this research was to determine the effect of learning models on mathematics achievement viewed from student's self-regulated learning. The learning model compared were discovery learning and problem-based learning. The population was all students at the grade VIII of Junior High School in Boyolali regency. The samples were students of SMPN 4 Boyolali, SMPN 6 Boyolali, and SMPN 4 Mojosongo. The instruments used were mathematics achievement tests and self-regulated learning questionnaire. The data were analyzed using unbalanced two-ways Anova. The conclusion was as follows: (1) discovery learning gives better achievement than problem-based learning. (2) Achievement of students who have high self-regulated learning was better than students who have medium and low self-regulated learning. (3) For discovery learning, achievement of students who have high self-regulated learning was better than students who have medium and low self-regulated learning. For problem-based learning, students who have high and medium self-regulated learning have the same achievement. (4) For students who have high self-regulated learning, discovery learning gives better achievement than problem-based learning. Students who have medium and low self-regulated learning, both learning models give the same achievement. t.

1. Introduction

Suherman [1] mentions that in mathematics learning, students need to get used gain understanding through the experience of existing or not existing properties a set of objects. Furthermore, with this abstraction students are trained to make predictions, conjecture or tendency based on experience or knowledge developed through specific examples. Concepts in mathematics are mutually sustainable, from easily material increases to difficult material. Students who don't know basic concepts will have difficulty when confronted with other concepts that related to the basic concept. The implications, student mathematics achievement was unsatisfactory.

Low mathematics learning achievement, especially for junior high school students in Boyolali Regency, can be seen as the result of National Examination year 2013/2014 [2]. One of the mathematics material has a low absorption rate is polyhedral material. Absorption rate for the ability to solve problems are relating determine the elements in the polyhedral material, resolve issues related to the geometrical surface area and resolve issues related to geometrical volumes respectively are 48.68%, 45.38% and 47, 03%. Absorption rate in Boyolali is lower than the absorption rate of the province at 49.95%, 47.75%, and 49.11%. The absorption rate for polyhedra material is lower than absorption rate other materials.

Many factors can affect mathematics achievement. These factors which came from outside or from within students themselves. Factors from within e.g. intelligence, learning interest, learning motivation, learning styles and etc. While external factors such as learning models used by teachers there is no significant change from year to year. Allegedly lower mathematics achievement of students is due to lack of precise used the learning model. In line with the opinion of Carol in Sudjana[3] that students' learning achievement influenced by the quality of teaching used by the teacher. The quality of a teacher's learning can be shown through the approach and learning model used. Lack of variation in learning causes the learning becomes monotonous and boring, so students became less enthusiastic about learning.

According to Permendiknas Number 20 of 2006 [4], one of the goals in mathematics learning is to solve problems that include the ability to problems understanding, design mathematics models, solve models and interpretation the solutions. Based on these objectives required learning model that can facilitate students to achieve that goal. Learning model that can facilitate to achieve learning goals is discovery learning and problem-based learning.

The characteristic of discovery learning is discovery itself. Each student must make an invention to discover the concept of the material to be studied. This model provides an opportunity for students to discover and construct their own knowledge. Wong [5] points out that "discovery learning is one of the pedagogic deals that reduce teachers' direct instruction and have students construct knowledge on their own. Guided discovery is superior to pure discovery in helping students learning and transferring". Discovery learning is a learning that develops teachers' pedagogic ability and enables students to construct their own knowledge. Discovery is part of the discovery learning, which helps students how to learn and transfer their knowledge. The role of the teacher is more determined as a learning coach and learning facilitator. Research of Balim [6] gives the result that discovery learning is one of the variations of learning model that can improve students' and teachers' activeness as their mentor. Discovery learning can improve student learning outcomes and student discovery skills compared conventional learning models. Alex and Olubusuyi [7] in their research found that there are significant differences in learning outcomes between students who used discovery learning and students who do not use discovery learning. Discovery learning has great potential to improve student learning outcomes.

Problem-based learning has characteristic learning based on problem-solving. The research results of Padmavathy and Mareesh [8] suggest that problem-based learning influences learning and improves student understanding, and increases the ability to used concepts already learned in real life. Krulik and Rudnick in Padmavathy and Mareesh [8] mentioned that problem-based learning is a problem-based mathematics learning and gives students more opportunities to think critically, present creative ideas and communicate with peers mathematically. The research results of Fatade et al [9] mentioned that there is a significant difference in mathematics learning achievement between problems based learning and traditional model. Problem-based learning provides better mathematics learning achievement. So problem-based learning recommended for teachers to use in classroom learning.

In addition to the approach and learning model, one of the factors that can affect student's mathematics learning achievement is self-regulated learning. Knowless at Scott [10] suggests that self-regulated learning is a process whereby individuals take their own initiative, with or without help from others, to diagnose learning needs, formulate learning objectives, identify learning objectives, select and define learning strategies, and evaluate learning outcomes. Research of Vrieling, et al [11] gives results that self-regulated learning has a strong relationship with the use of cognitive skills and student learning motivation. Students with high self-regulated learning have a stronger materials understanding when compared with medium and low self-regulated learning. According to the research results of Pintrich and De Groot [12] which states, students with high self-regulated learning will be easier to use their cognitive abilities and also maximize their learning outcomes.

The purpose of this research was to know the learning of mathematics of polyhedra material of grade VIII students of SMP Negeri in Boyolali Regency: (1) which one gives better mathematics achievement, discovery learning or problem based learning; (2) which one has better mathematics achievement, students who have high, medium, or low self-regulated learning; (3) for each learning model, which one has better mathematics achievement, students who have high, medium, or low self-regulated learning; (4) For each category of self-regulated learning, which one gives better mathematics achievement, discovery learning or problem based learning.

2. Methods

This research is a quasi-experimental study with 2 × 3 factorial design as shown in Table 1 below.

Table 1. Research Design

B \ A	Self-regulated learning		
	High (b_1)	Medium (b_2)	Low (b_3)
Discovery Learning	$(ab)_{11}$	$(ab)_{12}$	$(ab)_{13}$
Problem Based Learning	$(ab)_{21}$	$(ab)_{22}$	$(ab)_{23}$

The population in this research were students of class VIII SMP Negeri at Boyolali Regency. The population consists of 50 State Junior High Schools in Boyolali Regency. The sampling technique used in this research is stratified cluster random sampling. From the sample selection, the result for the sample is SMP 4 Boyolali, SMPN 6 Boyolali, and SMPN 4 Mojosongo.

There are two independent variables in this research, learning models and self-regulated learning, and one dependent variable that is mathematics learning achievement. Methods of data collection used achievement test, questionnaires, and documentation. The instruments used in this research are mathematics learning achievement test and self-regulated learning questionnaire. Mathematics learning achievement test consists of 25 multiple choice questions. Self-regulated learning questionnaire consists of 50 statements.

Data analysis techniques used to test the hypothesis in this study is unbalanced two-ways variance analysis with 0.05 significance level. Precondition data analysis of early mathematics ability data analysis using students' mathematics achievement test includes normality test using Lilliefors test and homogeneity test of variance using Bartlett test.

3. Results and Discussion

Normality data test result of student's early mathematics achievement test ability found that L_{obs} at discovery learning and problem based learning group are less than$L_{0,05;n}$. That means H_0 of both groups are accepted. In conclusion, each sample group came from a normally distributed population. Similarly, the homogeneity test of population variance on the student's early mathematics achievement test ability, obtained ${\chi^2}_{obs} = 0,38973$ with $DK = \{\chi^2|\chi^2 > 3,84146\}$, since ${\chi^2}_{obs} \notin DK$ then H_0 is accepted. Therefore it can be concluded that both groups of samples came from populations with equal variance.

Based on the result of equilibrium test on the student's early mathematics achievement test ability, obtained $F_{obs} = 0,62742$with $DK = \{F|F > 3,8872\}$. Then $F_{obs} \notin DK$, so H_0 was accepted and it was concluded that both groups of samples came from the population with a balanced mean early mathematics achievement test ability.

Before hypothesis test first tested the normality and homogeneity of the population on the mathematics learning achievement of students. Normality tests are done 12 times and Lobs is obtained for each group, it's smaller than$L_{0,05;n}$with $DK = \{L|L > L_{0,05;n}\}$, so $L_{obs} \notin DK$ and then H_o is accepted . It is concluded that all samples in this study came from normally distributed populations. For homogeneity test of population variance, it is found that ${\chi^2}_{obs}$ in each group is smaller than ${\chi^2}_{tabel}$with $DK = \{\chi^2|\chi^2 > {\chi^2}_{tabel}\}$, because ${\chi^2}_{obs} \notin DK$ then H_o is accepted. It was concluded that the population had the same variance.

Table 2 presents a summary of students' mathematics learning achievement mean data based on the learning model viewed from self-regulated learning.

Table 2. Mean of Data Students' Mathematics Learning Achievement On Each Learning Model and Self-Regulated Learning

Model	*Self-Regulated Learning*			Marginal Rate
	High	Medium	Low	
Discovery Learning	89,4194	75,3778	58,1538	75,2549
Problem based learning	74,3448	70,1395	60,0000	68,0377
Average of Marginal	82,1333	72,8128	89,2000	

Hypothesis test is conducted to find out whether there are differences in achievement between each learning models and self-regulated learning, its interaction on students' mathematics learning achievement

Table 3. Summary of the calculation unbalanced two-way analysis of variance

Source	JK	dk	RK	F_{obs}	F_{table}	Judgment
Model (A)	1893,5633	1	1893,5633	16,321	3,8881	H_{0A} Rejected
Self-regulated (B)	17557,734	2	8778,8671	75,669	3,0408	H_{0B} Rejected
Interaction (AB)	2405,7161	2	1202,8581	10,368	3,0408	H_{0AB} Rejected
Galat (G)	43667,225	205	116,0158			
Total	66795,443	206				

Table 3 is a summary of the results of unbalanced two-ways analysis of variance. Based on Table 3, it could be concluded as follows. (1) In the main effect (A), $F_{obs} = 16{,}321$with $DK= \{F|F > 3{,}8881\}$. Obtained $F_{obs} \in DK$, then H_{0A} rejected. This means that there are differences in students' mathematics achievement between discovery learning and problem-based learning. (2) In the main effect (B), $F_{obs} = 75{,}669$with $DK = \{F|F > 3{,}0408\}$. Obtained $F_{obs} \in DK$, then $F_{obs} \in DK$ rejected. That means that there are significant differences in the categories, high, medium, and low self-regulated learning in order to students' mathematics learning achievement. (3) In the interaction effect (AB), $F_{obs} = 10{,}368$ with $DK = \{F|F > 3{,}0408\}$. Retrieved $F_{obs} \in DK$, then H_{0AB} is rejected. It means there is interaction between learning models and self-regulated learning category on student's mathematics learning achievement.

The result of the Anava calculation shows that H_{0A} is rejected, So it must be tested the mean comparison between rows. The summary of the calculations is presented in Table 4 below.

Table 4. Summary of Intermediate Line Comparison Test

H_0	F_{obs}	$F = 2F_{table}$	Judgment
$\mu_{1.} = \mu_{2.}$	23,337	7,7763	H_0 Rejected

Based on Table 4 and the marginal rate in Table 1 it can be concluded that discovery learning provides better mathematics learning achievement than problem-based learning. In the discovery learning model students are given the opportunity to think experimentally, discovering and constructing their own knowledge. These results are in line with the study of Alex and Olubusuyi's [7] research, that there are significant differences in learning outcomes between students using discovery learning and students who do not use discovery learning.

The result of the analysis of variance calculation shows that H0B is rejected, So it must be tested the mean comparison between column. The summary of the calculations is presented in table 5 below.

Table 5. Summary of Inter-Column Mean Comparative Test

H_0	F_{obs}	$F = 2F_{table}$	Judgment
$\mu_{.1} = \mu_{.2}$	26,6830	7,7763	H_0 Rejected

$\mu_2 = \mu_3$	56,4960	7,7763	H_0	Rejected
$\mu_1 = \mu_3$	135,9999	7,7763	H_0	Rejected

Based on Table 5 and the marginal rate in Table 1 it can be concluded that students 'mathematics learning achievement with high self-regulated learning is better than students with medium and low self-regulated learning, and students' mathematics learning achievement with medium self-regulated learning is better than students with low self-regulated learning. These results are consistent with the hypothesis proposed research. Better students' mathematics learning achievement that has high self-regulated learning in line with the results of research Pintrich and De Groot [12] which states that students with high self-regulated learning will be easier in using their cognitive abilities and learning outcomes more leverage. Students with high self-regulated learning manage their behaviour and thoughts in learning so they obtain the information they need when learning.

Furthermore, students' mathematics learning achievement with medium self-regulated learning is better than those with low self-regulated learning. In line with Vrieling, et al [11] research, gives results that self-regulated learning has a strong relationship with the use of cognitive skills and student learning motivation. Students with high self-regulated learning have a stronger understanding of the material when compared with medium and low self-regulated learning. Students with medium self-regulated learning are had better learning motivation than students with low self-regulated learning, so students with self-regulated learning have better learning achievement.

The results of ANOVA calculation show that H_{0AB} is rejected, so we need a mean comparison test between cells in the same row. The summary of the calculations is presented in Table 6.

Table 6. Summary of Mean Comparison between Cells on the Same Line

H_0	F_{obs}	F_{table}		*Judgment*
$\mu_{11} = \mu_{12}$	31,1943	11,2950	H_0	Rejected
$\mu_{11} = \mu_{13}$	119,1445	11,2950	H_0	Rejected
$\mu_{12} = \mu_{13}$	42,1381	11,2950	H_0	Rejected
$\mu_{21} = \mu_{22}$	2,6400	11,2950	H_0	Accepted
$\mu_{21} = \mu_{23}$	27,7594	11,2950	H_0	Rejected
$\mu_{22} = \mu_{23}$	16,8258	11,2950	H_0	Rejected

Based on Table 6 and the marginal rate in Table 1 it can be concluded that in discovery learning, students 'mathematics learning achievement with high self-regulated learning is better than the students with low and medium self-regulated learning, students' learning achievement with medium self-regulated learning is better than low self-regulated learning students. Students with high self-regulated learning have a strong understanding of the material compared with medium and low self-regulated learning. In line with Vrieling, et al [11] research, gives results that self-regulated learning has a strong relationship with the use of cognitive skills and student learning motivation. Students with high self-regulated learning have a strong understanding of the material compared with medium and low self-regulated learning.

In problem-based learning, students' mathematics learning achievement with high and medium self-regulated learning as same as well. The factors caused by the understanding of student material with high and medium self-regulated learning is same. As a result students' ability in solving problems is same. Furthermore, the mathematics learning achievement of students with high self-regulated learning is better than those with low self-regulated learning. This is in line with the results of research Pintrich and De Groot [12] which states that students with high self-regulated learning will be easier in using their cognitive abilities and learning outcomes more leverage.

Furthermore, the mean comparative test between cells in the same column, the summary of the calculations are presented in Table 7 below.

Table 7. Mean Comparison Range Between cells in the same column

H_0	F_{obs}	F_{table}	Judgment	
$\mu_{11} = \mu_{21}$	29,3480	11,2950	H_0	Rejected
$\mu_{12} = \mu_{22}$	5,2006	11,2950	H_0	Accepted
$\mu_{13} = \mu_{23}$	0,4328	11,2950	H_0	Accepted

Based on Table 7 and the marginal rate in Table 1 it can be concluded that students with high self-regulated learning, discovery learning provides better mathematics learning achievement than problem-based learning. This is because students with high self-regulated learning have better cognitive and motivational skills. Students with high self-regulated learning viewed the problem as a challenge to overcome not a threat to avoid. In line with the results of the study of Alex and Olubusuyi [7] who found, there are significant differences in learning outcomes between students who use discovery learning and students who do not use discovery learning. This is also reinforced by high self-regulated learning so that mathematics learning achievement of students who used discovery learning is better than problem-based learning. In students with medium and low self-regulated learning, discovery learning and problem-based learning provide mathematics learning achievement as well. Students with high self-regulated learning have a strong understanding of the material compared with medium and low self-regulated learning. It is possible when students with medium and low self-regulated learning are confronted with a particular learning model they find it difficult and less motivated to learn. So in each learning model, students' mathematics achievement with medium and low self-regulated learning will be the same. In addition students with medium and low self-regulated learning when following the learning process need more guidance, but the guidance provided can't be maximized because of the limited time

4. Conclusion

From the results of the research can be concluded as follows. (1) Discovery learning gives better mathematics achievement than problem-based learning. (2) Mathematics achievement of students who have high self-regulated learning was better than students who have medium self-regulated learning, and both (high and medium) gives better mathematics achievement than low self-regulated learning. (3) For discovery learning, Mathematics achievement of students who have high self-regulated learning was better than students who have medium self-regulated learning. Both (high and medium) gives better mathematics achievement than low self-regulated learning. For problem-based learning, students who have high and medium self-regulated learning have the same mathematics achievement. Both (high and medium) gives better mathematics achievement than low self-regulated learning. (4) For students who have high self-regulated learning, discovery learning gives better mathematics achievement than problem-based learning. For students who have medium and low self-regulated learning, both learning models give the same mathematics achievement.

Suggestions from the results of this study are as follows, teachers should apply discovery learning as one of the references in learning on the polyhedral material in the classroom. Teachers should prepare them optimally, with the preparation of facilities, learning tools, and student conditioning that can support the learning process. So discovery learning really facilitates students' self-regulated learning. In addition, teacher guidance and encouragement are needed to help students with medium and low self-regulated learning.

References

[1] Suherman E 2003 *Strategi Pembelajaran Matematika Kontemporer* (Bandung: FPMIPA Universitas Pendidikan Indonesia)
[2] Badan Standar Nasional Pendidikan 2014 *Laporan Hasil Ujian Nasional Tahun Pelajaran 2013/2014* (Jakarta: Kemendikbud)
[3] Sudjana N 2004 *Dasar-Dasar Proses Belajar Mengajar* (Bandung: Sinar Baru Algensindo)
[4] Kemendiknas 2006 *Peraturan Menteri Pendidikan Nasional Nomor 20 Tahun 2006 tentang Standar Isi untuk Satuan Pendidikan Dasar dan Menengah* (Jakarta: Kemendiknas)

[5] Wong S I 2010 *Journal of ICCE*
[6] Balim A G 2009 *Eurasian Journal of Educational Research* **35** 1-30
[7] Alex A, Olubusuyi F 2013 *Journal of Education and Practice* **4** (12)
[8] Padmavathy R D, Mareesh K 2013 *International Multidiclipinary e-Journal* **II** (1)
[9] Fatade A O, Mogari D, Arigbabu A A 2013 *Acta Didactica Napocensia Journal* **6** (3)
[10] Scott K W 2006 *International Journal of Self Directed* **3** (3)
[11] Vrieling E, Bastiaens T, Stijnen S 2012 *Australian Journal of Teacher Education* **37** (8)
[12] Pintrich P R, De Groot E V 1990 *Journal of Educational Pscychology* **42** (1) 33-40

Acknowledgements

I would like to express my sincere gratitude to all the persons who have been helpful towards the successful completion ofthis paper. First and foremost I would like to thanks my thesis supervisor, Dr. Imam Sujadi, M.Si and Dr. Riyadi, M.Si for their support and assistance towards carrying out the research to complete this paper.

Enhancing students' mathematical problem posing skill through writing in performance tasks strategy

Kadir [1], R Adelina [1] and M Fatma [1]

[1] Department of Mathematics Education, Universitas Islam Negeri Syarif Hidayatullah Jakarta, Jl. Ir. Juanda 95, 15412 Indonesia

E-mail: kadir@uinjkt.ac.id

Abstract. Many researchers have studied the Writing in Performance Task (WiPT) strategy in learning, but only a few paid attention on its relation to the problem-posing skill in mathematics. The problem-posing skill in mathematics covers problem reformulation, reconstruction, and imitation. The purpose of the present study was to examine the effect of WiPT strategy on students' mathematical problem-posing skill. The research was conducted at a Public Junior Secondary School in Tangerang Selatan. It used a quasi-experimental method with randomized control group post-test. The samples were 64 students consists of 32 students of the experiment group and 32 students of the control. A cluster random sampling technique was used for sampling. The research data were obtained by testing. The research shows that the problem-posing skill of students taught by WiPT strategy is higher than students taught by a conventional strategy. The research concludes that the WiPT strategy is more effective in enhancing the students' mathematical problem-posing skill compared to the conventional strategy.

1. Introduction

In the Indonesian context, problem-posing plays an important role in the development of Kurikulum-2013. The standard process of Kurikulum-2013 emphasized a scientific approach to the implementation of learning method consists of five skills, namely observing, asking, reasoning, experimenting, and communicating in a consistent and persistent manner [1]. One aspect of the skills in Kurikulum-2013 is asking. In mathematics, asking or pose a problem is so-called problem-posing. The National Council of Teachers of Mathematics (NCTM, 1989) on the Curriculum and Evaluation Standards for School Mathematics explicitly states that students should "have some experience in recognizing and formulating their own problems, an activity that is at the heart of doing mathematics. Silver & Cai (1996) stated that the professional standards for teaching in mathematics underline the importance of teachers to provide opportunities for the students to pose their own problems. "Students should be given the opportunities to formulate problems in a given situation and create new problems by modifying the conditions of a given problem [2]. Problem-posing skill in mathematics can be defined as the skill to pose problems based on the tasks given by the teachers [3].

The problem-posing in mathematics learning has been widely recognized as an effective learning approach to train learners in problem-posing independently. Many researchers have studied the understanding of the students in problem-posing.[4, 5, 6, 7, 8]. However, only a few studies that specifically pay attention to the effect of learning strategy to measure the problem-posing skill in Indonesia.

As the fifth phase of the Polya's problem solving, problem-posing can be considered as an indicator of the relation between problem-posing and problem-solving [9]. The problem-posing skill could be enhanced in phases of Polya's problem-solving. The indicators of the problem-posing skills were developed based on classification by Stoyanova, namely reformulating the problem, reconstructing the problem, and imitating the problem. Reformulating the problem is conducted by restructuring the elements of the similar or identical problems using different sentences. Reconstructing problem is carried out by modifying or changing the properties of the problem in some ways with different contents. While imitating the problem is implemented by adding and expanding the purpose, as well as connecting other materials to the initial problems [10].

In general, the achievement in mathematics, which is related to the problem solving is low. The TIMSS (Trends in International Mathematics and Science Study) 2011, shows that achievement in mathematics of the Indonesian student's 8th Grade was ranked at 38th out of 42 countries with a score of 386. The results were grouped in 3 categories: score 400 or less (low), 401-475 (medium), 476-550 (high), and 551- 625 (advance).[11] Using these criteria, Indonesian students' achievement was in the lower category, focuses on memory rather than comprehensive understanding. Furthermore, the result of PISA study (Programme for International Student Assessment) 2015, shows that only about 10% of Indonesian students that able to answer the test at level 4, 5, and 6. In particular, level 5 contains problems that require the students to work in the complex situations, identifying constraint, selecting, comparing, and evaluating the suitable problem-solving strategies, using extensive reasoning, reflecting, formulating as well as stating the interpretation and reasoning [12]. Furthermore, the initial results of the study at the 7th-grade students of Junior Secondary School revealed that only 19.4% students achieved the average of problem-solving skill of 56.94 and achieve mastery level. In general, the level of problem-solving skills of the students is low. It is an indication of the low problem-posing skill among the students of Junior Secondary School in Indonesia.

One alternative to the learning strategies that can support the development of students' mathematical problem-posing skill is the WiTP strategy or the Writing in Performance Task. According to Junaidi, WiPT is a learning strategy that asks the students to demonstrate and communicate the mathematical understanding of the tasks. WiPT has two main stages, namely, writing and performing the task. The writing stage contains the activities of writing in accordance with the Bloom's taxonomy, namely: writing of knowledge, writing of comprehension, writing of application, writing of analysis, writing of synthesis, and writing of evaluation[13]. The performing stage contains a completion of the mathematical task, such as problem-solving, drawing an image and graph, developing tables, and preparing a summary. WiPT strategies can improve the mathematical disposition, confidence in problem-solving, delivering ideas, provide reasons, find mathematical ideas, find something new, reflecting the minds, appreciating the use of mathematics, and appreciate the role of mathematics in culture and language [14].

Based on the above problems, the purpose of this research is to examine the effect of WiPT strategy on the students' mathematical problem-posing skill. The improvement of the problem-posing skill is studied through an experiment.

2. Methods

This research used a quasi-experiment with Randomized Control Group Posttest Only design. The experimental group was treated with the WiPT Strategy, while the control group was treated with the Conventional Strategy. The population in this research is all students of the 8th Grade of the MTs Khazanah Kebajikan, an Islamic Public Junior Secondary School in Tangerang Selatan. The number of sample in this research is 64 students. The sample was taken through cluster random sampling technique. One class with 32 students was randomly selected from the experimental groups and another with 32 students was also randomly selected as a control group. The problem-posing skill data were collected using 7 essay test items. Content validity of problem posing skills test was determined based on expert evaluation using Lawshe method [15]. The reliability of the test was 0.82 and the validity of the test ranged from 0.45 to 0.89. Meanwhile, the level of difficulty of test items ranged from 0.24 to 0.75, and discriminating index test items ranged from 0.21 to 0.46. The test measures the indicators: the skill to carry out reformulation, reconstruction, and imitation of the problems.

Reformulation of the problem contains rearranging the elements in the structure of the initial problems to draw points and lines as well as determining the slope of the various forms. The reconstruction contains modifying the properties of a straight-line equation taken from a daily life. The imitation contains the addition and expansion of the purposes and determination of the slope based on the properties of the slope. The research uses an Algebra materials, which consist: 1) slope; 2) equation of a straight line; and 3) the graphs of straight lines. Data analysis was performed using test-t with SPSS program.

3. Results and Discussion

The students' mathematical problem-posing skill both for the experiment and control groups is provided in Table 1.

Table 1. Students' mathematical problem-posing skill between Experimental and Control Group

Statistics	Group	
	Experiment	**Control**
The number of students (n)	32	32
Average ($\bar{x}$)	60,68	53,26
Standard Deviation (s)	12,70	14.77

The indicators of the mathematical problem-posing skill in this research covering: problem formulating, problem constructing, and problem imitating. Base on the indicators, the percentage of both experimental and control group is provided in Table 2.

Table 2. Percentage skills of students base on indicators of mathematical problem posing for both experimental and control group

Indicators	Max score	Experiment		Control	
		$\bar{x}$	%	$\bar{x}$	%
Reformulation	4	3,91	97,66	3,68	92,19
Reconstruction	8	5,81	72,66	5,47	68,36
Imitation	12	4,81	40,10	3,62	30,21

Analysis results in Table 2 show that the indicator of problem-posing skill in mathematics for the reformulation is very high. On the reconstruction, the score for both of experimental and control group is moderately high. In contrast, the imitation scores low. The results of this analysis also show that the students treated by WiPT strategy have higher achievement in all indicators of reformulation, reconstruction, and imitation compared to the control group.

Based on the hypothesis testing, using program SPSS independent samples t-Test, shows that t_{abs} = 2.156; p-value = 0.035/2 = 0.018. Thus, H_0 rejected, in other words, the average of students' mathematical problem-posing skill for the experimental group is higher than the control group. The results of the t-test are provided in Table 3.

Table 3. Hypothesis Test Results with t-Test

t-test for Equality of Means						
t	df	Sig. (2-tailed)	Mean Difference	Std. Error Difference	95% Confidence Interval of the Difference	
					Lower	Upper
2.156	62	.035	7.422	3.442	.540	14.304

Analysis results in Table 3 show that the students' mathematical problem-posing skill with WiFT learning strategy is higher than of the one treated using the conventional strategy.

In general, the research findings indicate that the average of students' mathematical problem-posing skill for experimental group is significantly higher than the control group. This shows that the

WiPT strategy is more effective in improving the skill of students' problem-posing compared to the conventional strategy. This finding is consistent with all indicators including reformulation, reconstruction, and imitation, where students taught using the WiPT strategy tend to be higher than students who were taught with the conventional strategy. The findings of this study are similar to research by Suryana, which found that the average score of the capabilities for students with Problem-Based Learning is higher than the students with the conventional learning [16]. This research finding is slightly different from Esra and Unal, who found that the problem-posing skill is affected by the experience of the mathematics groups and learning motivation provided by the teacher [17].

The WiPT strategy emphasizing the group discussion, equipped with the student's activity sheet (LKS) that contains two stages, namely: task writing and task performing. The writing stage includes writing of knowledge, writing of comprehension, writing of the application, writing of analysis, writing of synthesis, and writing of evaluation. In the writing of knowledge activity, the students were asked to write down their knowledge, such as summarizing a theory or algorithm using their own language and formulating the concept that has been understood. On the writing of comprehension, the students were asked to re-explain the definition or summarize a paragraph using their own sentences. In the writing of the application, the students were given a task to write and explain how to solve the problem by applying mathematical concepts related to everyday life. In the writing of analysis, the students were asked to identify the parts and the relationship between parts, and apply the organizing principles in resolving the problem. Similarly, in the writing of synthesis, the student identify the relationships among problems systematically to get a pattern or certain formula. In the last activity, the writing of evaluation, the students solve the problems given by the teacher. In the next stage, task-performing, students were guided to create a mathematical statement to develop the earlier problems.

One of the writing the equation of a straight line created by the students in the Experimental Group is shown in Figure 1.

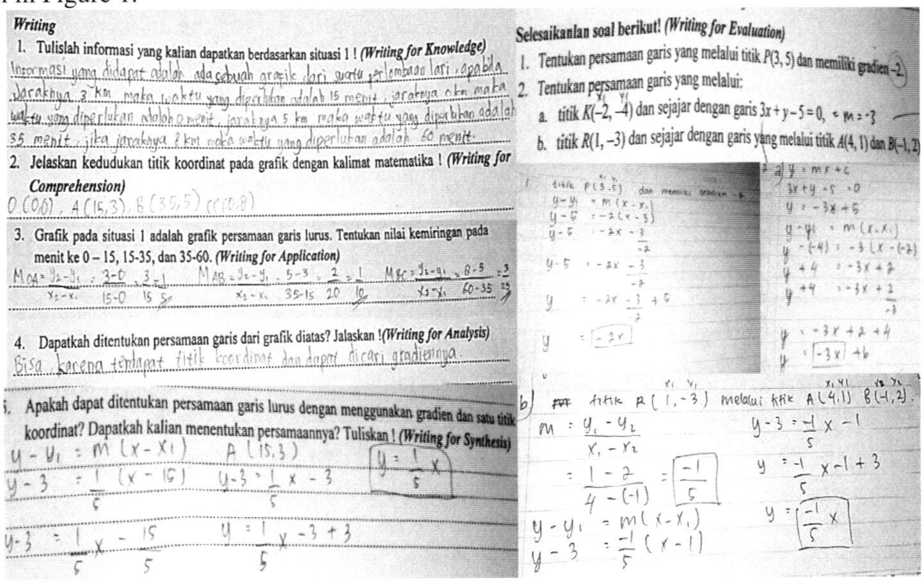

Figure 1. Determining the equation of a straight line in the writing stage

Figure 1 shows the problems presented in students' worksheet treated with a WiPT which are deliberately presented for exercise in order to assess the students' writing mathematical in the equation of a straight line chapter. In the experimental group, the majority of students both in individual lessons or group are able to answer the practice questions within students' worksheet WiPT. However, there are some questions that simply cannot be solved by all students and ultimately be discussed together with the teacher. In the control group, there are problems that can be done easily, but most cannot be solved by the students. This is because they are not taught the writing strategies, so they were not able

to understand the concepts that have been taught. Consequently, most of the mathematical writing problems were not able to be completed and finally discussed together with the teacher.
The answer of the students in the posttest as a tool to measure the skill of mathematical problem posing. Students are given a question to measure the skill to a reformulation of the problem as follows:
Question: "the following graph is the velocity a car,

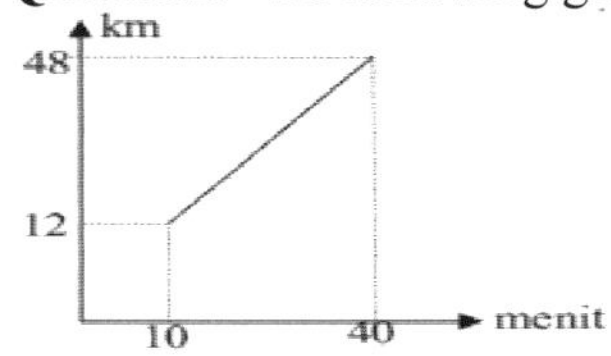

a. Write down the information that you should know based on the chart on the side!
b. Make at least 2 questions related the graph beside by adding other information!"

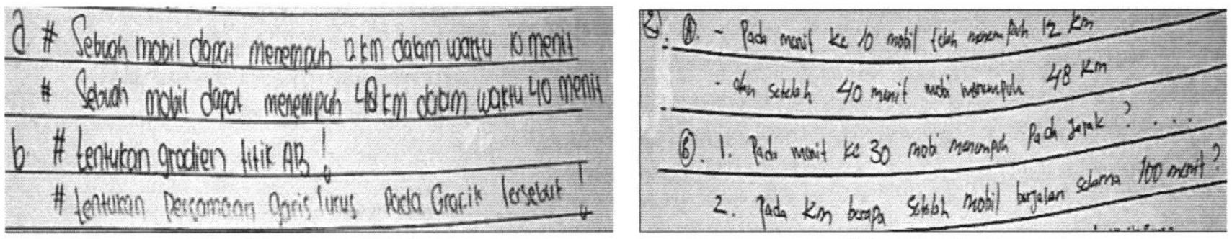

(a) Experimental (b) Control

Figure 2. Answers's Problem Reformulation of Experimental and Control Group

Based on Figure 2, shows that most of the students in the two groups were able to answer the question properly, but there is a difference in answers to the problem are given. Experimental group students could answer in more detail, so the students ask questions which related to the image fits the material equations of straight lines. Meanwhile, students' answer is as well as true of the control group students, but students ask questions not related to the material equations of the straight line. The learning process in the group stage of the experimental group with writing training. The students writing down information through understanding the graphs. While in the stage of performance tasks, students doing the presentation in front of the group. This stage train the students for expressing his opinion either in the form of a math question so that develop the problem-posing skill with reformulate problem is given.

Furthermore, the capabilities of problem posing on indicator reconstruction of the problem, students are given a question to measure the skill to as follows:
Question: " Point A (5-4), B (2,-8) and C (k-12) are in the same straight line. (a) Draw the problem in the form of charts!, and (b) Make at least 2 questions, then finish your question that you create!"

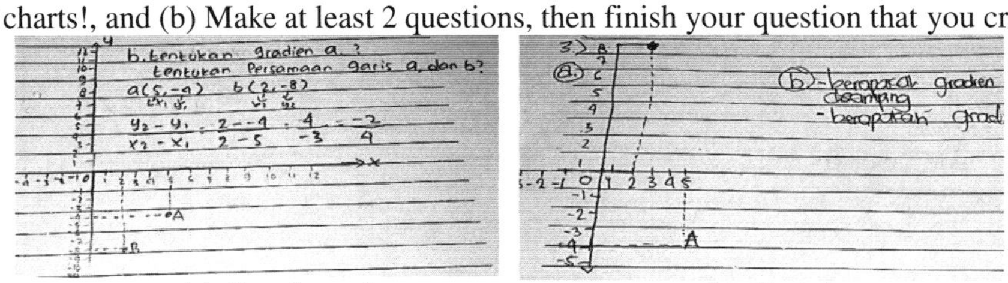

(a) Experimental (b) Control

Figure 3. Answers's Problem Recontruction of Experimental and Control Group

Based on Figure 3(a), Experimental group, students are able to solve the problem that has been made by themselves. Even though the students do not modify the early problem by finding the k-value in the graph. The experimental group is more trained in writing the information which associated situations in the graph given. A similar situation occurred in the group control (Figure 3 (b)), but in the control group, students haven't been able to resolve the problem they create.

Furthermore, the capabilities of problem posing on indicator imitation of the problem, students are given a question to measure the skill to as follows:
Question: "Three straight line l1, l2, and l3 have slope 3, 4, and 5. The three lines through Y-axis in the same point. If the amount of abscissa point through each line with the X-axis is 47/60". (1) Make the problem in form of charts!, (2) Make at least 2 question related the issue by changing the objective problem!, (3) Finish your question that you create by using more than one resolution procedure.

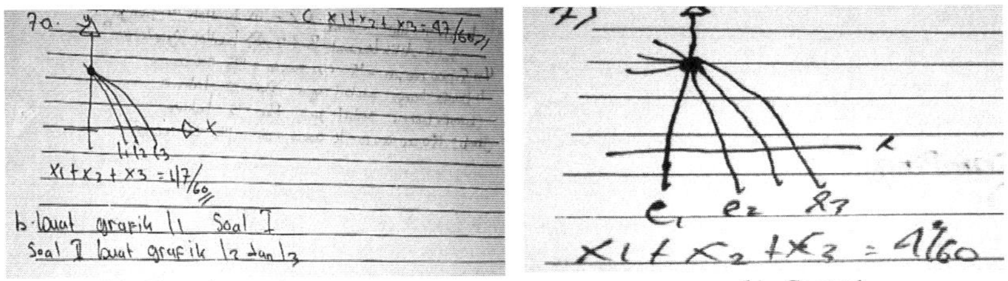

(a) Experimental (b) Control

Figure 4. Answers's Problem Imitation of Experimental and Control Group

4. Conclusion

Based on the results and discussion, it can be drawn conclusion that the students' mathematical problem posing skill who taught by WiPT learning strategy is good enough. Meanwhile, students who are taught strategy conventional still at less category. The gain of students' mathematical problem posing skills, including indicators of reformulating, reconstruction, and imitation of the problem. Students' mathematical problem posing skill on indicators of reformulating a problem, both for conventional and WiPT the learning strategy was already classified as very good in comparison with indicators of reconstruction and imitation of the problem. The high percentage of indicators of reformulating the problem because it is supported by the preparation of sentence that is identical to the earlier problem. Meanwhile, the low percentage in indicator of imitation the problems are caused by the students' difficulties in expanding, enriching, and associating with other materials with early problems.

In general, students' problem posing skill who were taught by WiPT learning strategy is higher than the students who are taught by conventional strategy. The intervention stage of writing in the learning process through the activities of the: writing of knowledge, writing of comprehension, writing of application, writing of analysis, writing of synthesis, and writing of evaluation is the main factor to enhancing the students' mathematical problem posing skill. Referring to the conclusion, it can be suggested that the teachers should have well-understood and could apply the stages of WiPT learning strategy. For further researchers are expected to develop research instruments which are most appropriate to measure imitation problem indicator.

References

[1] Nuh M 2014 *Menjawab Kompleksitas Masa Depan* (Jakarta: Tabloid Asah Asuh Edisi 7 Th 5)

[2] NCTM 1989 *Curriculum and Evalution Standarts for School Mathematics* (Reston Virginia)

[3] Silver EA and Cai J 1996 *Journal for Research in Mathematics Education* **27** pp 521-539

[4] Chen L, Van Dooren W, Chen Q and Verschaffel L 2011 *International Journal of Science and Mathematics Education* **9** pp 919-948

[5] Kulm G 1994 *Mathematics assessment: What works in the classroom?* (San Francisco, CA: Jossey Bass Inc.)

[6] Leung SS and Silver EA 1997 *Mathematics Education Research Journal*, **9** 5-24

[7] Yuan X and Sriraman B 2011 *An exploratory study of relationships between students' creativity and mathematical problem-posing abilities in The elements of creativity and giftedness in mathematics (*Rotterdam, the Netherlands: Sense) pp 5-28

[8] Rosli R, Goldsby D, and Capraro MM 2013 *Asian social science* **9** 54-60

[9] Kar T, et al. 2010 *Procedia social and behavioral sciences,* **2** pp 1577-1583

[10] Stoyanova E 2005 Extending and Exploring Student's Problem Solving via Problem Posing-A Study of Years 8 and 9 Students Involved in Mathematics Challenge and Enrichment Stages of Euler Enrichment Program for Young Australians pp 162-175

[11] Mullis IVS, Martin M O, Foy M, and Arora A 2012 *Trends in international mathematics science study (TIMSS) 2011 international results in mathematics* (Chestnut Hill, MA: TIMSS & PIRLS International Study Center, Lynch School of Education, Boston College)

[12] PISA 2015 *Results Excellence and Equity in Education volume* 1, OECD Publishing 2016/http://dx.doi.org/10.1787/9789264266490-en.
[13] Kennedy D 2007 *Writing and using Learning Outcomes: a Practical Guide* (Bolgna: Quality Promotion Unit) pp. 28-35
[14] Junaedi I 2010 *Jurnal Kreano*, **1** 2010
[15] Lawshe CH 1975 *A Quantitative Approach To Content Validity* (Personnel Psychology, INC)
[16] Suryana Y 2010 *Jurnal Pendidikan Dasar* ***1*** p 5
[17] Esra Arikan E and Unal H 2015 *International Journal of Research in Education and Science (IJRES)* pp 23-30
[18] Killen R 2016 *Effective Teaching Strategies: Lesson from Research and Practice,4th ed*, (Cengage Learning Australia: Thomson Social Science Press) pp 287-297

Acknowledgments
We thank all students and teacher for supporting us to collect the data in this paper. We really appreciate your support.

IOP Conf. Series: Journal of Physics: Conf. Series **948** (2018) 012023 doi:10.1088/1742-6596/948/1/012023

Developing entrepreneurship ability of pre-service mathematics teachers through GSSM

E E Rohaeti[1], M Afrilianto[1], R B Primandhika[1]

[1]STKIP Siliwangi, Jalan Terusan Jenderal Sudirman, Cimahi, Indonesia

E-mail: e2rht@stkipsiliwangi.ac.id

Abstract. This research aimed to describe mathematical entrepreneurship ability of 136 mathematics education students through Gerakan STKIP Siliwangi Mengajar (GSSM) that was conducted in 7 districts (of 17 villages) in West Java. GSSM was a programme that combines devotion to the society and college student internships activity at several schools within three months. The data was obtained through observation towards the activities performed by the students during GSSM. The questionnaire to measure the mathematical entrepreneurship ability of students. The results showed that 1) there were three activities that encourage the mathematical entrepreneurship ability of students; such as tutoring post, teaching practices in school and entrepreneurial activities in society, 2) through those three activities, students can develop their entrepreneurial spirit well and grow creativity, innovation and calculation take risk ability, 3) there was medium-association between student mathematical concept mastery that supports entrepreneurship with their mathematical entrepreneurship ability.

1. Introduction

For a country to be a prosperous country, there should be 2% of the citizens be an entrepreneur [1]. Therefore, the entrepreneurial spirit among the young generation should be raised remembering that the higher number of an entrepreneur in a country will gain the competitiveness of a country as well. A country with many entrepreneurs is certainly has: 1) Big income from tax sector for economic activities from the entrepreneurs 2) An independent economy (not relying on capitalist economic system) 3) Increasing Gross National Product (GNP) because with running a business, one has a chance to have higher earnings [2].

In fact, however, the number of entrepreneurs in Indonesia has only reached 0,18%. This makes a significant difference compared to Singapore that has reached 7%, Malaysia 5%, China 10%, and the United States 13% [2]. This would be a big challenge for the world of education in Indonesia to instill the knowledge and entrepreneurial spirit to the young generation so that there will be more learners that are motivated to improve their self-quality that appears to be creative, innovative, and calculated risk-taker entrepreneurs.

There are some different understandings related to the term "entrepreneur" entrepreneur refers to spirit, skills, attitude, and individual behaviour to handle a business and activities that leads to an effort to find, create, and apply the workings, technology, and new product with improving the efficiency to give a better service or obtaining the bigger profit [3]. Hence, not only referring to the business context but the term "entrepreneur" can also be used to the other field such as social, academic, governance et cetera.

IOP Conf. Series: Journal of Physics: Conf. Series **948** (2018) 012023 doi:10.1088/1742-6596/948/1/012023

STKIP Siliwangi, whose institutional vision is to generate reliable and professional teachers, has already conducted Gerakan STKIP Siliwangi Mengajar (GSSM) that combines devotion to the society and college student internships activity in several districts in West Java. One of GSSM's objectives is to set up the entrepreneurial soul of the educator, including the candidate of educators in mathematics education. This research is consequently aimed to describe the mathematical entrepreneurship ability of 136 students of mathematics education in 9 districts in West Java through GSSM.

Entrepreneurship is a realized value in behavior that becomes a resource, driving force, purpose, tactics, tips, processes, and business result [4]. In addition, entrepreneurship is also a process to apply creativity and innovation to solve a problem and looking for a chance to improve life quality. Mathematics, as the queen and servant of science, cannot be separated from the world of entrepreneurs [5]. The materials in it have the contents that support entrepreneurship; for instance, arithmetic, that contains the learning of profit and loss (P&L), taxes et cetera. Besides that, geometry also helps to calculate the shape of products' packaging. Linear Programming also teaches how to maximise the profit and reduce venture capital and cost. Accordingly, mathematical entrepreneurship in this research refers to entrepreneurial ability with applying mathematic materials within the activities process.

Gerakan STKIP Siliwangi Mengajar (GSSM) is a programme that combines devotion to the society and college student internships activity at some schools. The general ideas of this programme are (a) to prepare the SM3T Programme (*Sarjana Mengajar di daerah Terluar, Terdepan dan Tertinggal*) (b) Innovation that born from the heart/bottom-up (prioritise the and objective result) that tend to be more successful compared to innovation programme of the government (prioritise report and imaging); (c) STKIP Siliwangi, with a motto Innovative Campus: with quality and fitting cost (the innovation is based on optimism built through *istiqama, istikhara* and *istighfa*r developing the culture of action, not discussion) [6]. The achievement targets of GSSM is to create STKIP Siliwangi students and West Java people as learners. Piaget [7] holds that the acquisition of knowledge should be obtained by action and active interaction from students towards the environment.

To obtain the reliable mathematical entrepreneurship skill, the students are also provided with the general entrepreneurship material, planning and development service programme material of mathematics courses, economic mathematics, and financial mathematics.

2. Methods

This is research descriptively depicts mathematical entrepreneurship of 98 mathematics education students in 7 districts (17 villages) in West Java within 3 months along 2017 through Gerakan STKIP Siliwangi Mengajar (GSSM) that combines that combines devotion to the society and college student internships activity at a various level of school. Students in field practice experience are given various material beforehand. Not only the materials that are related to entrepreneurship but also a workshop to create educational devices.

The data are obtained through various students' activities during GSSM, practical teaching test to the students and filling the questionnaire to measure students' mathematical entrepreneurship ability. The data were performed using SPP and Microsoft Excel.

3. Results and Discussion

During three months of GSSM, the students conducted various activities. To measure students' mathematical entrepreneurship ability, however, the research focused on three activities. Those are tutoring at the post, teaching practices in school and students' participation in entrepreneurial activities at society.

3.1. Mathematics Tutoring Activities

Students conduct the tutoring at their post. As seen on Figure 1:

Figure 1. Tutoring at GSSM post with innovation and creativity.

Places to do the learning activities are provided by local people. Sometimes it was away from the post where students live. Students' struggle and determination are needed to reach the place, it could be district office, mosque veranda, and local people's house. Those can be seen in Figure 2 below:

Figure 2. Tutoring outside GSSM post with innovation and creativity.

Remembering that tutoring was free of charge, the participants came with high enthusiasm, even when they are far from the tutoring and the numbers of them are beyond predicted. The students are prepared with Mathematical Entrepreneurship III course to support their skill in organizing the tutoring and its curriculum. In addition, the students have already experienced to take part in teaching at free tutoring held by STKIP Siliwangi for junior to high school students. The knowledge they had were re-applied on this GSSM tutoring activities. The students faced various problems, including:

a. The lack of facilities and infrastructure
b. Age diversity of the participants
c. Diversity of interest towards mathematics lesson
d. Various level of mathematics material mastery
e. Society's support towards the implementation of the tutoring

All those problems require the creativity from the students to resolve the problem that the tutoring activities can be continued connecting with their mastery of mathematics learning method and mathematics material. Based on their mastery towards those two concepts, they tried to creatively think about teaching strategy to overcome the limited facilities and infrastructures, age diversity of the participants and diversity of participants' interest towards mathematics lesson. Some learning innovations are also applied to achieve the maximum results and to show to the public about the benefit of this activity. This was done to change the viewpoint of most society that is not welcoming about this kind of activity to be fully supportive. Students' creative ways and innovation are done through accurate calculation against the risk that will be inflicted. Thus, this tutoring activity explored the spirit of mathematical entrepreneurship well. This is in line with Rohaeti's [8] research declaring that when students were to face explorative situation, they will through three stages. The first one is educating stages, as we've done on Mathematical Entrepreneurship III lesson, where the students appeared to be unfamiliar with exploration and innovation. Then, when students took-part on free tutoring at the campus, they are in trial stage in which students create knowledge understandings and practiced. Until the GSSM held, students are accustomed to innovating and exploring several activities and calculating the risk from creativity and innovation that previously done.

3.2. The Tutoring Practices
From 17 locations of GSSM 2017, 98 undergraduate mathematics education students of STKIP Siliwangi (in their last semester) conducted teaching practices in elementary, middle-high, high (also vocational) school. The distance between the schools and their post are varied and the transportation is sometimes quite problematic.

The obligated number of lesson in school are 10 sessions. However, in fact, the total of meetings is sometimes exceeded, even more than 20 meetings. The obligatory for practicing students are to teach high school level, but again, it appeared that they teach more than merely one level of school. As found in GSSM 2016 [6], on GSSM 2017 are also found several problems:

3.2.1. Diversity of students' characteristic
There were students with harsh and rude intonation, shows no respect to the teacher(s), discipline problems, being lazy on doing the assignment, presence problems, and low mathematics skill. Passive students tend to perceive the information and following on what is being explained and exampled by the teachers. They generally dislike mathematics.

3.2.2. Diversity of school personnel characteristics
There was a nice and friendly teacher(s). However, there are also teacher(s) that shows seniority towards others, so busy and lack in presence that the practicing students receive less guidance. Monotonous and not innovative teaching method is also felt when some of the home teacher only give a note to their students. This causes problems for their understandings towards the material. Some of the practicing students had some difficulties as the school personnel uses very polite (tend to be old-fashioned) Sundanese language in which not familiar.

Figure 3. Learning practises at the school with innovation and creativity.

Understanding of mathematical concepts gives big contributions to create a creativity and innovation in a learning. This founding is according to Muskibin [9] that hypothesized that creativity and innovation is a process to create something involving current existing elements and experiences to be processed in the brain to generates something new.

3.3. Entrepreneurial Activities in Society

There are some entrepreneurial activities in the area where the students participated in GSSM. For example, a small business of *Gurilem* crackers in Desa Taman Jaya Gunung Halu, chicken farm in Ciamis, culinary business in Desa Ciburuy and more. In this activity, the students took part in those businesses. They initially obtained the knowledge and insights of the business that has run by doing some of practices, packaging, and selling. After some time, the students applied their mathematical knowledge in that activity, such as social arithmetic, financial mathematics that they already learned in Mathematical Entrepreneurship II lesson. They give some feedback related to profit and loss, innovation to be done with calculating the risk that possibly happened. Their knowledge in geometry was used when it came to packaging the product. Information and experienced assimilated by the students during the lecture can be utilized to initiate a creativity and innovation towards business in society. The figure depicts these below:

Figure 4. Students participation in entrepreneurial activity in society with innovation and creativity

Participating in entrepreneurial activities in the society means that they directly face the social environment connecting to their entrepreneurial ability. This made the students explored the skill and potential in themselves to arise innovation and creativity, as said by Coleman and Hammen [10] that stated that the factor that supports one's creativity and innovation, besides cognitive skill, is the open nature of internal and external stimulus, independent in thinking and expressing. It also needs independence of thought, unbounded to authority and existing social convention, and most of all, believe in self-ability.

3.4. The Association between Mathematical Concept Mastery with Mathematical Entrepreneurship

The connection between students' Mathematical Concept Mastery with their Mathematical Entrepreneurship is described in the table below:

Table 1. The number of students based on the criteria of mathematical concept mastery and their mathematical entrepreneurship

		Mathematical Entrepreneurship			Total
		High	Medium	Low	
Mathematical Concept Mastery	High	15	7	0	22
	Medium	15	31	22	68
	Low	1	1	6	8
	Total	31	39	28	98

Based on calculation performed on SPPS obtained sig=0,000 < 0,05 meaning that H_0 is rejected so that the conclusion obtained, there is an association between mathematical concept mastery with mathematical entrepreneurship ability. Also obtained C=0,57 C max, meaning that association is categorized as medium.

As shown in Table 1 we can see that between students' mathematical concept mastery with mathematical entrepreneurship ability has a close relationship. This is in line with the results of

Hendriana and Rohaeti's [11] research that declared that there is close relation between hard one's skill (of mathematics) and soft skill (characters). Therefore, those two abilities should be optimally developed and balanced through various of innovative learning with value and character nuances.

4. Conclusion

Based on the results and discussion, it can be concluded that: (a) There are three activities that encourage the mathematical entrepreneurship ability of students; such as tutoring post, teaching practices in school and entrepreneurial activities in society; (b) Through those three activities, students can develop their entrepreneurial spirit well and grow creativity, innovation and calculation take risk ability; (c) There is medium-association between student mathematical concept mastery that supports entrepreneurship with their mathematical entrepreneurship ability.

References

[1] Mc Clelland, DC 1961 *Achieving Society* (Princeton, N J: D Van Nostrand Co)
[2] Nurrokhman H A 2015 *Kewirausahaan* (Bandung: Kompas Cyber Media)
[3] Drucker P E 1985 *Innovation and Entrepreneurship* (New York: McGraw Hill Book)
[4] Sanusi, A 1994 *Strategi Operasional Peningkatan Mutu Wajar 9 Tahun dan Pendidikan Luar Sekolah di Desa Tertinggal* (Bandung: IKIP Bandung)
[5] Zimmerer 1996 *Entrepreneurship and The New Venture Formation* (New Jersey: Prentice International Hall Inc)
[6] Rohaeti, E E 2016, *Issues Formulating of Mathematics Teaching In Remote Area of West Java via "Gerakan STKIP Siliwangi Mengajar"* (Cimahi: STKIP Siliwangi Press)
[7] Suparno, P 1997 *Filsafat Konstruktivisme dalam Pendidikan* (Yogykarta: Kanisius)
[8] Rohaeti, E E 2009 Pembelajaran Melalui Pendekatan Eksplorasi Untuk Mengembangkan Kemampuan Berpikir Kritis Dan Kreatif Matematik Siswa SMP, *Proc. Int. Conf. on Mathematics Education* (Medan: Medan University)
[9] Musbikin, I 2006 *Mendidik Anak Kreatif ala Einstein* (Yogyakarta: Mitra Pustaka)
[10] Yudha, A S 2004 *Berpikir Kreatif Pecahkan Masalah* (Bandung: Kompas Cyber Media)
[11] Hendriana, H and Rohaeti, E E Values and Characters-Nuanced Innovative Learning to Develope Hard Skills and Soft Skills of Junior and Senior High Students' Mathematics, *Proc. Int. Conf. on ISSIMED* (Yogyakarta: Yogyakarta State University)

Affective strategies, attitudes, and a model of speaking performance development for engineering students

S Wijirahayu[1] and P Dorand[2]
[1]English Education Department Graduate School Universitas Muhammadiyah Prof. DR. HAMKA Jl. Buncit Raya 17 Pancoran Jakarta Selatan Indonesia
[2]Akademi Telkom Jakarta, Indonesia

E-mail: sucianawijirahayu@gmail.com

Abstract. Learning English as a Foreign language (EFL) as one of the challenges especially for students majoring in Telecommunication Engineering to develop their communication skill as a professional could be one of the chances for them to face a more global era. Yet, there are important factors that may influence the progress of the speaking performance and attitude is one of them. Therefore, a survey involving two main psychological variables in language learning namely attitude and affective strategies and the third variable is speaking performance was conducted and a model of affective strategies in language learning developing through the application of Content Language Integrated Learning and multimedia instruction was introduced. This study involved 71 sophomore students and two classes of university students majoring in Telecommunication Engineering and Electrical Engineering. The researchers used both survey and action research method with quantitative as well as qualitative in approach.

1. Introduction.

English as a foreign speaking performance can be frustrating to some students. The students who do not have positive perspective towards learning might not activate affective strategy features. However, along with positive attitudes, it is argued that they then would employ various kinds of affective strategies by activating affective strategy features. Assuming that those variables are interlocking each other in the human brain, such theoretical framework confirmed McGoarty in Bagheri and Andi [1] argued that attitudes will lead to greater language success if they are accompanied by effective learning strategies and that allowed the students to take an advantage of instructional opportunities. Strategy training is one of the ways to equip the university students to deal with problems they face in EFL and one-time strategy training could be applied in a short period of learning [2]. The shared effects of students' attitudes and affective strategy should be taken into account of contributing the students' speaking performances. Therefore a model of affective strategies training, in a form of the combination of both variables is expected to overcome the challenges and create autonomous learning that contributes greater effects on the students' speaking performance to an increase of the students' awareness and responsibilities to learn.

2. Methods

Fifty Affective strategy questionnaire items adapted from Oxford's SILL which were also used by previous researchers, such as; Shukri [3], Tabanlioglu [4], Riazi, [5], Wijirahayu [6] and others. Two sets of the questionnaire, than were employed to the 71 respondents after having moved out some invalid and unreliable items. By the help SPSS version 22.0, the data then was interpreted based on its mean score as the quality or level of the students' options. Meanwhile, speaking performance data was collected by a test instead of a questionnaire. The respondents were individually asked to retell a sequential story available in power point presentation. Such speaking test was evaluated and gathered with accumulative scoring by three raters using TOEFL speaking rubrics due to the broad coverage of speaking performance features. Jamieson [7] claimed that such speaking rubric is used to infer students' ability to produce spoken the language in real life situations.

Table 1. Mean Score Level Classifications of Affective Strategies

Quality	Respondents' Options	Mean Score Range
High/positive	Strongly Agree	4.5 to 5.0
	Agree	3.5 to 4.4
Moderate/Neutral	Neutral	2.5 to 3.4
Low/ Negative	Disagree	1.5 to 2.4
	Strongly Disagree	1.0 to 1.4

The t-test at multiple regressions analysis resulted from the finding that the positive effect of students' attitudes partially on students' speaking performance was significant.

3. Result

The t-test revealed that the positive effect of affective strategy partially on students' speaking performance was not significant since its sig. 0.140 is greater than 0.05. In other words, it can be concluded that attitudes determine speaking performance. The contribution effect of affective strategy did not demonstrate its significance unless accompanied by good attitudes. Based on the findings, a model of affective strategies application was developed. Students' progress and response related to speaking performance were reported in a form of both qualitative and quantitative findings. The influence of multimedia in building students attitudes and developing the affective language learning strategies was explored and reported in the further study.

This survey result in the students of Akademi Telkom Jakarta was quite a success as the students were able to manage themselves relax although only a small number of students practiced controlling their heartbeat while speaking. This result confirmed the similar success of a study conducted by Yunus and Singh [8] in which majority of the respondents claimed that they were strong users of affective strategy to lower their anxiety by relaxing when they had to speak English. They also stated that such strategies improved their confidence level and speaking performance.

3.1. Affective Strategy Analysis

The Affective strategies involved some variables and those are anxiety reduction, self-encouragement, and monitoring emotion. Each of the data from those categories was analyzed specifically in the explanation below.

3.1.1 Anxiety Reduction

There are three sub-affective strategies related to anxiety reduction namely using progressive relaxation, the use of music, and laughter or humor that are reviewed below based on the data analysis.

Using Progressive Relaxation. The survey showed that 52.1% claimed that they often visualized something to calm them down, while the rests considered as moderate users of such strategy. Meanwhile for the deep breathing sub-strategies, only 28.2% of the students who agreed to frequently use of controlling heartbeat when speaking English, while the majority was considered as moderate users.

Music. The survey result showed that 50.7% frequently listen to music to make them better at controlling emotions while 39.4% sometimes use this strategy. However, incorporating music into English activities really helped them reduce their anxiety level as the survey showed that 55% had high frequent use to do so.

Laughter. The use of humor or laughter seemed to be the highest frequency used to reduce anxiety as 87.4% agreed with such statement that it could release tension. The use of humor or laughter benefits from releasing anxiety and bring pleasure to the classroom.

3.1.2. Self Encouragement

Three sub-variables of self-encouragement are; making positive statement, taking risks wisely, and administering self-rewards highlighted as follows;

Making the positive statement.

The survey result revealed that the students were responsible enough to their own learning, and they believed in studying English for better future as majority claimed to do best on learning. Moreover, 59.1% encouraged themselves to speak in spite of making mistake.

Taking risks wisely.

Although majority realized that taking part in speaking activities allowed them to have the new learning experience, only 15.5% carried it out, causing the lowest mean score of 2.72. Such phenomenon is likely to occur as it was claimed that it is not easy to be wise risk-taker students since it needs the supportive use of other affective strategies, such as making positive statement or self-reward. Then, it was revealed that "there is a connection between risk-taking ability and self-rewarding sub-strategies." In this case, the low use of self-rewarding is commonly accompanied by the low use of risk-taking substrategy, and vice versa.

Administering self-reward.

It is defined as students appreciation of their own learning or achievement. The majority seemed to think that they appreciated their own performance. However, most of them did not give themselves a reward for doing well in performing language activities. It should be an important finding that the students should appreciate or giving self-reward as their good performance since it is useful to release learning tension and cultivate self-confidence. Similarly, the survey of the study result conducted only 30% respondents agreed to the statement of "I give myself a reward when I do well in English", represented the lowest mean score. In addition, the identical study result and stated that the low score of such statement was due to the practice of self-rewarding has not become a habit among students yet that same goes with the habit of writing the diary.

There have been a few studies conducted affective strategy use as a single variable related to the students speaking performance; therefore, this present research is expected to be a contribution in a field of education. The statistical calculation revealed that there is the positive effect of affective strategy on students' speaking performance. The survey conducted towards the students of Akademi Telkom Jakarta related to the affective strategy use discussed above revealed that the students are classified as high use of affective strategy. Similarly, the study conducted by Yunus and Singh, in which the findings indicated that there was a significantly positive relationship between affective strategy and the students' speaking performance. The students were found to be well used of the affective strategy as to lower their anxiety and control emotions. Of the different angle, Galti et al. [9] shared the same idea that there is a strong relationship between the use of affective strategy and speaking anxiety level. He emphasized that the students experiencing anxiety while performing speaking was mostly caused by the low use of affective strategy.

3.1.3. Monitoring Emotion Sub-variables

Listen to the body.
It seems that English is one of the compulsory subjects the students had to take in each semester, also the demand of English activity emphasizing speaking activities, the majority was found to force themselves too hard on their learning, even if they realized that their physical sign of stress might affect to their performance. As a result, the majority had more potential to have physical tension or stress since they less use of such strategy that might affect their body and mind. In fact, only 18.3% who often noticed if they were tense or nervous when they had to speak English. As Yunus and Singh said that this could be a good sign because the students were ready for oral presentations. Such strategy would help them control their nervousness. Yu argued that most students often ignore what the body says but physical state influences speaking performance as negative feelings tighten the muscles that affect all organs of the body. On the contrary, positive feelings can give a calming effect that enables to control emotions.

Writing diary.
In overall observation, the students seemed to find it is useful to write them down their mind in a language diary. They also realized that writing a diary could relieve tension, and minimize English mistake. In fact, it was contradictive as only 12.7% found it helpful to communicate feeling through their diary writing.

Discussing feeling.
It was the most successful strategy as it had the highest mean score of overall monitoring emotion sub-strategies. Such sub-strategy highlighted that the students like to share their worries both with their teachers and their peers. In fact, students preferred being straightforward discussing their feeling in a form of spoken to written. They also claimed that they felt more relieved to share their feelings, and a majority took some good advice about themselves while sharing their feelings.

Compared to discussing feeling orally, writing a language diary sub-strategy is less popular among the students; therefore, the teachers should familiarize the use of writing diary strategy that enables the students to manage their emotional aspects as well as practice the target language in a form of written work. With the high-frequent use of Affective strategy, the students agreed to deal with their emotional reaction towards learning that indicates to a greater opportunity to use language in forms of spoken. On the other hand, Oxford in Shukri [3] stated that the low frequent use of Affective strategy generally led the students to experience the high-level language anxiety.

3.2. Students' Attitudes Analysis
The survey findings discovered the students' attitudes as the responses of 33 valid questionnaire items within three aspects – behavioral, cognitive, and emotional. The survey result revealed that they had positive attitudes as the mean score range between 3.5 and 4.4, by referring to the mean score classification at table 1. In fact, emotional aspects contributed the highest mean score of 3.66, followed by cognitive 3.62, and behavioral of 3.57 as reported below.

3.2.1. Behavioral Aspects of Attitudes

Three sub-variables were investigated related to behavioral aspects were self-confidence, learning environment, and learning experiences. They are self-confidence, learning environment, and learning experience.

Self-confidence
The survey revealed that 70.5% did not feel embarrassed to speak English in front of the classroom. Also, almost half of the students were relaxed whenever they had to speak in front of the class. Then, such items represented that the students were confident enough to perform their English speaking.

Learning Environment
Although half of the students tended to be neutral, more than 40% agreed to the statement that their peers' English skills motivated them to be better, and 45.1% agreed that they would like to have many English-speaking friends. The students' success of learning is inseparable to their surroundings.

Learning Experiences
The percentage of 42.2% revealed that the students paid attention to the lesson while the rests mostly were neutral. Also, 51.8% of the students asked the homework or lesson taught while missing the class. Due to the importance of learning experiences as a basic thing that forms the attitude, the process of teaching and learning should give good impressions. For a non-English major classroom practice, the variety of prior knowledge and experiences may have a great influence on motivation [10]. Therefore, the classroom activities should be designed to minimize the students' negative experiences in learning.

3.2.2. Cognitive Aspects of Attitudes
Beliefs and Thought.
Half of the students believed that they had more knowledge and more understanding when studying English. The items revealed the majority believed that English is important for their lives, and found to have clear objectives of learning English. In fact, beliefs about language learning and language learning strategies influence not only learners but also lecturers in the EFL classroom practices Wijirahayu [11].

Skills, knowledge or abilities.
The survey disclosed that the students were in doubtful that they could summarize the important points in English lessons since 47.9% chose neutral, although the rests agreed to it. As a matter of fact, 49.3% of the students found it was not hard to apply English lessons in their real lives.

3.2.3. Emotional Aspects of Attitudes
Two sub-variables elicited were feeling-involved learning and learning preferences. Related to feeling-involved learning, the majority showed that they were enjoyable at studying English. Interestingly, 61.9% stated that they were looking forward to English speaking activities. Related to language learning preferences, the survey revealed that 62.2% of studying English better than other foreign languages.

This current survey finding is in line with Eshghinejad's work, which accidentally showed the same results that emotional aspects had the highest rank of mean score followed by cognitive, and behavioral aspects. In his study, Eshghinejad [12] emphasized that positive attitudes are essential factors influencing language performances. The researcher then argued that such condition should be at the advantages as the students who engage in learning activities emotionally would grow the other attitude aspects. Sharing similar study result that the respondents have positive attitudes in overall aspects, it was confirmed that the key to obtaining target language can be achieved if the teachers successfully define the importance of each activity and motivate the students to have learning responsibilities. Furthermore, the other item showed that most of the students felt embarrassed to speak that represented low mean score and negative behavioral aspect towards speaking.

In some other ways, the results of current research conducted to the students of Akademi Telkom Jakarta contrasted to some studies conducted by other previous researchers. For example, the study of Libyan Secondary School Students conducted by Abidin et al [13]. showed negative attitudes in overall aspects with the mean score of 2.61. The mean score of behavioral aspects was 2.57, the cognitive was 2.67, and the emotional mean score was 2.60. It is argued that the behavioral aspects showed the lowest mean score since the participants did not feel relaxed when they had to speak English. In the aspects of emotional, the participants preferred using their mother tongue to the English language. The students had negative attitudes towards their learning experiences. Still, in the same negative dimension of attitudes, the study conducted by Pakcr and Dcvrim [14] entitled "Attitudes and Perceptions of the Students and Instructors towards Testing Speaking communicatively" showed negative result towards their speaking performance that was mostly caused by the lack of preparation and speaking test experiences.

3.3. Speaking Performance Test Result

To highlight, topic development showed the lowest sum and mean score of overall aspects, revealing that some students seemed to find difficulty in providing relevant information by developing key ideas. Accumulation score revealed that 37 of 71 students demonstrated their communication satisfactorily in overall; however, the rests were categorized at level three, indicating that their communication is limited. The general descriptions of speaking rubrics notice that the group of students at level three was actually connected to the task, but some information was missing that obscure the meaning.

Table 2. Students' Speaking Performance Interpretation

Score Range	Delivery		Language Use		Topic _Dev		Accum. Score		Meaning of Score
	Fr	%	Fr	%	Fr	%	Fr	%	
4.50 – 5	5	7.0	4	5.6	2	2.8	4	5.2	Communicate Effectively
3.5 – 4.4	38	53.5	37	52.1	37	52.1	37	52.6	Communicate Satisfactorily
2.5 – 3.4	25	35.2	28	39.4	30	42.2	28	39.0	Communication is limited
1.5 – 2.4	3	4.2	2	2.8	2	2.8	2	3.3	Communication is weak
0 – 1.4	0	0	0	0	0	0	0	0	Need more practice
Total	71	100	71	100	71	100	71	100	

Delivery. It indicates the ability to give the clear speech, pronunciation, and the responses that easy to understand. 5 of 71 achieved the highest score and fulfilled the features of delivery effectively. Next, 38 students were placed at level four, meaning that their use of stress and intonation did not cause problems for the audience or listeners in spite of occasional error pronunciation and hesitancy. Other 25 students at level 3 as they had limited communication with frequent pauses; however, it was still understandable by sympathetic listeners, while the other 3 students had weak communication as they struggled to pronounce that caused difficulty for the listeners.

Language Use. Its components required an ability to demonstrate the control of both grammatical and vocabulary. 4 of 71 deserved to get the highest score at level 5, indicating that they were able to communicate effectively with good range use of grammar and vocabulary. Next, 37 students demonstrated communication satisfactorily with sufficient grammar and vocabulary, also some minor errors that caused difficulty to listeners. Other 28 students were at level three, revealed that they had limited communication in a term of grammar and vocabulary, while the other two students at level two, indicating that they had an inaccuracy of word choices, and only used basic structures with errors that unfortunately categorized them as weak communication.

Topic Development. The components were taken into account of the students presented the ideas coherently, the relationship between ideas was clear and easy to follow. Referring to the table above, 2 of 71 achieved the highest score at level five as they demonstrated their understanding the relationship of key ideas with clear progression and elaboration. Next, there were 37 students at level four as they were able to provide supporting reasons although with minor errors. Other 30 students deserved to get level three as their speaking performance were almost the same features of level four; yet, supporting information was limited with incomplete reasoning. Finally, the other two students were at level two as some key ideas were limited and too much repetition. In addition, some ideas were omitted and lack of coherence, identifying weak communication.

4. Discussion

4.1. Positive Effect of Affective Strategy on Students' Speaking Performance

With correlation value of 75.2%, the relationship between the variables of affective strategy and students' speaking performance is strong that goes to the same positive direction. As the F-test result is significant, the equation model allowed to estimate. This research finding corroborated Oxford in Hurd [15] who stated that affective is one of the biggest effects on success or failure by pointing out that affective and cognitive are interdependent aspects of human brain functioning. Thus, it is clear that affective strategy use becomes one of the factors influencing the students' speaking performance as this research figured out that affective strategy variable can explain as much as 56.50% towards speaking performance. Supported by the significance of t-test result, the hypothetical statements revealed that there is some positive effect of affective strategy on students' speaking performance. Thus, it confirms the study conducted by Xu [16]which figured out that there was a significant positive correlation between affective strategies and IELTS speaking performance test.

From a different angle, Galti [8] shared their study that there was a positive correlation between the students' speaking anxiety level and Affective strategy use during the speaking task. As a result, their study emphasized that the students' anxiety was caused by the lack use of Affective strategy. Such argument implies that the students could have performed better if they had relied upon the use of Affective strategy. In contrast to the study conducted by Galti, the students of Akademi Telkom Jakarta were reported as high users of reducing anxiety strategy. The majority of the students used the technique of visualizing something positive to calm them down. In addition, they relied upon practicing humor and incorporating music while learning. The study conducted by Yunus and Singh showed similar satisfying results that the students were strong users of lowering anxiety strategies. Within her great work of Affective strategy, Xu [16] highlighted in her study of Chinese context, the students used more self-encouragement over other affective strategy sub-strategies.

In fact, this study revealed that the students were feeling-oriented rather than motivational-oriented learners as the survey findings showed that the students relied upon the monitoring emotions rather than self-encouragement sub-strategies. This might be caused by the features of speaking performance activity allow them to share their feelings, either with the teacher or their peers as their preferences technique to release their tension. Indeed, sharing the feelings or showing care of each other are also needed for language learning community as learning environment might determine to learn success or failure. The effect of collaboration and task preference may influence the employment of the strategies and motivation (Wijirahayu [2]). To conclude, it has been proved by many empirical studies stating that there is some effect of affective strategy on students' speaking performance. The correlation value that goes in the same positive direction suggests that by increasing the use of Affective strategy will potentially increase the students' speaking performance score. In addition, the effect of affective strategy on students' speaking performance as the finding of this research is undoubtedly significant.

4.2. Positive Effect of Students' Attitudes on Students' Speaking Performance

The correlation value of 77.9% revealed that the relationship between the variables of students' attitudes and students' speaking performance is strong and goes in the same positive direction. Therefore, the equation model of regression Y = -15.583 + 0.401X1 indicated that the increased value of the students' attitudes will be followed by the increased value of the students' speaking performance. Since speaking performance is a product of the students' attitudes, this finding confirmed that attitude is considered as an essential factor influencing language performance. In line with this finding, the students' attitudes of *Akademi Telkom Jakarta* contributed the effect as much as 60.7% on their speaking performance. Supported by the significance of t-statistic test result, the positive effect of students' attitudes on students' speaking performance is significant. This current research is identical to the study conducted by Bagheri & Andi, which revealed that there was a significant positive correlation between medical students' attitudes towards their English proficiency. They argued that attitudes played a key role in language learning since attitudes determined the learners' behaviors during the learning process. They also revealed that attitude is one of the influencing factors since the students who have a positive attitude are more willing to take part in English speaking activities, while the negative ones are not. In accordance with similar studies and arguments above, it can be concluded that an increase of the positive attitude value of the students will increase the value of their speaking performance score as it is also supported by the significance of equation model. Therefore, it makes sense to say that the positive attitude students have greater achievement in speaking performance as they are willing to practice and use the language as consistent as the features of positive attitude aspects. As a whole, supported by some theories and empirical research, it is clear that there is some positive effect of students' attitudes on their speaking performance.

4.3. The Partial Effect of Students' Attitudes and Affective Strategy on Students' Speaking Performance

The independent t-test result, previously revealed that the partial effect of the students' attitudes on students' speaking performance is significant. It means that the students' attitudes still play great a contribution effect on the students' speaking performance, although the affective strategy is at rest. Thus, well-equipped attitude students will employ any other effective learning strategies they believe to conduct as the responsibility learning. To conclude, cultivating good learning attitudes is the most effective teaching and learning ways as the students who have good positive attitudes will find their own best way to enhance their progress that leads to autonomous learning.

The t-test result showed that the variance of Affective strategy significance value was 0.140 or greater than $\alpha = 0.05$, stating that the partial effect of affective strategy on students' speaking performance is not significant. Such inferential statistical finding implied that, while the attitude is at rest, the effect of affective strategy on students' speaking performance is unacceptable. Thus, it makes sense to say that unless underlying positive attitudes of the students, any learning strategies seem to be less effective. It can be an alternating speaking that the success of affective strategy as explanatory variables did not meet the expectation, while distracted by the variables of students' attitudes in multiple regression analysis. Thus, Dornyei [17] argued that the significant result of a research is not a matter as long as the results show meaningfulness. Furthermore, Sugiyono [18] suggested that identifying some possible factors that caused the variable changing is better than moving the insignificant variable out of the research.

The success of affective strategy use does not always depend on the students' roles; with some benefits lies, teachers should introduce the importance of using affective strategies, such as; progressive relaxation and deep breathing exercises, monitoring emotion, and making positive self-talk. According to MacIntyre and Noels in Barra [19], the affective strategy should be introduced as students benefit from the instruction if they understand the strategy itself. The teachers should help students identify their existing strategies and introduce them new or never-use strategies before. The survey findings revealed that the students seemed to have problems with risk-taking sub strategies use. The ability of risk-taking in language learning is identified as good language learners since it is a part of the learning process as Burgucu et al.[20]. argued that risk-taking ability is defined as being enthusiastic to get new information regardless of embarrassment. In this case, Wijirahayu [21] reported that teachers' beliefs in language learning strategy play important role in introducing suitable strategies to the students.

Therefore, teachers should create such a friendly learning atmosphere that students feel no worries for making mistakes to train the students' risk-taking ability. Minghe and Yuan [22] added that errors should also be allowed in the whole process of speaking performance to help the students relieve their anxiety. The teachers should also remind the students the importance of taking part in speaking classroom, and involve them in speaking activities that increase their learning ability to their performance. A study by Wijirahayu [23] revealed that teachers beliefs in integrating language learning strategies (LLSs) in their instructional plan will influence the way the strategies introduced classroom practices.

4.4.Positive Shared Effect of Students' Attitudes and Affective Strategy on Students' Speaking Performance

The contribution effect of students' attitudes and affective strategy variables on students' speaking performance is 61.90%, while the other 38.10% influenced by other variables out of this research. This shared effect contribution is greater than that of standing alone as a single explanatory variable. With correlation value of 78.70%, the relationship is strong that goes to the same positive direction, meaning that the increase of either students' attitudes or affective strategy will be followed by the increase of the speaking performance score. Since the F-test revealed that the relationship is significant, the equation model also allows estimating.

In connection with this significant result, Sadighi and Zarafshan [24] stated that the students who have positive attitudes will use affective strategy more frequently than those who have negative ones. Such argument emphasized that the higher positive students' attitude value, the more frequent affective strategies they use, resulting in the higher score of their speaking performance. Some extents to illustrate the significant the relationship, one hand, as Baker in Bagheri and Andi [1] shared the same ideas that the students who have good learning attitudes will use any possible strategies, such as; risk-taking, answering questions or volunteering information – as the features of affective strategies – to cope with their own learning. The employment of more variety of language learning strategies Wijirahayu [25], including Affective strategies influences the nature of language learning. Nevertheless, since the attitude is not a permanent state, the use of Affective strategy functions to assist students' mental and behavioral state balance to overcome the features of attitudinal problems such as anxiety, not-willing to speak, low motivation, and self-confidence that block their performance. In addition, McGoarty in Bagheri and Andi stated that unless accompanied by effective strategies that allow the students to take advantage of instructional chances obviously presented to them, even positive attitudes might be insufficient to lead language success. Thus, the presence of affective strategy as the other explanatory variable is in line to support students' attitudes to optimize students' speaking performance. Due to some reasons mentioned above, then the combination of attitudes and affective strategy resulted greater contribution effect on the students' speaking performance. As a whole, some positive shared effect of students' attitudes and affective strategy on students' speaking performance simultaneously is significant.

5. Conclusion

This research provides two main areas of discussions as they are intermingled to support each other. Inferential statistics aims at answering the research questions, while the survey finding discovers important information related to the variables of the research, the characteristics of the sample, and the tendency that emplies to. Besides, both of two main areas of discussions are compared, contrasted, and confirmed to the previous studies conducted by other researchers for the sake of significances

As the affective strategy functions to cope with the emotional state such as anxiety reduction and self-encouragement, it could have played greater effects on speaking performance if accompanied by orientation. However, what determines high or low of affective strategies use depends upon teaching and learning in a certain classroom, the people who deal with it, and individual differences. This might be special characteristics of Affective strategy that is different from other LLSs which can grow more naturally.

Thus, the study conducted in *Akademi Telkom Jakarta* noticed that the insignificances of the research might be caused by the inconsistency of the students' responses due to their unfamiliarity and lack of exposure to certain strategies. Although the survey revealed that the students were high-frequent users of Affective strategy in overall as some features appeared naturally, to move out the students' inconsistency should be done by conducting an orientation or training. A model of Affective strategy combined with suitable multimedia related to the content language integrated learning (CLIL) to reduce the anxiety in increasing autonomous learning could be built to have better development in EFL and part of the development have been shown in this study.

References

[1] Bagheri, Mehri, and Taghi, Andi 2015 *International Electronic Journal* **3** (I) 7–9
[2] Wijirahayu, S 2011 *Journal Penelitian Inovasi dan Perekayasa Pendidikan* **6** (2)
[3] Shukri, Kamarul 2013 *Journal of Islamic and Arabic Education* **5** (1) pp 11-18
[4] Riazi, Abdolmehdi, and Rahimi Mohammad Iranian EFL Learners' Pattern of Language Learning Strategy Use *The Journal of Asia TEFL* **2** (1) pp 103-129
[5] Tabanlioglu, Selime 2003 *Middle East Technical University* pp 21-112
[6] Wijirahayu, S. 2017 *Journal Penelitian Inovasi Pendidikan Dasar* **2** (2) pp 45-52
[7] Jamieson, Joan 2013 Developing Analytic Rating Guides for TOEFL iBT Speaking Tasks *TOEFL iBT Research Report*
[8] Yunus M, Norimah, and Singh Maktiar 2013 *Malaysia Procedia – Social and Behavioral Sciences* **12**(3) pp 204-214
[9] Galti A, Malah, Zanna, Tahiru and Grema A, Mustapha 2016 *International Journal of Multidisciplinary Research and Development* **3**(3) pp 319-322
[10] Wijirahayu, S 2008 *Majalah Mimbar Ilmiah UIJ* **18** (1)
[11] Wijirahayu, S 2009 The Relevance of Beliefs about Language Learning and Language Learning Strategy of University Students at UPI and UHAMKA (Jakarta: Research Report UHAMKA)
[12] Eshghinejad, Shahrzad 2016 *The Research Article of Curriculum and Teaching Studies Cogent Education* pp 1-15
[13] Abidin M J Zainol, Majid, Pour, and Alzwari, Hanan 2012 *Asian Social Science* **8** (2) pp 119-134
[14] Paker Turan and Devrim Hol 2012 *Pamukkale University Journal of Education* **32** (2) pp 13-24
[15] Hurd, Stella 2008 *Affect and Strategy Use in Independent Language Learning Second Language Acquisition The Open University Bristol* (UK: Multilingual Matters) pp 218-236
[16] Xu Jin 2016 *International Journal for 21st Century Education* **3** (2) pp 69-96
[17] Dornyei Zoltan 2003 *Questionnaires in Second Language Research: Construction, Administration, and Processing*. Lawrence Erlbaum Associates (New Jersey: Inc. Publishers)
[18] Sugiyono 2013 *Statistika untuk Penelitian* (Bandung: Alfabeta)
[19] Barra Carmen Gloria 2008 *Short Articles* **4** pp 1-12
[20] Burgucu, Assiye. Han, Turgay and Engin A, Osman 2010 *2nd Int. Symp on Sustainable Development* (Sarajevo)
[21] Wijirahayu, S 2013 *Int. Conf. on Computer in Education (ICCE)* (Bali: Uhamka Press)
[22] Wijirahayu, S 2017 *Proc. Konf. Nasional Pascasarjana PTM ke-5* (Sidoarjo: UM Sidoarjo)
[23] Minghe, Guo and Yuan, Wang 2013 *Higher Education of Social Science* vol. **5** (3) pp 57-61
[24] Sadighi, Firooz, and Zarafshan, Mehrdad 2006 *Journal of Social Sciences & Humanities of Shiraz University* **23** (1) pp 71-80
[25] Wijirahayu, S 2008 *Majalah Ilmiah Widya* **25** 271

IOP Conf. Series: Journal of Physics: Conf. Series **948** (2018) 012025 doi:10.1088/1742-6596/948/1/012025

Contextual approach using VBA learning media to improve students' mathematical displacement and disposition ability

Siti Chotimah[1,2], M Bernard[1,2] and S M Wulandari[2]

[1]Departement of Mathematics Education, Faculty of Mathematics Education, STKIP Siliwangi Jalan Terusan Jendral Sudirman, Cimahi 40526, Indonesia
[2]SMAN 4 Cimahi Jalan Kihapit Barat No. 323, Leuwigajah, Cimahi 40532, Indonesia

Email: chotimahchotie@gmail.com

Abstract. The main problems of the research were the lack of reasoning ability and mathematical disposition of students to the learning of mathematics in high school students in Cimahi – West Java. The lack of mathematical reasoning ability in students was caused by the process of learning. The teachers did not train the students to do the problems of reasoning ability. The students still depended on each other. Sometimes, one of patience teacher was still guiding his students. In addition, the basic ability aspects of students also affected the ability the mathematics skill. Furthermore, the learning process with contextual approach aided by VBA Learning Media (Visual Basic Application for Excel) gave the positive influence to the students' mathematical disposition. The students are directly involved in learning process. The population of the study was all of the high school students in Cimahi. The samples were the students of SMA Negeri 4 Cimahi class XIA and XIB. There were both of tested and non-tested instruments. The test instrument was a description test of mathematical reasoning ability. The non-test instruments were questionnaire-scale attitudes about students' mathematical dispositions. This instrument was used to obtain data about students' mathematical reasoning and disposition of mathematics learning with contextual approach supported by VBA (Visual Basic Application for Excel) and by conventional learning. The data processed in this study was from the post-test score. These scores appeared from both of the experimental class group and the control class group. Then, performing data was processed by using SPSS 22 and Microsoft Excel. The data was analyzed using t-test statistic. The final result of this study concluded the achievement and improvement of reasoning ability and mathematical disposition of students whose learning with contextual approach supported by learning media of VBA (Visual Basic Application for Excel) was better than students who got conventional learning.

1. Introduction

Education has an important role to prepare qualified human resources and to compete in the development of science. The education should be encouraged as well as possible. This can be achieved through a timely education in order to realize the learning objectives, which are implemented in the form of the learning process which is the implementation of the school curriculum through teaching activities. Mathematics is an important part that cannot be separated from human life. In the world of education, mathematics is often encountered in other subjects. Thus, mathematics is called an applied science. One of the goals of learning mathematics is to educate the life of the nation. Things to consider in math lessons are interest and willingness, and hard work in thinking.

According to Mathematics is one of those lessons closely related to real life. No little things and problems around we need the science of mathematics. Depdiknas [3] says, Mathematics has an important role in developing the human mind power so that mathematics becomes the foundation for the development of modern technology. Mathematics equips students to have logical, analytical, systematic, critical, and cooperative skills. Therefore, mathematics lessons should be given to all students for every level of education, from elementary to university level.

NCTM [9] recommends some general-purpose students of mathematics learning: 1) learning about mathematical values, understanding evolution and its role in society and science; 2) Have a high confidence, believe in the ability of mathematical thinking owned and sensitive to the situation and problems; 3) Being a problem solver, being a productive citizen and experienced in solving various problems; 4) Communicating mathematically, learning about symbols, symbols, and mathematical rules; 5) Mathematical reasoning is to make the conjecture, evidence and construct arguments mathematically.

The results of TIMSS and PISA over the past few years indicate that students' mathematics learning outcomes in Indonesia, especially about mathematical reasoning, are still low and are in positions below Malaysia and Singapore [10]. The facts encountered by researchers also showed the results of learning mathematics that is less satisfactory. Through the results of preliminary observations and interviews with mathematics teachers in SMAN 4 Cimahi obtained information that there are still many students who have not met the minimum completeness criteria (KKM) established by the school that is 66.5. Associated with the ability of mathematical reasoning that became the focus of research, researchers also get information from mathematics subject teachers that the ability of mathematical reasoning students SMAN 4 Cimahi also still low.

Based on the problems above, one of the objectives is how to overcome the problems faced by high school students who generally do not solve the problem of mathematics in the material opportunities. Another objective is solving the problem by providing an overview or by using a contextual approach in the form of examples. It improves the reasoning ability of high school students. This approach is very practical, fun and not boring because it focuses more on the real world of students. Thus, in learning the concept of mathematics more easily understood. It is also easier for the teacher to link the material taught to the real world of the students. Depdiknas in Hutagaol [7] says, there are seven main components in learning with contextual approach,

1. Constructivism
2. Questioning (ask)
3. Inquiry (find)
4. Learning community (learning community)
5. Modelling (modelling)
6. Reflection
7. Authentic Assessment

Keep in mind that Visual is more influential than listening and reading, this statement is in line with Hidayatullah [6]. De Porter's research results, humans are able to absorb a material as much as 70% of the work done, 50% of what is heard and seen, while from what he saw only 30%, then he heard only 20%, and from 10% read. Therefore, the basic knowledge of opportunity for high school required props that are able to describe the ICT-assisted mathematics associated with using VBA in Excel aimed at high school students able to make conclusions without having to memorize the formula of opportunity. In addition, the researchers also looked at the mathematical disposition of students.

2. Review of Literature

2.1. Mathematical reasoning abilities

The reasoning is a process or way of thinking of a person in taking a conclusion of knowledge, thinking activities that have certain characteristics to find the truth, and think reason has certain characteristics. Baroody in Bernard [2] says, reasoning is an essential tool for mathematics and the activities of everyday life. it means that students when learning math needs reasoning to spark ideas so that students understand the correct mathematical concepts.

The indicators of mathematical reasoning ability in Soemarmo [13] are: a) Drawing the conclusions of analogies, generalizations, and construct conjectures; B) Draw logical conclusions based on inference rules, check the validity of arguments, and construct valid arguments; C) Establish direct proof, and by mathematical induction. Where items a and b are reasoning capabilities by inductive step and item c is deductive reasoning ability. So with the learning of reasoning ability, students can be able to solve problems without having to memorize the mathematical concepts that become the students' burden.

2.2. Mathematical Disposition
According to Hendriana & Sumarmo [4], mathematical dispositions are attitudes that show: a) confidence; b) flexible; c) persistent, tenacious performing mathematical tasks; d) interest, curiosity; e) monitor, reflect on their own appearance and reasoning; f) passion for learning mathematics; g) implements mathematics to other situations; h) appreciate the role of mathematics; i) rape and metacognition; j) share with others. Polking in Herlina [5] argues that mathematical dispositions show: (1) confidence in using mathematics, solving problems; (2) flexibility in investigating mathematical ideas and seeking alternative methods of problem-solving; (3) diligently doing math tasks; (4) interest, curiosity in mathematics; (5) tend to monitor; (6) assessing mathematical applications to other situations; (7) appreciation (appreciation).

2.3. Contextual Approach
The contextual approach according to Trianto in Nuriadin [10] the main component of the contextual approach is constructivism; questioning, finding (inquiry), learning community, modeling, reflection, and actual assessment. The indicators can be used for teaching and learning in the classroom. The system of contextual learning includes the following eight [1] components: (1) meaningful associations (2) meaningful work (3) self-regulated learning (4) Working together (5) Critical and creative thinking (6) individuals to grow (7) reaching high standard (8) authentic assessment.

The principles of contextual learning according to Sanjaya [12] are (1) Constructivism, constructivism is the process of building or composing new knowledge in the cognitive structure of students based on experience, application of the principle of constructivism in contextual learning approach, students are encouraged to be able to construct their own knowledge through real experience; (2) Inquiry, the second principle in the contextual approach is the inquiry. that is, the learning process is based on the search and discovery through the process of thinking systematically. knowledge is not a number of facts the result of remembering, but the result of the process of finding itself; (3) Questioning, learning is essentially asking questions and answering questions. questioning can be viewed as a reflection of every individual's curiosity while answering questions reflects one's ability to think; (4) community learning, problem cannot be solved alone but requires the help of others. the concept of a learning community in a contextual approach suggests that learning outcomes be obtained through cooperation with others; (5) modelling (modelling), referred to as the modelling principle is the learning process by demonstrating something as an example that every student can imitate; (6) reflection is the process of settling the learned experience done by reordering the events or learning events that have been passed; (7) authentic assessment the process by which teachers collect information about student learning progress. authentic assessment is done in an integrated manner with the learning process. this assessment is done continuously during the learning activities take place. therefore, the pressure is directed to the learning process rather than to the learning outcomes.

2.4. VBA Helped Language Learning (Visual Basic Application for Excel)
Visual Basic Application For Excel is a programming language that gives commands needed in Microsoft Excel to speed up the operation automatically. In accordance with Winarno [15] that VBA (Visual Basic Application) or macros are functions and program commands in MS Office (including Excel) stored in the old Visual Basic, or Visual Basic before the .NET framework version is present. With VBA, a work in Office can be optimized. Microsoft Excel is generally used by most people in numerical processing because of the many mathematical functions that can be associated with Statistics, Economics, Engineering and more. But not limited to that alone, the fact that VBA in Excel can be utilized more usefulness. As for the usage of the props making using VBA in Excel first, the drawing

can be an interactive image, the students become active and fun to learn math because the students are more understanding than without using ICT, the second is not much big cost to make the props, And thirdly, that the use of Microsoft Excel Software is accessible for computer users.

3. Experimental Method

This study was designed in the form of experimental design with pre-test-post-test control group design that aims to examine the role of learning with the contextual approach to reasoning ability and mathematical disposition of high school students. The sample is randomly assigned to the experiment class and control class. The students' mathematical reasoning test is structured to refer to the characteristics of mathematical reasoning ability and good test preparation guidelines. Similarly, the preparation of students' mathematical disposition instruments is prepared on the basis of indicators on the student's mathematical disposition. The data will be analyzed by using statistical test t and test with statistic $\chi 2$ (for association test between variables).

Before the treatment is done, students are given an initial test of students' mathematical reasoning abilities. Furthermore, learning in the experimental and control classes begin at the end of the treatment will be done the final test to see the development of students' mathematical reasoning abilities. Then to find out the mathematical disposition of students conducted mathematical disposition tests of students by giving a questionnaire of the mathematical disposition of students in both classes. Thus the research design is as follows [11]:

A : 0 X 0
A : 0 0

Where:
A : Classroom random sampling
0 : Pretest = Posttest the ability of mathematical reasoning
X : Learning with Contextual Approach

The populations in this study were all of the high school students of class XI in Cimahi – West Java. The sample subjects were selected by one school at random based on school qualification. Of the three levels of existing classes in high school in Cimahi that is class X, XI, and XII then selected class XI with consideration in the first semester there is a subject of opportunity that will be used in research. Sampling in this study was randomly assigned to a class, in a random sampling technique of each sampling unit as a population element obtaining equal opportunity to be a sample or to represent a population.

The data in this study was collected through the mathematical reasoning skills test, the scale of mathematical disposition of the students. Data on mathematical reasoning ability were collected through pre-test and post-test, mathematical disposition data of the students was collected by dispersion of mathematical disposition scale after treatment. For the test data of the students' mathematical disposition the two groups were processed with the help of Microsoft Excel software. Furthermore, the test results data (preliminary and final tests) of both groups were processed using Microsoft Excel and SPSS 22 software assistance with the following steps:

1) Testing the normality of sample data.
2) Test the Homogeneity of Variance.
3) Average Difference Test.
4) N-Gain Test
5) Chi Square Test and Contingency Coefficient.

4. Results and Discussion

At the beginning of the study, the researchers conducted a second class observation which is a sample based on the average of classes in four classes XI and taken 2 classes. They got pre-test before doing learning and teaching activities.

Table 1. Description of Control and Experiment Classes during Pre-test

Class	Average (X)	Standard Deviation (S)
Control	4,49	0,78
Experiment	4,26	0,84

Maximum Score Index = 20

From the pre-test result, it is found that the control grade 4.49 and the experimental class 4.26 shows that the control class is higher than the experimental class and for the standard deviation that the control class with the value of 0.78 and the experimental class with the value of 0.84 means That the scattering of student grades in the experimental class is more diffuse against the average grade than the control class.

But it can be seen from Table 1 that the mean values of the two classes are not too far away, and the small average values will be used as experimental classes and other classes are the control classes. It gives the information to make a decision to determine whether there is a difference in average or not.

Table 2. Class Control and Experiment Normality Test

	Class	Kolmogorov-Smirnov[a]			Shapiro-Wilk		
		Statistic	Df	Sig.	Statistic	Df	Sig.
Value	Control	,306	41	,000	,828	41	,000
	Experiment	,260	41	,000	,867	41	,000

[a] Lilliefors Significance Correction

The first stage, the normality test of whether the two classes are normal data or not, seen from the value of sig Control 0.000 and sig experiments 0,000, the value of sig taken from the Kolmogorov-Smirnov normality test [11]. It turns out the two classes are not normal because the sig value is less than 0.05. This means that both data is continued with the test difference of nonparametric mean that Mann - Whitney test.

Table 3. Pre-test Mean Difference Test

Avg. Test	Value
Mann-Whitney U	698,000
Wilcoxon W	1559,000
Z	-1,415
Asymp. Sig. (2-tailed)	,157

Obtained a Mann-Whitney 698,000 value and a two-sig value to see the presence or absence of average difference and obtained sig0.157 value means greater than 0.05 means there is no average difference between the control class and experimental class during pre-test or test early. After that, they continue the activities by doing the post-test or final test after learning and teaching in class control and experiment class. From these results can be from the average grade and standard deviation of the two classes.

Table 4. Description of Control and Experiment Classes during Post-test

Class	Average ($\bar{X}$)	Standard Deviation(S)
Control	11,71	2,59
Exsperiment	12,85	2,81

Maximum Score Index = 20

The Post-test shows that both classes have an increase in the average value of post-test. It means that there is an increase in the results of each student at the time before doing student learning compared to after learning. For the control class having an average grade of 11.71 and for the experimental class of 12.85 means that the average value of the experimental class is higher than the average value of the control class and the experimental class. And for the standard deviation for the control class 2.59 and the experimental class 2.81 means that each student's data for the experimental class is greater than the control class against the average grade of each class. It can be said that the spread of control class data is closer to the control class average than the experimental class. And to see the decision whether the experimental class that is learning mathematics with Visual Basic For Excel is better than the class that uses the usual way.

The first stage is the second-class normality test using Kolmogorov-Smirnov and adapted sig value for the control class 0,015 and sig for the experimental 0,000 class means that the sig value is less than 0.05 or it.

Table 5. Normality Test of Post Control Class and Experiment

Class		Kolmogorov-Smirnov[a]			Shapiro-Wilk		
Value		**Statistic**	**df**	**Sig.**	**Statistic**	**df**	**Sig.**
	Postest Control Class	,155	41	,015	,955	41	,101
	Postest Experimental Class	,195	41	,000	,894	41	,001

It can be said that the control class and the experimental class are abnormal data. If the second is not normal to the second stage, test the difference in the mean of two nonparametric classes using Mann-Whitney test.

Table 6. Average Test of Experiment Class and Post-test Control

Avg. Test	Value
Mann-Whitney U	613,500
Wilcoxon W	1474,500
Z	-2,131
Asymp. Sig. (2-tailed)	,033

The table shows that if there is an average difference between the experiment class and the post-test controls where the Mann-Whitney test value is 613,500 with the sig value. 0.033. Because to see that the experimental class is better than the control class it is done by a one-party test by means of a value $sig\frac{0,033}{2} = 0,0165 < 0,05$ [14]. It means that there is an average difference between the experimental class and the control class and it is also known that the average of the experimental class is higher than that of the control class means that the learning class uses ICT to improve the ability of reasoning High school students are better than the class that is learning in the usual way.

As for the improvement of students for each class can be seen from the value of gain useful for the improvement of student work before learning or pre-test with after learning or post-test.

Table 7. Description of control and experiment gains during Pre-test

Class	Average ($\bar{X}$)	Standard Deviation (S)
Control	0,47	0,15
Exsperiment	0,54	0,18

From the average gain value of the control class 0.47 while the experimental class of 0.54 means that the gain value for the experimental class is higher than the control class although the standard deviation of the control class is smaller than the experimental class means that the gain value for the control class is more Closer to the mean than the experimental class.

Table 8. Test of gain class Normality of Experiment Class and Control Class

Class		Kolmogorov-Smirnov[a]			Shapiro-Wilk		
Value		Statistic	df	Sig.	Statistic	df	Sig.
	Postest Control Class	,104	41	,200*	,966	41	,260
	Postest Experimental Class	,157	41	,012	,934	41	,020

Normality test using Kolmogorov-Smirnov test, where for sig value control class 0,200> 0,05 mean data of control class is normal data. While the experimental sig of 0.012 <0.05 means that the experimental class data is not normal. Since there is one abnormal then proceed with Mann-Whitney Test.

Table 9. Difference Test Average gain of Control and Experiment Class

Avg. Test	Value
Mann-Whitney U	608,000
Wilcoxon W	1469,000
Z	-2,161
Asymp. Sig. (2-tailed)	,031

Mann-Whitney 608,000 test value and sig value 0.031, to see better the Mann-Whitney test for sig 0.031/ 2 = 0.0155 <0.05 means that the gain for the experimental class is better than the control class.

Table 10. Ability of reasoning and disposition

		Disposition			
		Low	Medium	High	**Total**
Experiment	Low	1	7	0	8
	Medium	8	14	7	29
	High	0	3	1	4
Total		9	24	8	41

For reasoning abilities, there is 1 student who has low disposition questionnaire and low reasoning ability also, for medium questionnaire and disposition ability there are 14 students possibility for the group is no relationship and there is one student who has high disposition questionnaire and high reasoning ability. But there is also a low reasoning capability but this moderate disposition questionnaire is not consistent that students with low reasoning abilities are also in a low disposition, as well as a group of high reasoning abilities there are 3 students who state that the moderate disposition questionnaire is high.

IOP Conf. Series: Journal of Physics: Conf. Series **948** (2018) 012025 doi:10.1088/1742-6596/948/1/012025

Table 11. Chi-Square test for reasoning and disposition reasoning associations

Test of the Association	Value	df	Asymp. Sig. (2-sided)
Pearson Chi-Square	5,417[a]	4	,247
Likelihood Ratio	7,722	4	,102
Linear-by-Linear Association	,735	1	,391
N of Valid Cases	41		

a. 6 cells (66.7%) have expected count less than 5. The minimum expected count is 78
b. The result of the association between reasoning ability and disposition can be seen from Pearson Chi-Square test has value 5.417 with sig value of two parties 0.247> 0.05 meaning there is no association between reasoning ability and disposition.
c. During classroom surveys, mathematics class XI classes on Opportunities and Statistics. The difficulties encountered by students are about understanding, calculating and applying into examples of opportunities such as enumeration, permutations, and combinations. With the problems faced by students, it is necessary is to improve students' reasoning abilities between the material opportunities in the example of daily life that can be analogous and provide some examples to get conclusions as well as from some sense to prove the notion of opportunities with the ICT-assisted contextual approach as a tool of mathematics Opportunities. The software used to create props is Microsoft Excel with the help of Visual Basic Application for Excel.

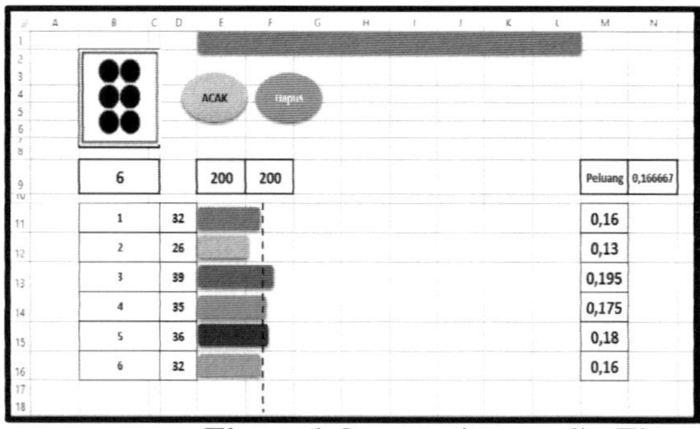

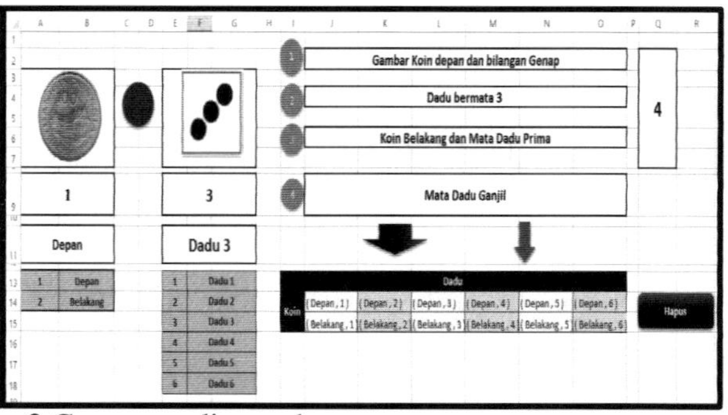

Figure 1. Opportunity one dice**Figure 2.**Game one dice and currency

Students are given a sense of opportunity with some basic examples such as selecting a sample of the eyes of the dice mentioned by the student towards the whole which will be randomized several times and the student will make the conclusions of the ICT-assisted props.

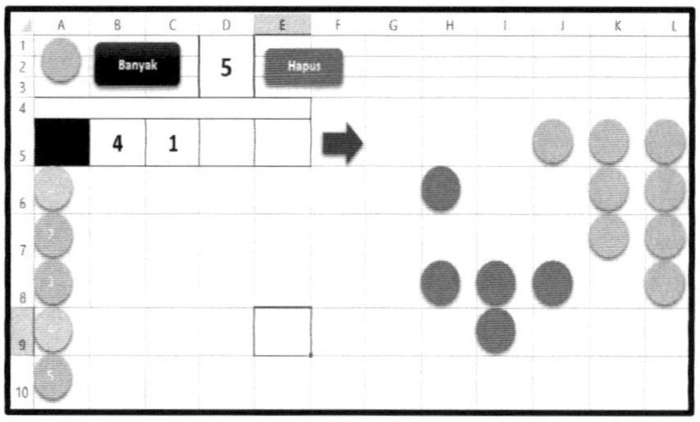

Figure 3.Combination and permutation with the concept of a worm and

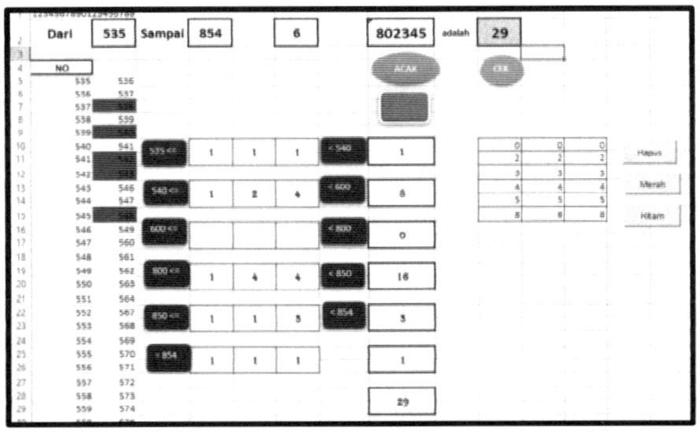

Figure 4.The game determines the number of numbers to explain permutations and combinations.

Students' understanding of the meaning of combinations and permutations is given some sample images of the button concept so that students blindly declare to distinguish combinations and permutations after which the students will be given an opportunity to try the truth and then given an application game of combinations and permutations by determining many numbers.

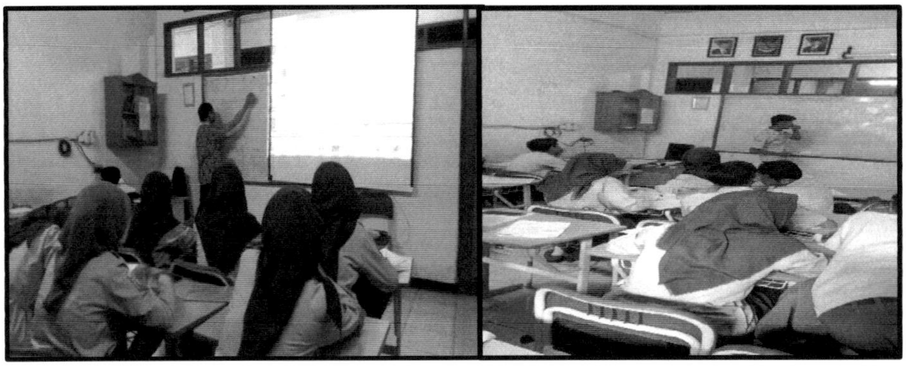

Figure 5. Teacher's explanation of opportunities using ICT

5. Conclusion

Based on the results of the discussion it can be concluded that Contextual Approach aided Application of Excel VBA Learning Media can Improve the Ability of Reasoning and Math Disposition of High School Students.

Mathematics learning does not always explain using ICT. ICT is only a tool to provide examples of drawing props and is explained on the basis of student conclusions from the observation of a sample image from ICT. This gives students more active and fun learning math.

References

[1] Baker Elaine DeLott 2009 Contextualized Teaching and Learning: A Faculty Primer (California:
[2] Benard M 2015 *Journal of Infinity* **4** (2) 197-222.
[3] Chotimah Siti 2015 *Didaktik* **9** (1) 26-32
[4] Hendriana H, Soemarmo U 2014 *Assessment of Mathematics Learning* (Bandung: Refika Aditama)
[5] Herlina S, Turmudi, Dahlan JA 2012 *Journal of Teaching MIPA* **17** (1) 1-7.
[6] Hidayatullah P, Akbar A M, Rahim Z 2008 *Making Education Animation Using Flash* (Bandung: Informatika Bandung)
[7] Hutagaol, K 2013 *Jurnal Infinity* **2** (1) 85-99.
[8] Hutajulu M 2010 *Jurnal IlmuPendidikan (JIP) STKIP Kusuma Negara* **3** (2) 45-49
[9] Mullis et al 2007 *TIMSS 2007 International Mathematics Report International Study CenterLynch School of Education Boston College.*

[10] Nuriadin I 2015 *Journal of Infinity* **4** (2) 168-181
[11] Ruseffendi E T 2010 *Fundamentals of Education Research and Other Non-Eksakta Fields.* (Bandung: Tarsito)
[12] Sanjaya W 2006 *Strategi Pembelajaran Berorientasi Standar Proses Pendidikan* (Jakarta: Kencana)
[13] Soemarmo U 2012 *Learning Materials Courses Mathematical Thinking Process* (Bandung: Unpublished)
[14] Uyanto S S 2009 *Guidelines for Data Analysis with SPSS* (Yogyakarta: Graha Ilmu)
[15] Winarno E, Zaki A 2014 *VBA Programming For Excel All Version* (Jakarta: Elex Media Komputindo)

Cloning of gene-encoded stem bromelain on system coming from *Pichia pastoris* as therapeutic protein candidate

Y Yusuf[1] and W Hidayati[1]

[1]Universitas Muhammadiyah Prof. DR. HAMKA, Jakarta, Indonesia

E-mail: yusnidaryusuf@uhamka.ac.id

Abstract. The process of identifying bacterial recombination using PCR, and restriction, and then sequencing process was done after identifying the bacteria. This research aimed to get a yeast cell of *Pichia pastoris* which has an encoder gene of stem bromelain enzyme. The production of recombinant stem bromelain enzymes using yeast cells of *P. pastoris* can produce pure bromelain rod enzymes and have the same conformation with the enzyme's conformation in pineapple plants. This recombinant stem bromelain enzyme can be used as a therapeutic protein in inflammatory, cancer and degenerative diseases. This study was an early stage of a step series to obtain bromelain rod protein derived from pineapple made with genetic engineering techniques. This research was started by isolating the RNA of pineapple stem which was continued with constructing cDNA using reserve transcriptase-PCR technique (RT-PCR), doing the amplification of bromelain enzyme encoder gene with PCR technique using a specific premiere couple which was designed. The process was continued by cloning into bacterium cells of *Escherichia coli*. A vector which brought the encoder gene of stem bromelain enzyme was inserted into the yeast cell of *P. pastoris* and was continued by identifying the yeast cell of *P. pastoris* which brought the encoder gene of stem bromelain enzyme. The research has not found enzyme gene of stem bromelain in yeast cell of *P. pastoris* yet. The next step is repeating the process by buying new reagent; RNase inhibitor, and buying liquid nitrogen.

1. Introduction

The use of protein as the disease therapy has rapidly grown for decades [9].That therapy is well known as therapeutic protein. The protein used as the therapeutic protein must have the high purity level. The enzyme, commonly protease, is a protein group which has the high purity level which will be used in both industry and therapy. More than 60% enzyme used in both industry and therapy is protease enzyme [7].

Bromelain is one of the protease enzymes which can be found on pineapple plant (*Ananas comosus*). Pineapple has been used to cure the digestion problem and reduce inflammatory [5]. In pineapple, this enzyme can be found in all parts. In the 1800s, bromelain enzyme was successful to be isolated from the stem and pollen of pineapple. There are two kinds of bromelain enzymes; stem bromelain with commercial code EC 3.4.22.32 and fruit bromelain with commercial code EC> 3.4.22.33.

In the food industry, bromelain enzyme is mostly used to make meat soft, in the food industry, cosmetics, skin, textile, and health field [10]. In the health field, this enzyme has many benefits as a therapeutic protein. In 1957, bromelain enzyme was firstly known as therapeutic compound [5]. In clinic application, this enzyme is used to heal some diseases, but the most effective effect is used to reduce the inflammatory because of infection, sinusitis, osteoarthritis, and cancer [8]. Bromelain enzyme which comes from rough extract is still mixed with protein, enzyme, and carbohydrates. This enzyme must have high purity level so that it can be used as therapeutic protein. This enzyme could be separated from those components with using some stages and methods; centrifugation, precipitation, ultra-filtration, electrophoresis, chromatography method and other procedures [4]. That conventional method caused the production and formulation of bromelain enzymes very difficult, it was almost impossible to be applied. It was often mixed with other compounds and the cost was expensive. Besides that, the process of downstream became one of the main factors the most expensive; more than 50%.

Along with the technological and scientific advancement and the existence of genetic modification technology, the enzyme production could be done in out of its original organism. Recombined enzyme production could be done by isolating that enzyme encoder gene and then being inserted into a plasmid and then being inserted into the host cell which was continually stimulated to produce that enzyme [2]. Some proteolytic enzymes which have been developed with genetic modification technology are papain enzyme using *Pichia pastoris* cell [3]. Genetic modification technology has successfully got recombinant bromelain protein which is fruit bromelain applied on a transgenic plant [6], and stem bromelain coming from *E. Coli* bacterium cell [1].

Currently, host cells that can produce enzymes or proteins have been developed extracellular. One is the *Pichia pastoris* cell. Extracellular proteins will make the process of isolating and purifying proteins easier and it can minimize protein degradation. In addition, *P. pastoris* cells are eukaryotic cells so it is expected that bromelain enzyme confirmation will not change due to the same protein synthesis process with pineapple plants.

There are many benefits to be gained from bromelain enzymes in the industry and health. Therefore, it is necessary to find a cheap and easy method to get this enzyme in a clean condition. Genetic modification technology can produce pure bromelain enzymes with cheap budget and easy methods when on a large scale. Because of the similarity of cell structure and protein synthesis processes that occur in *P. pastoris* and pineapple plants, bromelain enzymes can be produced with genetic modification in *P. pastoris* cells. Production of recombinant stem bromelain enzymes using *P. pastoris* yeast cells can produce pure bromelain rod enzymes and have the same conformation as conformation of enzymes in pineapple plants. This recombinant stem bromelain enzyme can be used as a therapeutic protein in diseases due to inflammation, cancer, and degeneration. This study was an initial step in several steps to obtain bromelain stem protein derived from pineapple made with genetic modification

2. Methods

Genetic modification technology, which is well known as DNA recombination, is a method used to combine some genes to make a component which will be inserted into the host cell which has been modified. This development technology gives the big change in protein and enzyme industry, both in industry and health field. This technology consists of some components; a wanted gene, vector/plasmid, DNA enzyme of polymerase, restriction enzyme, ligase enzyme, the host cell. Briefly, the main work of this technology is isolating the wanted gen from DNA molecule of the root organism, multiplying the wanted gene with in vitro using polymerase chain reaction technique (PCR), the wanted gene is cut and continued by combining those genes into a vector, combined vectors and genes are inserted into the host cell with transformation method [2].

Materials used in this experiment were pineapple, RNA isolation kit from pineapple, *Escherichia coli* enzyme for the first cloning, *Komagataella pastoris* yeast cell for yeast cell transformation, reverse transcriptase enzyme, agarose powder, loading dye, DNA ladder, Tris-Base, glacial acetic acid, EDTA, the primer specific forward and reverse, primer to construct cDNA forward and reverse, DNA enzyme of polymerase, an enzyme of T4 DNA ligase, restriction enzyme of BamH I and Xho I, DNA isolation kit, DNA purification kit from agarose gel, restriction enzyme to make recombined plasmid linear, ampicillin antibiotic, zeocin antibiotic, tryptone, peptone, yeast extract, yeast nitrogen base, biotin, jelly, glucose, dextrose, methanol.

Equipment used in this research was tools to have separation process of DNA and protein, the process of protein transfer, gene amplification, cloning process, and transformation of yeast *P. Pastoris*, protein extraction, and visualization of DNA fragment. To DNA separation used electrophoresis of agarose gel. To separate protein used Mini Protean (Bio-Rad). To protein transfer process from acrylamide gel to nitrocellulose membrane used semi-dry trans blot (BIO-RAD). To gene amplification used thermocycler PCR machine (Applied Biosystem) and Gradient thermocycler (Bio-Rad). To clone used the water bath and shaker incubator. To transform yeast of *P. pastoris* used electroporator EC100 (Thermo). To visualize DNA used a machine (Bio-Rad) with the software of quantity one. To extract the protein used vortex and water bat. The routine equipment which was used was centrifuged, micropipette, and tips to various volumes.

This research used pineapple, expression vector on *P. pastoris* cell, *Escherichia coli* bacterium, and yeast cell of *Komagataella pastoris*. To isolate pineapple RNA used a kit to isolate RNA. To make cDNA used *reverse transcriptase* enzyme, and to gene amplification was used with polymerase enzyme. To clone an expression vector, a restriction enzyme is used to limit the genes and vectors and ligase enzyme to ligase process.

3. Results and Discussion

3.1 Isolation of pineapple RNA

Isolation of pineapple RNA was used to get pineapple RNA molecule where pineapple RNA molecule kept an informative data of bromelain enzyme gene of pineapple stem. In this experiment, stem (head) of pineapple which was cut and measured for 100mg then was soaked in liquid nitrogen. Liquid nitrogen was kept in the special tube (See Figure 1).

Figure1. A tube to save liquid nitrogen

This submersion was done to make the hard sample in short term so that it could prevent the damage in next process. The final result which was gotten from pineapple RNA extraction which was RNA in the aquadest liquid which was free from RNase and DNase. The result of this step was not pictured because RNA could not be seen visually.

3. 2 Pineapple cDNA Construction

After obtaining the RNA molecules, then it was continued to synthesis process of cDNA and DNA using kit SuperScript III One-Step RT-PCR System with Platinum Taq DNA Polymerase (Invitrogen, USA). The construction of cDNA and DNA synthesis of specific bromelain enzyme on pineapple stem were done in one step only without stopping when RT-PCR on PCR machine process was running. The PCR machine can be seen in Figure 2. Because of the use of primary molecules (reverse and forward), this process produces DNA molecules of bromelain enzymes which designed by bromelain enzyme gene data residing within the bank genes. The result of this step was specific DNA molecules of bromelain enzyme whose successful process could be seen from electrophoresis process of agarose gel (see Figure 3).

Figure 2. PCR Machine

Based on the Figure-3B, it was known that DNA synthesis process of bromelain enzyme had not been successful to be synthesized. The occurred molecule on agarose gel was a primer reacting one another which signed that molecule RNA did not exist in RT-PCR reaction. This may have happened because the RNA degradation by RNase enzyme which could destroy RNA molecule happened. Kit, used for DNA synthesis, did not provide RNase inhibitor which could prevent RNase activity when construction process of cDNA and DNA synthesis was running.

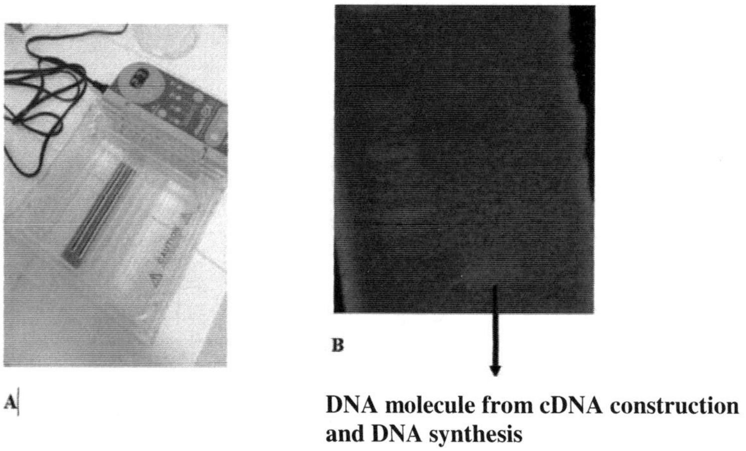

Figure 3. DNA molecule of cDNA construction and DNA synthesis

4. Conclusion

Until now, the research has not found enzyme gene of stem bromelain yet. The next step is repeating the process by buying new reagent; RNase inhibitor, and buying liquid nitrogen.

References

[1] Amid A, Ismail N, Yusof F, Salleh HM 2011 *Process Biochem* **46** 2232–2239
[2] Brown TA 1996 *Gene cloning an introduction 3rd edition* (London: Chapman & Hall)
[3] Dufour E, Storer A.C, Ménard R 1995 *Biochemistry.* **34** (50): 16382–16388
[4] Gautam SS, Mishra SK, Dash V, Goyal AK, Rath G 2010 *Thailand Journal of Pharmacy Science.* **34** 67-76
[5] Gregory S, Kelly N D 1996 *Alt. Med.Rev.* **1**(4) 243-257
[6] Jung Y, Choi C, Park J, Kang H, Choi J, Nou I, et al 2008 *Electronic Journal of Biotechnology* **1**(1) 1–9

[7] Lucia F, Tomas GV 2011 *Food and Bioprocess Technology* **4** (6) 1066-1088
[8] Maurer HR 2001 *Cell. Mol. Life Sci* **58** 1234-124
[9] Paul C 2004 *Protein purification Protocols* (New Jersey: Human Press)
[10] Tochi B. N, Wang Z, Xu S Y, Zhang W 2008 *Pakistan Journal of Nutrition* **7**(4) 513–520

Acknowledgments
Our gratitude to the Rector of UHAMKA and Ristekdikti who has provided assistance in the form of research fund may be useful for the development of science.

IOP Conf. Series: Journal of Physics: Conf. Series **948** (2018) 012027 doi:10.1088/1742-6596/948/1/012027

Laminar flow rate analysis on plate with stokes theorem

D Amirudin[1] and T I Hartini[1]

[1]Program Studi Pendidikan Fisika, Universitas Muhammadiyah Prof. DR. HAMKA, Jakarta, Indonesia

E-mail: Amirudin.dessy@yahoo.com

Abstract. This experiment as conducted to analyze velocity of the laminar flow on a plate by using Stokes theorem. The sample used was honey and oil which has a density of 643,4gr / cm³ and 762,2 gr / cm³. When the density of the two samples have known, it will be easier to obtain viscosity with compared both sample and test samples each having results as 0,00594 Ns/m² for honey and 0,00503 for the oil. The results of these calculations will be obtained pressure for each samples treated with different speeds the flow of each sample was different. The plate which flowing honey has a velocity 0,046 m /s² and velocity 0,002 for oil with spiral plate, while for honey and oil in zigzag plate have both of velocity 0,02 m /s²and 0,0058 m /s².

1. Introduction

In the middle of the development of science and technology, the study of the modification boundary layer very beneficial for development. Research modification of plate boundary layer directed to reduce the negative effects to fluid flow [1]. Usually, areas of the boundary layer flow near to the surface of the plate which influenced by force of viscous, therefore this layer often scrutinized for analyzing the variety of fluid flow, one of them is Laminar flow.

Laminar flow is turbulent fluid flow without current, moving relatively slowly and parallel [3]. This flow rare happening in the water and in the air, but a kind of this flow often occurs in the fluid which has a high relative degree of viscosity, such as oil, glycerin, and oil. The high viscosity of a substance can give many effects with a flow rate of the plate.

This experiment ever investigation [6], with using hot plat plate for analyzing influence turbulent flow of characteristic concerning boundary layer. The experiment only describes an effect of turbulence on the speed of the water, while research on laminar flow has not been known. Therefore it is important to make research about it. So that, this experiment created for to analyze the speed of laminar flow on the plate with using Stokes theorem.

2. Experimental Method

The experiment was carried out on 30 March 2015 until 30 June 2015 at the physics laboratory of UHAMKA. This experiment using many samples such as oswtald viscometer, pycnometer, digital balance, measuring cup, hollow pipe, oil X brand, honey X brand, and plate to type (zigzag and spiral). Pottery contain oil in measuring cup until 30 ml, weigh blank pycnometer using digital balance, and then poured the oil in pycnometer and weigh using digital balance, poured the oil into viscometer Ostwald with using tunnel, wait flow of oil until balance, after that suction the sample using a vacuum until through pipe B and up until through line M, then let the sample down to line N. Record this time when the sample through line M to N. The same treatment to honey [2].

The last treatment is putting the plate (zigzag plate) in the hollow pipe, pour the oil in this plate, observe the movement and view the time it takes for the flow get the bottom of the pipe using a stopwatch. the same treatment for honey with the different plate (spiral plate).

3. Results and Discussion

Table 1. Calculation of density and viscosity of materials

Description	Honey	Oil
ρ (g / cm³)	643,4	762,2
η (Ns/m²)	0,00594	0,00503

Table 2. Pressure and flow velocity (zigzag plate)

Material	P (Pa)	v (m/s²)
Honey	1,46	0,02
Oil	0,34	0, 0058

Table 3. The pressure and flow velocity (spiral plate)

Material	P (Pa)	v (m/s²)
Honey	2,46	0,0046
Oil	0,18	0,0023

From the table in J Waspakrik [5], both these fluids have the same volume with different mass, where the mass of honey is 38.110 gr and the mass of the oil is 32.700 gr. From table 1, the density of oil is bigger than honey, it can happen because density strongly influenced by the volume and mass, the greater mass of the fluid the greater density. From table 1, viscosity honey is greater than oil. It can happen because the viscosity affected by density. Determination viscosity of the fluid can be using comparison both this viscosity, between the viscosity of aquadest and viscosity which looking for.

From table 3, the pressure to honey greater than the pressure to oil, it can happen because pressure on the object very influenced by mass on the object, can look mass of honey greater than the mass of the oil, so that pressure too honey greater than the pressure to the oil. But the different case for table 2 with the zigzag plate, it can happen maybe oil common using machine with the spiral plate, so the movement oil greater than honey. From table 2, oil has the greater velocity than honey, it can happen because of the pressure of honey greater than oil. A velocity of flow very influenced by pressure.

The last part of the discussion is looking for the flow velocity using Stokes theorem. It using Stokes theorem who uses the surface integral to define a small area of a flow, the theorem Stokes in question is a blend of divergent functions with integral full on a formula of velocity, so that the calculation results can be seen, that's both types of fluid flowed to the plate (zigzag plate) is faster than the fluid flowing through the plate (spiral plate).

4. Conclusion

The conclusion from the experiment is that velocity of laminar flow can be analyzed using Stokes theorem, with different velocity for both of fluids.

References

[1] Tipler A Paul 1998 *Physics for Science and Engineering* (Jakarta: Erlangga)
[2] Dictionary of Development Team Language Center 2002 *The Indonesian General Dictionary* (Jakarta: Balai Pustaka)
[3] Ridwan *Basic Fluid Series KuliahMekanika* diktat (Jakarta: Gunadarma)
[4] Young and Freedman 2002 *University PhysicsVolume 1* (Jakarta: Erlangga)
[5] J Waspakrik, Hans 1993 *Fundamentals of Mathematics for Physics* (Jakarta: Gunadarma)
[6] Umar Faruk and Kamiran 2012 *Journal Of Science and Art Its* **1**

Acknowledgments

First I thank Allah that makes me in carrying out this study, do not forget I say thank you to laboratory assistants and experiment associates who helped me in the success of this study. On this occasion, the authors would like to express the appreciation and gratitude to LEMLITBANG UHAMKA as a sponsor and all the parties that have big impacts on the success and completion of this research.

IOP Conf. Series: Journal of Physics: Conf. Series **948** (2018) 012028 doi:10.1088/1742-6596/948/1/012028

The effectiveness of problem-based learning on students' problem solving ability in vector analysis course

R Mushlihuddin[1], Nurafifah[1] and Irvan[1]
[1]Universitas Muhammadiyah Sumatera Utara, Jl. Kapt. Mukhtar Basri No. 3 Medan Sumatera Utara

E-mail: rahmatmushlihuddin@umsu.ac.id

Abstract. The student's low ability in mathematics problem solving proved to the less effective of a learning process in the classroom. Effective learning was a learning that affects student's math skills, one of which is problem-solving abilities. Problem-solving capability consisted of several stages: understanding the problem, planning the settlement, solving the problem as planned, re-examining the procedure and the outcome. The purpose of this research was to know: (1) was there any influence of PBL model in improving ability Problem solving of student math in a subject of vector analysis ?; (2) was the PBL model effective in improving students' mathematical problem-solving skills in vector analysis courses? This research was a quasi-experiment research. The data analysis techniques performed from the test stages of data description, a prerequisite test is the normality test, and hypothesis test using the ANCOVA test and Gain test. The results showed that: (1) there was an influence of PBL model in improving students' math problem-solving abilities in vector analysis courses; (2) the PBL model was effective in improving students' problem-solving skills in vector analysis courses with a medium category.

1. Introduction

Mathematics learning consists of five content standard mathematics learning process [5]. The five content standard learning process is a mathematical ability that must be owned by a student. Nation Council of Teachers of Mathematics [5] they are: 1) Problem Solving; 2) Reasoning & Proof; 3) Communication; 4) Connections; 5) Representation;

The student's low ability in mathematics problem solving prone to the less effective of the learning process in the classroom. Effective learning is a learning that affects students math skills, one of which is problem-solving abilities. Polya in Mushlihuddin [4] said that problem-solving abilities consist of several stages: understanding the problem, planning the settlement, solving the problem as planned, Reviewing the procedures and results. Nation Council of Teachers of Mathematics [5] The instructional program of student problem-solving skills should enable each student, they are: 1) Build new mathematical knowledge through problem-solving; 2) Solve problems that arise in mathematics and in other contexts; 3) Apply and adapt a variety of appropriate strategies to solve problems; 4) Monitor and reflect on the process of mathematical problem solving.

One of the learning model that can solve the problem one of them is PBL (Problem Based Learning) model. Problem-Based Learning (PBL) model describes a learning environment where problems drive the learning. That is, learning begins with a problem to be solved, and the problem is posed in such a way that students need to gain new knowledge before they can solve the problem.

Rather than seeking a single correct answer, students interpret the problem, gather needed information, identify possible solutions, evaluate options, and present conclusions. Proponents of mathematical problem solving insist that students become good problem solvers by learning mathematical knowledge heuristically [2].

Arends [1] there are five stages in the syntax of problem-based learning, they are: 1) Provide orientation about the problem to students; 2) Organizing students to research; 3) Assisting independent and group investigations; 4) Develop and present artefacts and exhibits; 5) Analyze and evaluate the problem-solving process. The PBL model requires students to be more active. Because in the students' learning is directly involved in the investigation and find the solution of the problem, so in the end, the students are helped to become autonomous students who can help themselves, in solving the problems faced.

Therefore, the researcher was interested to conduct further research on problem-based learning model with the title of the effectiveness of PBL model in the student's ability mathematics problems solving on Vector analysis.

The purpose of this research was to know: 1) was there any influence of PBL model in improving ability Problem solving of student math in the subject of vector analysis?, 2) was the PBL model effective in improving students' mathematical problem-solving skills in vector analysis courses?

2. Methods and Materials

Determination of the sample by conducting preliminary tests to see the homogeneity of each class. Based on the preliminary tests the researcher assumes that the nine classes are homogeneous, so the researcher concludes taking only one class from eight classes at random (Cluster Random Sampling).

This research was a quasi-experiment research with two research variables. The research design is shown in Table 1.

Table 1. Research Design

Group	Pre-test	Treatment	Postest
Experiment	T_1	X_1	T_2
Control	T_1	X_2	T_2

Information :

T_1 = pre-test

T_2 = postest

X_1 = experimental group using PBL model

X_2 = control group using traditional model

The procedures in the study are as follows:

1. Early research was given preliminary or pre-test to see the homogeneity of each class of the population, in order to determine the sample class.
2. The researcher gave a pre-test to see the student's initial ability level.
3. The researcher provided a stimulus for the sample class using the problem-based learning model for X_1 and the conventional model for X_2.
4. Provided a post to see the level of success of students' math skills.
5. Conducting hypothesis test with the procedure as follows: 1) test description of data; 2) test data prerequisites; 3) statistical test that is ANCOVA-test to see the influence of PBL model in improving the mathematical ability of the student of Vector Analysis [7], and Gain-test Hake [6] to see the effectiveness of PBL model in improving students math skills Vector analysis.

$$N_{gain} = \frac{S_{post} - S_{pre}}{S_{maks} - S_{pre}} .$$

The scoring rate is categorized into three categories, shown in table 2, as follows:

Table 2. Category N-Gain Score

Category	Information
High	$N_{gain} \geq 0.7$
Medium	$0.3 < N_{gain} < 0.7$
Low	$N_{gain} \leq 0.7$

3. Results and Discussion

3.1. Research Result

Descriptive analysis of pre-test and score resulted from descriptive data. Some descriptive data of and from the results of this study are mean scores and standard deviations shown in Table 3.

Table 3. Descriptive Statistics

Statistics	Pre-test		Post-test	
	PBL	Traditional	PBL	Traditional
Mean	68.6000	67.5143	82.2571	77.4571
Standard Deviation	8.16088	6.21370	7.51212	6.63477

Based on the results of the descriptive analysis shows that the pretest averages for PBL model 68,60, pretest for Conventional model 67,51, post-test for PBL model 82,26, post-test for conventional model 77,46. As for standard deviation pretest model PBL 8.16, pretest for conventional model 6.21, post-test for PBL model 7.51, post-test for conventional model 6.64.

The normality test is a test of the spread of each score on the pre-test score and post-test score, both in the control group and in the experimental group. The scatter normality test scores were performed by Kolmogorov-Smirnov and/or Shapiro-Wilk statistical tests. The results of the scatter distribution normality test are shown in table 4.

Table 4. Tests of Normality

Model	Test	Kolmogorov-Smirnov[a]			Shapiro-Wilk		
		Statistic	df	Sig.	Statistic	df	Sig.
PBL	Posttest	.114	35	.200*	.970	35	.430
Konvensional	Posttest	.108	35	.200*	.970	35	.431

*. This is a lower bound of the true significance.
a. Lilliefors Significance Correction

Based on normality test results indicate that the value of sig. At Kolmogorov-Smirnov more than 0.05, so the conclusion is good post-test data by using normal distributed PBL model and traditional models.

The homogeneity test of variance was intended to determine the homogeneity of the variance of pretest score between the control group and the experimental group. Likewise, the homogeneity of variance scores post-test between controls and experimental groups. The homogeneity test of variance is shown in table 5.

Table 5. Tests of Homogeneity of Variance

		Levene Statistic	df1	df2	Sig.
Pre-test	Based on Mean	.587	1	68	.446
	Based on Median	.465	1	68	.497
	Based on Median and with adjusted df	.465	1	64.635	.498
	Based on trimmed mean	.610	1	68	.438
Post-test	Based on Mean	.243	1	68	.624
	Based on Median	.257	1	68	.614
	Based on Median and with adjusted df	.257	1	66.217	.614
	Based on trimmed mean	.252	1	68	.617

Based on homogeneity test results indicate that the value of sig. On Levene Statistic over 0.05, it was clarified that the variance of the pretest score between the control group and the experimental group was homogeneous. Likewise, the variance of posttest scores between the control group and the experimental group was homogeneous.

Test slope regression or interaction test is intended to determine whether there is the influence of learning model on learning outcomes. If there is an interaction ($p < 0.05$) between the regression line of the pretest score and posttest score in the control group with the regression line between the pretest score and posttest score in the experimental group, the student learning outcomes are influenced by other variables instead of the learning model variables. Conversely, if there is no interaction ($p > 0,05$) between regression line of pretest score and posttest score in control group with regression line between pretest score and posttest score in experiment group, student learning outcomes are only influenced by learning model variable, not by variable another. Interaction test results or homogeneity of the slope of the regression line are shown in Table 6.

Table 6. Tests of Between-Subjects Effects

Dependent Variable: Postest

Source	Type III Sum of Squares	df	Mean Square	F	Sig.
Corrected Model	2669.896[a]	16	166.868	7.699	.000
Intercept	156477.422	1	156477.422	7219.882	.000
Model	245.642	1	245.642	11.334	.001
Pre-test	2125.462	9	236.162	10.897	.000
Model * Pre-test	104.518	6	17.420	.804	.571
Error	1148.676	53	21.673		
Total	450220.000	70			
Corrected Total	3818.571	69			

a. R Squared = .699 (Adjusted R Squared = .608)

The significance figure for the Model line * Pre-test obtained from the interaction test is 0.571. This significance number is more than 0.05 so it can be concluded that the covariate variable has no significant effect on student learning outcomes. In other words, the variable of student learning outcomes is only influenced by the variables of applied learning model

After all assumption tests are met, hypothesis testing with ANCOVA statistics. The alternative hypothesis tested is "PBL model can improve students' mathematical problem-solving skills in vector analysis courses." The test results using ANCOVA statistic produce significance value as shown in Table 7.

Table 7. Tests of Between-Subjects Effects

Dependent Variable: Postest

Source	Type III Sum of Squares	df	Mean Square	F	Sig.
Corrected Model	2237.240[a]	2	1118.620	47.395	.000
Intercept	740.047	1	740.047	31.355	.000
Pretest	1834.040	1	1834.040	77.707	.000
Model	281.547	1	281.547	11.929	.001
Error	1581.331	67	23.602		
Total	450220.000	70			
Corrected Total	3818.571	69			

a. R Squared = .586 (Adjusted R Squared = .574)

The above output shows that the influence of student Pre-test and the difference of learning model to the Post-test value obtained by the students simultaneously can be seen from the number of signatures in the Corrected Model. It can be seen that the number of significance is 0.000. Because the value of significance is far below 0.05 then H0 is rejected. So at the level of 95% confidence can be concluded that simultaneously Pre-test students and learning models affect the value of post-test obtained by students.

The significance of the Pre-test variable is 0.000. Because of the value of Sig. <0,05 then H0 is rejected. This means that at the 95% confidence level there can be a linear relationship between the Pre-test and the Post-test value obtained by the student. This statement indicates that the ANCOVA assumption has been met. This test is done by eliminating the influence of differences from the learning model first.

Further testing is done to determine the effect of different learning models to the value of post-test obtained by students. This test is done by eliminating the influence of Pre-test of the learning model. From the results of processing, it can be seen that the significance of the learning model is 0.001. Because the value is far below 0.05 then H0 is rejected. Without the influence of Pre-test, at 95% confidence level there is the influence of learning model difference to the post-test value obtained by the student. So it can be concluded PBL model can improve students' math problem-solving skills in vector analysis courses. If we look at the posttest average score of the students, it appears that the mean score of the experimental group posttest is higher than that of the control group's posttest mean. From this post-test average score, the conclusion that can be drawn is the PBL model can improve students' mathematical problem-solving skills in vector analysis courses.

To test the second hypothesis using the test N-gain. The alternative hypothesis tested is "The PBL model is effective in improving students 'math skills in vector analysis courses." The test results using Again statistics is 0.44, it can be concluded that "PBL model is effective in improving students' math skills in vector analysis courses" with the medium category.

Based on the results of the first hypothesis test, it can be concluded that the PBL model can improve students' math problem-solving skills in vector analysis courses, shows that the pre-test averages for PBL model 68.60, and post-test for PBL model 82.26. There is an increase in learning outcomes from before applying the PBL model with after applying the PBL model. For the results of the second hypothesis test, it can be concluded that "PBL model is effective in improving students' math skills in vector analysis courses" with the medium category.

3.2. Research Discussion

The PBL model, the teacher gives students an orientation about the problem and motivates the students to engage in problem-solving activities. Orientation about the problem in question is the teacher gives a problem according to the topic to be discussed, the students are expected to solve the problem in the group. Teachers organize students to research, help students to define and organize learning tasks related to problems, teachers assist independent and group investigations, encourage students to get the right information, carry out experiments, and seek explanations and solutions. To organize the students in question is to form groups and make patterns or maps of thoughts on students looking for solutions to solve problems within the group. Teachers develop and present artifacts and exhibits, help students plan and prepare reports, and help them to communicate to others. Developing and presenting the intended ones is that students should be able to present the results of group discussions in front of other groups. Discussion results are the problem solving of the problems given by teachers within the group. Teachers analyze and evaluate problem-solving processes, help students reflect on their investigations and the processes they use. Based on the stages of learning above it is clear that learning requires more active students. Because in the students' learning is directly involved in the investigation and find the solution of the problem, so in the end, students are helped to become autonomous students who can help themselves, in solving the problems it faces.

The PBL model is a model of group learning, so the PBL model is part of the cooperative learning model. Mandal [3] said that cooperative learning can develop high-level thinking skills, create an active learning environment, improve student performance with less academic ability, and tolerate different learning styles among students.

Some principles in applying the PBL model, namely: (1) all participants listen well to what is delivered by students or other groups; (2) opinions must be based on sound evidence and problem-solving; (3) the discussion process should be in a dialogical setting. So that students are expected after learning PBL model not only have the cognitive ability but also effective or soft skill ability and skill or life skill.

IOP Conf. Series: Journal of Physics: Conf. Series **948** (2018) 012028 doi:10.1088/1742-6596/948/1/012028

Some of the obstacles encountered in applying the PBL model are: (1) the PBL model is an oral book study, each student or group must have many references on the topic being discussed. So that students have to collect many sources of reference, because at the beginning of the teacher's learning only provide topic material and problems to be solved in the group; (2) high-order thinking that causes students to take time to adjust to the PBL model. The solution given to overcome these problems is to direct and guide students using IT-based technology. Technological advances not only have a negative impact but actually, have a positive impact when used properly and correctly.

4. Conclusion

Based on the results of the results of this study, it can be concluded that the PBL model can improve the ability. The PBL model is effective in improving students' math skills in the medium category. This shows the application of the PBL model to the mathematical abilities of the vector analysis courses. Thus, the PBL model effectively improves the ability.

Based on the results achieved, suggestions can be made as follows. First, on the application of PBL model. Second, the PBL model can be used as an alternative learning model by the lecturer to improve the ability. Third, the use of PBL models is applied to higher learning. Fourth, the PBL model is an oral books learning model so that students can learn independently

References

[1] Arends I A 2008 *Learning to teach belajar untuk mengajar Edisi ketujuh buku kedua* (Yogyakarta: Pustaka Belajar)
[2] IAE-Pedia 2008 *Math Problem-based Learning* retrieved from http://iae-pedia.org/Math_Problem-based_Learning.html.
[3] Mandal Rita Rani 2009 *The Modern Journal of Applied Linguistic* **1**
[4] Mushlihuddin R 2017 *Seminar Nasional dan Workshop Matematika dan Pendidikan Matematika STKIP PGRI Sumatera Barat Padang* (Padang: STKIP PGRI Sumatera Barat Press)
[5] NCTM 2016 *Principles and Standards for School Mathematics* retrieved from http://www.nctm.org/Standards-and-Positions/Principles-and-Standards/Process/.
[6] Rahmaniati R et all 2015 *Jurnal Anterior* **14** (2)
[7] Redhana I W 2014 *Jurnal Cakrawala Pendidikan UNY* **XXXIII** (1)

Acknowledgments

Acknowledgments are sent to KEMENRISTEKDIKTI and UMSU which have funded this research. In addition, thanks are also conveyed to my beloved Family, Nur'afifah, and Irvan who helped the implementation of this research.

Students' thinking styles and their proof writing levels

Risnanosanti[1] and Ristontowi[1]

Department of Mathematics Education, Muhammadiyah University of Bengkulu, Indonesia

E-mail: nosantirisna@gmail.com

Abstract. This study was done to investigate the relationship between thinking styles and proof writing levels among university students. Participants in this study were 124 students who were selected via multi-stage cluster sampling. Sternberg and Wagners' thinking styles inventory was used for assessing students' thinking styles and their proof writing levels were assessed by the proof-writing scale. The results indicated that there was a significant relationship between thinking styles and proof writing levels and the level of significance was 0.95. That means students' thinking styles have the ability to predict their proof writing levels. There was also a positive and meaningful correlation between executive thinking style and the total scores of proof writing levels. The result was as follow: students who have external proofs schemes (44%), empirical proofs schemes (52%) and analytical proofs schemes (6%). There was a significant difference, in term of their thinking style, among students who had varying levels of proof-writing.

1. Introduction

Mathematics is a field of science that filled with various statements in the form of theorems or propositions. People do not know whether the statement about the propositions is true or not until they had proved it. Therefore, the proof is fundamental to develop mathematics. [1] Proof that provides assurance of mathematical information, in making and understanding of mathematics is a basic activity. Mathematical proof is an essential activity in doing and understanding mathematics, and it gives a warranty for mathematical knowledge [2]. The proof is not only the heart of mathematics education but also an essential tool for promoting students' mathematical understanding. According to [3] the teaching and learning of proof are recognized internationally as a key component of mathematics education and mathematics curricula. Therefore for anyone involved in mathematics, one important activity is reading a mathematical proof with the aim of determining whether or not it is valid.

With the increase in the role and significance of proof in mathematics, the developments and mental processes of students in various age groups when they prove have been the subject of research in mathematics education. [4] examined students' proof schemes through their considerations examples and counterexamples and reasoning in proof construction. Research shows that the student has difficulty in proving mathematics in general, especially in the field of geometry [5], and they show difficulties when trying to read proofs [6]. One of the important thing to develop students ability in proves can be done by looking at proof as an intellectual activity that involves the process of thinking.

When students do the proof, they should be able to understand and explain the facts in mathematics. It is explaining the facts of mathematics each person has different ways, some people explain by drawing sketches or graphics, while others use structures, patterns or formulas [7]. There are three aspects to considering to develop students proof and proving. The first aspect is a proof as a structural object, the second aspect a proof is an intellectual activity and the third aspect is the roles, function, and meaning of proof and proving [5]. Proof and proving as an intellectual activity is a support by students thinking styles. Thinking styles correlated positively with a student success in a variety of academic tasks. Successes and failure attributed to abilities often stem from styles [8].

1.1. Writing Proof

Harel and Sowder [9] conducted a study about thinking processes that classifies solution levels during proofing by interrogating the reasons behind them. The proof schemes can be grouped into three parts as follows.

Analytical or deductive proof schemes: students who hold analytical proof schemes prove and validate theorems by means of logical deductions. These schemes with involve assumptions with logical inferences and reasoning [10]. The characteristics of this scheme are that students can use various strategies. They can make generalizations and be able to utilize mathematical relationships between sections in mathematics.

Empirical proof schemes: this scheme describes student proof schemes who believed in evidence from examples or perceptions rather than logical reasoning. According to Harel and Sowder [9] students proof by giving number values to expressions or use similar instances in the confirmation process and prefer to account for some situations with their intuitions. There are two parts of proof in this category namely inductive and perceptual proof schemes. Inductive proof schemes held by students who are convinced by seeing a few examples. The perceptual proof is a scheme evidenced by a student who is convinced of a theorem's truth by a rudimentary, mental image or perception [11].

External proof schemes: students in this type are who believe in the truth of a theorem or statements that made by external factors such as mathematicians, teachers, or they classmates. According to [10], students perform proof based on the accuracy of their knowledge of books, rules that apply or to others who considered more expert. In general types of the proof scheme in this category are a ritual, non-referential symbolic, and authoritative. Students ritual proof schemes are convinced by the form or appearance of a proof. Non-referential symbolic.

On the other hand, Balacheff [12] divided mathematical proof into three levels: (1) Pragmatic Proofs: students who have pragmatic level of proof performance unrelated responses, proof completely based on numerical example, responses based simply on the writing of what is given; (2) Intellectual Proofs: students in this types proofs based on formulation, proofs are given through the use of incomplete mathematical language and demonstration, write proof with general argument; and Demonstration Proofs: in this type student completely accurate proofs that have to be organized with a theory or the use of information commonly accepted by a community. In this study level of proofs that be used is a combination of the two opinions the above.

1.2. Thinking Style

Research on the style of thinking has evolved since its emergence Sternberg's work. Some examples of research related to thinking styles include [13]. There is a significant relationship between adopted from students and their thinking styles [14]. Thinking styles have a relationship with problem-solving, decision-making, academic achievement, etc., and variables such as culture, gender, age, etc. [8]. Thinking styles can affect student achievement in mathematics [15].

Thinking styles is a particular procedure by which something done to a person. Strenberg [16] defines thinking styles is the way of thinking and describes how the individuals use their capacities. The theoretical basis for this research is based on Sternberg's theory. According to Strenberg [17] thinking styles dimensions divided into five categories, namely forms, functions, levels, learning, and scopes. The dimensions of functions are the basic of thinking styles and support students abilities in the study. In this dimensions there are three styles as follows:

- Legislative style: students who have legislative style prefer the problems that are requires them to devise new strategies and to create their own laws and they enjoy giving commands. [18]. Legislative Likes to create, discovery, design; do things using own method; and less structure.
- Executive style: students in this style prefer to use the ways that already exist to solve problems, application, and implementation of the rules. They will be able to start work if they know about the five 'W'- when, why, where, who and whose-. [19]. Executive Likes to follow instructions; does what is requested; structure must be given.
- Judicial Style: students of this method care about the stages of the work and the result. The question that they often asked such as why? What is the reason? [20] Judicial Like to criticise and evaluate people and things.

The questionnaire about ways of thinking by Strenberg's consist 65 item at an average 5 item for each thinking style. Table 1 shows the distribution of item in the ways of thinking:

Table 1. The distribution of items on the list ways of thinking by Strenberg

Types	Items	Types	Items	Types	Items
Legislative	1,14,27,40,53	Local	5,18,31,44,57	Monarchic	9,22,35,48,61
Executive	2,15,28,41,54	Liberal	6,19,32,45,58	Oligarchic	10,23,36,49,62
Judicial	3,16,29,42,55	Conservative	7,20,33,46,59	Anarchic	11,24,37,50,63
Global	4,17,30,43,56	Hierarchic	8,21,34,47,60	Internal	12,25,38,51,64
External	13,26,39,52,65				

The aims of this study is to investigate the relationship between thinking styles and proof writing levels among university students at Muhammadiyah University of Bengkulu based on Strenberg's ways of thinking.

2. Methodology

This study was both of qualitative and quantitative design with descriptive research. The participants in this study were 124 students from mathematics study program of Muhammadiyah University of Bengkulu in academic year 2016/2017. There were three tools for gathering the data: 1) questionnaire; 2) interview to obtain data on the student's thinking style; and 3) test to obtain data in student's writing proofs. The questionnaire made based on every indicator of thinking style.

Both a qualitative and a quantitative was use to collect the data through questionnaire, interviews and documents analysis. The questionnaire was based on an existing standardized instrument, which aimed to determine the different strategies used by the students to proofs the theorems or proposition, execute tasks and make decisions. The questionnaire consisted the functions dimensions of thinking style which three type. For each type, there were five questions on 1 – 4 point Likert scale. The score was then averaged over each type. The second phase of this study by conducting one-to-one interviews with the students based on their answer in the questionnaire. The aim was to ascertain their thinking style when doing the proof and proving. The data about students proofs levels taken from their work documents.

3. Result and Discussion

3.1. Thinking Style

Table 1 shows the distribution of item in the ways of thinking. Each of thinking style type, legislative, executive and judicial have 5 item. One of the item in the legislative type asked the students when they discussing or writing idea. Are the students have many ideas or not? And most of the students answer that they stick only to one idea. In general, the result of the questionnaire and one-to-one interview show in Table 2. Table 2 show students thinking styles based on the types of thinking style.

Table 2. Students thinking style

Type of Thinking Style	Frequency
Legislative	65
Executive	47
Judicial	12
Total	124

Table 2 shows that most of the students in mathematics study program of Muhammadiyah University of Bengkulu have legislative thinking style. Students who have judicial type thinking style only 12. From that we knew that most of the students stick to standard rules or ways of doing things. This result supported by [21] who claimed that mathematics requires that students be able to think in the terms, thus students should have a preference to create, discover and design which the indicator legislative thinking style.

3.2. Writing Proofs

Students writing proofs were determined based on the combination in [9] and [12]. The combination shown in Table 3.

Table 3. Levels of Students Proof

Levels of Proofs	Indicators/Students Response
Analytic or Deductive	completely accurate proofs that have to be organized with a theory or the use of information commonly accepted by a community
Empirical	1. based on formulation, 2. proofs are given through the use of incomplete mathematical language and demonstration, 3. write proof with the general argument
External	1. unrelated responses 2. proof completely based on the numerical example 3. responses based simply on the writing of what is given

Based on Table 3, we made students respond when they doing proofs. For examples when students asked about the set of the natural number is a group or not. They proving about the closure properties in natural number with the example $4 + 5 = 9$, 4 is a natural number, 5 is natural number and 9 is natural number so natural number is closed. Based on the answer we said the the students in external level because the response is unrelated. In general, the students level of proofs shows in table 4.

Table 4. Students Levels of Proofs

Level of Proofs	Frequency	%
Analytical	6	4.8
Empirical	54	43.6
External	64	51.6

From the Table 4, students who have an analytical level of proofs only 4.8%, it means most students have low abilities in proofs. Students in Muhammadiyah University have the external level of proofs. They often gave the unrelated respond when doing the proofs. The combination students thinking style and levels of proofs show in Table 5.

Table 5. Students Thinking Style and Their Writing Proofs Levels

	External	Empirical	Analytical
Legislative	37	28	0
Executive	14	29	4
Judicial	3	7	2

Table 5 shows that students who have legislative thinking style most have an external level of proof. It is clear because in legislative thinking style students does things using their own method even that is true or not. This result is also supported by [22] who said that the complexity of mathematics can make extensive demands on the reasoning, interpretive and strategic skills of students.

Therefore, based on the data tested correlational to see the relationship between student's thinking style with writing proofs levels. The results indicated that there was a significant relationship between thinking styles and proof writing levels with the correlation 0.67 and the level of significance was 0.95.

In this study, it was seen that students who have external type of writing proofs are 29.8% from legislative, 11.3% from executive and 2.41% from judicial thinking styles. Students who have empirical type of writing proof are from legislative 22.6%, executive 23.4% and 5.7% from judicial thinking styles. The analytical levels of writing proofs are 3.2 from executive and 1.6 from judicial thinking styles. No ones in analytical levels from legislative thinking styles. According to result, there are differences the levels of writing proofs with related to thinking style. The result of this study agreed with the result of a study by [5].

4. Conclusion

Generally, students have different thinking styles, and it is can affect their ability in writing proofs. However, students thinking styles could be taken into consideration as guidelines when teachers give a task writing proofs. Based on the finding of this research, further research is required to develop students writing proofs in the mathematics class.

References

[1] Sengul S and Guner P 2014 Relationship between Proof Schemes Used by Preservice Mathematics Teachers and Gender, Views Toward Proof *Procedia-Social and Behavioral Sciences* **116** p 617
[2] Ozdemir E and Ovez F T D 2012 A Research on Proof Perceptions and Attitudes Towards Proof and Proving: some Implication for Elementary Mathematics Prospective Teachers *Procedia-Social and Behavioral Sciences* **46** p 2121
[3] Hanna G and de Viller M 2012 Aspects of Proof in Mathematics Education. In G. Hanna & M. De Villier (Eds.), *Proof and Proving in Mathematics Education*: The ICMI Study, 1 – 10, (New York: Springer)
[4] Lee K 2016 *The Journal of Mathematical Behavior* **41** p 26
[5] Miyazaki M, Fujita T and Jones K 2017 *Educational Study Mathematics* **94** p 223
[6] Inglis M and Allock A 2012 *Journal for Research in Mathematics Education* **43** 4 p 358
[7] Borromeo F R 2010 *Journal fur Mathematikdidaktik* **31** 1 p 99
[8] Negahi M, Nouri N and Khoram A 2015 *Theory and Practice in Languange Studies*, **5** 8 p 1722
[9] Harel G and Sowder L 1998 Students Proof Schemes: Result from Exploratory Studies, In A. Schoenfeld, J. Kaput and E. Dubinsky. (Eds.), *Research in Collegiate Mathematics Education III*, 234 – 283, Providence, RI, American Mathematical Society
[10] Flores A 2002 *Teaching Children Mathematics* **8** 5 pp 269
[11] CadwalladerOlsker T 2011 *Journal of Humanistic Mathematics* **1** 1 p 201
[12] Balacheff N 1988 Aspects of Proof in Pupils' Practice of School Mathematics. In D Pimm (Ed.) *Mathematics, Teachers and Children*, 216-35. Trans. By D Pimm. (London: Hodder & Stoughton)
[13] Lun V M C, Fischer R and Ward C 2010 *Learning and Individual Difference* **20** p 604

[14] Gu X, Wang H and Mason J 2017 Are They Thinking Differently: A Cross-Cultural on the Relationship of Thinking Styles and Emerging Roles in Computer-Supported Collaborative Learning. *Educational Technology & Society*, **20** 1 p 13
[15] Risnanosanti., 2017. Mathematical Thinking Styles of Undegraduate Students and Their Achievement in Mathematics. *American Institute of Physics (AIP) Conference Proceedings* 1868.050018; America.
[16] Strenberg R J 2002 *Thinking Styles* Reprinted Edition (UK: Cambridge University Press)
[17] Strenberg R J 1990 *Phi Delta Kappa*, **71** 5 pp 366 – 71
[18] Zhang L F 2004 *The Journal of Psychology*, **138** 351- 70
[19] Turki J 2012 *American International Journal of Contemporary Research* **2** 3
[20] Bernardo A, Zhang Li and Callueng 2002 *The Journal of Genetic Psychology* **163** 2 p 149
[21] Spangenberg E D 2012 *Phytagoras* **33** 3
[22] Bohlmann C and Pretorius E 2008 *Pythagoras* **67** p 42

IOP Conf. Series: Journal of Physics: Conf. Series **948** (2018) 012030 doi:10.1088/1742-6596/948/1/012030

The application of vector concepts on two skew lines

F Alghadari[1,2], Turmudi[1] and T Herman[1]

[1]Department of Mathematics Education, Universitas Pendidikan Indonesia, Jalan Dr. Setiabudi No. 229, Bandung 40154, Indonesia

[2] Department of Mathematics Education, STKIP Kusuma Negara, Jalan Raya Bogor No.24, Jakarta 13770, Indonesia

E-mail: fikialghadari@student.upi.edu

Abstract. The purpose of this study is knowing how to apply vector concepts on two skew lines in three-dimensional (3D) coordinate and its utilization. Several mathematical concepts have a related function for the other, but the related between the concept of vector and 3D have not applied in learning classroom. In fact, there are studies show that female students have difficulties in learning of 3D than male. It is because of personal spatial intelligence. The relevance of vector concepts creates both learning achievement and mathematical ability of male and female students enables to be balanced. The distance like on a cube, cuboid, or pyramid whose are drawn on the rectangular coordinates of a point in space. Two coordinate points of the lines can be created a vector. The vector of two skew lines has the shortest distance and the angle. Calculating of the shortest distance is started to create two vectors as a representation of line by vector position concept, next to determining a norm-vector of two vector which was obtained by cross-product, and then to create a vector from two combination of pair-points which was passed by two skew line, the shortest distance is scalar orthogonal projection of norm-vector on a vector which is a combination of pair-points. While calculating the angle are used two vectors as a representation of line to dot-product, and the inverse of cosine is yield. The utilization of its application on mathematics learning and orthographic projection method.

1. Introduction

An issue about STEM (Science, Technology, Engineering, and Mathematics) still to be discussed until now and continues to be consistently a trending topic, especially in education since the beginning emergence of the last few decades. However, the focus of the discussion is on gender gaps which are the percentage of career opportunities for women in STEM areas are lower than men [8,11]. The reason isn't a historical heritage as a reflection where are women not to be encouraged to be a worker [7], but it is a spatial intelligence problem [2,10]. The occurrence of gender gaps has been caused more than the expectation of men careers related to physics, mathematics, or engineering, wherever more than the expectation of women careers in science or health-related professions [7,8,11], So that the professional areas of occupied bring out many gender similarities within a scope of work [7].

However, there is a development of the condition in Indonesia that the student who comes from a vocational high school still free to choose a faculty in the university program, which isn't rare to be found that the faculty of the students chosen was not relevant to the program or the vocational practiced at their senior high school (SMA) or vocational education. This condition is due to several factors including the maturity, self-regulated, and thinking skills are matured [3], so there is a tendency that student was realized about their talent and interest whichever they did, and the students will be brave to take a decision and choose a department on university program with their intelligence. This becomes interesting because still there a chance for a female student to get study at STEM areas which will become a focus of their career in the future. However, this chance will opponent with the founded at some case studies result, that is gender gap happened when an education system gives a freedom to students to choose majors [8,11].

Efforts to reduce the gender gap in STEM areas is recommended to get involved with parents for exploring their career [8] because parents who work in the STEM areas will reflect his talent even to his daughters. Nevertheless, the involvement of parents also should not exceed the limits of children abilities, because educational selected often as a selection of parents and not compatible with their child and then their child can't be responsible [3]. The gender gap in STEM areas is started on many options of high schools with science program as a possible selected so there are pursuing career gaps generated like in engineering area [7,8,11,12]. The problem of the gender gap marked by career expectation to an achievement of educational and working in STEM areas [11,12], this is shown by a little number of female representations [7]. The possible impact on gender gap because the option and working placement occur in old ages and then an ability and interest in STEM area to be down [12]. In addition, senior high school structural system can create to be bad of gender gap [11], so the initiative promote gender equality in STEM must be focused on senior high school structural system [8]

In addition to a focus on the gender gaps, the STEM is also influenced by dynamic of science development, so that should be accompanied by a demand on human resources with competence in particular areas of the STEM [5,12]. Mathematics has contributed ideas in the development of STEM. Application of mathematics in STEM areas give a view that mathematics is one of the scaffolding its development because science is closely related to mathematics and technology [1]. The advantages of mathematics learning are very well in daily life [9,10] because its integration involves particularly STEM subjects [9], so mathematics is one of the important subjects in the development of STEM [5,9,12]. Not only beneficial to the development of science and technology but also mathematics is beneficial to self-development. In this case, exemplified like the application vector concepts on 3D learning then mathematics has much beneficial and utilization. Vector is the findings by Josiah Willard Gibbs which has a great contribution to mathematics area [2,4]. Nevertheless, learning of 3D in senior high school by utilizing the geometric concept of the plane and involves of technology for learning assisted that is based on spatial intelligence. The effort to give a facility for students as a mean [2], whereas the problem on geometry can be solved easily by vector concepts [4].

The utilization of science and technology is one of the keys as a preparation to face times today [9]. Science development is important in factor variety and easiness especially in helping students overcome their learning obstacle. Students will not face an obstacle and difficulties on learning of 3D when they have a good spatial intelligence [10]. A part of all students may feel difficulties, it will impact on geometric idea development and learning achievement when was seen on gender views, so it needs more than understanding and strategy, one of all by applying the vector concept [2]. Writing this article aims to know how to apply vector concepts on two skew lines and its utilization. Two skew lines to be known the shortest distance between of them is one of the discussions in learning of 3D in senior high school. Mathematics curricula in Indonesia set vector as a chapter which is studied in basic mathematics, while the 3D chapter in additional mathematics and both have interrelated. Application of vector on 3D allows for a balance between students' mathematical abilities of male and female [2] so that the gender gap in STEM areas do not continue to be the focus of the discussions.

2. Methods

This study was focused on several articles which published at the end of the year from 2010, related to examples of the application of vector on STEM areas utilization, and vector concepts application. Next, reading and relating its article content. This study is describing how to applying vector concepts on two skew lines and the utilization of STEM area. This study is analyzing the calculation of the distance between two skew lines for three examples of the different geometrical object, that is a cube, cuboid, and pyramid, to find a common step calculation of the distance and how many degrees of the angle. Each a geometrical object is drawn on the rectangular coordinates of a point in space (3D coordinate) with the length of an edge is the one, so the vertex of a geometrical object can be named on the 3D coordinate, and two points of an edge can be represented in vector form. Next, vector concept will find both how long of the shortest distance and how many degrees of the angle on two skew lines.

3. Results and Discussion

Basically, the conventional 3D learning in senior high school by drawing 3D objects manually on the board in front of the class by a teacher and without helping of technology [2], is a part of practices as drawing, spatial thinking, spatial transformation, create a representation of a model, and arithmetic [10]. Its practices are directed as self-development of students in STEM areas. Because mathematics learning based on STEM education is a learning which the integrated design and expected a contribution to improving the competitiveness of Indonesian human resources in all areas [5,9]. Conventionally, mathematics learning is a study about drawing techniques that require 3D effects for adding magnificence of image and then getting the best outcome. Some technique of drawing like compatibility of orthogonal comparison and horizontal visual angle on the drawing orthogonal lines on 3D objects. The orthogonal comparison refers to the comparison of the length of an orthogonal line on the image and the actual. The orthogonal line is the line which perpendicular to the plane of the image. The angle of inclination is an angle between horizontal lines and orthogonal lines.

3D Object defined the object which has a spatial dimension on the 3D coordinate like on solid, the examples are a cube, cuboid, and pyramid [2]. On the 3D object, can be created two skew lines where have both how long of the shortest distance and how many degrees of the angle on two skew lines. Both of them can be calculated by applying vector concept. Application of vector is one of all alternative ways for determining how long distance and how many degrees of angle on two skew lines, for students who have a spatial disability problem on 3D learning conventionally [2].

3.1. Calculate Distance and Angle Between Two Skew Lines on 3D Objects

The means of two skew lines refers two lines which can be created on the 3D coordinate and it does not intersect [2]. Two skew lines are the example of the 3D object. On a cube, cuboid, or pyramid, can be created two skew lines within a distance between of them. The distance of between two skew lines is calculated by applying vector analysis concepts where the 3D object like a cube, cuboid, or pyramid is the first step was made on 3D coordinate with the length of an edge which is according to need [2]

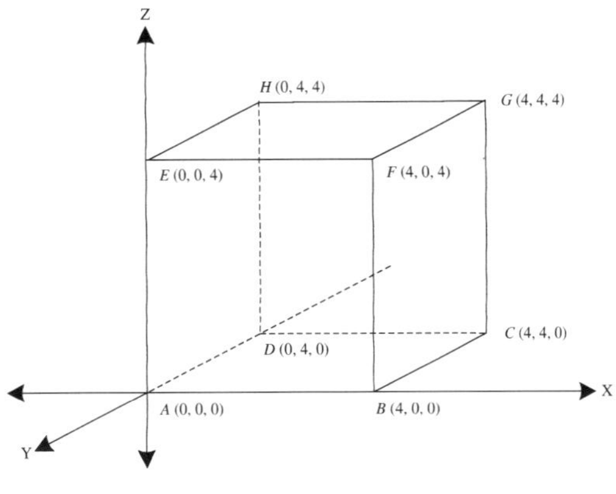

Figure 1. Cube *ABCD EFGH* on 3D coordinate axis with a length of the edge in 4 units.

Based on the coordinate points of the cube in Figure 1 above, the length of an edge of a 3D object is 4 units. On a cube can be created a variety of pair of two skew lines from a combination of four vertices. Not only on the vertices of figure 1, but also another vertex whose new can be made for one need of calculating the shortest distance between two skew lines, and a note that its vertices on a cube and the 3D coordinate can be determined. Wherever the position of coordinate points of two skew lines on a cube, vector concepts application still can be used in calculating the shortest distance on between of them. The application can apply also to the cuboid and pyramid [2]. For example of two skew lines on the cube is a line of *CE* and *AF*. The way of calculating the distance between its line with vector concept application is starting by making two skew lines in a representation of vector respectively [2].

$$\vec{CE} = \vec{AE} - \vec{AC} = (-4,-4,4) \text{ and } \vec{AF} = (4,0,4).$$

The shortest distance between two skew lines is the length of a line whose are perpendicular to its lines. On two skew line, can be determined one vector whose are perpendicular to its lines. The vector is commonly known as a norm-vector. Norm-vector is cross-product of two representation of vector respectively as a line of *CE* and *AF*. The following is the calculation of the norm-vector.

$$\begin{aligned}\vec{CE} \times \vec{AF} = \vec{n} &= (-4i-4j+4k)\times(4i+4k)\\ &= -16(i\times k)-16(j\times i)-16(j\times k)+16(k\times i)\\ &= -16(-j)-16(-k)-16i+16j\\ &= -16i+32j+16k\\ &= \lambda(-1,2,1).\end{aligned}$$

On figure 2, made the illustration of norm-vector ($\vec{n}$), the line of CE and AF as the line whose skew, points of P and Q as the point of intersection between the norm-vector with two skew lines, and then the line of PQ whose is perpendicular to two skew lines and collinear on norm-vector. Based on the figure, the shortest distance between two skew lines is the length of PQ [2].

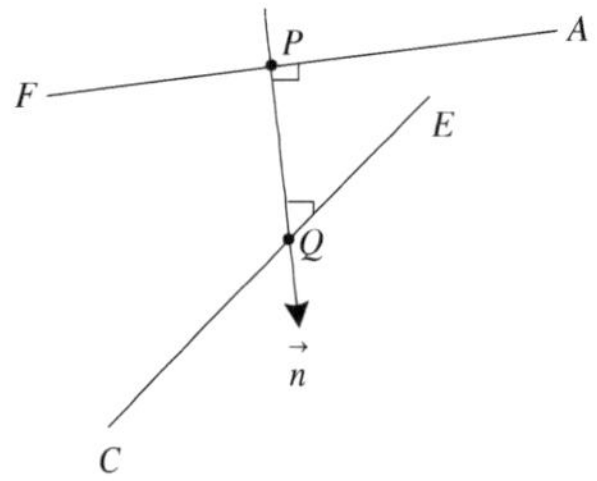

Figure 2. The intersection of two skew lines and norm-vector at points of *P* and *Q*.

Some of the vector concepts related to perpendicularity are dot-product, cross-product, norm-vector, scalar orthogonal projection, and vector orthogonal projection [2,4]. Because of the one which is calculated to be a scalar, then the shortest distance between two skew lines is a scalar orthogonal projection of vector *FC* on norm-vector, or vector *FE* on norm-vector, or vector *AE* on norm-vector, or vector *AC* on norm-vector. The scalar which is obtained by some calculating options is the same value [2]. So that one of all is chosen, it is a scalar orthogonal projection of vector *FC* on norm-vector, where

$$\vec{FC} = (0,4,-4), \text{ then } \left|\vec{PQ}\right| = \frac{\vec{FC}\bullet\vec{n}}{\left|\vec{n}\right|} = \frac{\lambda(0+8-4)}{\lambda\sqrt{(-1)^2+2^2+1^2}} = \frac{4}{\sqrt{6}} = \frac{2}{3}\sqrt{6} \text{ units.}$$

In addition, to calculating the shortest distance, the application of vector concepts on 3D objects also can be used to calculating of how many degrees of the acute angle between two skew lines. The degree of the angle is calculated with dot-product of vector concept [2,4], where are two vectors is meant refers to a representation of a line of *CE* and *AF*. So the angle is,

$$\theta = \cos^{-1}\left|\frac{\vec{CE}\bullet\vec{AF}}{\left|\vec{CE}\right|\left|\vec{AF}\right|}\right| = \cos^{-1}\left|\frac{-16+0+16}{\sqrt{48}\sqrt{32}}\right| = 90^0.$$

The way of calculation by vector concept to determine the shortest distance of two skew line isn't only on a cube but also applies to a cuboid and pyramid [2]. Here is a figure of cuboid *ABCD EFGH* with the length, width, and height respectively is 8, 4, and 3 units

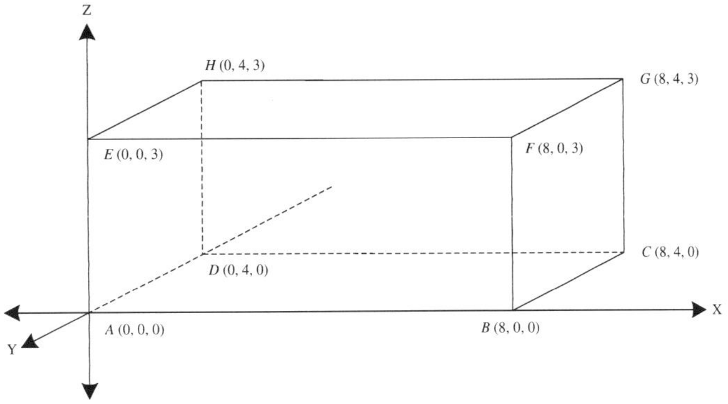

Figure 3. Cuboid *ABCD EFGH*, length, width and height are 8, 4 and 3 units respectively.

Suppose a sample taken of two skew lines are the same as on the cube in above, it is the line of *CE* and *AF*. In units of length, the shortest distance between the line of *CE* and *AF* on a cube is different on a cuboid, because of the length of an edge are also different, so a representation of vector on the lines is different exactly, as shown below.

$$\vec{CE} = (-8,-4,3) \text{ and } \vec{AF} = (8,0,3).$$

Calculating steps is same like on a cube, it is for finding a norm-vector in the cross-product way of two representation of vectors in above.

$$\begin{aligned}\vec{CE}\times\vec{AF} = \vec{n} &= (-8i-4j+3k)\times(8i+3k)\\ &= -24(i\times k)-32(j\times i)-12(j\times k)+24(k\times i)\\ &= -24(-j)-32(-k)-12i+24j\\ &= -12i+48j+32k\\ &= \lambda(-3,12,8).\end{aligned}$$

After that, the shortest distance between the line of *CE* and *AF* on a cuboid, and let *PQ* is it, it can be directly calculated by scalar orthogonal projection, with

$$\vec{FC} = (0,4,-3) \text{ and then } \left|\vec{PQ}\right| = \frac{\vec{FC}\bullet\vec{n}}{\left|\vec{n}\right|} = \frac{\lambda(0+48-24)}{\lambda\sqrt{(-3)^2+12^2+8^2}} = \frac{24}{\sqrt{217}} = \frac{24}{217}\sqrt{217} \text{ units.}$$

Next, calculate the angle between two skew lines were also in the same way like in a cube, using a dot-product, so obtained,

$$\theta = \cos^{-1}\left|\frac{\vec{CE}\bullet\vec{AF}}{\left|\vec{CE}\right|\left|\vec{AF}\right|}\right| = \cos^{-1}\left|\frac{-64+0+9}{\sqrt{89}\sqrt{73}}\right| = \cos^{-1}\left|\frac{-55}{\sqrt{6497}}\right| \approx 39{,}1^0.$$

IOP Conf. Series: Journal of Physics: Conf. Series **948** (2018) 012030 doi:10.1088/1742-6596/948/1/012030

On 3D learning conventionally, in abstraction process, there is a little of difficult to find a 2D representation of the results a spatial transformation form on two skew lines. Perpendicularity concept and the projection ways must be thoroughly understood well because it is the key to find a spatial transformation form and shortest distance. The similarity to another of a 3D geometric object like a pyramid, thus applying vector concepts could be one solution of all. For example, calculating the distance between the line of *AD* and *TB* at pyramid *T.ABCD* where is a square as a base, the length of an edge on base and the height is 4 units, is drawn on below.

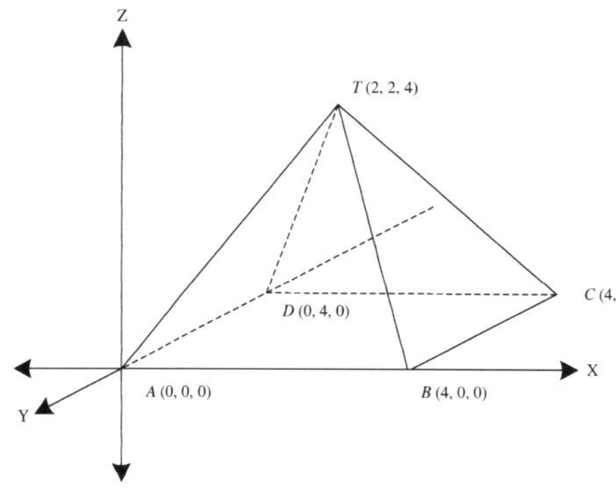

Figure 4. Pyramid *T.ABCD*, base form square, length of an edge on base and altitude are 4 units.

Based on figure 3 in above, the altitude of a pyramid is the line which is pull up from at intersection of base diagonals as an original point to top. According to 3D coordinate, the point of that intersection position at $(2,2,2)$. Because of the altitude is 4 units, so the distance from the vertex to the plane of *ABCD* is 4 units, on position of 3D coordinate at $(2,2,4)$, and then a representation of vector of line *AD* and *TB* are,

$$\vec{AD} = (0,4,0) \text{ and } \vec{TB} = (2,-2,-4).$$

Then, in the same way, like on cube and cuboid, calculating of the norm-vector of two vectors in above like at before, and obtained,

$$\begin{aligned}\vec{AD} \times \vec{TB} = \vec{n} &= (4j) \times (2i + -2j - 4k) \\ &= 8(j \times i) - 16(j \times k) \\ &= 8(-k) - 16(i) \\ &= -16i - 8k \\ &= \lambda(2,0,1).\end{aligned}$$

Therefore, the shortest distance between lines of *AD* and *TB* on pyramid *T.ABCD* if the length of *PQ* is supposed that, is

$$\vec{AB} = (4,0,0) \text{ and then } \left|\vec{PQ}\right| = \frac{\vec{AB} \bullet \vec{n}}{\left|\vec{n}\right|} = \frac{\lambda(8+0+0)}{\lambda\sqrt{2^2+0^2+1^2}} = \frac{8}{\sqrt{5}} = \frac{8}{5}\sqrt{5} \text{ units.}$$

Similarly to calculating how many degrees of the angle between lines of *AD* and *TB* on pyramid *T.ABCD*, isn't different with the way of calculating how many degrees between two skew line on a cube or cuboid, like below.

$$\theta = \cos^{-1}\left|\frac{\vec{AD}\bullet\vec{TB}}{\left|\vec{AD}\right|\left|\vec{TB}\right|}\right| = \cos^{-1}\left|\frac{0-8+0}{\sqrt{16}\sqrt{24}}\right| = \cos^{-1}\left|\frac{-8}{8\sqrt{6}}\right| \approx 65{,}9^0.$$

Thus, calculating how long of the shortest distance and how many degree of the angle between two skew lines on solid like a cube, cuboid, or pyramid, by applying vector concepts, it will help to be easy for student who have difficulties on spatial problem, because the calculation for three 3D objects is using the same of steps and ways, so this learning technique can be used on 3D learning at senior high school then learning achievement in this chapter especially for female student don't have any problem.

All of the 3D objects in the example, one vertex for each solid figure is always in the position of $(0{,}0{,}0)$. However, it shouldn't always be like that, because wherever a position coordinate of a 3D object and it is different of points, but the length of the edge is according to all, and then the vector of representation two skew lines will always be same, so the position of solid on 3D coordinates might be wherever. Anyhow, the application of vector on the calculation of the shortest distance and the angle isn't applicable to parallel lines, because two parallel lines have the same vector [2,4]. However, knowing the distance between two parallel lines may be not difficult because the level of abstraction and spatial thinking is not like the skew lines, and the angle of its projection shape 0 or 180 degrees certainly [4].

3.2. Utilization in STEM Area

In mathematics, arithmetic techniques utilization of shortest distance two skew lines can be applying on learning at senior high school in the 3D chapter for the additional mathematics subject group. Importance is an alternative learning to help students who tend to have problems in spatial thinking in the space object of 3D. Calculating for the third space object of 3D with the step and the same way, so it will facilitate learning achievement, especially for female students, and mathematics performance of male and female students will be possible to be balanced. By vector concepts, students neither need to imagine the position of two skew lines on 3D objects nor think to create a spatial transformation form that is generally a two-dimension figure as a spatial orientation two skew lines on a solid. Students only need 3D coordinate points that formed two skew lines and understanding of vector analysis concepts. However, this calculation technique isn't a priority on spatial intelligence training [2].

Beginning to applying this calculation technique, students who have a problem to imagine of two skew lines position on the 3D object, they will be led to find coordinate points of a 3D object at a figure in the classroom. If it does often repeatedly then students will be familiar with the position where the points on 3D coordinate so it will form a student imagination. This habit is a training method because a training can improve spatial thinking ability [2].

On technology area, vector application to calculating the distance between two skew lines is used for estimating a position of intersection point accurately of two virtual lines. This estimation is called by orthographic projection method where is more excellent than conventional perspective geometric method. Orthographic projection method used a mathematical vector calculation on a remote eye tracking system. This is an indication of mathematics development to continue showing its existence and facilitate of level achievement accurately better than conventional ways. It is important because human vision orientation on 3D way show the focus and attention of observing to objects that used a device with the perspective of the geometric conventional algorithm [6].

4. Conclusion

Based on results and discussion, application of vector on skew line is using to calculate the shortest distance and the angle between its lines. Three examples are in applying on solid , it is the way of applying vector concept in calculating the distance between two skew lines on a 3D object like a cube, cuboid, or pyramid, so that can be created a common steps of calculating: (1) two skew lines that each pass through at least two points on 3D coordinate was known, with those points, the line is created in vector representation respectively; (2) determining norm-vector by cross-product from two vectors was formed; (3) determining a vector from two points, the vector is formed by two points which was a combination of pair-points that was passed by two skew line; (4) the shortest distance between two skew lines is scalar orthogonal projection of norm-vector on a vector which is a combination of pair-points. Utilization application of vector on two skew lines as an alternative 3D learning in senior high school to help students who have spatial disabilities in solving a problem on determining the distance of a 3D geometrical object. On technology area, is used in orthographic projection method to estimate of intersection point position accurately of two virtual lines in 3D coordinates.

References

[1] Permanasari A 2016 STEM Education: Inovasi dalam Pembelajaran Sains *Prosiding Seminar Nasional Pendidikan Sains* ed M Masykuri (Surakarta: Sebelas Maret University, Indonesia) 23-34
[2] Alghadari F 2017 Menentukan Jarak pada Ruang Dimensi Tiga dengan Analisis Vektor *Konferensi Nasional Penelitian Matematika dan Pembelajarannya* ed S Rejeki *et al* (Surakarta: Muhammadiyah Surakarta University, Indonesia) 85-94
[3] Paramita G V 2010 *Humaniora* **1** p. 629-635
[4] Gibbs J W 1901 *Vector Analysis* (USA: Yale University Press)
[5] Milaturrahmah N, Mardiyana M and Pramudya I 2017 Science, technology, engineering, mathematics (STEM) as mathematics learning approach in 21st century *The 4th Int. Conf. on Research, Implementation, and Education of Mathematics and Sci. (4th ICRIEMS)* ed C Kusumawardani *et al* (Yogyakarta: Yogyakarta State University, Indonesia) p 050024
[6] Mujahidin S, Wibirama S, Nugroho H A and Hamamoto K 2016 3D gaze tracking in real world environment using orthographic projection *The 2016 Conf. On Fundamental and Appl. Science for Advanced Technol. (ConFAST)* ed D Y Kusuma *et al* (Yogyakarta: Ahmad Dahlan University, Indonesia) p 020072
[7] Sassler S, Glass J, Levitte Y and Michelmore K M 2016 *Soc. Sci. Research* **63** 192-208
[8] Han S W 2016 *Int. J. of Educational Development* **49** 175-187
[9] Milaturrahmah N and Pramudya I 2017 Mathematics Learning Process with Science, Technology, Engineering, Mathematics (STEM) Approach in Indonesia *Int. Conf. on Mathematics and Sci. Education (ICMScE)* ed A G Abdullah *et al* (Bandung: Indonesia University of Education; Indonesia) p 012030
[10] Riastuti N, Mardiyana M and Pramudya I 2017 Students' Errors in Geometry Viewed from Spatial Intelligence *Int. Conf. on Mathematics and Sci. Education (ICMScE)* ed A G Abdullah *et al* (Bandung: Indonesia University of Education, Indonesia) p 012029
[11] Eccles J S and Wang M T 2016 *Int. J. of Behavioral Development* **40** 100-106
[12] Wang M T, Ye F and Degol J L 2016 *J. of youth and adolescence* **46** 1805-1820

Acknowledgments

The Author would like to thank Department Mathematics Education Doctorate Program at Indonesia University of Education which has provided individual study subject until generating ideas to make an article. Furthermore, ICE-STEM which has provides an opportunity for publication

Enhancing students' critical thinking skills through critical thinking assessment in calculus course

Zulfaneti[1], S Edriati[1] and Mukhni[2]

[1]STKIP PGRI Sumatera Barat, Jl. Gunung Pangilun Padang, Indonesia
[2]Universitas Negeri Padang, Air Tawar Padang, Indonesia

E-mail: zulfaneti@gmail.com

Abstract. This study aims to determine the development of students' critical thinking skills through the implementation of critical thinking instruments in Calculus lectures. The instruments consist of observation sheets, critical thinking test, self-assessment, peer assessment and portfolio. The research was a qualitative research; with the participants were 53 first-year students who take Integral Calculus in Mathematics Education Department STKIP PGRI Sumatera Barat representing high-ability students, medium and low. The data in this study were collected by tests, interviews, observations and field notes. Data were analyzed descriptively; data reduction, data presentation, and conclusions. For testing the validity of data, it was used credibility test data by increasing persistence and triangulation. The results showed that in high-level students there is a change of ability from Critical enough to be Very Critical, in the students with moderate and low ability there is a change of ability from Uncritical to Critical. So it can be concluded that the assessment instruments have a good contribution and can improve the ability of critical thinking.

1. Introduction

Assessment is very important to measure achievement of learning objectives. Assessment and learning process is a unity in the achievement of learning objectives that reflected in the knowledge, skills, and attitudes that are not mutually exclusive. Especially for mathematics, the purpose of learning mathematics is achieved or skills expected, namely: (1) mathematical communication; (2) mathematical reasoning; (3) mathematical problem solving; (4) mathematical connections; and (5) the formation of positive attitudes to mathematics. In order for this learning objective to be achieved, assessment is a very important role to measure the achievement. Assessment or activity assessment is an important aspect learning [2]. In addition, assessment can improve the quality of the learning process and outcomes [12], so the quality of good learning should be followed by a good assessment strategy as well because the information from the assessment is useful for improving the quality of learning [13]. Similarly, delivered by [9] that in the field of mathematics education, we have growing tendencies that emphasize the development of thinking skills and is an essential skill for life in the Information Age.

Critical thinking is a synonym of decision making, strategic planning, scientific process, and problem-solving. Critical thinking is an attempt to deepen the consciousness and intelligence of comparing some of its problems to produce a conclusion and ideas that can solve the problem. Everyone has a different mindset. However, if everyone is able to think critically, the problems they face will be simpler and easier to find solutions. Critical thinking implies a process of evaluation or decision-making which is full of consideration and is carried out independently. Cabera in [6] states that teaching and developing critical thinking skills should be viewed as something that is urgent and cannot be underestimated. Mastery of critical thinking skills is not enough to serve as an educational goal alone, but also as a fundamental process that allows students to overcome future uncertainties

Critical thinking skills are among the higher-order thinking skills [1]. Critical thinking has several characteristics as suggested by Wade in [7] which identifies eight critical thinking characteristics, which include: (1) Activity formulating questions, (2) Limiting problems, (3) Testing data, (4)) Analyze various opinions and biases, (5) Avoid very emotional considerations, (6) Avoid oversimplification, (7) Consider various interpretations, and (8) Tolerate ambiguity. Based on the critical thinking criteria expressed by [7] it can be expressed that a person has been critical if they already have a clear flow of thinking that is a scientific thinking line that starts from formulating questions, limiting problems and avoiding errors as small as possible. So someone who is critical is someone who is able to argue with clear and definitive data and must be able to consider the decision to be taken.

One way to achieve the skills is by improving critical thinking skills. The basics teaching that can support that goal, one of them, are the Calculus. After this goal is achieved, students are expected to more easily study math on other subject based on Calculus as a tool. This goal is a unity reflected in the knowledge, skills, and attitudes that are not mutually exclusive. curriculum assessment should include the aspect of knowledge, skills, and attitudes intact and proportional [10]. This means that the learning assessment should be able to evaluate each student's knowledge, skills, and attitudes (cognitive, affective and psychomotor). So, aspects that focus only on cognitive judgments are deemed irrelevant to the curriculum. Therefore, an assessment is required that not only focuses on cognitive assessment but also touches affective and psychomotor assessments. Lecturers as evaluators in learning need to evaluate the process implemented in the classroom. The quality of good learning should be followed by a good assessment strategy as well because information from the assessment is useful for improving the quality of learning [13]. In addition, a few papers refer to using classroom assessment to establish a harmonious classroom environment and to promote students' confidence, in fact that the purpose is to find students' strengths during their learning process, to give encouragement, and to create a harmonious environment for teaching and learning, to find students' strengths during their learning process, to give encouragement, and to create a harmonious environment for teaching and learning [16]. Also [8] the classroom assessment in Mathematics is among an instructor's most essential educational tool. When properly developed and interpreted, assessments can help teachers better understand what their students are learning. So, assessment is a key component of all educational systems [11].

Assessment by [15] is defined as a process of collecting data and or information (including in the processing and documentation) systemically about an attribute, person or object either in the form of qualitative or quantitative data about the number, condition, ability or progress of an attribute, objects or persons/individuals assessed without reference to value judgment

Assessments conducted in conventional learning generally only reveal what students know, whereas class-based assessments aim to reveal what is known and what learners can do. Through this assessment, educators not only get a picture of students' understanding of mathematical concepts that have been studied. More than that educator can express students' attitudes and motivation to the lessons and mathematical abilities.

Efforts to improve the quality of learning can be done through the model of learning and how the applied assessment mathematics. This is actually one of the factors that determine student achievement. Another factor in learning mathematics is the factor of within each student in this case of critical thinking, and significantly improve 2nd graders' mathematics achievement, including for children living in poverty, by using assessment data to individualize the mathematics instruction they receive [5]. Therefore, the study aims to determine the development of students' critical thinking skills through the implementation of the critical thinking assessment instrument in the calculus subject.

2. Methods

This research applies mix method, that is a combination of quantitative and qualitative research. The instruments used were observation sheets, tests, interview guides, and field notes. The observed indicator refers to a critical thinking indicator consisting of 6 indicators. Observations, interviews, and field notes are used to support the data found through critical thinking test results.

Research begins by observing classes that use assessment instrument. Aspects focus in observation are 1) attention to teaching lecturers, 2) ability to ask and express opinions, 3) courage to correct, and 4) pay attention to word/criticism and respect others. In addition to that observation also noted a variety of unique things in the field notes. These four aspects are shown in Table 1.

Table 1. The Aspects of observation

Aspects	Indicators
Attentions	All attention is directed to material explanations
Asking and opinion	Ask friends or lecturers about how to work / solution a topic or problem, the questions asked are relevant to the topic of learning, answering questions according to the intent and purpose of the question, or giving an argumentative opinion/response
Correcting	Dare to correct other people's wrong answers
Giving and appreciate	Respond to what friends say, including positive criticism, pay attention to what a friend is doing (listening), or admitting mistakes

Techniques of collecting data through observation was observers doing observations in a place that may be able to see all the activities of students, every 5 minutes, observers observing dominant student activity, and 30 seconds later observers write the results of observations by providing assessment Good (B), Enough (C) or Less (K) on observed activity. Observers involved in the research are as many as 5 people each responsible for observing 5-7 students. In every implementation in the classroom, open interviews are also conducted to each subject involved especially to the representative student. This interview is useful to complement/ensure data observation activities in the classroom. The grid of the interview guide is presented in Table 2.

Table 2. Interview Guidelines

Aspects	Aims
Reason	Knowing the reason of the subject to do something about learning, for example why want to ask, why should pay attention to teacher explanations or why should write something.
Aim	Knowing the purpose of doing something both orally and in writing related to critical thinking during learning.
Opinion	Request opinions or suggestions about using an assessment instrument.

Every use of the device on learning ends with a critical thinking test. The test is done three times, with the test indicator shown in Table 3.

Table 3. The test indicators of critical thinking

Indicator	Aim
Identify assumptions	To know student critical

Formulate the main problem	thinking ability
Determining the consequences of a given provision	
Detect bias	
Uncover data / concepts / definitions / theorems in solving problems	
Evaluate the relevant arguments in solving the problem	

The results of the observations were analyzed by assigning values 3 for observations B, 2 for C and 1 for K. Tests The ability of critical thinking was analyzed with the rubric of critical thinking in Table 4.

Table 4. The Rubrics of Critical Thinking Test

Indicator	**Score**			
	0	**1**	**2**	**3**
Identify assumptions	Not at all able to identify assumptions.	Not able to identify assumptions but already leads	Be able to identify assumptions but there are still not exact ones	Be able to identify assumptions correctly
Formulate the main problem	Not at all able to formulate the main problem	Not able to formulate the main problem but already leads	Be able to formulate the main problem but are still not exact one	Be able to formulate the main problem correctly
Determining the consequences of a given provision	Not at all able to Determining the consequences of a given provision	Not able to Determining the consequences of a given provision but already leads	Be able to Determining the consequences of a given provision but are still not exact one	Be able to Determining the consequences of a given provision correctly
Detect bias	Not at all able to detect bias	Not able to detect bias but already leads	Be able to detect bias but are still not exact one	Be able to detect bias correctly
Uncover data / concepts / definitions / theorems in solving problems	Not at all able to uncover data / concepts / definitions / theorems in solving problems	Not able to uncover data / concepts / definitions / theorems in solving problems but already leads	Be able to uncover data / concepts / definitions / theorems in solving problems but are still not exact one	Be able uncover data / concepts / definitions / theorems in solving problems correctly
Evaluate the relevant arguments in solving the problem	Not at all able to evaluate the relevant arguments in solving the problem	Not able to evaluate the relevant arguments in solving the problem but already leads	Be able to evaluate the relevant arguments in solving the problem but are still not exact one	Be able to evaluate the relevant arguments in solving the problem correctly

This rubric (see Table 4) is developed again based on the needs of each given test item so as to provide an accurate description of the assessment. Scores obtained by students are converted into 4 criteria, i.e. are shown in Table 5.

Table 5. The Criteria of Critical Thinking

Observation score	**Test Score**	**Criteria**
2.51 - 3.00	2.26 - 3.00	Very Critical (VC)

2.51 - 2.50	1.51 - 2.25	Critical (C)
1.51 - 2.50	0.76 - 1.50	Enough Critical (EC)
1.00 - 1.50	0 - 0.75	Not Critical (NC)

The results of observation, interview and test are described and combined, and then look at the development of each critical thinking ability in each group (High, Medium, and Low) and taken conclusions.

3. Result and Discussion

Observation result of observation of critical thinking of student on calculus lecturing showed that by using the critical thinking evaluation instrument, the ability of high group increased from criteria of EV to VC, whereas in medium and the low group did not experience significant change that remained at criteria EC. Observation result of ability thinking the criteria is shown in Table 6.

Table 6. Observation result of critical thinking

Aspect	High Group			Medium Group			Low Group		
	I	**II**	**III**	**I**	**II**	**III**	**I**	**II**	**III**
Attentions	3.00	3.00	3.00	2.80	2.00	3.00	3.00	1.50	3.00
Asking and opinion	3.00	3.00	3.00	1.80	2.00	2.33	1.50	2.50	1.00
Correcting	1.50	1.50	3.00	1.20	1.00	-	0.75	-	-
Giving and appreciate	2.25	1.50	-	3.00	3.00	1.00	3.00	-	-
Average	**2.44**	**2.25**	**3.00**	**2.20**	**2.00**	**2.11**	**2.06**	**2.00**	**2.00**
Criteria	**EC**	**EC**	**VC**	**EC**	**EC**	**EC**	**EC**	**EC**	**EC**

From Table 6, it can be seen that the use of evaluation instrument of critical thinking can invite students to pay attention to learning that is all attention is directed to a material explanation. In High Groups, from the beginning of the use of the device, the attention has a value of 3.00 (Very Good), while in the medium and low groups there is little change but still on the EC hose. For the observation of the High Group "Asking and opinion" from the initial use of observations continues to contribute very well, while in the medium-use group the device provides increased activity to ask and dispel opinions although not significant, as well as in the low-use group of assessment tools has not provided a change Which means to improve the ability to ask questions and express opinions. For aspects of correcting, and giving and appreciating, the use of assessment instrument has not had a meaningful effect, this means that the device has not contributed well to improving the response to positive criticism, paying attention to what others are saying and admitting mistakes.

The result of the observation of critical thinking ability shows that based on the predicate of students 'critical thinking ability, students' critical thinking ability in High Group is categorized as Very Critical (VC), while in Medium and Low Group is categorized as Enough Critical (EC). This implies that in the Higher Group, the use of an assessment can improve attention, increase motivation to ask questions and express opinions, to correct errors and admit mistakes. While in the middle and low groups, the use of an assessment can only improve the ability of critical thinking in the category EC.

Table 7. The Test Result Critical Thinking

Indicator	High Group			Medium Group			Low Group		
	I	**II**	**III**	**I**	**II**	**III**	**I**	**II**	**III**
Identify assumptions	1.13	2.00	2.38	0.30	1.67	1.90	0.38	1.25	1.38

Formulate the main problem	0.25	0.00	2.88	0.00	0.00	2.00	0.13	0.50	1.75
Determining the consequences of a given provision	1.13	0.00	2.75	0.70	0.00	2.47	0.50	0.00	2.58
Detect bias	0.44	0.00	1.00	0.15	0.00	1.87	0.50	1.50	2.00
Uncover data / concepts / definitions / theorems in solving problems	2.08	0.00	1.75	1.27	0.00	1.80	1.50	0.00	1.00
Evaluate the relevant arguments in solving the problem	2.50	0.00	3.00	1.60	0.00	2.20	1.00	2.25	1.42
Average	**1.26**	**0.33**	**2.29**	**0.67**	**0.28**	**2.04**	**0.67**	**0.92**	**1.69**
Criteria	**EC**	**NC**	**VC**	**NC**	**NC**	**C**	**NC**	**EC**	**C**

The results of the I test show that in the High Group the average score of 1.26 which is included in the Enough Critical criteria (EC), the best value is given to evaluate the argument and the lowest value is the ability to formulate the main problem. In Test II, critical thinking has decreased and belongs to the category Not Critical (NC), it happens because almost all students in the High group are not able to complete the critical thinking skills test on the indicators formulate of the main problem, determine the effect of a provision, detect bias, reveal Data and evaluate arguments, only on identifying the assumptions of high-ability students capable of completing them. In Tests III, students' critical thinking skills of High students experienced a good improvement, almost all indicators can be solved well, the average obtained by the high group in TEs III is 2.29 which included in criteria Very Critical (SK). The average value of Test I in the Medium Group is 0.67 which is included in the Non-critical criteria. The lowest score is the indicator formulates the main issues and the highest in the evaluation of the argument, as well as in the second test, almost the same as the High group almost all indicators cannot be properly resolved. However, in Tests III, the average value of the moderate group was 2.04 which included in the Critical (K) criterion. This means that there is a good change to students' critical thinking skills. The test results in the low group changed. The test I, Tests II, and Tests III, respectively, the low group obtained a value of 0.67; 0.92 and 1.69 are included in the criterion Not Critical (NC), Enough Critical (EC) and Critical (C). So, in the low group, the use of critical thinking assessment tools provides a good influence on the development of students' critical thinking skills

Decreasing Test II results in High and Low groups are influenced by the level of questions given and also reduced exercise. This is consistent with that found by [9] which states that students' critical thinking can be successfully encouraged and developed in basic mathematics education with content selection (task), but this result is also influenced by the attitude of students in learning mathematics, it is relevant to result [4] that state is the Students` Emotional Life and Their Attitude toward Mathematics Learning.

The results of interviews with students, the use of assessment instrument make them more careful and increase accuracy to understand the problem. Especially must know the reason statement or argumentative opinion of the solution to the problem that is faced, so student in solving problem given more emphasis to focus attention on explanation, also student stated that in their work duty more raise attention and think deep to take action against Solving problems, and also better understand the purpose of doing something. This is certainly very influential on improving students' critical thinking in learning Calculus. Recognizing the weaknesses and the advantages of the assigned tasks gives a positive effect on the subsequent tasks, this is in accordance with the results of research [3] which states providing diagnostic feedback improves math test scores. In addition, Tran, Ho, Mackenzie and Le 14] findings indicate that the ACLC effectively supported the teachers in their preparation and implementation of teaching for creativity at these Vietnamese secondary schools. Findings also indicate that specifically guiding the teachers in how to use the ACLC and giving them prompt feedback on their teaching for creativity is helpful in promoting teaching for creativity.

4. Conclusion

The results showed that in high-level students there is a change of ability from Critical enough to be Very Critical, in the students with moderate and low ability there is a change of ability from Uncritical to Critical. So, it can be concluded that the assessment instruments have a good contribution and can improve the ability of critical thinking.

References

[1] Aktaş G S and Ünlü M 2013 Critical Thinking Skills of Teacher Candidates of Elementary Mathematics. *Procedia - Social and Behavioral Sciences* **93** 831 – 835

[2] Mahadaniar A G, Siswaningsih W and Nahadi 2013 *Jurnal Penelitian dan Evaluasi Pendidikan* ***18*** 2 pp 245–252

[3] Betts J R, Hahn Y and Zau A C 2017 *Journal of Urban Economics* **100** 54-64

[4] Colomeischi A A and Colomeischi T 2015 *Procedia - Social and Behavioral Sciences* **180** 744 – 750

[5] Connor C M, Mazzocco M M M, Kurz T, Crowed E C, Tighee E L, Wood T S and Morrison F J 2017 *Journal of School Psychology*

[6] Fachrurazi 2011 *Jurnal Penelitian Pendidikan Khusus* **1** 76–89

[7] Hassoubah Z 2004 *Developing Creatif & Critical Thingking Skill* Terjemahan Bambang Suriadi (Bandung: Nuansa)

[8] Junpeng P 2012 The development of classroom assessment system in Mathematics for basic education of Thailand *Procedia - Social and Behavioral Sciences* **69** 1965–1972.

[9] Maričića S and Špijunovićb K 2015 Developing Critical Thinking in Elementary Mathematics Education through a Suitable Selection of Content and Overall Student Performance, *Procedia Social and Behavioral Sciencies* **180** 653-659

[10] Mulyasa 2013 *Pengembangan dan Implementasi Kurikulum 2013* (Bandung: PT. Remaja Rosdakarya)

[11] Sangwin C J and Köcher N 2016 *Computers & Education* **94** 215-227

[12] Syahrul 2009 Keefektifan Penerapan Model Asesmen Autentik Terintegrasi Dalam Pembelajaran Praktikum Ft-Universitas Negeri Makassar *Jurnal MEDTEK **1*** 2

[13] Syahrul 2010 *Jurnal Penelitian Dan Evaluasi Pendidikan **10*** 2 pp 246–268

[14] Tran T T T, Ho T N, Mackenzie S V and Le L K 2017 *Thinking Skills and Creativity* **25** 10-26

[15] Yusuf M 2011 Asesmen dan Evaluasi Pendidikan (Padang: Kencana)

[16] Zhao, X Van den Heuvel-Panhuizen M and Veldhuis M 2017 *Studies in Educational Evaluation* **52** 42-54

Acknowledgments

This research project received a research grant from the Ministry of Research, Technology and Higher Education this research was supported by KOPERTIS Region X. in here, the researchers would like to express they're thanks to Coordinator KOPERTIS Region X, chair of STKIP PGRI West Sumatera and head of Research Unit and Community Service STKIP PGRI West Sumatera.

Promoting middle school students' mathematical creative thinking ability using scientific approach

A Istiqomah[1], K S Perbowo[2] and S E Purwanto[3]

[1] Department of Mathematics Education, Universitas Muhammadiyah Prof. DR. HAMKA, Jl. Tanah Merdeka, Jakarta Timur, Jakarta, Indonesia

E-mail: krisna_satrio@uhamka.ac.id

Abstract. This research aims to identify the strength of scientific approach in order to develop mathematical creative thinking in junior high school. Descriptive qualitative method is used in this research. 34 students in 7th grade are chosen using purposive sampling. For collecting data, this research uses test, observation, and interview. The test consists of 6 items which have been tested for their validity and reliability and used in pre-test and post-test. The pre-test shows that students average score in mathematical creative thinking is 43 (low), while in post-test it is 69 (middle). The N-gain in mathematical creative thinking point is 0.461, which is classified in the middle grade. Furthermore, the N-gain for each indicator, they score 0.438 for fluency; 0.568 for flexibility; and 0.382 for novelty. The N-gain for those indicators falls under middle grade. The research shows that scientific approach develops more flexibility, and, on the other hand, it develops less novelty.

1. Introduction

Mathematics is a course that students have been learning since elementary school aiming at providing students with logical, analytical, systematic, critical and creative thinking ability. As stated earlier, one of the mathematical thinking abilities is creative thinking. Creative thinking is a form of thinking to find new answers, methods, or ways in response to a problem in order to solve it [1][2]. Students' mathematical creative thinking ability is a habit which can be trained. Creativity will develop under the right circumstance, yet it will deteriorate when no support is given by those around. Schools can help students develop their problem-solving skills and develop their creativity at the same time [3][4].

Generally speaking, the teaching and learning at schools focus more on developing analytical ability. The teaching and learning emphasize more on memorization and finding the answers to the questions provided. The focus on and attention to the effort of improving creative thinking ability in mathematics have infrequently or never been developed [5]. Based on the research conducted by Fardah [6] out of the 30 students she studies only 7 are found to have high creative thinking ability. The result is not greatly different from Septianingrum's [7] research which finds that those students with high mathematical ability are not in any creative thinking ability level.

The indicators of creative thinking ability according to Torrance are *Fluency*, *Flexibility*, *Originality* and *Elaboration* [2]. Silver also adds that in order to assess one's creative thinking three key components in creativity in using TTCT need to be used, i.e. *Fluency*, *Flexibility*, and *Novelty* [8].

The Indonesian government in their effort of improving its education quality has amended their curriculum into Curriculum 2013. Curriculum 2013 refers to the process of teaching and learning which is executed with scientific approach. Scientific approach is a way or mechanism of teaching and learning to facilitate students in order for them to obtain knowledge or skill using a procedure which is based on a scientific method [9]. Scientific approach is designed to provide students with creative thinking ability. The teaching and learning processes in scientific approach are characterized by 5M activities, namely *mengamati* (observing), *menanya* (questioning), *mengumpulkan informasi* (collecting information), *mengolah informasi* (processing information), and *mengkomunikasikan* (communicating). The 5M activity will influence student's creative thinking mindset, where students are given a chance to observe the existing phenomena, to question, to reason, and, finally to communicate the results they have obtained. In addition, innovation is expected to surface through creative thinking ability to find various answers to a problem. Through these steps of scientific approach, it can then be determined students' mathematical creative thinking ability in scientific approach which is limited by such indicators as fluency, flexibility, and novelty. The fluency aspect includes students' ability in solving problems correctly and accurately. The flexibility aspect includes the ability to employ various problem-solving strategies. Finally, the novelty aspect includes students' ability to use new strategies.

Based on the description above, this research aims to analyze student's mathematical creative thinking ability in scientific approach, so that the strengths and weaknesses of scientific approach in building mathematical creative thinking ability can be found.

2. Method

This research is conducted in Junior High School 49 Jakarta, at Jalan Raya Bogor Km. 20 Kramat Jati, Jakarta Timur 13510. It uses descriptive qualitative method, intending to describe students' mathematical creative thinking ability in the scientific approach. The sample is taken using *Purposive Sampling* [10]. The research sample is class VII–F amounting to 34 students. The data are collected using tests, observation, and interview. The students' mathematical creative thinking ability test contains 6 subjective questions on triangle section the validity and reliability of which have been confirmed based on fluency, flexibility, and novelty indicators.

The observation aims at finding out students' mathematical creative thinking ability which is built during the teaching and learning process using the scientific approach as written down in a field record. The interview technique used here is unstructured interview aided with interview guideline. In this research, the researchers use the indicators and sub-indicators as adapted from Olson and Willian [11] as presented in Table 1.

Table 1. Indicators and Sub-Indicators of Students' Mathematical Creative Thinking Ability.

No.	Indicator	Sub Indicator
1.	*Fluency*	a. Stir/produce thoughts in large number. b. Determine problem-solving steps completely and correctly. c. Predict answers fluently and quickly to a problem.
2.	*Flexibility*	a. Provide varied answers/ideas. b. Be able to see a problem from different points of view. c. Finding many alternatives or different ways of solving a problem.
3.	*Novelty*	a. Provide new and unique expressions/responses. b. Possess uncommon ways of solving problems. c. Determine his/her own ways of solving problems.

3. Results and Discussion

This research aims at analyzing students' mathematical creative thinking ability in scientific approach. Pre- and post-tests are used to figure out students' mathematical creative thinking ability before and after the teaching and learning using scientific approach. Based on the pre-test results, the mean score is 43 (low) and their post-test score is 69 (medium). The fluency, flexibility, and novelty indicators during pre-test have such mean scores as 50 (low), 53 (low), and 26 (low) respectively, and the post-test results indicate that the fluency, flexibility, and novelty indicators have the following mean scores respectively 72 (medium), 80 (high), and 55 (low). Below is the table on student distribution in each of students' mathematical creative thinking ability indicators.

Table 2. Student Distribution in Students' Mathematical Creative Thinking Ability Indicators

Fluency Indicator

Category	**Number of Students**	
	Pre-Test	**Post-Test**
High	2	16
Medium	11	5
Low	21	13

Flexibility Indicator

Category	**Number of Students**	
	Pre-Test	**Post-Test**
High	6	20
Medium	4	6
Low	24	8

Novelty Indicator

Category	**Number of Students**	
	Pre-Test	**Post-Test**
High	0	3
Medium	0	1
Low	34	30

After the pre-test is administered, the researchers also perform some observation by attending the teaching and learning using scientific approach. Based on the observation results, it is found that the scientific teaching and learning process in building students' mathematical creative thinking ability does not fully involve students actively. This can be seen from the fact that during the teaching and learning process, those students sitting in the rear rows seem ignorant to the materials being delivered, and no student asks any question nor expresses their ideas even when they are working on the questions. The 5M activities are implemented quite well, except for 2 activities of collecting information and communicating the results, hence student's ability in expressing new and unique ideas remain unobserved.

Just as the observation results, the interview conducted by the researchers in relation to the teaching and learning process using scientific approach performed by teachers in building mathematical creative thinking ability finds that it is less optimal. The less attention that teachers give to students' preparedness to receive the materials to be tougher is seen from the interview where students fail to explain the results observed from the teaching and learning process on the materials being delivered. In addition, when interviewed students say they are shy to ask questions during the teaching and

IOP Conf. Series: Journal of Physics: Conf. Series **948** (2018) 012032 doi:10.1088/1742-6596/948/1/012032

learning process, yet they will ask when teachers go close to them. Students also say that they only work on the questions coming from package textbooks and they suffice themselves with just one answer. Later on, students' works are directly submitted and discussion (communication) is rarely held in the classroom.

To find out the improvement of Pre- and Post-Test scores of students' mathematical creative thinking ability, the N-gain formula according to Hake [12] is used.

$$\langle g \rangle = \frac{< S_f > - < S_i >}{100 - < S_i >}$$

The results of N-gain calculation is presented in table 3 below.

Table 3. Results of N-Gain Calculation

Indicator	Mean n-gain	Category
Fluency	0.438	Medium
Flexibility	0.568	Medium
Novelty	0.382	Medium

Based on the results of mean N-gain, it is found that students' mathematical creative thinking ability is 0.461 at medium category improvement. From the results of mean N-gain of each indicator, it is found that scientific approach can build students' mathematical creative thinking ability at the medium category. The sequence of strengths of scientific approach in building mathematical creative thinking ability is flexibility, fluency, and novelty.

4. Conclusion

The mean score of Pre-Test is 43 (low). And from the observation of teacher and student activities during teaching and learning using scientific approach, it can be concluded that 5M activities are implemented quite well, except for 2 activities of collecting information and communicating the results. The teaching and learning of mathematics using scientific approach fairly improve students' mathematical creative thinking ability as can be seen from the Post-Test results at a mean score of 69 (category medium). The students' improved mathematical creative thinking ability using scientific approach is also seen from the results of N-gain calculation at a medium improvement category at 0.461. From the results of N-gain calculation, it is also found that the strength of scientific approach in building mathematical creative thinking ability lies with flexibility indicator. However, scientific approach still fails to build the novelty indicator. Therefore, the teaching and learning using scientific approach should be performed optimally and by actively involving students in order for it to be able to improve students' mathematical creative thinking ability.

5. References

[1] Uno H B, Umar M K, and Panjaitan K 2014 *Variabel Penelitian Dalam Pendidikan dan Pembelajaran* (Jakarta: PT Ina Publikatama)
[2] Lestari K E and Yudhanegara M R 2015 *Penelitian Pendidikan Matematika* (Bandung: PT Refika Aditama)
[3] Sani R A 2014 *Pembelajaran Saintifik untuk Implementasi Kurikulum 2013* (Jakarta: PT Bumi Aksara)
[4] Riyanto Y 2012 *Paradigma Baru Pembelajaran* (Jakarta: Kencana)
[5] Fauziyah I N L 2013 *Jurnal Pendidikan Matematika Solusi* **1** p 75

[6] Fardah D K 2013 *Unnes Journal of Mathematics Education Research* **3**
[7] Septianingrum M F 2015 Analisis Tingkat Kemampuan Berpikir Kreatif Siswa dalam Mengajukan Masalah Matematika Berdasarkan Kemampuan Matematika Kelas XI-MIA G SMAN 1 Probolinggo (Jember: UNEJ)
[8] Siswono T Y E 2009 *Konstruksi Teoritik Tentang Tingkat Berpikir Kreatif Siswa dalam Matematika Retrieved from* https://tatagyes.files.wordpress.com/2009/11/paper07_jurnal_univadibuana.pdf
[9] Kemdikbud 2013 Materi Pelatihan Guru Implementasi Kurikulum 2013 (Jakarta: Kemdikbud)
[10] Moleong L J 2013 Metodelogi Penelitian Kualitatif (Bandung: PT Remaja Rosdakarya)
[11] Siswono T Y E 2008 Model Pembelajaran Berbasis Pengajuan dan Pemecahan Masalah Untuk Meningkatkan Kemampuan Berpikir Kreatif (Surabaya: UNESA Press)
[12] Hake R R 1999 Analyzing Change/Gain Scores *Retrieved from* http://www.Physics.indiana.edu/~sdi/Analyzingchange.Gain.pdf

IOP Conf. Series: Journal of Physics: Conf. Series **948** (2018) 012033 doi:10.1088/1742-6596/948/1/012033

Assessment of statistical education in Indonesia: Preliminary results and initiation to simulation-based inference

K V I Saputra[1], L Cahyadi[1] and U A Sembiring[1]

[1]Department of Mathematics, Universitas Pelita Harapan, Jl MH Thamrin Boulevard, Lippo Karawaci, Tangerang, Banten, Indonesia

E-mail: kie.saputra@uph.edu

Abstract. Start in this paper, we assess our traditional elementary statistics education and also we introduce elementary statistics with simulation-based inference. To assess our statistical class, we adapt the well-known CAOS (Comprehensive Assessment of Outcomes in Statistics) test that serves as an external measure to assess the student's basic statistical literacy. This test generally represents as an accepted measure of statistical literacy. We also introduce a new teaching method on elementary statistics class. Different from the traditional elementary statistics course, we will introduce a simulation-based inference method to conduct hypothesis testing. From the literature, it has shown that this new teaching method works very well in increasing student's understanding of statistics.

1. Introduction

In the undergraduate education in Indonesia, statistics is one of the few courses that are taught in almost every department whether the department is in engineering faculty, natural science faculty, social science faculty or in business school. Statistics course sometimes is renamed as something else such as Elementary Statistics, Introductory Statistics, or even Biostatistics in biology department but the contents of those are just the same. There have been a lot of books and resources on Elementary Statistics that first-year undergraduate students can have whether it is in Indonesian or in English but the contents of such books are also conceptually the same. The fact that many students in Indonesia are still afraid of these statistics courses means that the way of teaching statistics both conceptually and pedagogically has to be evaluated. However, there has been an effort neither from the government nor from academia to assess the statistical education in undergraduate institutions.

We believe that the Elementary Statistics course that has been taught in Indonesian Institution follows the algebra-based introductory statistics course that has been implemented in a number of popular textbooks (see for instances [1, 2, 3, 4]. These books tend to the same four parts (although some books mentioned three parts as the first two are considered to be in one part). The parts are descriptive statistics; data collection and design; probability and sampling distributions and some knowledge about basic random variables; and inferential statistics. In some literatures (see [5, 6]) they found that by the time students get to inferential statistics (confidence intervals and hypothesis testing), which is the pivotal point of this course, they end up in "survival" mode, in which they "just" memorize formulas without really understanding the underlying concepts of how to draw valid and reliable conclusions from data. With this sequencing, Cobb [7] argues that student felt comfortable learning the first two parts, however, their understanding somehow dissipated as they landed in the third part as it conceptually difficult, and technically disconnected with real data analysis that was learned in the beginning. By the final fourth of the course, many students were not able to connect the dots in the previous part and were not able to grasp the idea of inferential statistics. This problem was also combined by the "stress" students had to experience by the end of the semester and resulted in a little understanding of what is statistics really about.

2. Methods

2.1. Recent Development of Elementary Statistics in United States

As far as the author knows, development of elementary statistics course probably started in 1997 when Moore [8] challenged all statistical educators to change their teaching pedagogy from the old model to the new model. The old model, as we also know, is more teacher-centered where students learn by absorbing information, while the new model is more student-centered, in which students learn through their own activities. Then there was relatively no result until 2005 when the American Statistics Association published the Guidelines for Assessment and Instruction in Statistical Education, also known as GAISE (see [9]). This guideline gave recommendations which at that time had not been implemented by either any instructor in their elementary statistics classes or in many statistics textbooks. Some recommendations in GAISE are: (1) to develop statistical thinking, (2) to use real data, (3) to emphasize statistical concepts rather than procedures, (4) to focus on student active learning, (5) to use technology in analyzing data, and (6) to use assessment to evaluate and to improve the learning process. However, Cobb [7] again argued that this guide along with its recommendation is not enough and urged statistical educators to reconsider the pedagogy and the content of the elementary statistics course.

Along with the above guide, in 2007 Delmas et al. [10] developed a comprehensive assessment of student understanding in the elementary statistics course, also known as the CAOS test. This assessment was the first standard, comprehensive assessment instrument for the introductory statistics course. When it first implemented before class and after class, the result was startling as the students only gaining average 9% difference, which implying most students didn't really understand about statistics. It also implied that the statistical educator must do something about this. Other assessments that have been used to evaluate elementary statistics classes are SATS (Survey Attitudes Toward Statistics) [11] and also GOALS-2 (The Goals and Outcomes Associated with Learning Statistics) [12]. Both have been successfully evaluated the impact of introductory statistics classes in U.S. institutions.

Tintle et al. [5] as far as the author knows was the first to develop a new pedagogy in teaching statistics. They not only took a student active learning approach which also is in line with the GAISE recommendation but also completely re-ordering, re-emphasizing, adding and subtracting content of the traditional elementary statistical course. The result is quite significant as it showed that there was a strong evidence (p-value < 0.001) that the student average in the new pedagogy is higher than that in the current pedagogy. Recently, Chance et al. [13] also showed a promising evidence that the simulation-based inference (SBI) curriculum works better than the traditional curriculum. The article analyzed multi-institution data which are pre/post of student attitudes and conceptual understanding of topics in statistics course in those institutions.

2.2. Development of THE simulation-based inference (SBI) CURRICULUM

The need of changing the pedagogy also is motivated by the fact that some statistical concept was taught in high school. High school education in US has already seen much of the material in the first part of the traditional statistics course [9]. This does not only happen in US but it also happens in many countries such as Indonesia [14]. Indonesia high school education currently uses the 2013 curriculum which exposes students to learn statistics descriptive. They will learn as much as center data (mean, median), variability, histogram and other concepts of presenting data. This means that the elementary statistics material taught in first-year college or first-year university has to be reshaped.

Another motivation of changing the pedagogy is the number of exceptions given in the traditional statistics course. As an example, suppose we wanted to test of hypothesis for the population mean. We use the z distribution to perform a test if the number of samples is more than 30. However, if the number of samples is less than 30 we must use the t distribution to perform such a test. But sometimes the population distribution is known normal and the population standard deviation is given, so even if the number of samples is less than 30, we can still use the z distribution to perform this test. This exception is one of many that not only confuse students but also drive the student to just memorize the rule and the formula.

In our statistics course with simulation-based inference (SBI), we shall teach students data collection, formulation research question, hypothesis testing, and statistical inference in the first week. We believe students have to know the statistical process from the beginning until making conclusion. We also emphasize the logic of statistical inference by using simulation test, and therefore we will integrate computer technology into our curriculum. The students have to be active in our class, because we will expose them some real problems that are aligned with the topic we are discussing. We highlight the following:

1. We only review topics on descriptive statistics but do not spend explicit time discussing this topic
2. After establishing the logic of inference, students are asked to draw a connection between simulation-based and theory (asymptotic) based
3. We present confidence interval as a result of test of significance, not the other way around
4. We underline the difference between observational study and experiment
5. We utilize statistical process in a large-scale research project

We now give an example of the syllabus of the traditional elementary statistics course which one can see from Table 1. We assume that the course will be taught in 16 weeks with mid-semester exam at week 8 and final semester exam at week 16. As we have stated before, typical syllabus of traditional elementary statistics course consists of four parts, i.e. data analysis (week 1-2), descriptive statistics (week 3), probability theory and random variables (week 4-7), and inferential statistics (week 9-15). The students first learn about hypothesis testing in week 10, which is near the end of the course. Some students may be able to understand the concept of inferential but most of them will not be able to understand the first three parts and how to use them in inferential statistics.

We would like to give also an example of the syllabus of the elementary statistics with simulation-based inference. As one can see from Table 2, we already introduce statistical inference the first two weeks. We then expose the student to the logic of statistical inference every week, adding more and more statistical concepts as time goes on. The descriptive statistics will be reviewed as needed but we do not dedicate a time slot for it. We also replace the probability theory and random variables part in the traditional statistics course with simulation-based. We also will compare the simulation-based inference with the theory-based but we shall explain to the students that the theory behind the theory-based inference will be taught in the next statistics course.

Table 1. Typical syllabus of the traditional elementary statistics course. This may vary depending on the textbook used by lecturer and also depend on the faculty this course taught

Week	Topics	Learning Outcomes
1	Introduction to Statistics	• Understand statistics and type of statistics • Understand difference between sample and population • Understand type of variables
2	Organizing and Graphing Data	• Understand how to organizing and graphing qualitative and quantitative data • Able to make cumulative frequency distributions • Able to make graphical representation (stem and leaf, dot-plots) of data

3	Descriptive Statistics	• Understand measurement of central tendency of data • Understand measurement of dispersion of data • Able to make box-plot data
4	Probability Theory	• Understand the theory of probability • Understand addition and multiplication rule to compute probability • Understand permutation and combination
5	Some Discrete Distributions	• Understand probability distribution of discrete random variables • Understand binomial, Hyper-geometric and Poisson distributions
6	Some Continuous Distributions	• Understand normal distribution and standard normal distributions • Able to determine probability of normal distribution using normal table and its inverse • Understand normal distribution approximation using binomial distribution
7	Sampling Distributions	• Understand sampling distribution • Understand sampling error and non-sampling error • Understand mean and standard deviation of $\bar{x}$ • Understand mean and standard deviation of $\hat{p}$
8	Mid Semester Exam	
9	Estimation of mean and proportion	• Understand estimation of μ when σ is known • Understand estimation of μ when σ is not known • Understand estimation of population proportion
10	Hypothesis testing about mean and proportion	• Hypothesis testing of μ when σ is known • Hypothesis testing of μ when σ is not known • Hypothesis testing of population proportion
11	Estimation and Hypothesis Testing: Two Populations	• Hypothesis testing and estimation of two population when $\sigma 1$ and $\sigma 2$ are known • Hypothesis testing and estimation of two population when $\sigma 1$ and $\sigma 2$ are unknown but equal
12	Estimation and Hypothesis Testing: Two Populations	• Hypothesis testing and estimation of two population when $\sigma 1$ and $\sigma 2$ are unknown and unequal • Hypothesis testing and estimation of two population for paired sample • Hypothesis testing and estimation of two population proportion
13	Chi-Square Tests	• Understand chi-square distribution • Understand a goodness of fit test
14	Analysis of Variance	• Understand F distribution • Understand one-way analysis of variance
15	Simple Linear Regression	• Able to compute simple linear regression • Understand inferences of slope • Able to compute correlation • Understand application of regression model
16	Final Semester Examination	

Table 2. Example syllabus of the elementary statistics with simulation-based inference. This is one example of syllabus that has been implemented in the Department of Mathematics,

Universitas Pelita Harapan.

Week	Topics	Learning Outcomes
1	Introduction to Statistics (P1 - P3)	• Understand six step statistics process • Understand how to explore data • Understand random process • Small Project
2	Significance: How strong is the evidence? One proportion test (1.1 - 1.5)	• Understand chance model in one proportion test • Understand measuring strength of evidence; p-value • Understand factors that affect strength of evidence • One proportion; theory-based approach • Small Project
3	Generalization: How Broadly Do the Results Apply? One mean test (2.1 - 2.3)	• Understand chance model in one mean test • Understand sampling in finite population • Understand error and significant • Small Project
4	Estimation: How Large is the Effect? Confidence Interval (3.1 - 3.3)	• Understand the concept of confidence interval • Understand 2SD method for estimation of proportion and mean populations • Small Project
5	Estimation: How Large is the Effect? Confidence Interval (3.2 - 3.5)	• Understand the theory-based method for estimation of proportion and mean populations • Understand factors that affect confidence interval • Small Project
6	Causation: Can We Say What Caused the Effect? (4.1-4.2)	• Understand definition of causation and confounding • Understand observational studies and experiments • Small Project
7	Comparing Two Proportions (5.1 - 5.3)	• Understand hypothesis testing in more general situation • Understand how to compare two proportions: simulation-based • Understand how to compare two proportions: theory-based • Small Project
8	Mid Semester Exam	
9	Comparing Two Means (6.1 - 6.3)	• Understand how to compare two means: simulation-based • Understand how to compare two means: theory-based • Small Project
10	Paired Data: One Quantitative Variable (7.1 - 7.3)	• Understand pair design vs two mean test • Understand how to do a paired test: simulation-based • Understand how to do a paired test: theory-based • Small Project
11	Generate population/chance model (Review Chapter 5,6,7)	• Understand how to generate simulation for two proportion test • Understand how to generate simulation for two mean test • Understand how to generate simulation for a paired test • Small Project
12	Comparing More Than Two Proportions	• Understanding simulation-based approach to compare multiple proportions

	(8.1-8.2)	• Understanding the theory-based approach to compare multiple proportions • Small Project
13	Comparing More Than Two Means (9.1-9.2)	• Understanding simulation-based approach to compare multiple means • Understanding the theory-based approach to compare multiple means • Small Project
14	Two Quantitative Variables (10.1-10.2)	• Able to compute correlation between two quantitative variables • Inference of correlation: simulation-based approach • Small Project
15	Two Quantitative Variables (10.3-10.5)	• Able to compute simple linear regression • Inference of slope: simulation-based approach • Inference of slope: theory-based approach • Small Project
16	Final Semester Examination	

We now would like to give an example of simulation-based inference on hypothesis testing. Suppose we want to do a hypothesis testing on population proportion. We want to check whether the null hypothesis (H0: $\pi = 0.5$) is true or the alternative hypothesis is true (Ha: $\pi > 0.5$). From the sample, we are given that there are 8 correct trials out of 10 trials. In this case, we could not use the theory-based inference, as our sample is not large enough. In case the sample is large enough, we need to compute the statistic z to determine whether we can reject H0. However, using simulation-based inference, we could easily do this. When the null hypothesis is true, we have $\pi = 0.5$, which is equivalent to a coin toss process. Therefore, we are going to toss a coin ten times and compute how many times the coin landed on head. We shall repeat this simulation many times (say 500 times) and we shall compute a number of simulations that landed on head more than 8 times. This proportion (relative to 500) is an approximation of p-value. Using this logic of inference, students will do some experiments on tossing coins. Each of them will have to toss a coin ten times and record how many times it landed on head. They will understand that if p-value is small then the sample (8 out of 10) provides a strong evidence against the null hypothesis. But if the p-value is large then the sample gives us not enough evidence to reject the null hypothesis. The student will use computer to simulate this many times. We will also introduce the theory-based inference as an alternative.

3. Results and Discussion

In this preliminary stage, we conducted a pre and post survey on students of batch 2015 from Mathematics Department at Universitas Pelita Harapan in the Even Semester 2015/2016. In this semester the traditional elementary statistics was taught. Students would complete our survey that consists of their demography, their attitude toward statistics, and their understanding of some statistical concepts, in the first week and again during the last week of the semester. We are going to compare the result of these two surveys and we want to see the difference between the two. In our next paper, we will conduct the same analysis of pre and post survey on students of batch 2016 from Mathematics Department in the Odd Semester 2016/2017. This batch, however, will use the elementary statistics using simulation-based inference (SBI).

The first analysis that we are going to do is about the student's attitude toward statistics. One can see the result of our survey in Table 3. The survey about student's attitude toward statistics was using Likert-type scale. Before the class started, student's attitude toward statistics is rather high, that is about 62%. It shows that on average students understand that statistics is important. The standard deviation is also small, meaning that there is not a lot of variation here. When we conducted the post-survey, the result is about the same. The average student's attitude toward statistics is 63% but the variation is now increasing. We suspect that some of the students find statistics difficult and decided to have a non-statistics career.

Table 3. Comparison of student's attitude toward statistics before and after class.

	n	Average (%) student's attitude toward statistics	SD
Pre Test	32	62.7	4.8
Post Test	24	63.7	6.2

Next, we would like to analyze the student's understanding of some statistics concept. One can see the result in Table 4. In this survey, we test some concepts in statistics, such as data collection, descriptive statistics, confidence interval, scope of conclusion, significance and simulation. On average, the score of the students before the class started is 43.2% however, after the semester ended the average score of students is 40.8%. This is the thing that we expected. After the semester ended, we ought to have an increasing value but instead, we could not see the improvement of student understanding. The variation is not so much changed.

Table 4. Comparison of student's understanding to some statistical concepts before and after class.

	n	Average (%) student's understanding	SD
Pre Test	32	43.2	9.4
Post Test	24	40.8	10.4
Change	-	-2.4	1

We can also break down the concept to see which of the concept students can easily understand. One can see in Table 5. It is somehow interesting that the score after the semester ended is lower than the score before the semester started. This could be because of many reasons. The important point is that the students are not getting better in statistics.

Table 5. Comparison of student's understanding to each of statistical concepts before and after class.

	Average Percent Correct Pre	SD	Average Percent Correct Post	SD	Average Change	SD
Data collection	46.1	29.5	35.9	19.3	-10.2	-10.2
Descriptive Statistics	41.5	20.6	39.8	19.7	-1.7	-0.9
Confidence Interval	31.8	20	32.6	24.3	0.8	4.3
Scope of conclusion	54.7	34.5	41.7	28.2	-13	-6.3
Significance	51.8	16.9	51.3	12.6	-0.5	-4.3
Simulation	36.3	17.9	32.5	18.4	-3.8	0.5

Our last analysis is a pair test. In previous analysis, we assume that the data is independent, so that we can conduct a two mean test. However, it is obvious that the data is not independent. Since the number of participants is not the same, we have to only consider participants that are active in both pre and post survey. In all cases, we also find that p-value is greater than 10%. Therefore we cannot conclude that there is a significant difference between the pre and the post.

Table 6. Pair data analysis of student's understanding to each of statistical concepts before and after class.

	Average Percent Correct Pre	SD	p-value
Mean difference	-1.598	29.5	0.52
Standard deviation	10.9	20.6	

4. Conclusion

In our analysis above, we assume that the data is reliable. However, this might not be true. We have to consider the situation of students who conducted the test. When they conducted the pre-survey, they might consider this seriously, but when the semester has ended, they might just do it recklessly. We do not give them benefit for writing this survey. On top of it, the survey itself is a long survey and it consists of more than 50 questions.

Next thing we want to do is to conduct another survey on students, but in this case, we are going to use the simulation-based inference curriculum. We also want to spread this new teaching method through the website, workshop and other publications (see details on Fig. 1).

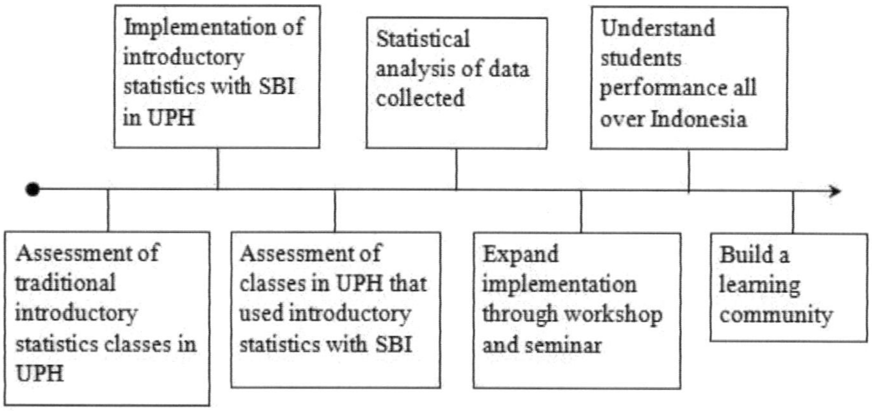

Figure 1. Timeline of our project on implementation of elementary statistics with simulation-based inference.

References

[1] Mann P S 2007 Introductory Statistics (Hoboken, NJ: John Wiley & Sons)
[2] Newbold P, Carlson W and Thorne B 2012 Statistics for business and economics (New Jersey: Pearson)
[3] Wild C J and Seber G A F 2000 Chance encounters: The first course in Data Analysis and Inference (Hoboken, NJ: John Wiley & Sons)
[4] Walpole R E et al. 1993 Probability and statistics for engineers and scientists **5** (New York: Macmillan)
[5] Tintle N, VanderStoep J, Holmes V-L, Quisenberry B and Swanson T 2011 *Journal of Statistics Education* **19** 1 pp 1-25
[6] Tintle N, Topliff K, VanderStoep J, Holmes V-L and Swanson T 2012 *Statistics Education Research Journal* **11** 1 pp 21-40
[7] Cobb G W 2007 *Technology Innovations in Statistics Education* **1** 1 pp 1-16
[8] Moore D S 1997 *International statistical review* **65** 2 pp 123-165

[9] Aliaga M, Cuff C, Garfield J, Lock R, Utts J and Witmer J 2005 Guidelines for Assessment and Instruction in Statistics Education (GAISE): College Report. American Statistical Association.
[10] Delmas R, Garfield J, Ooms A and Chance B 2007 *Statistics Education Research Journal* **6** 2 pp 28
[11] Schau C and Emmioğlu E 2012 *Statistics Education Research Journal* **11** 2 pp 86-94.
[12] Sabbag A G and Zieffler A 2015 *Statistics Education Research Journal* **14** 2
[13] Chance B, Wong J and Tintle N 2016 *Journal of Statistics Education* **24** 3 pp 114-126

Acknowledgments
Funding for this project is provided by the USAID through Partnership for Enhanced Engagement in Research (PEER) with grant no: 2000006278. We acknowledge all of our students who became our respondents of these activities. We also thank Nathan Tintle et al. for willing to have us adapt materials from their projects.

IOP Conf. Series: Journal of Physics: Conf. Series **948** (2018) 012034 doi:10.1088/1742-6596/948/1/012034

Improving students' understanding of mathematical concept using maple

Y L Ningsih and R Paradesa

Universitas PGRI Palembang, Indonesia

E-mail: yunikalestari@univpgri-palembang.ac.id

Abstract. This study aimed to improve students' understanding of mathematical concept ability through implementation of using Maple in learning and expository learning. This study used a quasi-experimental research with pretest-posttest control group design. The sample on this study was 61 students in the second semester of Mathematics Education of Universitas PGRI Palembang, South Sumatera in academic year 2016/2017. The sample was divided into two classes, one class as the experiment class who using Maple in learning and the other class as a control class who received expository learning. Data were collective through the test of mathematical initial ability and mathematical concept understanding ability. Data were analyzed by t-test and two ways ANOVA. The results of this study showed (1) the improvement of students' mathematical concept understanding ability who using Maple in learning is better than those who using expository learning; (2) there is no interaction between learning model and students' mathematical initial ability toward the improvement of students' understanding of mathematical concept ability.

1. Introduction

The first goal of learning mathematics is students have the ability to understanding the mathematical concept. This statement was mentioned in the Regulation of the Minister of National Education of Indonesia, number 22 in year 2006 [1]. Having ability to understanding mathematical concept is very important to students when they learn mathematics, and when they solve their daily life problems. In learning mathematics, the ability to understanding the mathematical concept is the main thinking that can lead students to have other mathematical ability [2, 3].

The NCTM describes that the ability to understanding the mathematical concept is the ability to (1) define the concept, (2) identifying the examples and not examples, (3) using the models, diagrams, and symbols to present a concept, (4) changing one form into representation to another form, (5) recognize the various meanings and interpretations of the concept, (6) identifying the properties of the concept, and compare and contrast the concept (as cited in [2]).

Considering the importance of students' understanding mathematical concept ability, the learning process of mathematics in the classroom should be conducted to support the development of this ability. But the fact, the mathematics ability of students in Indonesia at middle school is still low. It can be seen from the result of PISA (Programme for International Student Assessment) evaluation in year 2015. The mean score of mathematical ability of students in Indonesia is 386, meanwhile the mean score in PISA is 490 [4].

The lack of understanding the mathematical concept is not only happened to students at the middle level, but also for students at the higher level of education. For example, many of the students in undergraduate education whose learn mathematics got the low score in Calculus. Tall as cited in [5], stated that the lack is caused because of the learning process of Calculus is focusing on procedural not conceptual.

Based on the discussion, to develop the students' understanding of mathematical concept ability, teachers should use the innovative approach to learning mathematics. NCTM [6] suggested the using of computer technology in the learning process, such as Computer Algebra Systems (CAS). The using of computer technology in learning mathematics supported students to explore and to identify the concepts in mathematics and its relation. One of a mathematical computer program which is used as assisted in learning mathematics is Maple.

Maple software gives many benefits for students in undergraduate education. When they learn Linear Algebra or Calculus, it is helping students to do the verifying computation and drawing the complicated graph [7]. In addition, students' understanding of mathematical concept could improve by using Maple in the learning process [8, 9].

Another factor that can impact the improvement of students' understanding of mathematical concept ability is the mathematical initial ability (MIA). According to [10] students' prior knowledge before the learning process will influence the learning goal. The students' MIA divided into three categories, high, medium and low.

The purposes of this study are: (1) to know if students' mathematical concept understanding ability taught by using Maple in learning is higher than students taught by expository learning? (2) to know if students' mathematical concept understanding ability taught by using Maple in learning is higher than students taught by expository learning based on the students' mathematical initial ability (high, medium and low)? And (3) to know if there is an interaction between learning model and students' mathematical initial ability to the improvement of students' mathematical concept understanding ability?

2. Methods

This research is a quantitative research model of quasi-experimental design with pretest-posttest control group (pretest-posttest control group design). Research conducted at Mathematics Education of PGRI Palembang University, South Sumatera in the second semester of the Academic Year 2016/2017. The samples are two groups of classroom: the experiment class (E) in which learning with Maple was applied, and control class (C) that was treated with expository mathematics instruction. To know the improvement of students' understanding of mathematical concept ability, the data of student's n-gain for each level of students' MIA was analyzed between those who received the Maple instructions and the expository instructions. Meanwhile, to know the interaction between the learning approach and the students' MIA toward the improvement of students' understanding of mathematical concept ability, the two ways ANOVA was conducted.

3. Results and Discussion

3.1. Description of students' MIA

The MIA of students is divided into three categories levels (high, medium and low). The distribution of research sample based on MIA is shown in Table.1.

Table 1. Distribution of Research Sample.

Sample Research Group	Students' MIA Category		
	High	Medium	Low
Experiment Class	7	18	6
Control Class	6	18	6

To see if there is any different of students' MIA between the experiment and control class before the research conducted, the data of students' MIA is tested. Before using the statistic test, the test of normality and homogeneity for students' MIA is addressed.

Table 2. Data Normality Test for Groups and Each Level of Students' MIA.

MIA Category	Class					
	Experiment			Control		
	N	K-S	Sig.	N	K-S	Sig.
Overall	31	0.832	0.492	30	0.717	0.682
High	6	0.701	0.710	6	0.547	0.926
Medium	18	0.864	0.444	18	0.714	0.688
Low	7	0.151	0.141	6	0.998	0.272

The result of data normality test on Table 2, shows that the probability value (sig.) for every level of students' MIA on both of class is higher than 0,05. It means that the data distribution of students' MIA for every level in every class is normal.

Table 3. Variance Homogeneity Test for Both Groups and Each Level of Students' MIA

MIA Category	Lavene Statistic	Sig.
Overall	0.416	0.521
High	0.701	0.710
Medium	0.864	0.444
Low	0.151	0.141

The result of variance homogeneity test on Table 3, shows that the significant value for every level of students' MIA is higher than 0.05. It means that the variance data of students' MIA for every level is homogenous. Because data is normal and homogenous, so to see if there is the different mathematical ability between both classes the testing is continuing by using the t-test.

Table 4. Significant Different Test.

MIA Category	*t*	Sig.
Overall	-0.262	0.795
High	0.000	1.000
Medium	-0.273	0.787
Low	0.171	0.867

The significant value on Table 4 for every level of students' MIA is higher than 0.05. It means that the students' MIA of experiment class and control class are not different.

3.2. Description the improvement of students' understanding of mathematical concept ability.

One of the aims of this study was to describe the difference of students' improvement in understanding mathematical concept ability between those who got the Maple in learning and those in the expository instruction, based on the students' initial mathematics ability. The study was also conducted to see the difference of the interaction between those approaches. The data analyzed were the n-gain score of understanding mathematical concept ability between students' pretest and posttest score.

From the data analyzed, it could be concluded that there was an improvement of students' understanding mathematical concept ability for every level of students' MIA both in Maple-based classroom and expository-based classroom. Descriptive result of the n-gain score of students' understanding of mathematical concept ability on both groups is shown in Table 5.

Table 5. Description of Students' Understanding of Mathematical Concept Ability Improvement (n-gain score).

MIA Category	Class					
	Experiment			Control		
	N	Mean	SD	n	Mean	SD
Overall	31	0.68	0.13	30	0.47	0.14
High	6	0.83	0.07	6	0.65	0.18
Medium	18	0.66	0.12	18	0.42	0.94
Low	7	0.62	0.12	6	0.42	0.08

Based on Table 5, known that the mean n-gain for every level of students' MIA of experiment class is higher than control class. According to Hake (1999), the n-gain score of experiment class for high level of students' MIA is in high category, and for medium and low level is in medium category. Meanwhile, the n-gain score of control class for every level of students' MIA is medium.

Next step is testing the different of students' understanding of mathematical concept ability based on students' MIA level. The result of the data normality test n-gain of the students' understanding of mathematical concept ability had shown that probability score (sig.) for every level of students' MIA in each class is more than 0.05. From this result, it can be concluded that the sample in both groups is normally distributed. The result of data normality test is in table 6.

Table 6. Data Normality Test N-gain of Students' Understanding of Mathematical Concept Ability.

MIA Category	Class					
	Experiment			Control		
	N	K-S	Sig.	N	K-S	Sig.
Overall	31	0.666	0.767	30	1.075	0.198
High	6	0.454	0.986	6	0.489	0.971
Medium	18	0.453	0.986	18	0.529	0.942
Low	7	0.434	0.992	6	0.368	0.999

The result of the variance homogeneity test n-gain of the students' understanding of mathematical concept ability had shown that probability score (sig.) for every level of students' MIA in each class is more than 0.05. From this result, it can be concluded that the variance of both groups is homogenous. The result of variance homogeneity test is in Table 7.

Table 7. Variance Homogeneity Test N-gain for Each Level of Students' MIA.

MIA Category	Lavene Statistic	Sig.
Overall	0.346	0.559
High	1.697	0.222
Medium	1.732	0.197
Low	0.537	0.479

From the test of normality and homogeneity, it is known that the data n-gain is normal and homogenous, so for the test of signification using the t-test. The null hypothesis of this test is there no differences between n-gain of understanding of mathematical concept ability of students learning with Maple and those learning with the expository. The criterion for this test is if the probability value (sig.) is more than 0.05, the null hypothesis will be accepted. The result of t-test can be seen in Table 8.

Table 8. Signification Different Test of Students' Understanding Mathematical Concept Ability Improvement.

MIA Category	*T*	Sig.
Overall	6.203	0.000
High	2.314	0.043
Medium	6.855	0.000
Low	3.224	0.008

Based on Table 8, the probability value (sig.) for every level of students' MIA is less than 0.05. It means that the null hypothesis is rejected. So the result of this test is there is the differences between the understanding mathematical concept ability of students learning with Maple and those who learning with the expository one.

3.3. Description the interaction of learning process and students' MIA level toward the students' understanding of mathematical concept ability improvement.

To find out if there is the interaction between learning process and level of students' MIA toward the improvement of students' understanding mathematical concept ability, the two ways ANOVA was conducted. The null hypothesis of this test is there is no interaction between learning process and level of students' MIA toward the improvement of students' understanding mathematical concept ability. The criteria of this test is accepted the null hypothesis if the probability value (sig.) between learning process and level of students' MIA is more than 0,05.The result of two ways ANOVA test is can be seen in Table 9.

Table 9. Two Ways ANOVA Test of Students' Understanding Mathematical Concept Ability Improvement.

Variance Resource	Type III Sum of Squares	df	Mean Square	F	Sig.
Learning Process	.504	1	.504	39.952	.000
Level of Students' MIA	.403	2	.202	15.987	.000
Learning Process * Level of Students' MIA	.011	2	.006	.449	.641
Error	.693	55	.013		
Total	22.168	61			
Corrected Total	1.833	60			

Based on Table 9, the probability value (sig.) is more than 0.05. It means that there is no interaction between the learning process and level of students' MIA toward the improvement of students' understanding mathematical concept ability. So, in another word, we can conclude that the improvement of students' understanding mathematical concept ability is affected by the learning process only. This test result is also could be seen from the plot in Figure 1.

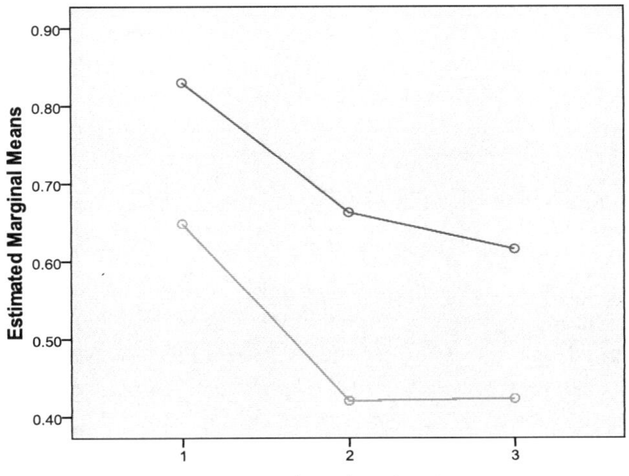

Figure 1. The Interaction between learning process and level of students' MIA

This result study shows that the improvement of students' understanding mathematical concept ability students which is learning with Maple was higher than those who learning with the expository one, for every level of students' MIA. It means that using Maple in learning support the improvement of students' understanding mathematical concept ability [9].

The biggest achievement n-gain score of students' understanding mathematical concept ability improvement is obtained in the high level of students' MIA of experiment class. The n-gain score is 0, 83, high category. It means that students in high level of MIA could learn with Maple very well. They could understand how to operate the Maple Program, as well as they learn to understanding the mathematical concept. Besides that, students in high level of MIA felt that Maple was very useful for them in learn to understand the mathematical concept, such us when they wanted to draw the difficult graph, the drawing process become much easier and more attractive with Maple program. This statement is accordance to [7].

4. Conclusion

The results of this study show (1) students' ability of understanding the mathematical concept taught by using Maple in learning is higher than students taught by expository learning, (2) students' ability of understanding the mathematical concept taught by using Maple in learning for every level of students' mathematical initial ability (high, medium and low) is higher than students taught by expository learning, and (3) there is no interaction between learning model and students' mathematical initial ability to the improving of students' understanding of mathematical concept ability.

Maple is very useful to development the students' understanding of mathematical concept ability. From the result of this study known that the achievement of students' understanding mathematical concept ability of students which is taught by using Maple is higher than the expository one. For the future, Maple can be used to development others mathematical ability, such as mathematical communication ability and mathematical problem solving ability.

References

[1] BSNP 2006 *Permendiknas No. 22 Tahun 2006 tentang Standar Isi untuk Satuan Pendidikan Dasar dan Menengah* (Jakarta: Depdiknas) pp. 140
[2] Lambertus 2016 *International Journal of Education and Research* **4** 7

IOP Conf. Series: Journal of Physics: Conf. Series **948** (2018) 012034 doi:10.1088/1742-6596/948/1/012034

[3] Lestary L and Surya E 2017 *International Journal of Sciences: Basic and Applied Research (IJSBAR)* **34** 1 pp. 91-100
[4] OECD 2016 *Country Note – Results from PISA 2015* available at https://www.oecd.org/pisa/PISA-2015-Indonesia.pdf
[5] Ningsih Y L, Darmawijoyo, and Hartono, Y 2015 Developing students' worksheet of derivative based on APOS theor *The 3rd SEA-DR International Conference* edited by R. I. I. Putri *et al.* (Palembang: Sriwijaya University) pp. 184-194
[6] NCTM 2015 *Strategic Use of Technology in Teaching and Learning Mathematics : A position of the National Council of Teachers of Mathematics* available at http://www.nctm.org/uploadedFiles/Standards_and_Positions/Position_Statements/Strategic%20 Use%20of%20Technology%20July%202015.pdf
[7] Kilicman A, Hassan M A, and Husain S K S 2010 Teaching and learning using mathematics software : the new challenge *Procedia Social and Behavioral Sciences* **8** pp. 613-619.
[8] Klasa J 2010 *Linear Algebra and its Appliations* **432** pp. 2100-2111
[9] Salleh T, and Zakaria E 2013 Enhancing students' understanding in integral calculus through the integration of maple in learning *Procedia-Social and Behavioral Science* **102** pp. 204-211
[10] Surya E, Putri F A and Muktar 2017 *Journal on Mathematics Education* **8** 1 pp. 85-94

Acknowledgements
This work was supported by Minister of Research, Technology and Higher Education of Republic Indonesia, in HIBAH PENELITIAN DOSEN PEMULA schema, year 2017.

IOP Conf. Series: Journal of Physics: Conf. Series **948** (2018) 012035 doi:10.1088/1742-6596/948/1/012035

Pre-service teachers' challenges in presenting mathematical problems

R Desfitri

University of Bung Hatta, Indonesia

E-mail: rdesfitri@bunghatta.ac.id

Abstract. The purpose of this study was to analyzed how pre-service teachers prepare and assigned tasks or assignments in teaching practice situations. This study was also intended to discuss about kind of tasks or assignments they gave to students. Participants of this study were 15 selected pre-service mathematics teachers from mathematics education department who took part on microteaching class as part of teaching preparation program. Based on data obtained, it was occasionally found that there were hidden errors on questions or tasks assigned by pre-service teachers which might lead their students not to be able to reach a logical or correct answer. Although some answers might seem to be true, they were illogical or unfavourable. It is strongly recommended that pre-service teachers be more careful when posing mathematical problems so that students do not misunderstand the problems or the concepts, since both teachers and students were sometimes unaware of errors in problems being worked on.

1. Introduction

Mathematicians as well as mathematics educators have turned their attention to teacher's knowledge. In the last decade, researchers in mathematics education have begun to specify and assess teachers' knowledge in new ways [1]. At the same time, there is a general expectation that teachers have enough mathematical knowledge to teach. In mathematics teaching, knowledge for teaching is necessary because it will direct a teacher in deciding what to do in class, including considering examples to be given, the form of presentation to be used. This knowledge will also help teachers determine what relation or connection will be used during the learning process, including how to respond to students' thinking. Knowledge for teaching will also affect teachers' preparation in organizing class.

For mathematics teachers, teacher development is an extensive and thriving research area around the world. It faces multidimensional challenges as regards both pre-service and in-service teachers' programs [2]. Focus on pre-service mathematics teachers, professional development can be supported by all parties, and university mathematics education play an important role when teaching their students as prospective mathematics teachers. One aspect that can be developed during their teaching practice is their skill of posing mathematical problems.

2. Mathematical Problems

Problems abound in mathematics education at all level [3]. There are many kinds of mathematical problems described in the literature. Many researchers also classified problems in different ways. Souviney in [4] classified as routine story or word problems and non-routine problems. If word or story problems can be solved by applying previously learned concepts and skills, non-routine problems, on the other hands, cannot be solved by only selecting and applying one operation as world problems. Solving non-routine problems requires flexible thinking.

Another way to classify mathematical word problems is to group them into standards and non-standard problematic (parallel) problems. Standard word problems are those that can be solved by applying the most obvious arithmetic operations using given numbers. Non-standard word problems can be solved by using arithmetical operations, but it requires the problem solvers to use their real-life knowledge into account in problems' context. For non-standard word problems, appropriate mathematical model are less obvious and mathematical modeling assumption is problematic [4].

Students can only be asked to solve mathematical problems if the problems to be solved exist. Even though almost all mathematics textbooks for any level provide mathematical problems, it does not mean that it should depend on textbooks based problems. One aspect of problem solving that school mathematics teachers need to engage in is the art of problem posing [5]. Teachers are also suggested to think of self-generated mathematics problems. If students need to practice how to solve mathematics problems, teachers need to provide good mathematical problems. It leads to the fact that school mathematics teachers need to develop their problems posing skill.

3. Challenge in Posing Mathematical Problems

It has been quite difficult to literarily find the exact definition of problem posing. Some sources describe posing problems in indirect ways. Problem posing is recognized as an important component of mathematics teaching and learning [6]. Similar to this, problem posing has been identified by some distinguish leaders in mathematics and mathematics education as an important aspect of mathematical education [7]. Stoyanora and Ellertonin [8] described problem posing as a process of creating meaningful mathematics problem from interpretation established throughout concrete cases and based on mathematical experiences.

Problem posing referred to both generator of new problems and the re-formulating of given problems. Thus, posing can occur before, during and after of the solution of the problems. Problem posing has long been viewed as a characteristic of creativity or exceptional talent. For this reason, Gonzales in [9] noted that teachers have an important role in the implementation of problem posing in the curriculum. Since problem posing has been embedded in the assessment of creativity or mathematical talent, it is reasonable to assume that there are some links between posing and creativity [7]. Crespo in [10] identified three different approaches that pre-service teachers used in posing problems to pupils: i) making problems easier to solve, ii) posing similar problem, and iii) posing problems blindly.

If we referred to Crespo, problems which are easy to solve, could be kind of forms where hints were provided to lead someone to as solution strategy or answers. Familiar problems could be kind of problems as quick translation story problems or computational exercise, and blindly problems refer to problems without solving beforehand or deeply understanding mathematics.

Problem solving can also be considered as necessary part for pre-service teachers because teaching entails posing good question would impact students' development of mathematical understanding. It means that one excellent way for pre-service teachers to practice posing good questions is to apply problem posing in their teaching practice. Many researchers in mathematics education have also been highlighted on relationship between pre-service teachers and their classroom practices. However, for pre-service teachers who have been prepared to be mathematics teachers, practicing to pose mathematical problem could be kind of challenge of learning to teach mathematics. Pre-service teachers may learn this essential practice when they took part on practice teaching during teaching preparation program. Pre-service mathematics teachers can get many benefits when generating problems. First, since problem posing can be viewed as a window into mathematical understanding, pre-service teachers will be more motivated to master the concept better before posing problems in their class. Second, in posing mathematical problems, pre-service teachers are required to be able to create problem related to the facts or conditions that exist around. Third, problem posing might also help pre-service teacher in reducing their dependency on their textbooks.

It has been asserted in several studies that pre-service teachers have some issues related to problem posing activities [4]. In order that pre-service teacher will gain the knowledge and required confident for incorporating problem posing activities in their next classes, they have to experience it first [9]. However, the challenges faced by pre-service teachers in posing mathematical have not been widely discussed. Considering that posing mathematical problems are not always easy, it is important to find out the challenges pre-service mathematics teachers face in posing mathematical problems during teacher education as a means of educating them. Within this context, the study about how pre-service mathematics teachers prepared to give mathematical problems during their teaching practices, are getting important. In this sense, the answers of the following questions were searched.

1. What was the main sources of the mathematical problems the pre-service teachers provide in the classroom?
2. What approaches pre-service mathematical teachers used to pose mathematical problems
3. What types of tasks of posing problems the pre-service teachers during their teaching practice?
4. What are difficulties and challenges facing the pre-service students in posing mathematical problems?

4. Methods

This was a descriptive research to identify how pre-service teacher posed what challenges they faced in posing mathematical problem. Participants of the study were 15 pre-service mathematics teachers enrolled in mathematics teaching practice or micro-teaching class at University of Bung Hatta during February – June 2017 term. All participants had already completed coursework in mathematics teaching method course the earlier semester. Each of 15 pre-service mathematics teachers had a chance to practice twice. For each class, one mathematical problem provided by teachers was selected for analysis. The total 30 of mathematical problems were picked up. Data of the study were collected during mathematics teaching practice course. The main sources of the data for the study were: (i) the collected mathematical problems the pre-service teachers posed during their practice; (ii) the observation sheets described how pre-service mathematics teachers posed mathematical problem during their practice teaching; (iii) discussion's note written by mathematics educator as course's supervisor.

5. Results and Discussion

By asking those pre-service teachers about the sources from which of the selected mathematical problem provided during the class, their response can be classified as follow:

Table 1. The sources of mathematical problems pre-service teachers gave to students.

Sources of mathematical problems	Frequency	Percentage
Original from textbooks	3	10
Slightly revised from textbooks	7	23.3
Self-generated	20	66.67

Based on the information provided in Table 1, it was found out that the sources pre-service teachers referred when posing problems could be derived from two main sources, course textbook and self-generated. However, part of the problems was taken from textbooks have been slightly revised to be adjust into appropriate situation in the class.

In order to be more focused on the posing problems, the only self-generated problems to be analyzed in more detail. The approach pre-service teacher used in the total of 20 posed mathematical problems can be described as follows

Table 2. The approaches pre-service teachers used in posing mathematical problems.

Approach used	Frequency	Percentage
Posing problems easy to solve	9	45
Posing familiar problems	6	30
Posing problems blindly	5	25

From Table 2 above, majority of pre-service teachers tend to use easy problem. They also preferred to provide simple algorithm so it will be easy to check students' answers. Based on our discussion with pre-service teachers as research participants, it was known that even though some of them were eager to pose familiar problems, but in majority, pre-service teachers tended to pose easier problem for two reasons: i) lack confidents to pose more complicated mathematical problems, and ii) to avoid their own mistake in front of class.

By categorizing as standard and non-standard problems, distribution of mathematical problems which were posed by pre-service teachers can be described in Table 3 as follow:

Table 3. Types of tasks (mathematical problems) given

Types of Problems/Tasks	Frequency	Percentage
Standard Problems	15	75
Non-standard Problems	5	25

Based on data provided in Table 3, if we referred to the standard/non-standard problems classified by Kiliç discussed earlier, it was figured out that majority of posed problems were categorized as standard, since those problems could be solved by applying simple arithmetic operations. It also showed that most mathematical problems post by pre-service mathematics teachers reflected that mainly, lower mathematical thinking level needed to solve those problems. These standard problems could be solved by applying the most obvious arithmetical operations using the given numbers.

The language (text) pre-service teacher used in posing problems was analyzed regarding to level of easiness to understand by reading the problems. The result was described in the following table.

Table 4. The language levels of mathematical problems provided by pre-service teachers.

Language Level	Frequency	Percentage
Not clear	4	30
Relatively clear	9	45
Clear	7	35

Table 4 told us that most pre-service teachers were able to pose problems which were relatively understandable. However, some posed problems were incomplete and need to be more explained since they do not specifically indicate what was being asked. The example of the case was noted in the following example of posed problem.

Solve the following problem: $\sin\ 2x = \cos x, 0\ \leq x\ \leq 360$

There was no specific clue was given by pre-service teacher regarding to problem $\sin 2x = \cos x$ above, so students had no idea how to solve problem since there was no specific question to declare.

Based on the observation and note discussion, pre-service teachers faced a plenty of challenge in posing mathematical problems. There were mainly four challenges they faced were related to various aspects of which can be described in the following table:

Table 5. The challenges pre-service teachers faced in posing mathematical problems.

Challenges	Number of pre-service teachers face it	Percentage
Lack of experience	11	73
Lack of mastering content knowledge	10	67
Eager to find problem issues	15	100
Time allocation	7	46

The Table 5 above describes the fact that majority of students did not face single challenge in posing problems. Data also showed us that all pre-service teachers in this study admitted that finding related issues is part of their challenges; they even argued that the process on how to find related issue was the biggest challenge. They also admitted that other challenges took into account to pose problems. 11 out of 15 pre-service teachers thought of their lack of experiences and lack of mastering content knowledge also influenced their ability in posing mathematical problems.

One unconsidered factor that came up was that time allocation factor. This factor also arose due to the time limitation of teaching practice provided to them. Almost half of them pointed out that their short time in front of class to teach could also be limiting them to be creative in posing problems. This situation should be overcome by having enough preparation before they come to teach.

By looking at more deeply on posed mathematical problems, there were some interesting facts that need to be discussed since there were various errors in the problems.

One interesting thing that appeared what was called a 'hidden error', it referred to the existence of the posed problem that did not look like a mistake, so all the works on automatically according to the formula that might apply. However, it was not likely to happen. That kind of posed mathematical problems had the unaware error. When problem was seen there was no problem at a glance, in fact, there was a hidden error in posed problem. The examples of posed problem given by one pre-service teacher were restated as follows:

Determine the circumference of the symmetric trapezoid in the next figure

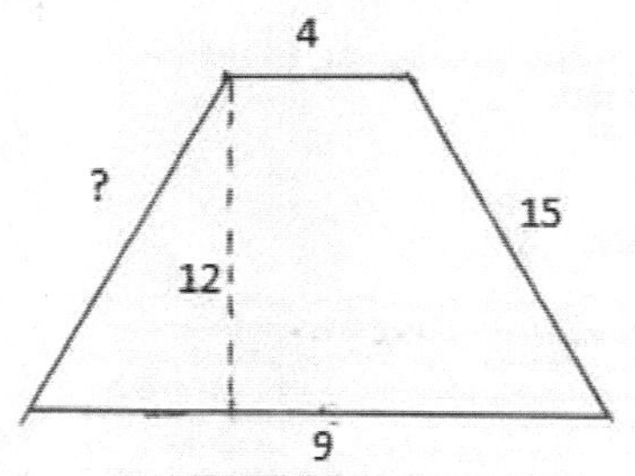

Figure 1. Trapezoid posed by pre-service teacher to find its circumference.

Figure 1 describes the posing problem which was, at first glance, seemed obvious. The language was clear and understandable as students would easily understood what they were supposed to do. The problem also looked pretty good because it required students to think how to use Pythagorean's theorem to determine the length of the unknown side before adding all the side to get its circumference. However, it the problem was examined more carefully, it would be recognized that according to Pythagorean's theorem, the trapezoid could in fact, not exist since the length of the opposite side was 15 instead of 13. When the particular pre-service teacher who posed that problem was asked whether he was careless about the side which length was 15, he said that it did not affect anything, because circumference was just the sum of all sides of trapezoid that were already known. It also showed us that pre-service teachers who have inadequate meaningful mathematical content knowledge often posed problems with contextual error. The measurement had also no specific dimension or unit, which was not suggested in general mathematics, because circumference was supposed to have its unit.

Another thing that came up from the data was the presence of posed mathematical problems that impressed less rational/realistic. This was seen from the following problem:

If one paving block has its thickness of 8 cm, what is the height of 55 paving blocks which are arranged vertically?

At a glance, there was nothing wrong with the problem above. It was just kind of simple problem with simple calculation. The language was also clear that students were asked to do direct multiplication. But if viewed further, this problem seemed less realistic, since it will be difficult to arranged 55 paving blocs vertically in the form of single line

In posing mathematical problems, pre-service teachers sometimes also less concerned with things in real world that seems trivial. For example, the following bathtub problem

A bath in a shape of cube with the outside wall has its length of 50 cm. If only half-filled, how much volume of water in the tub?

There was one issue to be addressed in the problem above. One might think that this problem seemed easy to solve directly by calculating the volume of the tub and do some other simple calculations. However, it would seem less logical if the bathtub did not have a thickness or just as thick as a piece of paper. This data also showed that pre-service teachers were sometimes less concerned with the essence of real world problems.

Regarding to posing world problems, it was also found that pre-service teachers were less confident in posing world problems. Majority of posed problems posed by pre-service teachers were not world problems, even if the subject being though was geometry, pre-service teachers were still eager to find issues in real world problems. This because one factor or more combination of challenges the faced we discussed earlier, including lack of experience and/or did not master subject content knowledge.

Some errors in mathematical problems above related to the challenges the pre-service teachers faced in posing problems. Since problem posing could be used as a task to asses students' mathematical thinking, then this approach would be a good alternative for pre-service teachers to realize their mathematical knowledge and learning from their experiences to resolve and reduces their errors in posing next mathematical problems.

6. Conclusion

This study investigated pre-service mathematics teachers' challenge and how they pose mathematical problems. Based on the data and discussion, some important aspect could be picked up. Many pre-service teachers felt lack of experiences, and lack of realizing their lack of mastering content knowledge which had influences to their posing problems' ability. Some of them experienced difficulties in finding related issues so most of them tended to pose non-real world problems. This was also because of what many teachers and textbook perceived to be real life mathematics were not always the same as what people experienced in their daily life. Their focused on how to pose problems were still constrained by what type of questions should be given regarding to the subject, as the result, they were not aware of the errors appeared in their posed problems. Pre-service teachers still needed to some guidance from their mathematics educators on how to pose mathematical problems that they would use as task or assignment in their class practices.

The main goals of teachers' preparation and professional development program are in helping pre-service teachers master mathematics content they teach, understand how their students learn, and know how they facilitate their students to learn. For these reasons, pre-service teachers' ability in posing problems need to be the main concern. This because in posing mathematical problems, pre-service teachers' venture may have multiple approaches and solutions can also be adjusted to real world situations. The pre-service teachers may pose open-ended tasks which need some explorations by themselves before giving it to their students.

Since it is impossible for someone to teach what he/she does not know, it reminds us that the only way to achieve better mathematics education is to improve mathematics teachers' capabilities. Professional development of pre-service mathematics teachers becomes necessary in helping them be better prospective mathematics teachers.

References

[1] McCrory R 2006 Mathematicians and mathematics textbooks for Prospective elementary teachers *Notices of the AMS* **53** pp 20-29
[2] Naik S Mathematics Teachers Education in India – demanding change and reform in teachers'professional development Preprint gr-qc/0401010
[3] Dindyal J 2009 Mathematical Problems for Secondary Classrooms *Mathematical Problem Solving* ed Kaur and Kapur (Singapore: Association of Mathematics Educator,Word Scientific) pp 208-225
[4] Kiliç 2015 The Tendency of turkish pre-service teachers to pose word problem *Turkish J. of Computer and Mathematics Education* 6 pp 163-178
[5] Yu V 2009 Using innovation technique to generate new problems *Mathematical Problem Solving* ed Kaur and Kapur (Singapore: Association of Mathematics Educator Word Scientific) pp 185-207
[6] National Council of Teachers of Mathematics 2000 Principles and Standards for School Mathematics (Reston: NCTM)
[7] Silver E et al 1996 *J. Research in Mathematics Education* **27** pp 293-309
[8] Şengul S and Katranci Y 2015 Free problem posing cases of prospective teachers: difficulties and solutions *Proc. Social and Behavorial Science* **174** pp 1983-1990
[9] Lavy I and Shriki A 2007 Problem posing as a means for developing mathematical knowledge of prospective teachers *Proc. 31st . Int. Conf. for the Psychology of Mathematics* **3** pp 129-136
[10] Crespo S 2003 *Educational Studies in Mathematics* **52** pp 243-270

IOP Conf. Series: Journal of Physics: Conf. Series **948** (2018) 012036 doi:10.1088/1742-6596/948/1/012036

Analysis of students' mathematical reasoning

Sukirwan[1], Darhim[2], and T Herman[2]

[1]Universitas Sultan Ageng Tirtayasa, Jl. Jakarta Km. 4, Serang 42121, Indonesia
[2]Universitas Pendidikan Indonesia, Jl. Dr. Setiabudi 229, Bandung 40154, Indonesia

Email: sukirwan@untirta.ac.id

Abstract. The reasoning is one of the mathematical abilities that have very complex implications. This complexity causes reasoning including abilities that are not easily attainable by students. Similarly, studies dealing with reason are quite diverse, primarily concerned with the quality of mathematical reasoning. The objective of this study was to determine the quality of mathematical reasoning based perspective Lithner. Lithner looked at how the environment affects the mathematical reasoning. In this regard, Lithner made two perspectives, namely imitative reasoning and creative reasoning. Imitative reasoning can be memorized and algorithmic reasoning. The Result study shows that although the students generally still have problems in reasoning. Students tend to be on imitative reasoning which means that students tend to use a routine procedure when dealing with reasoning. It is also shown that the traditional approach still dominates on the situation of students' daily learning.

1. Introduction

Reasoning is a very important aspect of mathematical ability in teaching and learning of mathematics. The National Council of Teachers Mathematics [1] revealed that mathematics is reasoning that any activities in mathematics will not be detached from reasoning. Consequently, reasoning becomes a much-needed basic capability to improve general mathematical abilities. Program for International Student Assessment [2] mentioned the reasoning as one of the fundamental mathematical capabilities continually would be a strategic issue in the future.

Reasoning, in fact, always be a major concern of a mathematician. Its role in solving mathematical problems [3] but the implications of reasoning is a crucial issue that always attends in daily learning. Some researchers have revealed the low level of mathematical reasoning achievement in junior high school [4], even Arwinie [5] reported that from the 40 students of the junior high school who he studied, only one being able to achieve with good reasoning. While Wardhani and Rumiati [6] said that the lack of imagination and creativity into the causes of students are not able to use reasoning skills well. Finding Wardhani and Rumiati [6] provides an overview of the condition explicitly reality problems faced junior high school students at the time associated with reasoning. It also provides illumination that issues regarding the reasoning are a complex issue, not only just a basic mathematical skill but the ability that requires students to think comprehensively. On the other hand, Rizta, Zulkardi and Hartono [7] next give a more detailed analysis of the obstacles students in reasoning, including the ability of the analyzing, generalizing, integrating, giving a reason, and the solving non-routine problems. They argued that the mathematical reasoning abilities of the students in the fourth such capabilities only reached an average of 13.33% for the category of very high and 14.82% for the higher categories. In other words, only 28.15% of the students achieving good categories.

Bjuland [8] revealed that the complexity of reasoning regarding the process of the five mathematical processes; namely: sense-making, conjecturing, convincing, reflecting, and generalizing. Sense-making is the ability to build a schema problems and represent knowledge. Sense-making is the adaptation and association process of new information acquired with prior knowledge. This process occurs when the mathematical situation is understood and then try to be communicated in mathematical symbols or languages. Conjecturing has the sense to predict theoretical activities and conclusions based on incomplete facts. Products from the conjecturing are done through the completion strategy of arguing and communicating mathematical activity. The completion strategy is then implemented through an activity called convincing. While reflecting are activities undertaken to re-evaluate the three processes earlier. These five abilities are in principle a series of capability processes in relation to the quality of mathematical reasoning.

As Bludjan pointed out, the activity of arguing and communicating mathematically became the core of the quality of mathematical reasoning. Basically, these two activities are closely related to the activity of expressing ideas or ideas in justifying the claims submitted. The idea or the idea could arise from the process of meaning that comes from prior experience and prior knowledge [8]. Lithner [9] has a similar opinion to Bludjan by classifying two types of reasoning based on their quality [10], namely: imitative reasoning and creative reasoning. The imitative reasoning is solely reasoning based on the experience before and effort that are original. Activities copying example of textbooks or remembering Algorithm certain mathematical reasoning fall into the category of this kind. Reasoning Imitative is superficial, meaning reasoning based solely on the surface only, based on mathematical characteristics deeply. The Imitative reasoning consists of two categories, namely: memorized reasoning and algorithmic reasoning. Both are distinguished from strategies used in solving mathematical reasoning. The characteristics of memorized reasoning, first, the strategy built on complete answer remembered by rote (memory); second, strategy built by writing each step that has been memorized before. While the characteristics of algorithmic reasoning, first, the strategy built by remembering an algorithm that guarantees that the solution can be achieved correctly, second, implementing a strategy consisting of nontrivial transformation or actions follow a set of rules. Although algorithmic reasoning includes imitative reasoning, but in certain cases (e.g when solving non-routine problems), algorithmic reasoning is very reliable, even by maths. The creative reasoning is reasoning that is based on creativity, the ability to produce work that is original and meaningful [9]. Lithner [9] described four characteristics of reasoning creatively, (1) novelty, refers to the new fact, that is the order of reasoning created or recreated, (2) flexibility, means the ability to use approaches and different adaptations from the specific problems, (3) plausibility, means that there are arguments in favour of the chosen strategy and explain why the conclusion gathering of the true or plausible, (4) a sound mathematical foundation, meaning that the argument is built based on a deep mathematical characteristics.

Study of the quality of reasoning becomes a very important in order to see the influence of the learning environment on improving mathematical reasoning abilities. This study is also a key reference in this paper, where an overview of the quality of Lithner perspective will provide valuable information for teachers in creating an approach and a pedagogical framework aimed to improve mathematical reasoning abilities.

2. Methods

The research approach in this study is a qualitative research. The goal of the research is to uncover naturally on students' mathematical reasoning based Lithner perspective. The research was conducted in the second year of the academic year 2016/2017 in one junior high school in Tangerang city. The research sample was set at 33 students of class IX. Research begins with the provision of mathematical reasoning tests. The test results in the form of student's answer then analyzed and categorized using the criteria of reasoning, namely: Memorized Reasoning (MR), Algorithmic Reasoning (AR), and Creative Reasoning (CR). The completeness of student answers is not taken into account in this analysis; but if the student's answers do not appear at all, giving wrong answers or beyond the proposed problem, then the students' reasoning is deemed to appear (NA). Details of such a criteria can be seen in Table 1.

Table 1. Criteria for Grouping Mathematical Reasoning

Code	Explanation
NA	Answer wrong /blank/ not suitable
MR	Explain the answers without giving a supportive reason, mentioning and defining supporting terms against the answers given
AR	Outlining the answers in detail, give reasons for the answers given
CR	Answer contains components *fluency, flexibility, originality,* and *elaboration*

The next stage is to conduct an interview. The purpose of this interview is to confirm the student's answers in order to verify the grouping of mathematical reasoning done in the previous stage. Interviewees are determined on the basis of theoretical sampling representing high, middle, and low-ability students. The results of these interviews it is possible to change the grouping when grouping different interpretation of the results by means of students who are actually in answer to the problems posed.

3. Result and Discussion

The study results show data on the quality of mathematical reasoning, as seen in the table below.

Table 2. Data on The Quality of Mathematical Reasoning

Question	Quality of Reasoning			
	NA	MR	AR	CR
1	26	0	1	8
2	0	19	16	0
3	32	0	0	3
4	32	0	1	2
5	28	4	3	0

In Table 2, the number of students in the NA category (except number 2) is dominant compared to other categories. This shows that in general, students still experience constraints when dealing with mathematical reasoning. In the question 1, students experiencing difficulties in making a conjecture that meets the conditions of figure 1 and 2. Since students are unusual with this kind of problem, most students forget to use the answer option to create conjectures that satisfy both conditions. One student can determine the answer correctly by utilizing the choice of answers so categorized as algorithmic reasoning. Eight other students make conjectures directly by giving reasons for unfulfilled conditions if the load is worth 8 grams. Although these eight students do not explain for other unfulfilled conditions, at least raising the idea of conjecture is an original thought that is not easy for other students. Because of this thinking, the eight students are categorized as creative reasoning.

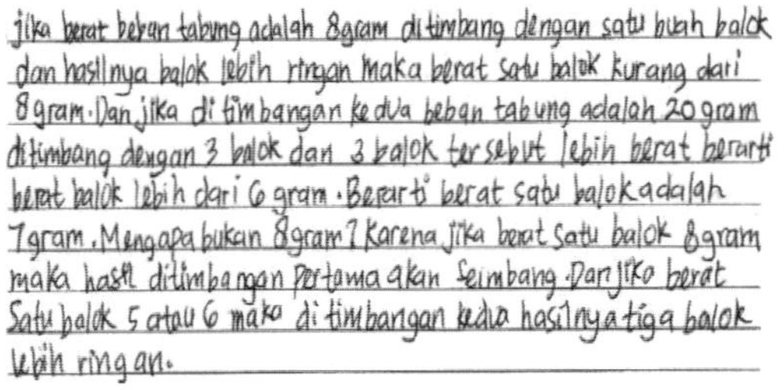

jika berat beban tabung adalah 8gram ditimbang dengan satu buah balok dan hasilnya balok lebih ringan maka berat satu balok kurang dari 8gram. Dan jika di timbangan kedua beban tabung adalah 20gram ditimbang dengan 3 balok dan 3 balok tersebut lebih berat berarti berat balok lebih dari 6 gram. Berarti berat satu balok adalah 7gram. Mengapa bukan 8gram? Karena jika berat satu balok 8gram maka hasil ditimbangan pertama akan seimbang. Dan jika berat satu balok 5 atau 6 maka di timbangan kedua hasilnya tiga balok lebih ringan.

Figure 1. Student Answer to Question 1, Category CR

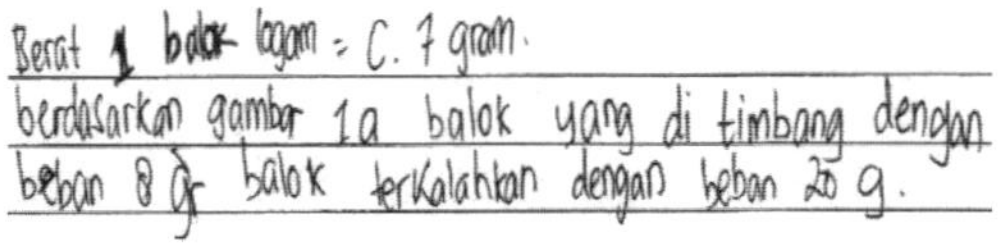

Berat 1 balok logam = C. 7 gram.
berdasarkan gambar 1a balok yang di timbang dengan beban 8 gr balok terkalahkan dengan beban 20 g.

Figure 2. Student Answer to Question 1, Category AR

Excerpts of the following interviews provide reinforcement for CR category on question 1.

Teacher : Can you explain the purpose of this answer? (Shown answer Student 20 on question 1)
Student 20 : At that time, I took 8 grams, lost the same that 20 grams
Teacher : So...
Student 20 : Because picture 1a is not the same? So I make sure it is 7 grams.
Teacher : Why not 6 grams or 5 grams?
Student 20 : It is too small, sir?
Teacher : Of which student 20 believe that 6 grams or 5 grams do not satisfy?
Student 20 : (Somewhat thinking) Hmm ... Yeah well yes ... But, figure 2 unfulfilled, must be heavy to the right. (Pointing to the load image in figure 2)

On question 2, students commonly can determine the extent of the field with the right, but wrong in drawing conclusions. In other words, students cannot utilize preliminary calculations for proper inference. Nevertheless, MR and AR category remains can be identified by looking at the pattern of the students' answers on the use of formula for the area of a rectangle.

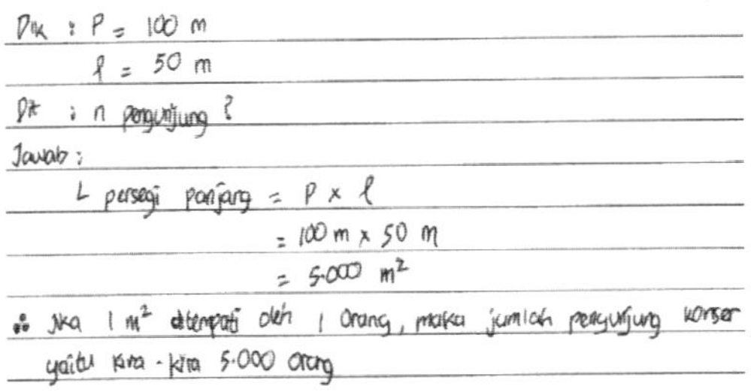

Figure 3. Student Answer to Question 2, Category MR

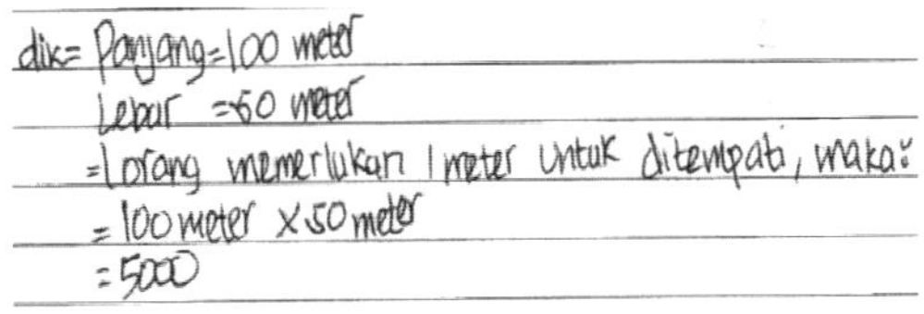

Figure 4. Student Answer to Question 2, Category AR

A total of 19 students were supposedly using the rectangular area formula as a routine procedure when dealing with rectangular objects. In addition, there are 16 students allegedly using the rectangular square formula as a step to be taken to find out the field capacity of known size in the matter. In figure 3, the students immediately calculate the area of a rectangle using a mathematical formula that has been previously known. The way students do such a settlement is categorized as MR. Meanwhile, in figure 4, students use the rectangular formula as the idea (source illumination) in determining the field capacity according to rectangular matter. The way students do such settlement is categorized as AR.

In addition, there are also 3 students first draw a rectangle shape as the initial idea to understand the issues in question explicitly.

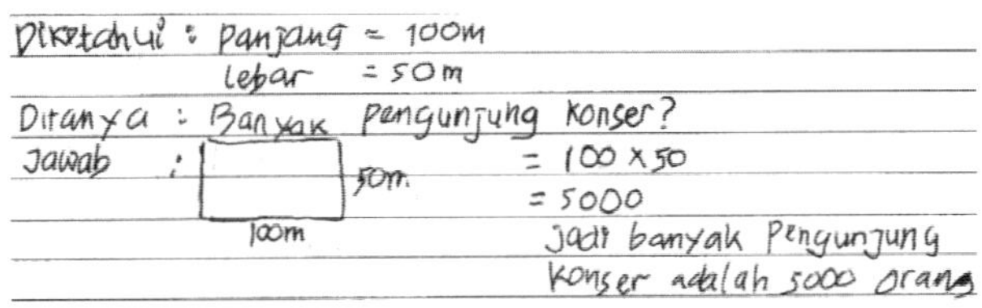

Figure 5. Students Answer to Question 2, Category AR

These three students allegedly only used the rectangular images to illustrate problems in the matter, then using the illustrations to solve the problem. An excerpt of the interview below provides clarity regarding the students in answering the question.

Teacher : *Try to look at your test! Do I want to ask, why did you write the formula for the rectangular area here? (Pointing to answer Student 24 in question 2)*
Student 2 : *If it is wrong, this is a rock concert in the square-shaped field.*
Teacher : *Are you sure!*
Student 20 : *Yes, because it is a rectangle, but I count area of its, sir?*
Teacher : *How about Student 18?*
Student 18 : *I use the rectangular formula as an aid to know the capacity of the concert audience.*
Teacher : *Oh, I see... How about Student? Why draw a rectangle first?*
Student 5 : *The same as Student 20. Because there is a long and wide, I multiply*

Excerpts of the interview above provide a clearer picture of how students in answering question 2. Since student 5 using the image merely as a routine procedure in resolving problems related to the rectangle, then student 5 is categorized as MR. So, for question 2, there are 22 people categorized as MR, the rest as many as 13 people are categorized as AR.

For question 3, students are required to solve problems with non-routine procedures. Students can actually take advantage of the ratio of the inner square area to the outer square area, but because the student's calculation focuses on the same right-angled triangle on all four square corners, it makes it difficult for students to calculate the area of the right triangle on the square in it. Only 3 students are capable of performing non-routine procedures so CR is categorized.

In question 4, students also experienced difficulties in utilizing the formula for the area of the circle. Most students guessed that a small pizza is cheaper, because of the smaller diameter stuck up with great pizza. A student tries to make use of the circle formula, but he does not relate to other conditions (in this case the price for each pizza), so the final answer becomes imprecise. Meanwhile, 2 students were able to use the exact circle formula. Although initially using the circle formula, but the creativity of the settlement appears from the division operation with the price of each pizza which is then multiplied by 1 zed.

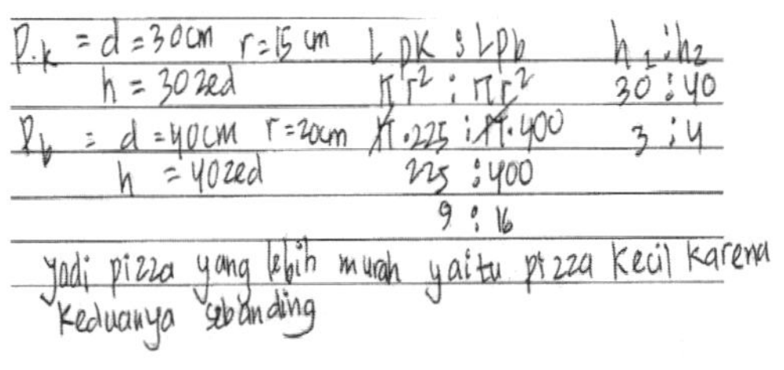

Figure 6. Student Answer to Question 4, Category AR

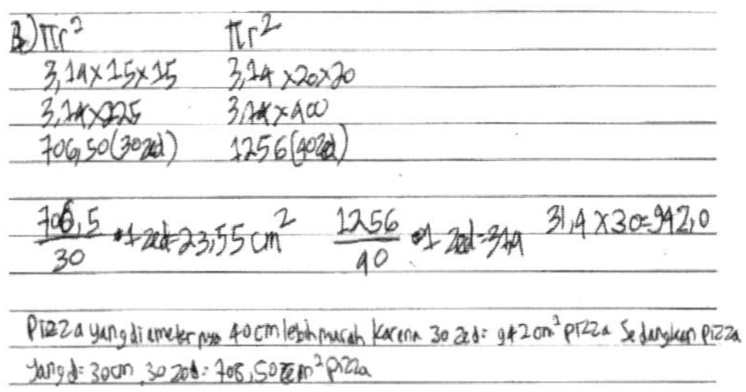

Figure 7. Student Answer to Question 4, Category CR

In question 5, the student is only able to solve the problem for the D option. There are 4 students who answer the problem by using the formula around the rectangle as a routine procedure. These four students are categorized as MR. In addition, there are 3 students who take advantage circumference of a circle, by way of sum up the line segments into others forming the circumference. These three students are categorized as AR.

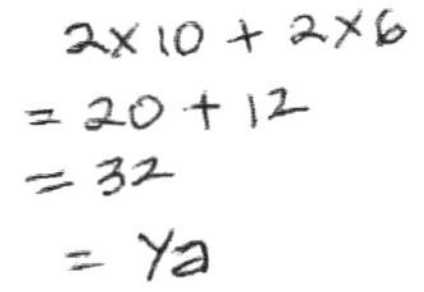

Figure 8. Student Answer to Question 5, Category MR

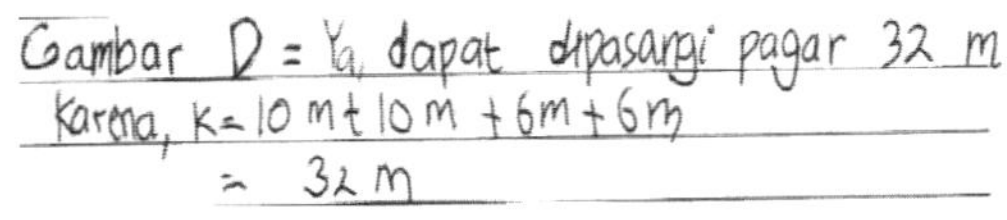

Figure 9. Student Answer to Question 5, Category AR

By taking a close look the various responses of the students from the five questions above, it appears that the use of imitation reasoning still dominates if compared to creative reasoning. It is not certainly independent of the students' learning environment of everyday life, where the traditional approach is still to be excellent teachers in presenting the study of mathematics. As a result, students stop at the computational results but do not arrive at the actual solution. Lithner [9] has ever revealed four-step solution regarding reasoning; (1) a task-solving situation is entered, in which students are exposed to problematic situations to determine how to solve the problem, (2) strategy choice, try to choose strategies that are believed to be able to solve the difficulties encountered, (3) strategy implementation, running a strategy, if the chosen strategy does not fit, go back to the previous step, (4) conclusion, any solution must come to the conclusion, that the conditions in which the answer to the problems was obtained.

4. Conclusion

Based on the founded study and the results, the conclusion that can be delivered from this research is as follows. The student still runs into constraints when dealing with the reasoning in general. The quality of students' mathematical reasoning is still dominated by imitative reasoning, where the faced situation problematics of the student is fixated on the application of routine in daily lessons.

References

[1] National Council of Teachers Mathematics 2000 *Principles and standards for school mathematics* (Reston V A: National Council of Teachers of Mathematics)

[2] Programme for International Student Assessment 2015 *Draft mathematics framework* (http://www.oecd.org/pisa/pisaproducts/pisa2015draftframeworks.htm)

[3] Rohana 2015 The enhancement of students' teacher mathematical reasoning ability through reflective learning *Journal of Education and Practice* **6** 108

[4] Pratiwi W I 2014 Penerapan metode penemuan terbimbing untuk meningkatkan kemampuan pemahaman dan penalaran matematis serta self-confidence siswa SMP *Tesis* (Bandung: Universitas Pendidikan Indonesia)

[5] Arwinie N 2014 Meningkatkan kemampuan penalaran dan komunikasi matematis serta self-concept siswa MTs melalui pembelajaran berbasis masalah *Tesis* (Bandung: Universitas Pendidikan Indonesia)

[6] Wardhani S and Rumiati 2011 *Instrumen penilaian hasil belajar matematika SMP: Belajar dari PISA dan TIMSS* (Yogyakarta: P4TK Matematika)

[7] Rizta A, Zulkardi and Hartono Y 2013 Pengembangan soal penalaran model TIMSS matematika SMP *Jurnal Penelitian dan Evaluasi Pendidikan* **17** 230

[8] Rosita C D 2014 Kemampuan penalaran dan komunikasi matematis: Apa, mengapa, dan bagaimana ditingkatkan pada mahasiswa *Jurnal Euclid* **1** 33

[9] Lithner J 2008 A research framework for creative and imitative reasoning *Educational Studies in Mathematics* **67** 255

[10] Bergqvist T 2005 How students verify conjectures: Teachers' expectations *Journal of*

Mathematics Teacher Education **8** 171

IOP Conf. Series: Journal of Physics: Conf. Series **948** (2018) 012037 doi:10.1088/1742-6596/948/1/012037

Promoting students' mathematical problem-solving skills through 7e learning cycle and hypnoteaching model

H Saleh[1], D Suryadi[2] and J A Dahlan[2]

[1]Universitas Muhammadiyah Tangerang, Jl. Perintis Kemerdekaan I/33 Tangerang Banten Indonesia
[2]Indonesia University of Education, Jl. Dr. Setiabudhi 229 Bandung 40154 West Java Indonesia

E-mail: hairulsaleh@umt.ac.id

Abstract. The aim of this research was to find out whether 7E learning cycle under hypnoteaching model can enhance students' mathematical problem-solving skill. This research was quasi-experimental study. The design of this study was pretest-posttest control group design. There were two groups of sample used in the study. The experimental group was given 7E learning cycle under hypnoteaching model, while the control group was given conventional model. The population of this study was the student of mathematics education program at one university in Tangerang. The statistical analysis used to test the hypothesis of this study were t-test and Mann-Whitney U. The result of this study show that: (1) The students' achievement of mathematical problem solving skill who obtained 7E learning cycle under hypnoteaching model are higher than the students who obtained conventional model; (2) There are differences in the students' enhancement of mathematical problem-solving skill based on students' prior mathematical knowledge (PMK) category (high, middle, and low).

1. Introduction

Mathematics deals with abstract ideas that are hierarchically arranged and deductive reasoning. Such a thing will certainly lead to the occurrence of the process of learning mathematics that leads to mastery of student mathematics. Lack of mastery of mathematics, especially in solving math problems such as students are not learning and not accustomed by educators solve mathematical problems that appeal to students of the phenomenon encountered. In the curriculum, it is emphasized that the problem-solving approach is the focus of mathematics learning covering closed problems with single solutions, open issues with non-sole solutions, and problems with various solutions. This is in line with the standard process based on the National Council of Teachers of Mathematics [1] which states that the standard process in learning mathematics include problem-solving skills, reasoning ability, communication skills, make connections (connections), and the ability of representation (representation).

Educator as facilitator should have the ability to create learning situations involving students actively so that leads to improving students' mathematical skills, especially problem-solving skills. Efforts that can be done, among others, by applying the appropriate model of learning and more meaningful learning model that involves students in learning, one of the learning model is the learning cycle.

The learning cycle is one of the learning models that adopt constructivist in learning. The rationale of the constructivist is an effective learning process requires that educators know how the students bring the facts and phenomena that become the subject of learning. Implementation of this model in learning activities can help students understand the concept through the stage of data collection (exploration), the introduction of the concept (concept introduction), and the application of the concept (concept application). Three cycles (phases) have been developed into five stages consisting of engagement, exploration, explanation, elaboration, and evaluation, and now developed into seven cycles (phase) consisting of elicit, engagement, exploration, explanation, elaboration, evaluation and extend [2]. In this research will be used 7E learning cycle that is elicit, engagement, exploration, explanation, elaboration, evaluation and extend.

In the 7E learning model, students are trained to identify a mathematical problem, connect their own mathematical understanding, express opinions, ask questions, present and explain answers, reformulate mathematical concepts with their own language, solve problems of challenge, evaluate and develop answers - more innovative answers. A teacher is student facilitators in learning, guiding and directing students during the learning process. Opinions or ideas raised by students, well appreciated by lecturers, so that students feel themselves have contributed to learning in the classroom.

Hypnoteaching is a unique, creative and imaginative way of teaching that can make students feel comfortable in the teaching and learning process, since the lesson begins until the ending lesson is done by using subconscious language which gives students the impression to concentrate fully on the learning process. Hypnoteaching is the art of communicating by giving suggestions that students become smarter. With the suggestion given they are expected to be awakened and enlightened that there is tremendous potential that they have never optimized in learning. By utilizing hypnoteaching expected barriers or difficulties faced by students in every cycle of the 7E learning cycle can be overcome.

Based on the above description, the authors are encouraged to conduct research entitled " The Enhancement of Students' Mathematical Problem-solving Skill Through 7E Learning Cycle Under Hypnoteaching Model ". In this course, students' mathematical problem-solving abilities are reviewed as a whole based on the student's prior mathematical knowledge (PMK).

2. Mathematical problem-solving skill

Problem-solving in the context of mathematics is seen as a very important activity in learning, because it demands critical and creative thinking. Solving mathematical problems is a complex mental process such as creating new ideas in determining problem-solving, or finding new techniques or products.To develop problem-solving skills, Polya [3] provides an overview of the framework for solving mathematical problems as: (1) Understanding on the problem (identification of purpose); (2) Create a problem-solving plan; (3) Implement the plan; and (4) Looking back.

Furthermore, in this research, the ability to solve mathematical problems developed against students are: (1) Understanding the problem, ie the ability to identify known elements, being asked, checking the adequacy of the data and creating a mathematical model of the problem; (2) Plan for problem-solving, ie the ability to choose or develop problem-solving strategies; (3) Carry out calculations, namely the ability to solve mathematical problems according to plan; and (4) Examines the truth of the results, ie the ability to interpret the results according to the original problem.

3. Hypnoteaching

According to Dale [4] hypnosis is a state of mind that shows high-level suggestions. Kurniawati [5] concluded that hypnoteaching is a learning method that seeks to decrease the frequency of brain waves by making the students focus (focusing), either with video shows, inspirational stories, and yelling and using relaxation and imagination so that students' attention becomes centered, students become relaxed and more suggestive in capturing the positive values of a learning process. According to the steps of learning hypnoteaching among others as follows: identify the needs of the students first; plan learning by using hypnosis media such as voice, picture, writing and symbols; start teaching by making students focus well with games, singing together, inspiring stories, and yeling (focusing); perform relaxation and imagination; divide the students into groups (grouping) where each group discusses, shares knowledge and teaches and praises; declare something positive from the student (affirmation); repeating the words of inspirational motivation (repeating); evaluate, and before the learning ends, the student reflects on the lesson learned.

4. 7E learning cycle under hypnoteaching model

The learning cycle is an active teaching approach based on Piaget's theory of mental development and constructivism. Through development by researchers for a long time, the learning cycle model comes in three types, depending on the number of phases it contains (4, 5 or 7). The latest version has the following 7 stages: elicit, Engage, explore, explain, elaborate, evaluate and extend. Research has shown that learning cycle models have a positive effect on specific learning products, proving to be an important tool for teaching concepts, influencing students 'academic performance in a positive way, enabling students to develop or improve positive attitudes toward learning and improve students' ability to think scientifically, experimental and practical skills [6].

In the next research, researchers will combine the learning of 7E cycles with hypnoteaching. The steps of the 7E cycle according to Eisencraft [7] are as follows:

1) Elicit phase, intended to identify the initial knowledge of the students, to ensure whether the students already know the lesson to be learned.
2) Engage phase, intended to attract attention or arouse students' interest and motivation towards concepts to be taught by asking questions, telling stories, giving demonstrations, or by showing an object, image or video.
3) The explore phase is intended to provide students with the opportunity to acquire knowledge with immediate experience related to the concepts to be studied, where students observe, isolate variables, plan inquiry, interpret results and develop hypotheses and organize conclusions and learning materials that have been provided previous.
4) Explain phases, students give explanations of concepts, introduce concepts, terms and summarize the results obtained in the exploration phase.
5) Elaborate phases, students are given the opportunity to apply their newly discovered knowledge to problems related to the example of the lesson learned, for example, providing practice questions with a deeper level of analysis to the students.
6) Evaluate phase, is an advanced cycle to evaluate students' knowledge. By using formative assessments to see student progress, the desired developments in the students' understanding of concepts, principles and abilities apply the concept.
7) Extend phase is the phase of students encouraged to link and explain examples of the application of concepts they learn with everyday life.

5. Method

The research method used in this research is quasi-experiment with the design of Nonequivalent group pretest-posttest design [8] consisting of two groups: experimental group and control group. The experimental group is a group that gets 7E learning cycle under hypnoteaching model (LCH), while the control group is a group that gets conventional learning (CL). The data to be analyzed in this research is N-Gain data (normalized gain) from students' mathematical problem-solving test. This normality and homogeneity test used the Kolmogorov-Smirnov and Levene tests. The difference test and interaction analysis using the t-test and ANOVA. All statistical calculations use the IBM SPSS 21 computer program.

6. Results and Discussion

6.1 Description of mathematical problem-solving ability

Data analysis of students' mathematical problem-solving ability is done by data which is derived from mathematical problem-solving test. The result of descriptive statistical improvement of students' mathematical problem-solving abilities is presented in Table 1.

Table 1. Descriptive statistics of improved mathematical problem-solving ability

Learning	PMK	Mean	N	Learning	PMK	Mean	N
LCH	high	0.6000	7	CL	high	0.4225	8
	medium	0.6311	35		medium	0.5239	44
	low	0.6163	8		low	0.5050	6
	Total	0.6244	50		Total	0.5079	58

Based on Table 1, the PMK category (high. medium. low) the average N-Gain of the students' problem-solving ability in the class by LCH learning is greater than in the class taught by CL. This can be seen from the total N-Gain average of LCH and CL is 0.6244> 0.5079. This result shows that the students' problem-solving abilities improvement based on the overall data and CL categories PMK (high, medium, low) in the classroom with the students by LCH learning better than the control class with the students by CL.

6.2 Mathematical problem-solving skill improvement based on learning analysis and PMK

Based on the normality analysis data, the category PMK each data is normally distributed. Further analysis to determine the improvement of the ability of PM based on the overall data and category PMK (high, medium, low) between the experimental class and the control class is done by using the t-test. The different test results between the experimental and control classes are presented in Table 2.

Table 2. Different test enhancement mathematical problem-solving

Category of PMK	t-test	Sig (2-tailed)
Overall	2.901	0.005
High	1.149	0.285
Medium	2.403	0.020
Low	1.031	0.323

Based on Table 2, the difference test between the two classes of experimental class and control class in overall data and medium category of PMK with the probability result is $0.0025 < \alpha = 0.05$ and $0.010 < 0.05$, means reject the H_0 hypothesis. These results show that the improvement of students' mathematical problem-solving abilities based on the overall data and PMK medium category in the experiment class with LCH learning better than the control class with CL learning.

While to know the difference test between experimental class and control class in high and low category with result of probability value is $0.285 > \alpha = 0.05$ and $0.165 > 0.05$ so Ho accepted. These results show that the improvement of students' mathematical problem-solving abilities based on LCH learning less than problem-solving learning in category high and low PMK.

6.3 Interaction analysis of mathematical problem-solving ability

Interaction analysis between learning model and PMK was done by ANOVA Analysis. The ANOVA Analysis results are presented in Table 3 below.

Table 3. Interaction between learning model and PMK

Source	Type III Sum of Squares	df	Mean Square	F	Sig.
Corrected Model	.440[a]	5	.088	2.179	.062
Intercept	17.815	1	17.815	440.892	.000
Learning	.257	1	.257	6.355	.013
PMK	.055	2	.028	.686	.506
Learning * PMK	.016	2	.008	.193	.825

Table 3 above shows that the value of significance 0.825 > 0.05, so that the Ho hypothesis is accepted. It means that there is no interaction between learning model (LCH, CL) and PMK category (high. medium. low). This means that the difference in mean score of mathematical problem-solving ability of mathematical knowledge students (high, medium, and low) who treated by LCH does not differ significantly with students who have treated by CL.

7. Conclusion

Based on the analysis of results obtained in this study it can be concluded several things as follows: (1) The students' achievement of mathematical problem-solving skill who obtained 7E learning cycle under hypnoteaching model are higher than the students who obtained conventional model; (2) There are differences in the students' enhancement of mathematical problem-solving skill based on students' prior mathematical knowledge (PMK) category (high, middle, and low).

References

[1] NCTM 2003 *Program for Initial Preparation of Mathematics Specialists* (http://www.ncate.org/ProgramStandards/NCTM/NCTMStandards.pdf)
[2] Eisenkraft A 2003 A proposed 7E model emphasizes "transfer of learning" and the important of eliciting prior understanding *The Science Teacher* 70 (National Science Teachers Association NSTA)
[3] Polya G 1985 On solving mathematical problems in high school In S Krulik eds *Problem Solving in School Mathematics* (Reston, Virginia: NCTM) p 5
[4] Dale R A 2014 *Hypnosis and Education* (http://eric.ed.gov/?id=ED087710)
[5] Kurniawati L 2014 Meningkatkan kemampuan berpikir intuitif-reflektif, pembuktian matematis dan disposisi mahasiswa melalui pembelajaran berbasis masalah dengan metode hypnoteaching *Disertasi* (Bandung: Program Pascasarjana Universitas Pendidikan Indonesia)
[6] Celik H 2013 The Effect of the computer-aided 7E teaching model on students' science process skills, *Mediterranean Journal of Educational Research* **14a** 926
[7] Pitriati 2014 Pengaruh penerapan model learning cycle 7E terhadap peningkatan kemampuan penalaran dan kemampuan komunikasi matematis siswa SMP *Tesis* (Bandung: Program Pascasarjana Universitas Pendidikan Indonesia)
[8] Ruseffendi E T 2010 *Dasar-dasar penelitian pendidikan dan bidang non-eksakta lainnya* (Bandung: Tarsito)

Improving students' mathematical representational ability through RME-based progressive mathematization

Warsito[1], Darhim[2] and T Herman[2]

[1]Tangerang Muhammadiyah University, Jl. Perintis Kemerdekaan I/33 Tangerang Banten Indonesia
[2]Indonesia University of Education, Jl. Dr Setiabudhi 229 Bandung 40154 West Java Indonesia

E-mail: warsito@umt.ac.id

Abstract. This study aims to determine the differences in the improving of mathematical representation ability based on progressive mathematization with realistic mathematics education (PMR-MP) with conventional learning approach (PB). The method of research is quasi-experiments with non-equivalent control group designs. The study population is all students of class VIII SMPN 2 Tangerang consisting of 6 classes, while the sample was taken two classes with purposive sampling technique. The experimental class is treated with PMR-MP while the control class is treated with PB. The instruments used are test of mathematical representation ability. Data analysis was done by t-test, ANOVA test, post hoc test, and descriptive analysis. The result of analysis can be concluded that: 1) there are differences of mathematical representation ability improvement between students treated by PMR-MP and PB, 2) no interaction between learning approach (PMR-MP, PB) and prior mathematics knowledge (PAM) to improve students' mathematical representation; 3) Students' mathematical representation improvement in the level of higher PAM is better than medium, and low PAM students. Thus, based on the process of mathematization, it is very important when the learning direction of PMR-MP emphasizes on the process of building mathematics through a mathematical model.

1. Introduction

Mathematics becomes one of the important knowledge to be mastered by students from elementary school to college level to achieve mathematical ability. In the Indonesian national curriculum, both in 2006 curriculum and the 2013 curriculum, mathematical ability is the target to be achieved in mathematics learning.

Although this mathematical ability is very important, the students' achievement level in mathematical ability in Junior High School is still low. This is seen from the results reports on student achievement of Indonesia in the international scale TIMSS and PISA from the first participate until now, there are no significant change outcomes ranking and always occupy the 10 lowest among other participating countries [1]. The low mathematical ability of students according to [2] is influenced by several factors such as students accustomed to acquire and use formal mathematics knowledge in the classroom that is the teacher give without knowing how to get the formula, and evaluation system in Indonesia still use level low.

It is important to understand various approaches to learning mathematics that can be used in order to develop students' skills, especially the ability of mathematical representation. One of the mathematical approaches that provide ample space for students to construct and discover mathematical concepts is realistic mathematics education. In Indonesian version, known as Pendidikan Matematika Realistik (PMR). Realistic learning is a learning approach that places mathematics as a human activity derived from the real world or the world that is close to the student and can be imagined by the students [3]. In this case, students are actively given the opportunity to reinvent a mathematical concept through teacher guidance [4]. The process of student activity is mainly to build a mathematical model or concept. In RME, the process of mathematization consists of horizontal and vertical mathematization.

The interesting thing about the mathematization process is that students should not be considered as passive receivers of ready-made mathematics, but students are considered as active individuals who are able to develop their own mathematical potential ([5]; [6]). In the process of mathematization, students are required to build or develop a mathematical model of a model that has been formed previously with the knowledge possessed. This is in line with the opinion of Freudenthal [4], mathematization in learning mathematics is as a process of improvement and development of mathematical ideas progressive mathematization process occurs gradually when the process of vertical mathematization forms formalization process of mathematical concepts of mathematical models derived from horizontal mathematization [1]. Based on the background of the problems that have been stated above, then his study aims to determine the differences in the improving of mathematical representation ability based on progressive mathematization with realistic mathematics education (PMR-MP) with conventional learning approach (PB).

2. Progressive Mathematization

The meaning of representation based on [7] mentions that the student's mathematical representation is a form of ideas expressions or mathematical ideas that are displayed as an attempt to find a solution of the problem he was facing. To construct mathematical concept as a form of mathematical representation of a contextual problem hence required step or process of mathematization that consists of the process of horizontal mathematization and vertical mathematization. Freudenthal in [4], in the learning of mathematics, is a process of improvement and development of mathematical ideas gradually. In accordance with De Lange's opinion [1] which defines mathematization as organizing activities in finding order, relationships, and mathematical structures using the prior knowledge and skills built.

The importance of mathematization is also used by PISA to develop students' mathematical abilities. In PISA's explanation [8], states that mathematical processes are an activity that describes what someone does to connect the contextual problems with mathematics. The process of mathematizing in PISA's version related to building mathematical models and concepts begins by identifying the contextual problems, looking for relationships with symbols, patterns relating to the problem so as to form a mathematical model. The staging process from contextual problem to the mathematical problem is done by the process of horizontal mathematization. While finding a mathematical problem-solving process in form of mathematical representation, formal mathematics and generalization is the process of vertical mathematization or model for and formal model.

The constructing or finding mathematical process concepts using mathematical models that have been formed before is called progressive mathematization process. For Trafer in [9], schematization was not the only argument in the aforementioned progressive schematization course, but be also emphasized other characteristics of informal strategies such as mental calculation, flexible arithmetic, and estimation. This is in accordance with Wijaya's statement [1] that, using models, schemes, diagrams, and symbols for progressive mathematization is important for mathematical discovery and development concepts by students. The model, in this case, relates to the model of the situation and the mathematical model developed by the students themselves. [4] and [10] mentions four levels or stages in the development of mathematical models, which are: 1) situational level, 2) referential level, at this level, 3) the general level, 4) formal level. The mathematization process is a key in the RME so that it is not only the main activity of math but also introduces students to situations close to the students' daily lives in searching for mathematical models. In addition, at the final stage of mathematization is the formalization through the axiomatization mathematical concepts so that with this process the students' mathematical ability, especially students' mathematical representation can increase.

3. Method

The research method used in this research is quasi-experiment with the design of nonequivalent control group design [11], consisting of experimental group and control group. An experimental group is a group that gets PMR-MP, while the control group is a group that gets PB. The data to be analyzed in this research is N-Gain data (normalized gain) from students' mathematical representation test. This normality and homogeneity test used the Kolmogorov-Smirnov and Levene tests. The difference test and interaction analysis using the t-test and ANOVA. All statistical calculations use the IBM SPSS 21 computer program.

4. Results and Discussion

4.1 Mathematical Representation Ability Description

The students' mathematical representation improvement data analysis was done by using N-Gain data from mathematical representation test. The descriptive statistical is presented in Table 1

Table 1. Descriptive Statistics Improving Ability of Mathematical Representation

	Average N-Gain on PAM Category							
	High	n	Medium	n	Low	N	Total	n
PMR-MP	0.53	7	0.39	21	0.30	8	0.39	36
PB	0.41	6	0.27	24	0.12	6	0.27	36

Based on Table 1, the PAM category (high. medium. low) the average N-Gain of the students' mathematical representation ability in the class by PMR-MP learning is greater than in the class taught by PB. This can be seen from the total N-Gain average of PMR-MP and PB is 0.39> 0.27. This result shows that the students' mathematical representation abilities improvement based on the overall data and PAM categories (high. medium. low) in the classroom with the students by PMR-MP learning better than the control class with the students by PB.

4.2 Mathematical Representation Improvement Based on Learning Analysis and PAM

Based on the normality analysis data, the category PAM each data is normally distributed. Further analysis to determine the improvement of the ability of mathematical representation based on the overall data and category PAM (high, medium, low) between the experimental class and the control class is done by using the t-test. The different test results between the experimental and control classes are presented in Table 2.

Table 2. Different Test Enhancement Mathematical Representation In PAM Category Data

Category of PAM	t-test	Sig. (2-tailed)
Overall	3.262	0.002
High	2.392	0.036
Medium	2.653	0.011
Low	2.633	0.026

Based on Table 2, the difference test between the two classes of experimental class and control class with the probability result is $0.001 < \alpha = 0.05$, means reject the Ho hypothesis. These results indicate that students who have received PMR-MP learning have better mathematical representation than students with PB learning in overall data. While to know the difference test between experimental class and control class in PAM (high. medium. low) category with result of probability value is 0.0175 $< \alpha = 0.05$, $0.0056 < \alpha = 0.05$. and $0.013 < \alpha = 0.05$ means reject Ho. These results indicate that students who have received PMR-MP learning have better mathematical representation ability than students receiving PB learning in the high PAM category, medium, and low PAM category.

The conclusion of different test analysis based on learning in the PAM category is there is an improvement of students' mathematical representation ability in the experimental class where the students treated by PMR-MP learning better than the students in the control class who treated by PB.

4.3 Interaction Analysis of Mathematical Representation Ability Based on Learning and PAM

Interaction analysis between learning and PAM was done by two-path ANOVA analysis. The ANOVA Analysis results are presented in Table 3 below.

Table 3. Interaction between Learning and PAM with ANOVA Test

Source	Type III Sum of Squares	df	Mean Square	F	Sig.
Corrected Model	.750[a]	5	.150	8.346	.000
Intercept	5.823	1	5.823	324.134	.000
Learning	.218	1	.218	12.144	.001
PAM	.492	2	.246	13.687	.000
Learning * PAM	.003	2	.002	.097	.907
Error	1.186	66	.018		
Total	9.913	72			
Corrected Total	1.935	71			

From table 3 above, It shows that the value of significance $0.907 > 0.05$, so that the Ho hypothesis is accepted. It means that there is no interaction between learning (PMR-MP. PB) and PAM (high. medium. low). In other words, this means that the difference in mean score of mathematical representation ability of mathematical knowledge students (high, medium, and low) who treated by PMR-MP does not differ significantly with students who have treated by PB. Graphically, the interaction can be seen in Figure 1 below.

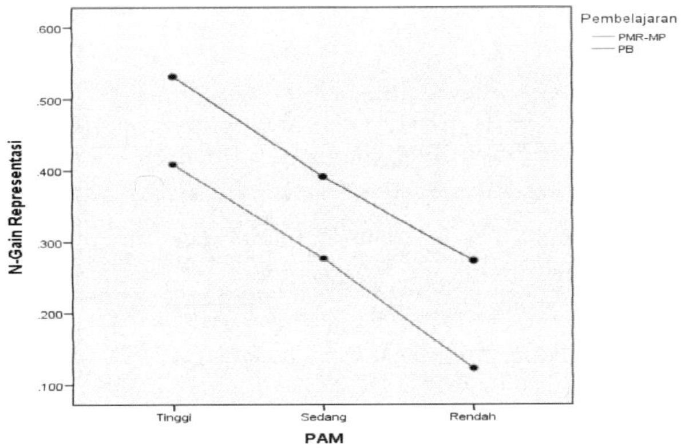

Figure 1. Interaction between learning and PAM

Based on figure 1 above, it can be explained that the learning with PMR-MP is appropriate for all categories of PAM (high, medium, low) in improving mathematical representation ability. This can be seen from the average students' mathematical representation ability score who treated by PMR-MP higher than students who received treated by PB. From figure 1 above also indicates that students with low mathematical ability gain the greatest benefit in learning based on PMR-MP rather than students with high and medium mathematical skills. This can be demonstrated through the difference of the average score of N-Gain ability of mathematical representation between students who learn through PMR-MP and PB in successive high-ability students (0.12), moderate (0.12), low (0.18).

From Table 3, it can be seen that the value calculation of F for the learning approach is 12.144 with a significance level of 0.001 < 0.05, so the Ho hypothesis is rejected. This means that there is a mathematical representation skill improvement difference based on the learning approach. In other words, there is a difference in the improvement of the significant mathematical representation ability among students who get the learning approach of PMR-MP and PB learning.

Referring to the research and discussion results above, it can be concluded that Freudenthal [6] further mentions that mathematization is a key process in RME. The reason is the first mathematization is not just a math-centered activity but also introduces students to situations close to the student's daily life. Second, the final stage of mathematics is formalization through axiomatization. This final stage is not a starting point when mathematical concepts are considered as found in traditional mathematical learning. Therefore, mathematics learning should be delivered in the form by giving students an opportunity to reinvent ideas and mathematical concepts along with teacher guidance through exploration of various contextual issues. If we hold fast to the objectives of learning mathematics and apply the process standard properly with the RME approach then the students 'mathematical abilities development in particular students' mathematical representation may increase

5 Conclusion

Based on the analysis results of the findings obtained in this study. it can be concluded several things as follows: 1) students' mathematical representation improvement ability who received learning with the PMR-MP approach is better than students who received learning with PB 2) students' mathematical representation improvement ability who received learning with the PMR-MP approach are better than students who received learning with PB in terms of PAM category (high. medium. low). 3) there is no interaction between learning approaches (PMR-MP and PB) with PAM (high. medium. low) to mathematical representation ability. and 4) student responses on learning by PMR-MP approach students give a positive response they feel happy with the implementation experienced learning.

References

[1] Wijaya A 2012 *Pendidikan Matematika Realistik* (Yogyakarta: Graha Ilmu)

[2] Hayat B and Yusuf S 2010 Mutu Pendidikan (Jakarta: Bumi Aksara)

[3] Hadi 2005 *Pendidikan Matematika Realistik Indonesia dan Implementasinya* (Banjarmasin: Tulip)

[4] Gravemeijer 1994 *Developing Realistic Mathematics Education* (Utrecht: Kluwer Academic Publishers)

[5] Hadi S 2002 Effective teacher professional development for implementation of realistic mathematics education in Indonesia *Dissertation* (Netherland: University of Twente Enschede)

[6] Fauzan 2002 Applying realistic mathematics education (RME) in teaching geometry in Indonesian primary schools *Dissertation* (Netherland: University of Twente)

[7] NCTM 2000 *Principles and standards for school mathematics* (Reston VA: National Council of Teachers of Mathematics)

[8] PISA 2015 *Draft mathematics framework*: (http://wwwoecdorg/pisa/pisaproducts/pisa2015draftframeworkshtm)

[9] Putten C M V, Snijder P A V and Beishuizen M 2005 Progressive Mathematization of Long Division Strategies in Dutch Primary Schools *Journal for research in Mathematics Educations* **36** **1** 44

[10] Kairuddin and Darmawijoyo 2011 *Journal on Mathematics Education* **21** 67

[11] Ruseffendi E T 2010 *Dasar-dasar penelitian pendidikan dan bidang non-eksakta lainnya* (Bandung: Tarsito)

[12] Gravemeijer K 2011 *Journal on Mathematics Education* **21** 1

IOP Conf. Series: Journal of Physics: Conf. Series **948** (2018) 012039 doi:10.1088/1742-6596/948/1/012039

Interactive design for self-study and developing students' critical thinking skills in electromagnetic radiation topic

D Ambarwati [1] and A Suyatna[1]

Physics Education of Graduate Program, Lampung University, Jl. Sumantri Brojonegoro No. 01, Lampung, Indonesia.

E-mail: ambarwati_desih@yahoo.com

Abstract. The purpose of this research are to create interactive electronic school books (ESB) for electromagnetic radiation topic that can be used for self-study and increasing students' critical thinking skills. The research method was based on the design of research and development (R&D) model of ADDIE. The research procedure is used limited the design of the product has been validated. Data source at interactive requirement analysis phase of ESB is student and high school teacher of class XII in Lampung province. The validation of interactive ESB designs is performed by experts in science education. The data of ESB interactive needs were collected using questionnaires and analyzed using quantitative descriptive. The results of the questionnaire obtained by 97% of books that are often used in the form of printed books from schools have not been interactive and foster critical thinking of students, and 55% of students stating physics books are used not meet expectations. Expectations of students in physics learning, teachers must use interactive electronic books. The results of the validation experts pointed out, the design of ESB produced is interactive, can be used for self-study, and increasing students' critical thinking skills, which contains instruction manuals, learning objectives, learning materials, sample questions and discussion, video illustrations, animations, summaries, as well as interactive quizzes incorporating *feedback* exam practice and preparation for college entrance.

1. Introduction

Learning science has an important role in preparing students to be able to think critically, creatively, logical and initiatives. This is in line with the goals of the curriculum in 2013, which is to encourage the students to be able to better observation, questioning, reasoning, and communicate what they earn or they know after receiving learning materials.

Physics is one branch of science that is closely related to the phenomena that occur in the natural environment. Not all physics phenomena can be observed directly, one of which is the material of Electromagnetic Radiation. Learning all this time, the teacher in teaching physics is still limited explanation abstract concepts through lectures, because the material is difficult in practice Electromagnetic Radiation directly, because to do the lab work on the matter requires powerful tools that are not available in schools. Physics is a unique and interesting science, because the subject of physics is difficult to teach and difficult to understand [1]. In the learning process, 83% of students sometimes experience difficulties in learning Physics materials, and 29% of students say teaching methods used by teachers are less appropriate.

Another difficulty is to bring scientific learning in teaching Electromagnetic Radiation, due to limitations of equipment and time owned lab. Limitations of time in discussing this matter because these materials are in the second semester of class XII in which schools use the time to prepare for the National Examination (UN). With the limited time of study that is owned by students of class XII in the even semester, not all class XII material in the even semester complete taught. Though thoroughly learned, the teacher only explains the material briefly or the important material that will be about to come out on the exam and the teacher only gives the exercise questions. Problems given by teachers are limited to the choice of plural and essay/descriptions that only require students to use related formulas, without making students think critically.

Physics learning can be used as a vehicle to train students to think critically. Based on preliminary research that is conducted by researchers to physics teachers and students of grade XII high school in Lampung province. The results of the questionnaire obtained by 97% of media often used in learning in the form of printed books from schools have not been interactive and foster critical thinking of students, as much as 70% of students stated the physics questions provided by teachers have not made students think critically. At 64%, students stated that the physics book should be useful in the learning process that is to facilitate the study of physics, and 55% of students stated that the physics book used in the lesson is less hopeful. Student expectations in physics learning, teachers must use electronic interactive school books (ESB) that can be used independently, which can facilitate the study of physics that contains instructions on the use, learning objectives, learning materials, sample questions and discussion, video illustration, animation, as well as interactive quizzes with feedback for exam exercises and preparing for admission to college.

In the learning process, in addition to using printed books from schools, there are students who seek other learning resources, one of them by using ESB. ESB can be used as a learning resource. Electronic books have a coherent presentation format, both the language, the high level of scholarship, and the extent of the discussion [2]. An electronic book is a book in electronic format containing information that can be texts or images; another advantage of ESB is to be able to display multimedia illustrations, such as animation [3].

Some Physics phenomena on certain subject matter cannot be observed directly by the students, because it is an abstract review, one of them is on Electromagnetic Radiation material. Innovative process of physics learning, which can facilitate students in observing the phenomena that are difficult to observe directly, one of which uses ESB featuring multimedia illustrations. Multimedia is an instrument that can create dynamic and interactive presentations by combining text, graphics, animation, audio, and video [4]. This multimedia-based ESB learning aspect supports a learner-centered learning strategy where each learner is responsible in his/her own learning process [5]. The media can be arranged so that learners can independently learn the material, and gain more knowledge, so that students are able to learn independently.

Physics can be regarded as a science that encourages students to think high-level, it makes enormous academic demands on students in learning [1,6]. Knowledge gained through higher-order thinking processes is easier to transfer, so students with deep conceptual understanding. High-level thinking skills include the ability to think critically. Students who are able to think critically can solve problems effectively [7]. By seeing the demands of the current learning curriculum, students' understanding of concepts can be improved through the development of students' critical thinking skills.

The purpose of this research is to make electronic school book (ESB) interactive electromagnetic radiation material that can be used for self-study and improve students' critical thinking ability. Until recently, a book like this for electromagnetic radiation matter yet, because the material electromagnetic radiation is very abstract, so to solve this problem required electronic school books (ESB) interactive.

2. Methods and Materials

The research method is based on research and development (R&D) design, which uses the ADDIE development model, consisting of analysis, design, development, implementation, and evaluation. The research procedure used is limited to the validated design stage of the expert. The first stage is the analysis. The ESB needs analysis phase uses interactive data sources of 87 high school students and 2 physics teachers drawn at random from 2 different SMAs in Lampung Province. ESB needs data were collected using a questionnaire. Questionnaire needs analysis is done to get information about real conditions in the learning process in school. Data were analyzed quantitatively descriptively. After performing the data analysis, then made an interactive ESB design followed by design validation stages. This design validation test is performed, to know concepts, formulas, images, animations, simulations, and videos, as well as critical thinking types contained in interactive ESB. Interactive ESB design validation is performed by experts in science education by using expert test instrument in the form of ESB interactive design score sheet. Expert validation test assessment guides are presented in Table 1.

Table 1. Rubric Assessment Test Validation Expert

Choiches	Score
Very Important	5
Important	4
Quite Important	3
Less Important	2
Umimportant	1

The results of average score are searched by using below formula:

$$\text{Average Score} = \frac{Total\ Score}{Number\ of\ experts}$$

Having obtained the average, then converted to the assessment statement. The conversion of the score into an assessment statement is presented in Table 2.

Table 2. Score Convertion

Average Score	Decision
4,20 – 5,00	It is very important to be put in an interactive ESB
3,40 – 4,19	It is important to be put in an interactive ESB
2,60 – 3,39	It is quite important to be put in an interactive ESB
1,80 – 2,59	It is less important to be put in an interactive ESB
1,00 – 1,79	It is not important to be put in an interactive ESB

3. Results and Discussion

Based on the results of questionnaire of requirement analysis, the preparation of ESB interactive design of Electromagnetic Radiation material for self-study and improving students' critical thinking skill (Figure 1). Interactive ESB is designed as a learning resource that can be used independently by students containing instruction manuals, learning materials, sample questions and discussion, video illustrations, animations, summaries and interactive quizzes with feedback.

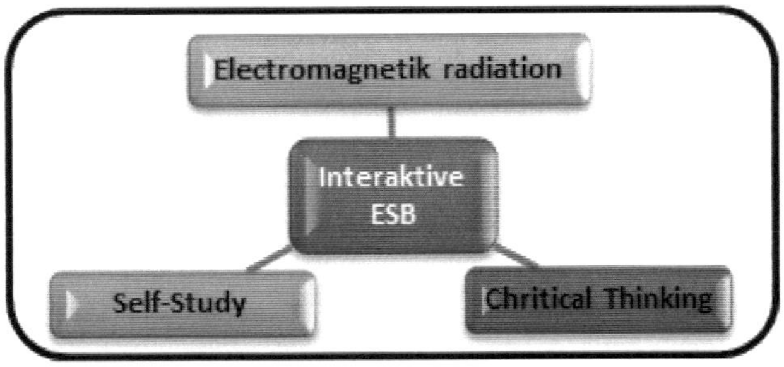

Figure 1. Design chart of interactive ESB

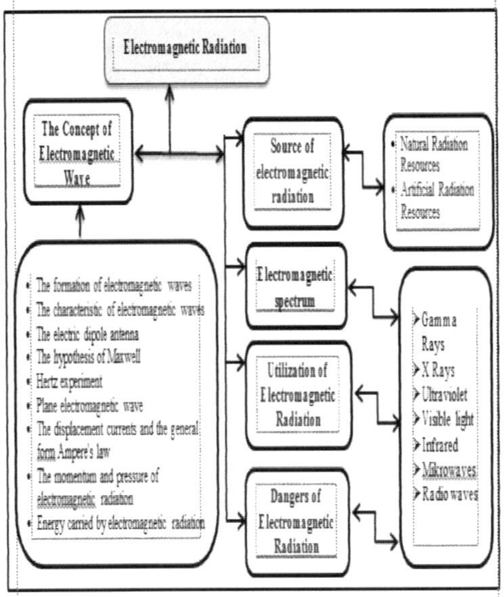

Figure 2. Electromagnetic Radiation Topic

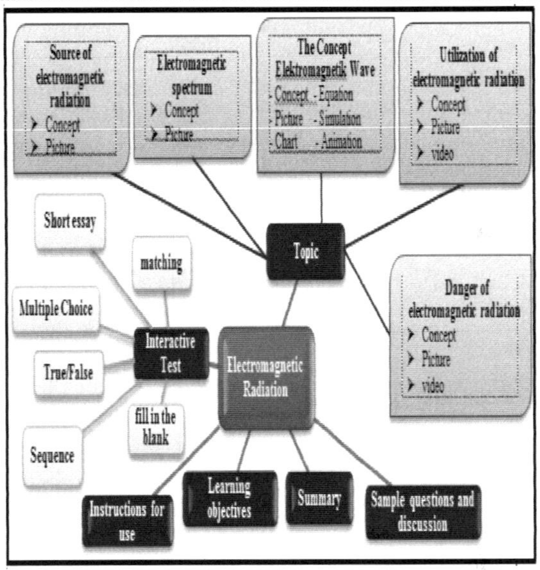

Figure 3. Interactive ESB Design of Electromagnetic Radiation

Based on the results of expert validation tests, the scope of electromagnetic radiation material consists of the concept of electromagnetic waves consisting of the properties of electromagnetic waves, the formation of electromagnetic waves, electric dipole antennae, the Maxwell hypothesis, Hertz experiments, current displacement and general form of Ampere law, electromagnetic waves, momentum and pressure of electromagnetic radiation, as well as the energy that electromagnetic radiation carries. Subsequent material coverage is the spectrum of electromagnetic radiation, electromagnetic radiation sources, and the utilization of electromagnetic radiation and the danger of electromagnetic radiation (Figure 2).

Based on the expert validation results, it is obtained ESB interactive design of electromagnetic radiation material (Figure 3). This interactive ESB design contains usage instructions, learning objectives, learning materials, sample questions and discussion, and interactive quizzes with feedback. Learning materials in this interactive ESB design are equipped with pictures, animations, simulations, and videos. Electromagnetic radiation material is an abstract material, or cannot be observed directly by students. Then it is needed an innovation in learning. There are many innovations in education and knowledge, from face to face with virtual education, different technologies have played a big role at different times [8]. The advancement of computer technology has attracted the attention of many educators. The use of technology in education provides a more suitable learning environment for learning, and helps to increase student motivation. Computer-based teaching has had an impact on the development of educational technology for the most part in the twenty-first century, and has resulted in the production of computer-based instruction software [9].

Table 3. Expert Experiment Test Results of ESB Interactive Electromagnetic Radiation Design

No	Topic	The result of expert validation	Note
A	**The Concept Electromagnetic Wave**		
	1. The formation of electromagnetic waves		
	a. Concept The formation of electromagnetic waves	*4*	*important*
	b. Picture lines of electric field	*3,33*	*important*
	c. Animation lines of electric field	*3,33*	*important*
	2. Characteristic of electromagnetic waves		
	a. Concept characteristic of electromagnetic waves	*4*	*important*
	3. The displacement currents and the general form Ampere's law		
	a. Concept The displacement currents and the general Form Ampere's law	*3,83*	*important*
	b. Picture The displacement currents	*3*	*quite important*
	c. Equation Ampere's law	*3,67*	*important*
	d. Simulation the displacement currents	*3*	*quite important*
	4. The electric dipole antenna		
	a. Concept the electric dipole antenna	*3,33*	*quite important*
	b. Picture current the electric dipole antenna	*3,33*	*quite important*
	5. The hypothesis of Maxwell		
	a. Concept electromagnetic Maxwell	*3,67*	*important*
	b. Equation speed of visible electromagnetic waves	*3,67*	*important*
	6. Hertz experiment		
	a. Concept Hertz experiment	*3,33*	*quite important*
	b. Picture Hertz experiment	*3,33*	*quite important*
	c. Simulation Hertz experiment	*3,67*	*important*
	7. Plane electromagnetic wave		
	a. Concept Plane electromagnetic wave	*3,33*	*quite important*
	b. Equation Faraday's law and electric flux	*3,33*	*quite important*
	c. Picture Plane electromagnetic wave	*3,67*	*important*
	d. Chart Plane electromagnetic wave	*3,56*	*important*
	8. The momentum and pressure of electromagnetic radiation		
	a. Concept the momentum and pressure of Electromagnetic radiation	*3*	*quite important*
	b. Equation the momentum and pressure of Electromagnetic radiation	*2,33*	*less important*
	c. Picture the momentum and pressure of Electromagnetic radiation	*2,67*	*quite important*
	9. Energy carried by electromagnetic radiation		
	a. Concept energy carried by electromagnetic radiation	*3,67*	*important*
	b. Equation total instantaneous energy density	*3,67*	*important*
B	**Electromagnetic Spectrum**		
	a. Concept electromagnetic spectrum	*4,33*	*very important*
	b. Picture electromagnetic spectrum	*4,33*	*very important*
C	**Source of electromagnetic radiation**		
	a. Concept Source of electromagnetic radiation	*3*	*quite important*

	b. Picture Source of electromagnetic radiation	*3*	*quite important*
	c. Simulation Source of electromagnetic radiation	*2*	*less important*
	d. Video Source of electromagnetic radiation	*2,44*	*quite important*
D	**Utilization of electromagnetic radiation**		
	a. Concept Utilization of electromagnetic radiation	*4,33*	*very important*
	b. Picture Utilization of electromagnetic radiation	*4,33*	*very important*
	c. Video Utilization of electromagnetic radiation	*5*	*very important*
E	**Danger of electromagnetic radiation**		
	a. Concept Danger of electromagnetic radiation	*5*	*very important*
	b. Picture Danger of electromagnetic radiation	*4,33*	*very important*
	c. Video Danger of electromagnetic radiation	*4,67*	*very important*

Based on expert validation, component concepts, graphics, formulas, images, animations, video illustrations are required in ESB interactive electromagnetic radiation materials (Table 3). Along with the development of information and communication technology (ICT), e-learning has emerged as an innovative approach to promote learning to students for continuing to higher education. E-learning provides an alternative to traditional classroom education and allows students to access information only without time constraints [10]. One of the E-learning that can be used is Electronic School Book (ESB). Electronic books are defined as the electronic form of a book with features similar to traditional print books including pages that can "change," and digital features can help readers like word pronunciation, text highlighting, and hypermedia (e.g. video, animation, and sound) [11]. One of the innovations of physics learning that can associate technology into learning is by using ESB featuring interactive multimedia illustrations. The term 'interactive multimedia' is catching all phrases to describe a new wave of computer software that is primarily concerned with the provision of information. The 'multimedia' component is characterized by the presence of text, images, sound, animation and video [8]. In line with technological advances: technological devices, in particular computers began to be used in educational environments to develop audio-visual materials such as animation and simulation, resulting in the development of computer-based instructional techniques [9].

Based on the data of expert validation results, the images to be contained in the interactive ESB are for the concept of electromagnetic wave formation consisting of electric field lines drawings, GEM sketch drawing, and current drawings in electric dipole antennas, and radiation intensity images produced by oscillating electric dipoles. In the spectrum of electromagnetic materials, the images of the spectrum of electromagnetic waves are gamma rays, x-rays, ultraviolet light, visible light, infrared rays, microwaves and radio waves, and contain images of electromagnetic radiation, electromagnetic radiation drawings and electromagnetic radiation hazards. The animations that will be contained in the interactive ESB are the electric field lines out of the positive charge toward the negative charge. The simulations contained in interactive ESBs include Hertz experimental simulations, and simulations of *Phet Simulation* about electromagnetic radiation depicting electrical charge lines, radio waves depicting electron position graphs.

Videos contained in ESB are interactive utilization of electromagnetic radiation, and the danger of electromagnetic radiation. E-learning has recently become a promising alternative to classroom learning. Video is a rich and powerful media used in e-learning. It can present information in an interesting way. Interactive video improves learning interactivity, thus potentially improving students' effectiveness and student motivation consistently [8]. One of the features that make learning interesting for learners is a real-life concept that makes learning more meaningful when the latest technology is used unlike traditional methods. In addition, most of the knowledge related to natural phenomena is now available in the computer environment. That is why, when teachers use computers and other technologies as teaching tools by means of students can visualize in the form of three dimensions, this will help students to find ways to relate learning to their environment [13]. Interactive test types that are considered to be very suitable in interactive ESB are the multiple choices, true/false, fill in the blank, matching, short essay, and sequence types. The type of interactive test used should adjust the material characteristics. Computer Based Instruction (CBI) allows students to learn by themselves by evaluating and reflecting on their learning. CBI motivates children to learn better by providing direct feedback and reinforcement by creating a fun and exciting atmosphere [9].

The use of computers in education has a positive effect to increase students 'attention and curiosity, and help provide in conceptual learning, so as to enhance students' critical thinking skills. Critical thinking is a learned skill that must be developed, practiced, and continuously integrated into the curriculum to engage students in active learning. Instructions that support critical thinking use interrogation techniques that require students to analyze, synthesize, and evaluate information to solve problems and make decisions (think) not just to repeat information (memorization) [7]. Critical thinking is a group of skills that help to achieve information and outperform difficulty dealing with ease. Critical thinking is not a congenital characteristic, it can be taught, explained and easily applied in the learning process [14]. The most important role of teachers in developing students 'critical thinking skills is to create instructional applications and environments that support critical thinking students, allowing students to comment on lessons, incidents, conceptions and events from a variety of perspectives vital to improving students' critical thinking [15]. Critical thinking skills are an educational goal because through this learner's skill move away from learning to think about what they are learning and not what others are learning [16]. The development of critical thinking as a result of desirable education requires teaching methods that help learners improve their ability to think critically [17]. On the one hand, e-learning is not limited to geographic barriers. Students can engage in self-learning, and learning resources can be used repeatedly [18].

4. Conclusion

Expert validation results show that the resulting ESB design can improve students' critical thinking skills, which include instruction manuals, learning objectives, learning materials, sample questions and discussion, video illustrations, animations, summaries and interactive quizzes with feedback, and can be used independent and interactive.

References

[1] Ramos J L S, Dolipas B B and Villamor B B 2013 *International Journal of Innovative Interdisciplinary Research* **1** 4 pp. 48-60
[2] Prabowo A, Heriyanto S and IM M 2013 *Jurnal Ilmu Perpustakaan* **2** 2 pp. 1-9
[3] Eskawati S Y and Sanjaya I G 2012 *Unesa Journal of Chemical Education* **1** 2 pp. 46-53
[4] Wu M 2016 *International Journal of Modern Education and Computer Science* **8** 5 39
[5] Teoh B S P and Tse-Kian N E O 2007 *TOJET: The Turkish Online Journal of Educational Technology* **6** 4
[6] Ramos J L S, Dolipas B B and Villamor B B 2013 *International Journal of Innovative Interdisciplinary Research*, **1 4** 48-60
[7] Peter E E 2012 *African Journal of Mathematics and Computer Science Research*, **5(3),** 39-43.
[8] Mishra S and Sharma R C (Eds.) 2004 *Interactive multimedia in education and training*. Igi Global.115-122
[9] Serin O 2011 *TOJET: The Turkish Online Journal of Educational Technology* **10** 1
[10] Al-Samarraie H, Teng B K, Alzahrani A I and Alalwan N 2017 *Studies in Higher Education* pp. 1-17
[11] Moody A K 2010 *Journal of Literacy and Technology* **11** 4 pp. 22-52
[12] Zhang D, Zhou L, Briggs R O and Nunamaker J F 2006 *Information & management* **43** 1 p 15-27
[13] Aremu A and EFUWAPE B M 2013 *American Journal of Educational Research* **1** 2 p 41-48
[14] Semerci C 2005 *Pakistan Journal of Social Sciences* **3** 4 p 598-602
[15] Yenice N 2011 *European Journal of Social Sciences* **20** 4 p 593-603
[16] Alotaibi K N 2013 *Education Journal* **2** 4 p 176-185

[17] Karami M, Pakmehr H and Aghili A 2012 *Procedia-Social and Behavioral Sciences* **46** p 3266-3270
[18] Wu W and Hwang L Y 2010 *International Journal of Electronic Business Management* **8** 4 p 312

Acknowledgments
The author would like to thank the validator, which has been providing advice and improvements to the author.

Introducing geometry concept based on history of Islamic geometry

S Maarif[1,2], Wahyudin[2], A Raditya[3] and K S Perbowo[4]

[1]Department of Mathematics Education, Universitas Muhammadiyah Prof. DR. HAMKA, Indonesia
[2]Department of Mathematics Education, Universitas Pendidikan Indonesia, Indonesia.
[3]Department of Mathematics Education, Universitas Muhammadiyah Tangerang, Indonesia
[4]Department of Mathematics Education, Universitas Muhammadiyah Prof. DR. HAMKA, Indonesia

Email: samsul_maarif@uhamka.ac.id

Abstract. Geometry is one of the areas of mathematics interesting to discuss. Geometry also has a long history in mathematical developments. Therefore, it is important integrated historical development of geometry in the classroom to increase' knowledge of how mathematicians earlier finding and constructing a geometric concept. Introduction geometrical concept can be started by introducing the Muslim mathematician who invented these concepts so that students can understand in detail how a concept of geometry can be found. However, the history of mathematics development, especially history of Islamic geometry today is less popular in the world of education in Indonesia. There are several concepts discovered by Muslim mathematicians that should be appreciated by the students in learning geometry. Great ideas of mathematicians Muslim can be used as study materials to supplement religious character values taught by Muslim mathematicians. Additionally, by integrating the history of geometry in teaching geometry are expected to improve motivation and geometrical understanding concept.

1. Introduction

Since geometry has an important role in mathematics and sciences history, so the process of teaching and learning geometry should be done innovatively. Innovative learning in geometry can also develop a positive attitude of students towards geometry. Therefore, the teaching of geometry should be done in an environment that motivated students to discover knowledge independently. It is the reason that makes the development of geometry should be accompanied by activities to improve the quality of learning. Revealed that learning mathematics should be continuously developed accordance the technology development so that it can improve student learning outcomes [1].

Revealed that learning mathematics (including geometry) effectively can be done by integrating the history of mathematics into mathematics lesson [2]. History of geometry may be as old as a human civilization. A brief history of geometry is important to explore the development of geometry and forming geometry curriculum in schools, it is important that we know a brief history of geometry from time to time. Teacher requires knowledge of the history of geometrical concept so that the student can learn not only the geometry knowledge but also the philosophical values of geometry concepts.

Liu presented five reasons why the history of mathematics should be included in mathematics lessons. These reasons are: 1) historical knowledge increases students' motivation, 2) helps them to develop a positive attitude towards mathematics, 3) seeing the obstacles experienced in the development of mathematics in the past allows them to see difficulties encountered at present, 4) to solve problems from history helps the development of students' mathematical thinking, 5) history brings out the human side of mathematical knowledge and it is a guide to teachers [2]. On the other hand, Katz reveals that usage wisely history of mathematics (including geometry) can be a medium of learning and can assist teachers in creating more effective learning. In particular, the integration of the history of mathematics (including geometry) in learning not only can improve students' understanding but also can stimulate communication, connection skill and improve students' appreciation in mathematics [3].

Jones revealed that teaching geometry effectively involves, amongst other things, appreciating the history and cultural context of geometry, knowing how to recognize interesting geometrical problems and theorems, understanding the many and varied uses to which geometry is put, and incorporating all these things into the practice of teaching in the classroom [4]. Zebrowski revealed that by teaching what has been learned in the past, allow students to apply the past knowledge on the knowledge in the future. We cannot teach or learn if we did not believe that the universe represents the continuity of history [5].

For teachers, history of geometry is an important part to understand geometry. If teachers know the origins and context of a geometrical concept, then it will be easier for them to make a learning material more related to their everyday life. According to Gursoy, pre-service teachers have developed a positive attitude towards the history of mathematical teaching and learning process. There are many things that can be learned from the history of mathematics to prepare a lesson plan that will be implemented in the classroom [2].

The importance of teaching geometry with attention to the history of the development, also with regard to the students' thought process. In learning, geometry either as a product or a process of thinking. According to Castagnola, there are two ways of thinking about mathematical knowledge: either as a product or a process. Thinking about mathematical as a product means being concerned with the results and the structure of that knowledge, that is to say, with mathematical discourse [6]. Thinking about mathematical as process means being concerned with mathematical activity. A history of mathematics centered on problems brings to the fore the process of the construction and rectification of knowledge arising out of the activity of problem-solving.

In Indonesia, history of Islamic geometry is less popular in the mathematics curriculum. One of the reasons for this situation because the reference about Islamic geometry is not widely distributed. Primary to high school textbook rarely place the Muslim mathematicians work. In the college level, study of the history of Islamic geometry are less studied. In fact, the history of Islam in particular geometry and the history of Islamic mathematics, in General, is not included in the existing curriculum in Indonesia.

The objectives of educational activity that contained in Constitution No.20/ 2003, revealed that the purpose of education is to bring about a human being who has the power of religious, spiritual self-control, personality, intelligence, morals, as well as the necessary skills themselves, the community, the nation and the State. Then, the geometrical teaching and learning by integrating historical development of Islamic geometry are very precise. Students not only learn geometry but also have a positive attitude of Muslim mathematicians who have contributed to the development of geometry.

In the development of geometry, there are some geometric concepts invented by Muslim mathematicians. As has been expressed by Maarif that by an analogy to the case in the present study, so students can be trained to see the extent to which the concept can understand, see the microscopic structure of a concept by researching the connections between the cases presented with concepts that are learned and open the minds of students about the applications and benefits of the concepts he had learned [7].

The idea of a Muslim religious studies material can be used to add the values of the characters. By integrating the history of geometry in a geometry learning is expected to add to the learning motivation of students to understand the concepts of geometry. As has been expressed by Maarif, the geometry is a branch of mathematics that studies the form of objects and their characteristics [8]. Geometry is a representation of the universe in mathematics. Surely the universe contains many meanings that must be revealed by geometric approach in mathematics. The existence of the people of the world on the basis of the existence of a God with nature *rakhman* and *rakhim*.

From above explanation, it's important to understand geometric concept and history from the perspective of Islamic mathematician. Hope this new perspective will improve the quality of teaching and learning geometry.

2. Methods

The research using literature research methodology. Literature research methodology is to read through, analyze and sort literatures in order to identify the essential attribute of materials [9]. In this research, literature material is containt a various research finding about history and concept of geometry that developed by Islamic mathematician.

3. Results and Discussion

3.1. History of Geometry Periods

Jones revealed that history of geometry indicates the following three distinct periods: Intuitive, Classical, and Modern geometry [10].

3.1.1. Intuitive Geometry (8000 B.C - 500 B.C)

The periodical inundation of the Nile which swept away the landmarks in the valley of the river, and by altering its course increased or decreased the taxable value of the adjoining lands, rendered a tolerably accurate system of surveying indispensable, and thus led to a systematic study of the subject by the priests [11]. Smith writes that the earliest geometry "was intuitive in its nature; that is, it sought facts relating to mensuration without attempting to demonstrate these facts by any process of deductive reasoning" (p.270) [10]. In earliest geometry nature was the main problem to solve intuitively. That experiences not only transferred but also gradually develop from one generation to another. Around 1700 BC, the very important manuscript called "Ahames" papyrus was written. This manuscript consists of many techniques to solve arithmetic problem and mensuration, areas of rectangles and circles [12].

3.1.2. Classical Geometry (500 B.C -1600 A.D)

In this period, Greek mathematicians more dominant than Babylonian and Egyptian mathematicians. Because of that reason the geometry was not focused to solve a practical problem but as a mathematical concept. The Greek mathematician using geometry as a concept to understand the universe. The Greek mathematicians such as: Thales, Pythagoras, Eudoxus and Euclid produced a masterpiece in geometry. Yazdani writes that in this period the geometry idea was demonstrated and formalized systematically as a mathematical concept [10]. Thales in early sixth century was the first Greek mathematician that using a deductive reasoning to prove a geometrical problem. The contribution of Thales is not lies in the discovery of theorem, but in their proof. In 300 B.C, Euclid writes the "Element". It is a collection of definitions, postulates (axioms), propositions (theorems and constructions), and mathematical proofs of the propositions. After over 2000 years this book is still the most influential textbook ever written [13].

While Europe endured its "Dark Ages" in 6th to 14th centuries, the Middle East became Muslim and in high point of Islamic civilization. In early 9th century under Caliph Harun al-Rashid and (continued by his son) al-Ma'mun, The House of Wisdom (bayt al-hikma) was established and became a center for the study of science and humanities. The scholar translating a large number of Greek, Hindu, Syriac-Persian, and Hebrew texts into Arabic. Al-Khwarizmi, Abu Wafa, and Omar Khayyam were great names that joined and worked in the House of Wisdom.

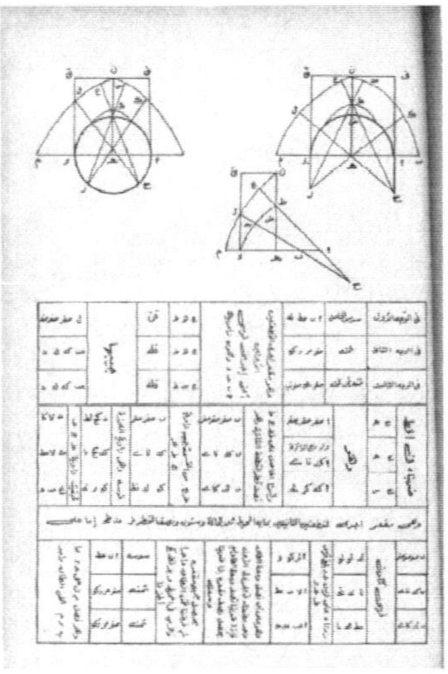

Figure 1. Page from *Al-Kashi* [15]

3.1.3. Modern Geometry (1750 M- now)

Eves revealed that the modern geometry period began when the attempted to prove Euclid's fifth postulate. Euclid established five axioms and five postulates for geometry but in the fifth postulate, better known as Euclid's parallel postulate, has become one of the most controversial statement in mathematical history. For roughly two millennia there were attempts to prove the postulate [8]. Recorded efforts were made by Proclus (fifth century), Thabit ibn Qurra (ninth century), ibn al-Haytham (tenth century), Khayyam (eleventh century), Nas.ir al-Din al-Tusi (thirteenth century); and, in 'modern' times like Playfair, Legendre, Gauss, Bolyai, Lobachevsky, Beltrami, Klein, Poncelet, Poincare, and Riemann. When many mathematicians tried to prove the Euclid's parallel postulate, few of them (such as: Saccheri, Beltrami and Lambert) using alternative postulate developed non-Euclid geometry [14,10].

3.2. Muslim mathematician and Geometry

3.2.1. Mohammad Abu'l-Wafa

Mohammad Abu'l-Wafa al'Buzjani (940 – 988) called al-Wafa born at Buzjan in Khorasan region where Iran is today. He was one of the most productive and well-known Islamic mathematicians and astronomers. He did most of his work at Baghdad observatory or the "House of Wisdom". His known for his translation and commentators on Greek works such as Diophantus, Euclid and al-Khwarizmi. He also was developed concept of trigonometry, arithmetic, study orbit of the moon and introduced the sec and cosec and studied the interrelations between the six trigonometric lines associated with an arc [15].

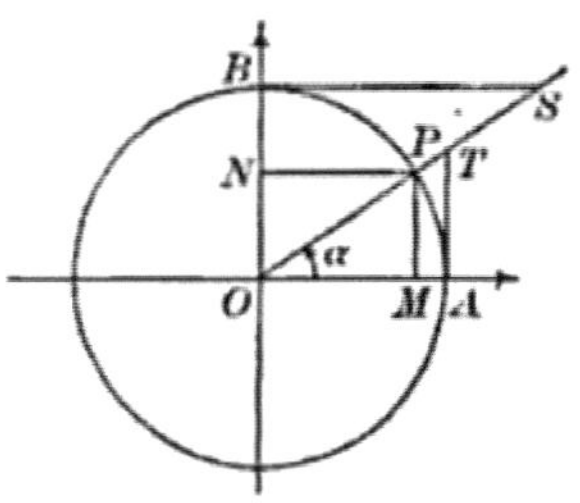

Figure 2. Page from *Al-Kashi*

PM= Sin; TA= tangent; OT = Secant; NP= Cos; RS = Cotangent and OS= Cosecant

Al-Wafa's contributions to trigonometry also proving of trigonometry sum-difference identities [16]:

$$\sin(\alpha+\beta)=\sin\alpha\cos\beta+\cos\alpha\sin\beta$$
$$\sin(\alpha-\beta)=\sin\alpha\cos\beta-\cos\alpha\sin\beta$$
$$\cos(\alpha+\beta)=\cos\alpha\cos\beta-\sin\alpha\sin\beta$$
$$\cos(\alpha-\beta)=\cos\alpha\cos\beta+\sin\alpha\sin\beta$$

And from which we can get the double angle identities:

$$\sin(2\alpha)=2\sin\alpha\cos\alpha$$
$$\cos(2\alpha)=\cos^2\alpha-\sin^2\alpha$$

Another contribution from al-Wafa to geometry was found in Kitab Al-Hindusa. This book consisted of solutions of a large number of geometrical problems on the fundamental construction of plane geometry with the opening of the compass [17]. Al-Wafa used his dissection and construction method to create a novel geometric proof of the Pythagorean theorem. Two unequal squares are added together to make a third square [18].

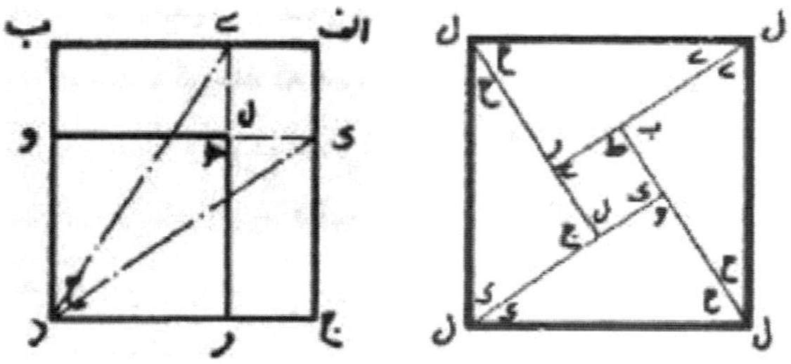

Figure 3. Drawing from original Persian text

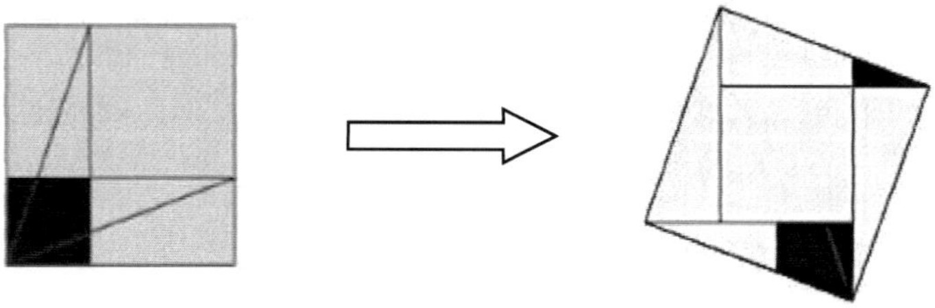

Figure 4. Geometric proof of the Pythagorean theorem

IOP Conf. Series: Journal of Physics: Conf. Series **948** (2018) 012040 doi:10.1088/1742-6596/948/1/012040

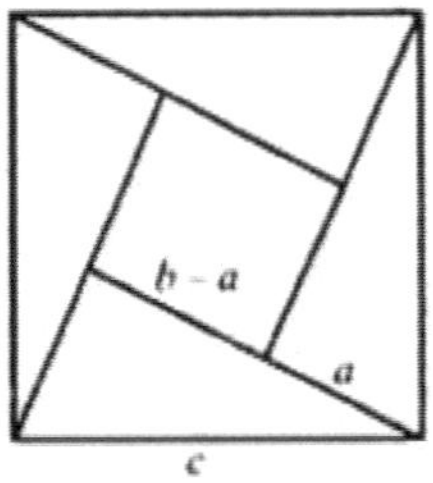

Figure 5. Geometric proof of the Pythagorean theorem

In al-Wafas' proof, a small square is placed to share the corner and two side segments of a larger square. Then a larger triangle is built up by dissecting both the large square and the small square and adding the large square to the small square to make a bigger square. In Figure 4, the area of small black square is equal to a^2, the area of large grey square superimposed behind the small black square is equal to b^2, and the area of larger square made from adding a^2 and b^2 is equal to c^2 (See Figure 4). In the final square, as shown in Figure 5, the edge a, b, and c can be viewed as the hypotenuse and other sides of a right triangle taking up part of the square [18].

3.2.2. Al-Khwarizmi

Al-Khwarizmi from Baghdad (ca. 780 – 850) is probably the most famous Muslim mathematician. His contribution not only in mathematics concept such as (number of mathematical concepts, geometry, arithmetic, algebra) but also in astronomy. In his book Al-kitāb Al-Mukhtaṣar fī Hisāb Al-Gabr Wa'l-muqabala, Al-Khwarizmi created the method of algebra equation hand in hand with the area of the square. Al-Khwarizmi also demonstrated a geometry approach to solve the algebra problem, this method now we call "completing the square" [19]. If he had an equation:

Figure 6. Page from Al-kitāb Al-Mukhtaṣar fī Hisāb Al-Gabr Wa'l-muqabala

$$x^2 + 10x = 39$$

He would create a small square on each side has a length that equal to x.

IOP Conf. Series: Journal of Physics: Conf. Series **948** (2018) 012040 doi:10.1088/1742-6596/948/1/012040

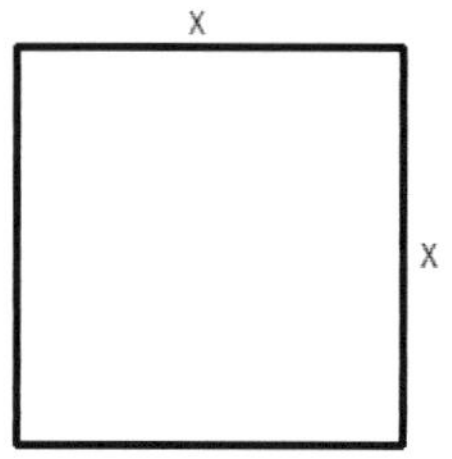

Figure 7. Square with x length

This square has an area equal to

$$x^2 = 39 - 10x$$

Then he added a rectangle to each side of the square, with a width of $\frac{10}{4} = 2.5$ unit.

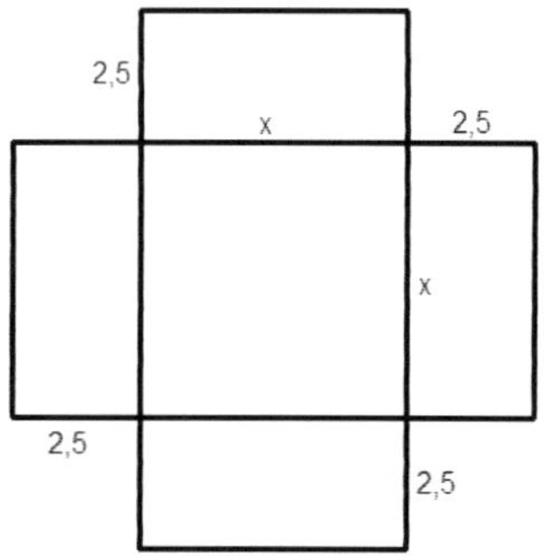

Figure 8. Modified Rectangle

So, the area of a rectangle's in Figure 8 has an area equal to 39 units. In other word:

$$x^2 + 4\ (2.5x) = x^2 + 10x = 39$$

Next, he added a line in corner between both rectangles, so each corner became a square with side length equal to 2.5 units.

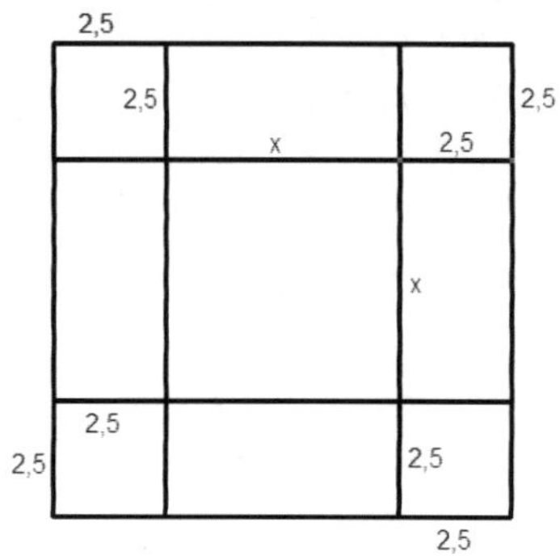

Figure 9. Modified Rectangle

The rectangles has a length equal to $x + (2,5 + 2,5) = x + 5$ unit. Hence, the area of the rectangle was $(x + 5)^2$ units. In the other hand, the area of rectangle in picture 8 is 39 units and the square in corner (picture 9) has an area equal to 6.25 units each. Hence the total area for rectangle in picture 9 equal to 39 + 4 (6.25) = 64 units. From both statement we can assume that $(x + 5)^2 = 64$ units, using simple arithmetics the x value equal to 3 units.

3.2.3. Abu Nasr al-Farabi

The Center for South Asian and Middle Eastern Studies, University of Illinois at Urbana-Champaign revealed that Islamic geometry also produced some concept in geometry. Abu Nasr al-Farabi (870 – 950 CE) was one of the Islamic Mathematician that concern about geometry, philosophy and astronomy. He wrote a treatise called "A Book of Spiritual Crafts and Natural Secrets in the Details of Geometrical Figures", in which he talked about several concepts of geometry and how to construct it [20].

Few problems that solved by Aby Nasr al-Farabi were 1) To construct at the endpoint A of a segment AB a perpendicular to that segment, without prolonging the segment beyond A and 2) To divide a line segment into any number of equal parts (for example, three equal parts).

Problem 1: "To construct at the endpoint A of a segment AB a perpendicular to that segment, without prolonging the segment beyond A".

1) Create segment AB and put point C anywhere in segment AB
2) Construct a circle in A with radius AC and construct a circle in C with radius AC. Hence an intersection from both circle in D.
3) Construct a circle in D with radius BD. Create a line CD, intersection between line and circle is E.
4) Hence AE is perpendicular with AB in A.

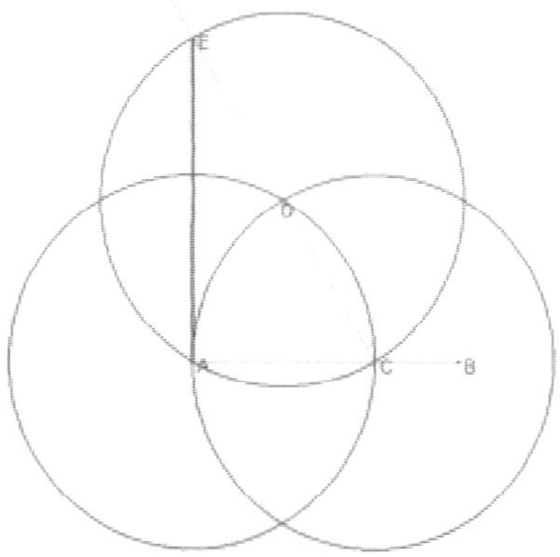

Figure 10. Problem 1 Abu Nasr al-Farabi

Problem 2: Divide a line segment into any number of equal parts (for example, three equal parts).

1) Create segment AB and put point C as a midpoint segment AB
2) Construct a circle in A with radius AC
3) Construct a circle in B with radius BC
4) At both endpoint creates perpendicular AD = BE
5) Add midpoint F and G hence AF=FD=BG=GE
6) Create segment DG and FE. Both segment intersecting AB in H and I.
7) Hence segment AB divided into three equal part in which segment AH, HI and IB.

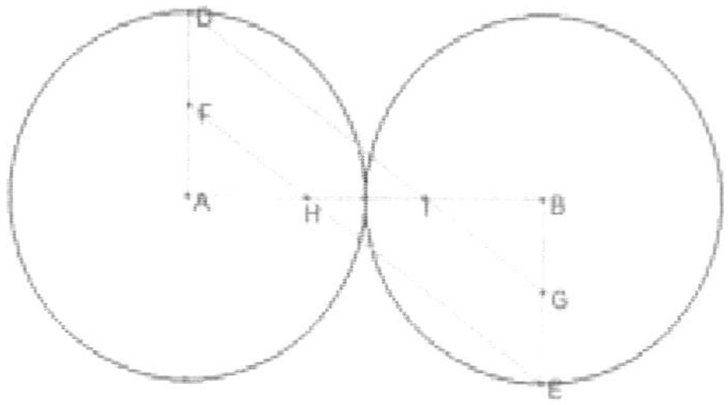

Figure 11. Problem 2 Abu Nasr al-Farabi

3.2.4. Ibrahim Ibn Sinan

Ibrahim ibn Sinan from Asfhana (in 980 CE) was another Islamic mathematician. He was known as a translator of Archimedes. He also constructed a nice and simple step to construct a parabola curve with this following method [21].

1) Create segment AG
2) Put point B between segment AG, on BG put as many points as you want (in this case 5 points: H, D, Z, X, and Y)
3) Create a circle with diameter AH and create a segment BE that perpendicular from segment AG
4) Put point T as an intersection between circle and segment BE.
5) Create line parallel to AG through point T and create line parallel to BE through H, hence the intersection was point K
6) Create a circle with diameter AD
7) Put point L as an intersection between circle and segment BE.
8) Create line parallel to AG through point L and create line parallel to BE through D, hence the intersection was point M
9) Follow the same construction method for point Z, X, and Y
10) Create K', M', ... on the extension of the lines KH, MD and ... so that KH = HK', MD = DM' .... Then K', M', ... Also lie in parabola.

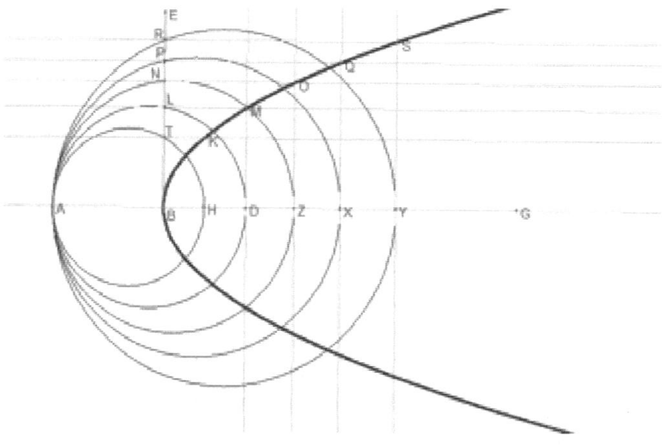

Figure 12. Parabola's construction from Ibn Sinan

4. Conclusions

Geometry lesson with historical perspective not only will provide a deep understanding for student but also can motivate them to learn geometry. The various example and story from historical perspective is good way for students to understand that geometrical problem is also everyday problem. Then students realise that problem can be solved with logic, persistence and an understanding of concept that invented before. Problems from the history of mathematics may be included as different methods in teaching and learning environments. For example, said that history of mathematics could be used in the method of the invention [22].

However, for the purpose of this paper, it suffices that the Muslim mathematicians also had a great part in development of geometry. Indonesian, as a biggest Islamic community in the world, needs to consider embedding history of Muslim mathematician in mathematics (especially geometry) curriculum.

References

[1] Maarif S 2015 *Pembelajaran Geometri Berbantu Cabri II Plus: Panduan Praktis Mengembangkan Kemampuan Matematis* (Jakarta: In Media)
[2] Goktepe S and Ozdemir S O 2013 *European Journal of Science and Mathematics Education* **1** 3
[3] Michael A B 2001 *ZDM* **33** 5
[4] Jones 2000 *Teaching Knowledge and Professional Development in Geometry*. Proceedings of the British Society for Research into Learning Mathematics 20(3) November 2000
[5] Zebrowski E 1999 *A history of the circle: Mathematical reasoning and the physical universe* (New Brunswick, New Jersey: Rutgers University Press)
[6] Castagnola E 2009 *The Use of The History of Mathematics in The Learning and Teaching of Algebra: The Solution of Algebraic Equation a Historical Approach* Proceding ICTM II 2009. Proceedings of 2nd International Conference on The Teaching of Mathematics
[7] Maarif S 2016 *International Journal of Research in Education and Science* **2** 1 pp 114-124
[8] Maarif S 2015 *Infinity Journal **4*** 2 pp 223-236
[9] Lin, G. 2009. *International Education Studies* **2** 4 pp.179-181
[10] Yazdani M A 2007 *Journal of mathematical Sciences & Mathematics Education* p 30-40
[11] Malek M 2008 *Classic Persian Poets Hakim Omar Khayyam Philosopher, astronomer, mathematician, and poet Part 2: Khayyām's Scientific Works.* [Online] available at http://www.cs.stevens.edu/~mmalek/IC5-Khayyam2.pdf accessed 11:00 AM, date June, 26th 2016.
[12] Nemati A 2008 *Persian Poet, Astronomer and Mathematician Omar Khayyam*. [Online] available at http://www.dusharm.com/content/view/16/2/ accessed at 11:45 AM, date, June 26th 2016.
[13] Greenberg M J 1993 *Euclidean and Non-Euclidean Geometry (Development and History)* (New York: W. H. Freeman and Company)
[14] Sickle J V 2011 *A History of Trigonometry Education in the United States: 1776-1900.* Disertation Columbia University.
[15] Hudgkin L 2005 *A History of Mathematics from Mesopotamia to Modernity* (Oxford: Oxford University Press)
[16] Mancillas K 2014 *Abu'l-Wafa al'Buzjani*. [Online] available at http://kobotis.net/math/MathematicalWorlds/Fall2014/131/Presentations/pdf/MancillasK_p2.pdf accessed at 9:33 AM, date July, 14th 2016
[17] Nowlan R A 1999 *Chronicle of Mathematics People* [Online] available at http://www.robertnowlan.com/contents.html accessed at 1:09 AM, date June, 25th 2016
[18] Nielsen J L 2010 *The Heart is a Dust Board: Abu'l Wafa Al-Buzjani, Dissection, Construction, and the Dialog Between Art and Mathematics in Medieval Islamic Culture*. [Online] available at http://historyofmathematics.org/wp-content/uploads/2013/09/2010-Nielsen.pdf accessed 11:34 AM date July 14th 2016
[19] Allaire R and Bradley R E 2001 *Mathematics Teacher* **94** 4 pp. 308

[20] Center for South Asian and Middle Eastern Studies, University of Illinois at Urbana-Champaign. *Islamic Mathematics*. [Online] available at http://www.csames.illinois.edu/documents/outreach/Islamic_Mathematics.pdf accessed at 10:05 PM, date June, 27th 2016
[21] Rogers E 2008 *Islamic Mathematics* (Illinois: University of Illinois at Urbana-Champaign)
[22] Yevdokimov O 2006 *Using Materials from the History of Mathematics in Discovery-based Learning*. Proceedings of the First International Congress on Mathematical Education

Measuring metacognitive ability based on science literacy in dynamic electricity topic

Warni[1], Sunyono[2] and Rosidin[1]

[1]Physics Education Study Program, Universitas Lampung, Indonesia
[2]Chemical Education Study Program, Universitas Lampung, Indonesia

E-mail: warniuuf@yahoo.co.id

Abstract. This study aims to produce an instrument of metacognition ability assessment based on science literacy on theoretically and empirically feasible dynamic electrical material. The feasibility of the assessment instrument includes theoretical validity on material, construction, and language aspects, as well as empirical validity, reliability, difficulty, distinguishing, and distractor indices. The development of assessment instruments refers to the Dick and Carey development model which includes the preliminary study stage, initial product development, validation and revision, and piloting. The instrument was tested to 32 students of class IX in SMP Negeri 20 Bandar Lampung, using the design of One Group Pretest-Postest Design. The result shows that the metacognition ability assessment instrument based on science literacy is feasible theoretically with theoretical validity percentage of 95.44% and empirical validity of 43.75% for the high category, 43.75% for the medium category, and 12.50 % for low category questions; Reliability of assessment instruments of 0.83 high categories; Difficulty level of difficult item is about 31.25% and medium category is equal to 68.75%. Item that has very good distinguishing power is 12.50%, 62.50% for good stage, and medium category is 25.00%; As well as the duplexing function on a matter of multiple choice is 80.00% including good category and 20.00% for medium category.

1. Introduction

The learning process that requires students to engage directly or actively participate can be applied in learning on science materials. Students are not only required to understand the concept of science, but also can apply the concepts in daily life so that the learning becomes more meaningful. Meaningfulness in science learning for students can be obtained if students have good science literacy skills [1].

Literacy of science is one's ability to solve problems by using science knowledge [2]. Literacy of science is very important to be developed because everyone needs information and knowledge to make choices and to solve problems faced every day [3]. Based on PISA 2015 results, students' literacy skills in Indonesia are still low. The literacy skills of Indonesian students are ranked 62nd out of 69 countries that follow PISA [4]. The low average science literacy of Indonesian students in PISA 2015 can be one of the illustrations that science learning conducted in Indonesia still needs improvement, so it is expected that the average science literacy of Indonesian students will be better.

Science is a human activity characterized by a thought process that occurs in the mind of anyone involved in it [5]. Tawil and Liliasari [6] states that in general thinking is a cognitive process, a mental activity to gain knowledge. Thinking is produced from metacognition that is possessed by every human being [7]. Metacognition plays an important role in the learning process because metacognition is crucial in regulating and controlling the cognitive processes of a person in learning and thinking, so that the learning and thinking done by a person becomes more effective and efficient [8]. It may be stated that metacognition is a person's awareness of the monitoring process as well as maintaining and controlling his own thoughts and actions. Thus, metacognition is necessary for the thinking activity.

Metacognition skills need to be developed, so that students are trained to always design the best strategy in choosing, remembering, re-recognizing, organizing the information it faces, and in solving problems [9]. The importance of metacognition to be developed in learning because metacognition is a fundamental factor in learning and problem solving that can ultimately create a conducive learning experience [10], through the development of metacognition awareness, students are expected to always monitor, control and evaluate what he has done [11], besides, Rahman and Philips [12] proves that metacognition awareness has a significant relationship with academic achievement.

The development of metacognition capabilities in the learning process will only be effective if the teacher has a good and continuous monitoring mechanism. This may only be possible through an assessment process capable of providing meaningful feedback loops during the learning process [13]. Thus, the assessment process needs to be integrated throughout the learning process. At the time of metacognition assessment process required metacognition assessment instrument.

There have been many studies that examine the instrument assessment of metacognition ability. The instrument of metacognition ability first developed by Schraw and Dennison [11] is MAI (Metacognitive Awareness Inventory), the instrument is a questionnaire to measure adult metacognition. The MAI instruments were used further by researchers for various purposes [14, 15, 16, 17]. In fact, [8] of the MAI instrument was developed into an instrument of Jr MAI (Junior Metacognitive Awareness Inventory) for students in primary education, especially grades 3-9.

Regarding the above description, a research on the development of metacognition ability assessment instruments based on science literacy on dynamic electrical materials was conducted. This research aims to produce metacognition ability assessment instruments based on science literacy on dynamic theoretical material that is valid theoretically and empirically, and describe the validity of theoretical and empirical assessment instruments. The benefits of this research are to contribute to the teacher as one alternative assessment instrument that can be used as a tool for collecting data and tools to obtain information about students' metacognition ability; to help students know their metacognition ability; and to be a reference for other researchers to conduct a research on the development of metacognition ability assessment instruments.

2. Methods

This study used the research method adopted from Dick and Carey [18]. The research and development activities undertaken were assessing the needs to analyze objectives, perform learning analyzes, analyze the characteristics of learners and the context, formulate specific objectives of learning, develop assessment instruments, design and implement formative evaluations. The instrument was tested on 32 students of class IX in SMP Negeri 20 Bandar Lampung, using One Group Pretest-Postest Design. The sample was chosen through purposive sampling. Data collection used validation sheet, metacognition ability test, and student response questionnaire. Theoretical validity is obtained from 3 validators. The data of theoretical validity results were calculated by using the equation (1):

$$\%Ji = \sum_{i=0}^{n} \frac{\text{Ji}}{\text{N}} \text{ X } 100\% \quad (1)$$

Information:
$\% Ji$ = percentage of achievement of the ideal score for each of the i aspects.
ΣJ = Percentage of achievement of the ideal score for each of the i-th aspects provided validator.
N = maximum score (ideal score).

Empirical validity was analyzed by calculating the validity of the item and the reliability of the item. The validity of the item was calculated by Product Moment Pearson correlation formula by correlating between the scores obtained by the students on an item with the total score obtained. The equation (2) used:

$$r_{xy} = \frac{N\sum XY - (\sum X)(\sum Y)}{\sqrt{(N\sum X^2 - (\sum X)^2 . (N\sum Y^2) - (\sum Y)^2}} \quad (2)$$

Information:
r_{xy} = correlation coefficient between variable X and variable Y
N = many test takers
X = value of variable X
Y = value of variable Y

The reliability of the item was calculated using the Alpha Cronbach formula. The equation (3) used:

$$r_{1i} = \left[\frac{n}{n-1}\right]\left[1 - \frac{S_i^2}{S_t^2}\right] \quad (3)$$

Information:
r_{1i} = instrument reliability coefficient
n = number of items.
S_i^2 = the number of variance scores per item.
S_t^2 = total score variance.

The degree of difficulty, distinguishing factor, and distractor index of the items were analyzed by the ITEMAN formula.

3. Results and Discussion

The development of metacognition skills assessment instruments based on science literacy on dynamic electrical materials applied Dick and Carey development model with the data collection instrument were in the form of validation sheet, metacognition ability test, and student response questionnaire. This study produced primary and secondary data. The primary data were assessment instrument development process, theoretical validity, empirical validity, reliability, difficulty level, distinguishing factor, and distractor index. The secondary data were in the form of student responses to instrumental assessment of metacognition ability developed.

Primary data were obtained from the result of instrument development in the form of metacognition ability assessment instrument, theoretical validity data obtained from the validation sheet of expert, empirical validity data obtained from the students' metacognition ability test which then were calculated the validity, reliability, difficulty level, distinguishing factor, and instrument distractor index developed. Secondary data in the form of student response were obtained from questionnaire of student response to metacognition ability assessment instrument developed. The theoretical validation results of the assessment instruments are presented in Table 1.

Table 1. Expert rating results

Validation	Aspect	Validator I, II, III	%	Category
Content / material	Compatibility of content/material with KI-KD	Average	91.11	Very valid
Construction/ Instruments	Construction in accordance with indicators of metacognition and science literacy	Average	95.83	Very valid
Language/ Legibility	The language is in conformity with Indonesian grammar	Average	99.38	Very valid

Based on the data in Table 1, the theoretical validity for the assessment instrument was 95.44% which belongs to a very valid category. Then the metacognition ability data of assessment instrument was tested. Based on the results of the experiment the obtained empirical validity were presented in Table 2.

Table 2. Percentage of category of validity item problem.

Number	The average moment person correlation coefficient	Category Validity	Number of Problems	%
1.	0.70	High	7	43.75
2.	0.46	High enough	7	43.75
3.	0.37	Low	2	12.50
		Amount	16	100,00

Based on the data in Table 2, there is an equation of the number of questions that has high category validity with high enough and has the same percentage, there is a high degree of category validity level and high enough level of 0.24. The figure is obtained from the average high coefficient correlation moment person of high category.

Table 2 shows that most of the questions are of high validity and are quite high category. That is, the developed problem can be used to measure students' metacognition and literacy skills. Based on the measurement results, the highest average score of students' metacognition ability is found in the declarative knowledge component and the lowest score is in the conditional knowledge, moreover, the average score of the highest science literacy ability is at nominal level, while the lowest is at the multidimensional level.

The empirical validity of most of the high and high categories in some of the assessment instruments was caused by the manufacturing process that has met the theory of problem-building steps and is based on basic competence indicators, metacognition indicators, and science literacy indicators adapted to dynamic electrical material concepts, it is also supported by the use of language that is easily understood by students. In line with [20], in order that the instrument has a high validity, then in making the instrument must be based on basic competence indicators and indicators of student competence to be achieved. Most developed instruments have high and high empirical validity, meaning that the developed instruments can be used to measure students' metacognition and literacy capabilities validity.

Based on the measurement result, the highest metacognition ability of students is found in the component of declarative knowledge and the lowest value is in the component of conditional knowledge. The high value of declarative knowledge of students is caused by the effect of learning, so that students have already known and understood about the facts (phenomena) associated with electricity. This is in accordance with the opinion of [21] who states that in the learning process, students experience a change in behaviour in the form of newly acquired knowledge, understanding, skills, and habits.

The lowest metacognition of students is in conditional knowledge. The lack of conditional knowledge of students because the students cannot provide the right reasons to apply the knowledge and skills possessed. Here is an example of students, answer in providing unappropriated reasons in applying the knowledge they have.

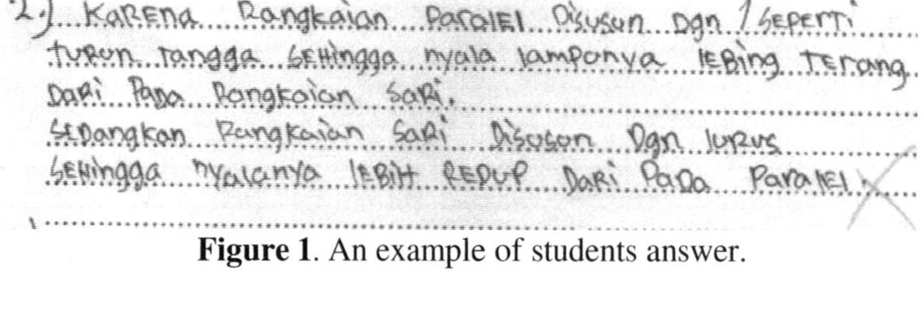

Figure 1. An example of students answer.

Based on the results of the measurements the average literacy ability of the highest student science was at the nominal level (level 1). Similarly, the results of the study of [22] on the analysis of students 'science literacy skills on the concept of IPA, it turns out that students' science literacy skills are mostly at the first level of the nominal. At this level, students have only vocabulary, concepts, context, and cognitive ability to identify questions scientifically. This is in accordance with the results of PISA 2015 which states that the average literacy skill of Indonesian students is at level 1.

The lowest level of students' literacy ability is at a multidimensional level (level 4). The results of the study [23] produce the same thing, it turns out the ability of science literacy students at multidimensional level was included in the low category. The cause of the low level of multidimensionality is that students have not been able to combine a broader understanding of science beyond the subject concept and scientific inquiry procedures. In line with [3], which states that only a small proportion of students answer multidimensionality, students have difficulty in integrating material in a global context. In addition, students also have difficulty in applying science in everyday life.

Based on the data in Table 2, there are 12.50% of questions that have low category validity. This low validity can be influenced by factors from the problem itself, that is the problem including difficult categories so that students cannot answer the question. [24] has similar opinion which states that some factors that affect the low validity of the item are the factors of the problem itself, among others, too easy, so the answers can be predicted by the students, the allocation of time is too short to do the problem, has a bad distinguishing power, and the degree of difficulty of the problem does not match with the material already taught.

Reliability is the level of consistency of the developed instrument. Based on the calculation result, the reliability assessment instrument metacognition ability of 0.83 are included in the high category. The degree of reliability of the high category instruments, indicates that the students were consistent in providing a relatively high response [25]. This means that the developed instrument has a high degree of consistency to measure students' metacognition ability. In line with the opinion of [26] which states that the instrument is said to be reliable, if it is used to measure the same object/subject by the same or different people will produce relatively similar data.

The high reliability of the instrument is influenced by the student's response to the language aspect and guidance on the problem. According to most students, telling the language and instructions on how to work on the assessment instrument is clear and easy to understand, so that students can correctly answer the questions on the instrument.

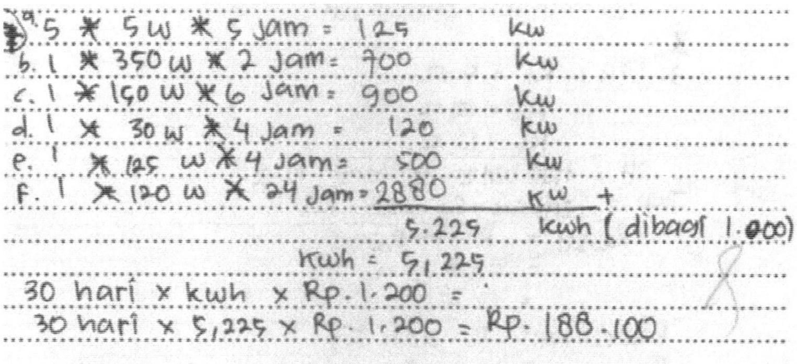

Figure 2. An example of students answer on the problem of electrical problems in a household.

Based on the description of student answers, it indicates that students easily understand the language in the problem, so students can determine the steps that must be done to solve the problem correctly. This is what causes students to answer the problem correctly. The difficulty level of the item is the proportion of test participants who answered correctly [27]. Description of the degree of difficulty and distinguishing factor of the item can be seen in Table 3.

Table 3. Percentage of different levels and distinguishing power.

Number	Category of Tribunal	%	Distinct Power Category	%
1.	Difficult	31.25	Very good	12.50
2.	Medium	68.75	Good	62.50
3.	Easy	0.00	Enough good	25.00
4.			Ugly	0.00
	Amount	100.00	Amount	100.00

Table 3 shows that the difficulty level of the items developed was mostly in the medium category and a few difficult categories. This means that most students can answer correctly. The problem-solving power of the item is the ability of a matter to distinguish between high-ability students and low-ability students [13]. The differentiating grains of questions that developed most of the categories are good and few are very good and good enough category (Table 3). This means that the items developed can be used well to distinguish the students who can answer correctly the items tested with students who cannot answer correctly.

In the multiple-choice items, there is an alternative answer that functions as a liar. Based on the results of functionality analysis of 80.00% good category and 20.00% category good enough. This indicates that the choice of answers works well, so it can outwit the students who do not understand the concept of choosing an answer. Similarly, the opinion of [13] who says that a deception can be said to work well if the deceiver has a great appeal for test followers who lack understanding of the concept or lack of control of the material.

4. Conclusion

The metacognition ability assessment instrument is considered feasible theoretically with theoretical validity percentage of 95.44% and the empirical validity of 43.75% for the high category, 43.75% for the medium category, and 12.50% for low category question; Reliability of assessment instruments of 0.83 high categories; Difficulty level of difficult question item is about 31.25% and medium category is equal to 68.75%; Items that has very good distinguishing power is 12.50%, 62.50% for good, and medium category is 25.00%; As well as the duplexing function on the multiple choice 80.00% including good category and 20.00% for medium category

References

[1] Arlianovita D, Rachmadiarti F and Setiawan B 2016 *Jurnal Mahasiswa Teknologi Pendidikan* **4** 3 pp 1-7
[2] Imaningtyas C D, Karyanto P, Nurmiyati N and Asriani L 2016 *Jurnal BIOEDUKASI* **9** 1 pp 4-10
[3] Soobard R and Rannikmae M 2011 *Science Education International* **22** 2 pp 133-44
[4] OECD 2016 PISA Results in focus. [Online] Tersedia http://www.oecd.org/pisa/keyfindings/pisa-2015-results-overview.pdf
[5] Maryuningsih Y 2013 *Jurnal Scientiaeducatia* **2** 1 pp 1-19
[6] Tawil M and Liliasari 2013 *Berpikir Kompleks dan Implementasinya dalam Pembelajaran IPA* (Makasar, Badan Penerbit UNM) p 4
[7] Nurmahanani I 2015 *Jurnal Metodik Didaktik* **10** 1 pp 57-68
[8] Muhali 2013 *Jurnal Kependidikan Kimia "Hydrogen"* **1** 1 pp 1-7
[9] Haryani S 2015 Pengembangan model pratikum kimia analitik instrumen berbasis masalah untuk meningkatkan metakognisi mahasiswa calon guru *Disertasi* Universitas Pendidikan Indonesia pp 3-4
[10] Urena S S, Cooper M M, and Stevens R H 2011 *International Journal of Science Education* **33** 3 pp 323-40
[11] Schraw G and Dennison R S 1994 *Contemporary educational psychology* **19** 4 pp 460-75
[12] Rahman S and Philips J A 2006 *Jurnal Pendidikan* **31** 2 pp 21-39
[13] Arikunto S 2016 *Dasar-Dasar Evaluasi Pendidikan* (Jakarta, Bumi Aksara) p 281
[14] Gassner L 2009 Developing metacognitive awareness -a modified model of a PBL-tutorial *Thesis* Malmo University p 7

[15] Young A and Fry J 2008 *Journal of the Scholarship of Teaching and Learning* **8** 2 pp 1-10
[16] Balcikanli C 2011 *Electronic Journal of Research in Educational Psychology* **9** 3 pp 1309-32
[17] Yildirim S and Ersozlu Z N 2013 *Eurasia Journal of Mathematics, Science & Technology Education* **9** 4 pp 411-5
[18] Sperling R A, Howard B C, Miller L A, and Murphy C 2002 *Contemporary educational psychology* **27** 1 pp 51-79
[19] Dick W and Carey L 2002 *The Systematic of Decision of Instruction Third Edition* (New York, America Harper Collins Publisher) p 6
[20] Noviyanti L, Indriyanti D R and Ngabekti S 2015 *Lembaran Ilmu Pendidikan* **43** 1 pp 32-9
[21] Trianto 2009 *Mendesain Model Pembelajaran Inovatif Progresif* (Jakarta, Kencana Prenada Media Group) p 9
[22] Odja A H and Payu C S 2014 Analisis kemampuan awal literasi sains siswa pada konsep IPA *In* Prosiding *Seminar Nasional Kimia* 20 September pp 40-7
[23] Ardiansyah A A I, Irwandi D and Murniati D 2016 *Educhemia (Jurnal Kimia Dan Pendidikan)* **1** 2 p. 149-61
[24] Syamsiah S, Puspitawati R P dan Widodo W 2016 *Jurnal Pendidikan* **4** 3 pp 1-5
[25] Azizah U, Suyono dan Suyatno 2015 Desain dan validasi instrumen untuk mengukur keterampilan metakognitif mahasiswa dalam materi larutan *Prosiding Seminar Nasional Kimia* Jurusan Kimia FMIPA Universitas Negeri Surabaya, 3-4 Oktober 2015 pp 59-65
[26] Sugiyono 2015 *Metode Penelitian & Pengembangan* (Bandung, Alfabeta) pp 179-80
[27] Jihad A dan Haris A 2013 *Eveluasi Pembelajaran* (Yogyakarta, Multi Presindo) p 182

IOP Conf. Series: Journal of Physics: Conf. Series **948** (2018) 012042 doi:10.1088/1742-6596/948/1/012042

Cabri 3D - assisted collaborative learning to enhance junior high school students' spatial ability

Muntazhimah[1] and A Miatun[1]

Mathematics Education Department, Universitas Muhammadiyah Prof. DR. HAMKA, Indonesia

E-mail: muntazhimah@uhamka.ac.id

Abstract. The main purpose of this quasi-experimental study was to determine the enhancement of spatial ability of junior high school students who learned through Cabri-3D assisted collaborative learning. The methodology of this study was the nonequivalent group that was conducted to students of the eighth grade in a junior high school as a population. Samples consisted one class of the experimental group who studied with Cabri-3D assisted collaborative learning and one class as a control group who got regular learning activity. The instrument used in this study was a spatial ability test. Analyzing normalized gain of students' spatial ability based on mathemathical prior knowledge (MPK) and its interactions was tested by two-way ANOVA at a significance level of 5% then continued with using Post Hoc Scheffe test. The research results showed that there was significant difference in enhancement of the spatial ability between students who learnt with Cabri 3D assisted collaborative learning and students who got regular learning, there was significant difference in enhancement of the spatial ability between students who learnt with cabri 3D assisted collaborative learning and students who got regular learning in terms of MPK and there is no significant interaction between learning (Cabri-3D assisted collaborative learning and regular learning) with students' MPK (high, medium, and low) toward the enhancement of students' spatial abilities. From the above findings, it can be seen that cabri-3D assisted collaborative learning could enhance spatial ability of junior high school students .

1. Introduction

There are many ways to facilitate mankind acquire knowledge. One of the most common ways is through education. Through education, learners are expected to be personally ready and to compete properly and optimally in the global world. Mathematics is a subject that can address the challenges of the global world because it is a universal science that underlies the development of modern technology, have an important role in a variety of disciplines, and develop the power of human thought trough the purpose of studying mathematics that show attitude logical, critical, analytical, meticulous and thorough, to be responsible, responsive, and not quitter in solving problem. Have a sense of want to know, spirit continuous learning, confidence, and interest in mathematics. Have confidence on power and usability mathematics, formed through learning experience, Have an attitude open, objective in interaction groups or daily activities [1].

Mathematics teaches us how to be able to think logically and systematically. It also teaches us to be more disciplined, obey rules and follow agreements. Besides, mathematics is also one of the subjects that have been taught at every level of education systems and has a very important role in the development of students' abilities. This is in line with the goal of learning mathematics formulated by National Council of Teachers of Mathematics (NCTM) which are: "(1) mathematical communication; (2) mathematical reasoning; (3) mathematical problem solving; (4) mathematical connection and; (5) mathematical representation" [2].

Since there are many benefits of mathematics for life, mathematics should be the important subject to be taught in schools, and also should be one of the most favorite subjects for the students. However, it turns out that mathematics is not the students' favorite subject after all. The results of a preliminary study by interviewing a few Junior High students, some students think that mathematics is an appalling subject, hard to understand, and taught in monotony. They believe that studying mathematics was boring.

Geometry is a part of mathematics which deals with the shape and size of an object that has regularity. The purpose of learning geometry at school as reported by Thomas [3], such (a) to develop spatial awareness, intuition and the ability to visualize geometry; (b) To provide the breadth of experience both in the geometry of two-dimensional space and 3-dimensional; (c) To develop knowledge and understanding and the ability to use the properties and geometry theorem; (d) to encourage the development and use of conjecture, deductive reasoning and evidence; (e) To develop the application of geometry skills through modeling and problem solving in the real world; (f) to develop the use of ICT skills in the context of geometry; (g) to ensure a positive attitude towards mathematics and; (h) to develop an awareness of the historical and cultural heritage of geometry in society and contemporary applications of geometry. However some testimony from the field showed that the results of learning geometry in Indonesia remain low because there are many students who have difficulty in learning geometry, ranging from elementary to university level [4] while Syahputra's [5] statement that students find many difficulties to understans object or image wake geometry.

The purpose of learning geometry is that the students will gain confidence in their math ability, be a good problem solving, can communicate mathematically, and think mathematically. Meanwhile, according to Budiarto stated that "the purpose of learning geometry is to develop the ability to think logically, develop spatial intuition, imparting knowledge to support other materials, and able to read and interpret mathematical arguments" [4]. Therefore, developing the students' spatial ability is important because it is one of the goals of learning geometry.

Wardhani and Rumiati [6] in their research revealed that "the analysis of the results of the PISA study, there are still many Indonesian students who have difficulty in solving these problems. Only 33.4% of the participants (Indonesian students) were able to answer correctly and the rest answered incorrectly". This condition illustrated that the level of spatial ability of the Indonesian students is still low, especially at 15 years old teenager. Furtheremore, Faradhila, Sujadi and Kuswardi [7] in their research found "at the junior level, in the daily test of the students in grade VIII about solid geometry is very unsatisfactory and far below the minimum completeness criteria". It is because of the students' disability in interpreting the images in a visual form for their low learning achievement.

The study of Syahputra [5] revealed that the low spatial ability of the students is caused by the teachers' pressure of learning geometry by giving and forcefully memorizing mechanical terms (e.g. to determine the ribs, the field side area, and the volume of the cubes) without adequate emphasis on the geometry spatial aspects they have studied. It is important for the teacher to design a lesson that stimulates the students to become more active in the classroom by having the group discussion or class discussion so that students' thinking skills and communication skills can be formed. An alternative solution to improve the spatial and mathematical communication ability of students is collaborative learning.

Besides designing an appropriate learning in order to get the maximum learning outcomes, the infrastructure also has a significant role. A learning process is also expected to visualize mathematical ideas clearly in order to help the students understand mathematics, especially in geometry. The visual demonstration has a very important role in learning geometry. Both teacher's demonstration and through technologies demonstration (e.g. software designed to convey geometry concepts) are surely helpful. The learning that combines face-to-face with teacher and technology is very effective. According to Syahputra [5], a dynamic mathematics software that can be used in geometry is Cabri-3D. This program will assist the teacher in explaining geometry. Cabri-3D will help the students to visualize two-dimensional and three-dimensional geometric shapes. It will help them to uncover the geometric shapes easily and it surely will help them to understand the concept of geometry and to train their spatial skills. With Cabri-3D students can explore, observe, and see the geometry shapes in detail. This application is expected to help the students to visualize geometry concept clearly.

Based on the description that has been presented about the importance of the efficiency and effectiveness of mathematics learning, the authors propose a research study of mathematics, especially in geometry by using Cabri-3D to improve spatial skills of the Junior High students.

The specific purposes of this research are:

1. To figure out the differences in the spatial abilities of the student learning with collaborative learning assisted by Cabri-3D and the students who get regular learning;
2. To figure out the differences of the spatial abilities of the student learning with collaborative learning assisted by Cabri-3D and the students who get regular learning in terms of the students' prior knowledge in math IMK (High, Medium, Low);
3. To figure out the effects of the interaction between learning model and the students' prior knowledge of mathematics to increase their spatial abilities.

The definition of spatial ability in this research is the ability of the students to visualize, compare, suspect, determine the suitable object, construct, present and find information from visual stimuli in the spatial context, which is the indicators as follows:

1) The ability to visualize the position of a geometric object after it undergoes rotation, reflection, or dilation.
2) The ability to compare the logical interrelationships of the elements of 3D shapes.
3) The ability to accurately guess the actual shape of the geometric shape that is seen from a certain perspective.
4) The ability to determine the suitable object in a certain position of the 3D geometric object.
5) The ability to construct a model of an object associated with the 3D shapes.
6) The ability to specify a simple object embedded in a more complex image.

2. Methods

This research is a quasi-experimental research with the nonequivalent control group design form. In this design, the subjects of study are grouped into two, namely the experimental group and the control group. The experimental group was treated by Cabri-3D assisted collaborative learning and the control group was treated by regular learning, in addition to the two classes grouped based on mathematical pior knowledge (MPK) into three groups: high, medium, and low.

The data were obtained using a test instrument types; a test for spatial ability. It was given before treatment (pretest) and after treatment (posttest). The test was conceived and developed by the researchers based on the procedure of preparation instrument properly.

The analysis of the data used in this study is called quantitative data analysis. The steps in the analysis of spatial ability test instruments are as follows:

a. Give the score to the students' answers according to the answer key and scoring guidelines.
b. Make the pretest and posttest tables scores of the students of experimental group and the control group.
c. Examine the similarities mean of the pretest from both groups.
d. Grouping the data of samples in the experimental and control class based on the classification of the MPK.
e. Determine the spatial ability with N-gain formula.

After the increasing data obtained, the next step is to do a statistical test. Before the statistical test, data were checked with normality and homogeneity of variance test as a condition for the use of parametric statistics called the two-way ANOVA test.

3. Results and Discussion

To answer the formulation problem proposed, the spatial ability of students were analyzed based on spatial ability pretest and posttest are given. Pretest data were analyzed to see whether the two groups have the common mean and then based on data pretest and posttest analyzed whether there is an increase or not. Furthermore, do the calculation of the N-gain. N-gain data are then analyzed to test the hypothesis by the previous normality and homogeneity test as a condition for using parametric statistics.

Other data values for the mean and standard deviation score data pretest, posttest, the gain and the normalization gain of spatial ability of students based on prior knowledge of mathematical categories are : The average value obtained an increase in spatial ability to both classes, the group of high MIK students obtained higher average than the control class for 0.87 and 0.79, with a variation of the spread of a more balanced spread also in the experimental class which can be seen from differences in the two classes, for standard deviation of 0.10 and 0.08. Similarly, the increase in spatial ability for groups of students MIK scores were obtained for the experimental class higher than the control class is 0.72 and 0.58 with a similar deployment on 0.13 and 0.11. The same result is also increasing the ability of lower MIK spatial student group class experiment is still superior to the control class for 0.68 and 0.39. With a more diverse spread of scores well in class experiments, standard deviation obtained from the experimental class larger than the control class for 1.12 and 0.65. Total mean score of the increasing of spatial abilities in experimental class is certainly higher than the control class. Note that the average N-gain experimental class is 0.74 (High), and the average N-gain control class is 0.58 (middle). To see the description of an increase in the spatial ability of students by the prior mathematical knowledge of students, presented by the following diagram.

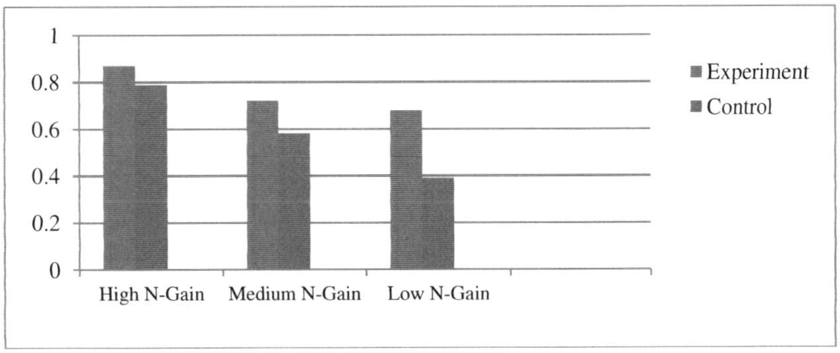

Figure 1. The mean of spatial ability of students based on MIK

The average value of the N-largest gain was the experimental class for highly capable students, amounting to 0.87. The average value of the smallest N-gain is also in the control class for low-ability students, for 0.39. Furthermore, the category of an increase in the spatial ability of students based on MIK presented in the following table:

Table 1. The Category of Increasing in Student's Spatial Ability Based on MIK

MIK	Experiment			Control		
	n	N-Gain	Category	n	N-Gain	Category
High	5	0.87	High	5	0.79	High
Medium	10	0.72	High	12	0.58	Middle
Low	5	0.68	Middle	5	0.39	Middle

IOP Conf. Series: Journal of Physics: Conf. Series **948** (2018) 012042 doi:10.1088/1742-6596/948/1/012042

Total	20	0.74	High	22	0.58	Middle

According to Table 1 found that a group of students of high MIK and medium MIK in the experimental class gained increased spatial ability with the high category, only the low group are upgrading spatial ability in the average category. In contrast to the control group which spatial enhancement to the group of students who entered high MIK with higher categories, while the medium and low MIK have only low increase of spatial ability. It is also included in the Middle category.

The first thing to do in analyzing the improvement score of spatial ability is to do normality test. If the data are normally distributed then proceed to test the homogeneity of variance. Normality and homogeneity test scores N-gain spatial skills performed in order to determine what types of statistical tests to be used in hypothesis testing. If the data are normally distributed and the variance of data is homogeneous, then the hypothesis test performed using two-ways ANOVA. If not, then the hypothesis test is done by using Freidmann's Nonparametric Test. Data spatial enhancement of students coming from a normally distributed population, and also from the population that has a homogeneous variance, so the statistical test used was two-way ANOVA statistical test. Here is the result of N-gain two-way ANOVA test scores:

Table 2. The Result of Two-Way ANOVA of N-Gain

Factor	F	Sig.	Conclusion	Description
Class	17.467	0.000	H_0 Rejected	There was significant difference
Category of IMK	13.997	0.000	H_0 Rejected	There was significant difference
Class*IMK	1.87	0.169	H_0 Accepted	There was no significant difference

Based on the Table 2, it can be seen that the value of class significance, i.e. learning model = 0.000 < 0.05 = α, so that H_0 is rejected. It means that there are differences in spatial abilities enhancement among students who learn by collaborative learning using Cabri-3D and regular learning. This indicates that the class factor is the type of learning has the significant influence on the improvement of students' spatial abilities.

Based on table 2, it can also be seen that the value of significance to MIK = 0.000 < 0.05 = α, so that H_0 is rejected. It means that there are differences in spatial abilities increase among students who received collaborative learning with Cabri-3D and regular learning in terms of MIK students. This indicates that MIK also has a significant influence on the enhancement of students' spatial abilities.

Furthermore, to determine the category of MIK which provides the significant difference, then the advanced ANOVA test, Scheffe Post Hoc test, must be done. Post Hoc test is an advanced ANOVA test which aims to determine which data group had significant differences among the six groups of data. Post Hoc test used is Scheffe Post Hoc test. Here is the results Scheffe Post Hoc test:

Then, based on Table 2 above, also found that the significance of learning approaches and MIK category to the improvement of spatial ability students = 0.169 > 0.05 = α, H_0 is accepted. It means that, there is no significant interaction between learning models and the categories of MIK to enhance students' spatial abilities. It also means that, learning models and the categories of MIK together do not have a significant influence on the enhancement of students' spatial abilities.

4. Conclusion

Based on the research and discussion, the findings of this study can be concluded such as: (1) There is a difference between the enhancement in spatial ability of students who learn by collaborative learning Cabri-3D and regular learning; (2) There is a difference between the enhancement in spatial ability of students who learn by collaborative learning Cabri-3D and regular learning in terms of MIK of the students (High, Medium, Low); and (3) There is no interaction between learning model with MIK of students to enhance students' spatial abilities.

IOP Conf. Series: Journal of Physics: Conf. Series **948** (2018) 012042 doi:10.1088/1742-6596/948/1/012042

References

[1] Badan Standar Nasional Pendidikan 2016 *Standar Isi Pendidikan Dasar dan Menengah* Jakarta: BSNP 116-117
[2] National Council of Teacher of Mathematics 2000 *Handbook of Research on Mathematics and Learning,* edited by A. Douglas (Macmillan Library Reference, USA).
[3] N Thomas *2001 Report of A Royal Society/Joint Mathematical Council Working Group*. 2001-06, SW1Y5AG
[4] Abdussakir 2009 *Jurnal Madrasah*. **2, 1** (http://dx.doi.org/10.18860/jt.v2i1.1832)
[5] Syahputra E 2013 *Cakrawala Pendidikan* **3** p 353-357
[6] Wardani, S., & Rumiati 2011 Instrumen penilaian hasil belajar matematika SMP : Belajar dari PISA dan TIMSS. Kementerian Pendidikan Nasional. Yogyakarta : Badan Pengembangan Sumber Daya Manusia Pendidikan dan Penjamin Mutu Pendidikan. PPPPTK Matematika.
[7] Pratiwi Y A, Sujadi I and Kuswardi Y 2017 *JPPM Solusi* **1** 2 p 130

IOP Conf. Series: Journal of Physics: Conf. Series **948** (2018) 012043 doi:10.1088/1742-6596/948/1/012043

Differences between quadratic equations and functions: Indonesian pre-service secondary mathematics teachers' views

T A Aziz[1], P Pramudiani[1] and Y W Purnomo[1]

[1]Program Studi Pendidikan Matematika, Universitas Muhammadiyah Prof. DR. HAMKA, Jalan Tanah Merdeka, Rambutan, Pasar Rebo, Jakarta Timur

E-mail: tian_aziz@uhamka.ac.id

Abstract. Difference between quadratic equation and quadratic function as perceived by Indonesian pre-service secondary mathematics teachers (N = 55) who enrolled at one private university in Jakarta City was investigated. Analysis of participants' written responses and interviews were conducted consecutively. Participants' written responses highlighted differences between quadratic equation and function by referring to their general terms, main characteristics, processes, and geometrical aspects. However, they showed several obstacles in describing the differences such as inappropriate constraints and improper interpretations. Implications of the study are discussed.

1. Introduction

A plethora of studies indicates that there is relevance of teachers' mathematical knowledge to high quality instruction and students' mathematics achievement [1, 2]. Teachers' knowledge of mathematics has become a remarkable headline and ongoing trend in mathematics education research for several decades now. Frameworks have been developed to understand teachers' knowledge of mathematics. It was the work of Ball and colleagues who firstly acknowledged the term of mathematics knowledge for teaching which has become one of the fundamental constructs in an effort to comprehend knowledge required for teaching mathematics [3]. According to them, mathematics knowledge for teaching consisted of three types of subject matter knowledge (common content knowledge, specialized content knowledge, and horizon content knowledge) and three types of pedagogical content knowledge (knowledge of content and students, knowledge of content and teaching, and knowledge of curriculum). Even though they are categorized into distinct knowledge, sometimes certain cases could not be elucidated independently.

Teaching involves various activities such as showing students the way to address mathematical problems, answering students' posed questions, and monitoring their works [3]. It is common in mathematics classrooms that in dealing with their puzzlements, students pose questions to their teachers. The questions posed occasionally could not be predicted initially by teachers. For that reason, ability to anticipate what students are likely to think and ask and provide best responses to students' why questions are necessary for teachers. Subsequently, teachers are suggested to provide responses by leveraging decompressed mathematical knowledge that facilitates students to comprehend, refine, and develop their knowledge. Many demands of teaching activities necessitate knowledge at the combination of student and contents. Therefore, teachers' mathematical knowledge for teaching is crucial for teaching.

The rationale behind the study is that there is scarcity in literature with respect to pre-service secondary mathematics teachers' views relative to discrepancies between quadratic equations and quadratic functions. The ground of why the author considered the topics was an interesting event occurred in peer teaching session. The peer teaching was one of the activities designed for the mathematics teaching development course. There was a pre-service teacher who tried to teach lesson segment of the concept of quadratic functions. At the beginning of the lesson, he was surprised by his friends who wondered about the differences between quadratic function and quadratic equations. Unfortunately, he could not make any significant and clear explications. The author then took this question into account.

As a matter of fact, the topics of quadratic equation and quadratic functions are presented in distinct area in secondary school mathematics curriculum. Quadratic equations precede quadratic functions as algebra is followed by function. Both are part of polynomials in which a large percentage of the middle and secondary mathematics curriculum is focused on by carrying out the operation and determining solutions or roots. The topics are cornerstone of subsequent mathematics topics such as trigonometry, limit, integral, and so forth. It is prevalent that in learning quadratic equation or quadratic function majority of students just memorize its rules devoid of reasoning and thinking deeply about them [4]–[6]. Previous studies highlighted students' lack of knowledge and misconceptions about factorization and the use of root statements [5]. In addition, students encounter adversity in articulating notion of variables and determining solution to a quadratic equation [7].

Students' difficulties and misconceptions in catching on both concepts are likely to be brought about by teachers' inability to transform and present mathematical knowledge properly. Brousseau, Sarrazy, and Novotná [8] acknowledged didactical obstacles, a term of which teachers transform a body of knowledge that might be different from the origin due to practical consideration. It appears to be the case that students' puzzlements in learning quadratic equations and quadratic functions might be attributed to several factors, one of which is pedagogical content knowledge of secondary mathematics teachers. Hence, teacher education is responsible for refining and developing pre-service teachers' mathematical knowledge in order to ensure high quality teaching.

The present study is substantial for several accounts. Recognizing pre-service secondary mathematics teachers' views about the differences could provide good opportunity for instructors to improve their performance in effort to improve and develop pre-service teachers' mathematical knowledge for teaching. In addition, findings of the present study could contribute to enriching body of knowledge of how pre-service secondary mathematics teachers' views concerning the differences between quadratic equations and quadratic functions. Besides, in-service mathematics teachers could benefit from pre-service teachers given responses regarding differences between quadratic equations and quadratic functions.

On the ground of the aforementioned discussion, the purpose of this study is to explore Indonesian pre-service secondary mathematics teachers' views about the differences between quadratic equations and quadratic functions. The study seeks explanation concerning their opinions as well as obstacles that pre-service teachers made when revealing their opinions.

2. Methods

Purposive sampling method was drawn upon to select participants. Participants of this study included fifty five pre-service secondary mathematics teachers who were enrolled as students at one private university in Jakarta City. They were in the third year at fourth-year secondary mathematics teacher education program. Data for the present study were collected in the spring of 2016-2017 academic year. At the time of the study, they have completed several teaching-related courses such as mathematics teaching and learning course and development of mathematics teaching course. Eight of the participants were male students, meanwhile the rest were female students.

2.1. Data Collection

Two sources of data were leveraged for this investigation: (1) participants' written responses; and (2) interviews. Participants' written responses were students' responses towards presented questions about the differences. The analysis of written responses is used to grasp participants' arguments. Subsequently, further investigation of participants' written responses was conducted by means of interviews. In this case, the author interviewed participants to gain deep information and justification directly concerning what they think and to conduct triangulation. Out of all participants, five of whom were interviewed whom the author thought would provide significant information. At the outset, protocol interview questions were prepared to guide the author during the interview.

The open-ended question that was asked to the participants, that is: *In case of a mathematics teacher introduces the concept of quadratic function (one-variable case), he/she reminds students about quadratic equations. Before continuing to subsequent explanation, one student interrupts and asks, "Sir/Mrs. what are the differences between quadratic equations and quadratic functions?" Suppose that you are the teacher, what will you explain to your students so that they understand its differences.* The question was given to the participants, and they were provided 100 minutes to write their argument concerning the differences. In addition, the question essentially required participants to describe the differences between quadratic equation and quadratic functions by means of their own so that students in secondary school level could grasp it easily.

2.2. Data Analysis

As an open-ended question was drawn on in the search of pre-service secondary mathematics teachers' views regarding the differences between quadratic equations and quadratic functions, qualitative approach was selected for the purpose of analyzing data. Categories and subcategories were constructed based on participants written responses. With respect to the reliability of categorization procedures, a second coder was involved. Agreement between coders for categorizing participants' responses was 94%. Triangulation method was conducted by interviewing five participants whose responses are interesting and in need of further and deep exploration.

3. Results and Discussion

In the following section, the findings will be presented in two major categories based on participants' responses concerning the question. The first category portrays participants' views about differences of quadratic equations and quadratic functions. This category consists of four sub-categories. The second category explicates participants' obstacles in describing the differences. In this category, there are four sub-categories constructed.

3.1. Participants' views about differences

Students' responses were likely to be influenced by their conceptions about algebra as Usiskin [9] described, namely, algebra as generalized arithmetic, algebra as the study of relationships among quantities, and algebra as the study of structures. In addition to it, we tried to describe participants responses in effort to shed light on the differences based on their written response by adapting conceptual framework for students understanding of logarithms proposed by Berezovski and Zazkis [10] and [11], namely, logarithm as object, logarithm as process, logarithm as function, and logarithm in contextual problems. The subcategories constructed are the differences based on the general terms, process, geometrical aspects, and main characteristics.

3.1.1. The differences based on its general forms

Out of 54 participants, 25 participants referred the differences to the general form of quadratic equation and quadratic functions. It means that the majority of responses fall into this subcategory. This response could be associated with the conception of algebra as generalized arithmetic. In this case, variables are wielded as generalizers in lieu of standing for specific unknown numbers [4]. According to Tanton [12], the general form of quadratic equations is and quadratic functions have general form in which a, b, and c are either real or complex number and a is not zero. In addition to the general forms, several participants exemplified it by using numbers directly. Referring it to the general forms was highlighted on account of practical consideration in which students could easily discern the differences.

It is important to note that focusing solely on the general forms might lead students to develop procedural or superficial knowledge devoid of conceptual or relational knowledge. Procedural knowledge, according to Hiebert and Lefevre [13], was described as knowledge concerning mathematical formal language or symbol representation system. In Indonesia, current mathematics curriculum is no longer centered solely on procedural skills; rather, it considers integration between procedural and conceptual development. Nevertheless, in fact procedural skills are still likely to be focus in most mathematics classrooms. Therefore, referring the differences based on the general forms is not enough and in need of being integrated with concept-based explanation.

3.1.2. The differences based on its main characteristics

Characteristics of equation and function were uncovered by twenty one participants in effort to differ the quadratic equations and functions. Tanton [12] described quadratic as any expression, function, or equation comprising variables raised to the second power, but no higher power. Several participants relied the difference on definition of quadratic similar to what Tanton [12] described. However, they made some additions such as explaining definition of equation and function. The equation was perceived by several participants as similarity between expression found in right side and left side. It is worth noting that participants' responses regarding equal sign converge with prior research [14] showing that students might develop various conceptions concerning equal sign, one of which is relational meaning. Relational meaning refers to various subcategories, such as formal equivalence describing equivalent terms [15]. Besides, several of them highlighted the definition of equation as an open sentence. The rationale behind the response is that a quadratic equation is a mathematical statement that can be either true or false relying on substituted values.

With respect to quadratic equation, several participants referred it to components of functions and related terms such as domain, codomain, range, dependent variables, and independent variables. They also emphasized function as a relation in which each component of the domain was mapped to exactly one component in the codomain. A participant expressed that quadratic function consisted of two distinct variables, that is, x and $f(x)$ or y.

3.1.3. The differences based on process

There were five teen participants who viewed that quadratic equations and quadratic functions as processes. According to them, quadratic equations referred to process to determine its roots. It means that the participants saw that quadratic equation automatically led them to determine the value of x. It fits with conception of algebra as study of procedures for solving certain kinds of problems [9]. They were likely to consider that quadratic equation and quadratic functions as problem that should be addressed. Just as logarithm [16], quadratic equation and quadratic functions had its operational characteristics and complex process.

Factorization and quadratic formula were two methods that were mostly mentioned by participants in attempt to determine the value of x or solving quadratics equation. Only two participants mentioned completing the square as one method to determine the roots of quadratic equation. Several participants completed their explanation by presenting direct examples of how to solve quadratic equations. To solve it effectively and correctly, mastering topics such as algebra, fractions, negative numbers and algebraic expansions are required [17]. In short, algebraic manipulations are perceived as the most important aspect in dealing with quadratic equations.

Meanwhile according to the participants, quadratic functions referred to process of fixing certain value of x or domain of function and determining the value of function or range of functions. This conception is likely to be in line with what Usiskin [9] shed light on the conception of variables as argument or parameter that have distinct values and relate to other variables. In this case, the value of y as dependent variable is determined by the value of x as independent variable. Therefore both values are related each other.

3.1.4. The differences based on geometric aspects
Thompson and Chappell [18] revealed that representation is a substantial component in the elaboration of mathematical literacy. Therefore, in addition to dealing with symbolic manipulation, geometric aspects were considered by twelve participants as quadratic functions referred to constructing graphic or curves. The form of curve that is constructed is parabola. One participant wrote that in attempt to draw graphic, the initial step was substituting dependent variable (x) by specific number and subsequently determining points on Cartesian coordinates. Another response was determination of intersection point with x or y axis. Concerning quadratic equation, one participant claimed that there was no graphic or curve that could be constructed to stand for quadratic equation. Nevertheless, another participant drew quadratic equation as one or two straight lines that intersected x axis. Referring to the general form of quadratic equation which consists of one unknown or variable solely and disregarding another variable, argument expressed that there was no representation of quadratic equation could be retained. Considering visual representation of quadratic functions might lead students to learn mathematics in a meaningful manner.

3.2. Participants' Obstacles
In attempt to uncover the discrepancies between quadratic equations and quadratic functions, several participants indicate obstacles in describing it, that is, inappropriate constraints and improper interpretation of the general forms.

3.2.1. Inappropriate constraints
Even though the majority of the participants highlighted the differences based on its general forms, several of them were unable to show appropriate constraints. The following are participants' deficiencies in articulating general forms of quadratic equations and quadratic functions properly.

- Confusions between real and constant number in describing the value of a, b, and c in general form of quadratic equation and functions.
  There were several participants expressed that in quadratic equation the value of a, b, and c were real numbers, yet in the following explanation he/she considered that in quadratic function, the value of a, b, and c were constants or vice versa. Some of them argued that the value of a, b, and c were constant real numbers. It is not surprising since literature also indicate discrepancies in illuminating the value of a, b, and c in which most of literature refer it to real numbers [12], meanwhile the others refer it to constants [19]. Even though they are expressed in a distinct manner, as a matter of fact, a a, b, and c are known or given. In addition, a constant is a specific real number. Nevertheless, consistency is required in describing both quadratic equation and quadratic functions. Besides, confusion relative to the value of x was demonstrated by several participants. They argued that the value of x was not equal to zero. It seems that they have puzzlement in determining which variables whose value is zero.
- Incomplete constraints
  Several participants did not get across the constraints of the general terms completely. The most two common mistakes are that the value of a is not determined and the value of a, b, and c are not specified succinctly. Determining the value of a is important in describing quadratic equation and function.

This obstacle might be attributed to mathematics teaching process in which teachers might focus solely on its general form of quadratic equation and function, yet they might not write completely with the constraints or explain it further. Getting rid of constraints and ignoring deep and further explanation of its general terms seems to be frequently found in mathematics classrooms.

3.2.2. Improper Interpretations
Improper interpretation is a common and persistent problem in algebra learning. In this case, for example, several participants interpret the general form of quadratic equation and quadratic functions improperly such as:

a. Quadratic equation is algebraic statement whose result is zero, meanwhile the result of quadratic function is not zero.
b. Quadratic equation is a mathematical statement which has equal sign and has value of zero, meanwhile the value of quadratic function can be zero and non-zero.
c. Quadratic equation has equal sign, whereas quadratic function does not have.

These obstacles might stem from participants' inability to give other explanations concerning the differences. In addition, it might occur as a result of teaching strategies that present mathematics concept or topics separately. Therefore, they made comments based on the general forms by their own. Mathematics concepts or topics should be taught in an integrated manner. At the beginning of lesson, teachers are suggested to review previous concepts and make connection to new concepts. Ignoring related previous concepts might lead students to interpret the concept improperly. According to the above participant's responses, the statement a and statement b are very succinct that participants make improper interpretation. Direct interpretation devoid of considering other mathematical concepts lead participants to make over-generalizations. On the other hand, the 3rd statement stems from participants' understanding of another written form of function, that is, $f: x \rightarrow f(x)$ and she concluded that the general form of quadratic functions did not have equal sign.

4. Conclusion

A study conducted by Vaiyavutjamai and Clements [20] showed that common problem encountered in middle grade was that even though students were able to obtain correct solution in determining roots of quadratic equations, they had serious puzzlements concerning the actual meaning of quadratic equations. It might be due to teaching strategies which emphasize heavily on procedural knowledge in lieu of integration of procedural and conceptual knowledge or on certain type of representation instead of multiple representations. Therefore, when mathematics teachers are asked by students who have difficulty in grasping certain concept, the teachers should provide appropriate responses by considering various aspects or representations. In addition, teachers are in need of decompressing mathematical knowledge in such that students grasp it easily and properly.

The written responses of the participants relative to differences between quadratic equation and quadratic function provided sufficient insights for us to consider various ways to deal with it. The results thus far show that participants referred the differences to their general forms, main characteristics, processes, and geometric aspects. Use of multiple aspects and representations in unpacking the differences might help both students and teachers in dealing with the topics. Besides, in expressing their responses several participants demonstrated obstacles such as inappropriate constraints and improper interpretation. The reason might lay in the fact that instruction or learning strategies emphasize heavily on symbolic manipulation in lieu of concepts and meaning.

The present study will assist mathematics teacher educators in obtaining information on pre-service secondary mathematics teachers' view and obstacles concerning quadratic equations and quadratic function. For the purpose of coming to grips with the obstacles, teacher educators should be more aware of pre-service teachers' mathematical knowledge and learning. The findings could be a reference for teacher educators to seek alternative methods in dealing with pre-service teachers' obstacles especially in the topics. Further studies might focus on investigation of pre-service secondary mathematics teachers' views on similarity and relations between quadratic equations and quadratic function.

References

[1] Ball D L, Lubienski S T and Mewborn D S 2001 Research on teaching mathematics: The unsolved problem of teachers' mathematical knowledge *Handb. Res. Teach.* **4** pp 433–456
[2] Hill H C, Rowan B and Ball D L 2005 Effects of teachers' mathematical knowledge for teaching on student achievement *Am. Educ. Res. J* **42** 2 pp 371–406
[3] Ball D L, Thames M H and Phelps G 2008 Content Knowledge for Teaching: What Makes It Special? *J. Teach. Educ.* **59** 5 p 389–407

[4] Chua B L and Wood E 2005 Working with logarithms: students' misconceptions and errors *Math. Edu.* **8** 2 p 53–70.
[5] Memnun D S, Aydın B, Dinç E, Çoban M and Sevindik F 2015 Failures and Inabilities of High School Students about Quadratic Equations and Functions *J. Educ. Train. Stud* **3** 6 p 50–60
[6] Emre E and Argun Z 2012 Talking About Mathematics: Quadratic Equations *Procedia - Soc. Behav. Sci.* **46** 1996 p. 5306–5310
[7] Vaiyavutjamai P, Ellerton N F and Clements M A 2005 Students' attempts to solve two elementary quadratic equations: A study in three nations *Build. Connect. Res. theory Pract. Sydney, Aust. Math. Educ. Res. Gr. Aust.*
[8] Brousseau G, Sarrazy B and Novotná J 2014 Didactic contract in mathematics education *Encyclopedia of Mathematics Education* Springer pp. 153–159
[9] Usiskin Z 1988 *Conceptions of school algebra and uses of variables* **5**
[10] Berezovski T and Zazkis R 2006 Logarithms: Snapshots from two tasks *Int. Gr. Psychol. Math. Educ* 145
[11] Williams H R A 2011 *A Conceptual Framework for Student Understanding of Logarithms* Brigham Young University
[12] Tanton J 2005 Quadratic *Encyclopedia of Mathematics* Facts On File, Inc. pp. 428–430
[13] Hiebert J and Lefevre P 1986 Conceptual and procedural knowledge in mathematics: an introductory analysis, in J Hiebert (Ed.) *Conceptual and procedural knowledge: the case of mathematics* Xiii (Hillsdale, NJ: Lawrence Erlbaum Associates) p 1–27
[14] Alibali M W, Knuth E J, Hattikudur S, McNeil N M and Stephens A C 2007 A Longitudinal Examination of Middle SchoolStudents' Understanding of the Equal Sign and Equivalent Equations *Math. Think. Learn* **9** 3 p 221–247
[15] Prediger S 2010 How to develop mathematics-for-teaching and for understanding: The case of meanings of the equal sign *J. Math. Teach. Educ.* **13** 1 p 73–93
[16] Aziz T A, Pramudiani P and Purnomo Y W 2017 How do college students solve logarithm questions? *Int. J. Emerg. Math. Educ* **1** 1 p 25–40
[17] Zakaria E, Ibrahim and Maat S M 2010 Analysis of Students' Error in Learning of Quadratic Equations *Int. Educ. Stud.* **3** 3 p 105–110
[18] Thompson D R and Chappell M F 2007 Communication and Representation as Elements in Mathematical Literacy *Read. Writ. Q* **23** 2 p 179–196
[19] Satyadevi C 2006 *Quantitative Technique* (New Delhi: S. Chand Publishing)
[20] Vaiyavutjamai P and Clements M A 2006 Effects of classroom instruction on students' understanding of quadratic equations *Math. Educ. Res. J* **18** 1 p 47–77

IOP Conf. Series: Journal of Physics: Conf. Series **948** (2017) 012044 doi:10.1088/1742-6596/948/1/012044

Dienes AEM as an alternative mathematics teaching aid to enhance Indonesian students' understanding of algebra concept

S Soro[1], S Maarif[1], Y Kurniawan[1] and A Raditya[2]

[1]Departement of Mathematics Education, Universitas Muhammadiyah Prof. DR. HAMKA, Indonesia.
[2]Departement of Mathematics Education, Universitas Muhammadiyah Tangerang, Indonesia.

E-mail: slamet.soro@yahoo.co.id

Abstract. The aim of this study is to find out the effect of Dienes AEM (Algebra Experience Materials) on the ability of understanding concept of algebra on the senior high school student in Indonesia. This research is an experimental research with subject of all high school students in Indonesia. The samples taken were high school students in three provinces namely DKI Jakarta Province, West Java Province and Banten Province. From each province was taken senior high school namely SMA N 9 Bekasi West Java, SMA N 94 Jakarta and SMA N 5 Tangerang, Banten. The number of samples in this study was 114 high school students of tenth grade as experimental class and 115 high school students of tenth grade as control class. Learning algebra concept is needed in learning mathematics, besides it is needed especially to educate students to be able to think logically, systematically, critically, analytically, creatively, and cooperation. Therefore in this research will be developed an effective algebra learning by using Dienes AEM. The result of this research is that there is a significant influence on the students' concept comprehension ability taught by using Dienes AEM learning as an alternative to instill the concept of algebra compared to the students taught by conventional learning. Besides, the students' learning motivation increases because students can construct the concept of algebra with props.

1. Introduction

Education is a dynamic process and always required to adjust to the needs of society and technological developments. It also has a major contribution in preparing quality human beings and directing students to understand their role as human beings who are responsible for continuing and improving development. To improve the development of education is needed a system that can affect the system of teaching and learning such as curriculum, books, learning facilities, teachers and students, and evaluation of learning outcomes. Therefore, it is required adjustment in all aspects that affect the implementation of the education system in order it can be achieved optimally.

Mathematics is a universal science underlying the development of modern technology in education in Indonesia, it has an important role in various disciplines and the development of human mind power.

Mathematics can develop logical, systematic and careful thinking. This is because of the hierarchical, dynamic, deductive, and generative nature of mathematics.

The ability to understand students' math concepts in schools should always be improved. This is based on the findings of research conducted by PISA (Program for International Student Assessment) in 2012 which states that Indonesia is ranked 64 out of 65 countries, while in the previous year Indonesia ranked 61 out of 65 countries, it shows Indonesia has more than 80% of students who do not like learning math at school [1]. It can happen because students think that math is difficult, both difficulty in calculating, understanding and solving mathematical problems so they do not like mathematics. In addition, mathematics too many symbols and terms, sometimes make students confused in interpreting it. It is common for students to take math lessons in the classroom without knowing what they are learning, because they are rarely asked to communicate what they have learned.

Difficulties experienced by students especially in understanding the concept of algebra. As in the settlement of algebraic forms $x + x = 2x$; $x \times x = x^2$; $(x + 1)(x + 2) = x^2 + 3x + 2$; and $(x - 2)(x - 3) = x^2 - 5x + 6$. The completion of the algebraic form is only understood that the rules are abstract by the students. This is because the concept of algebra is often presented in abstract in classroom learning. This difficulty occurs because the ability of students in understanding abstract mathematical concepts is very limited.

According to The *Great Dictionary of the Indonesian Language of the Language Center* "understanding" means "really understand". Understanding is an important aspect of learning. Sumarmo states that understanding is the second cognitive level in Bloom's taxonomy that describes mastery using the relevant rules without connecting it with other ideas and all its implications [2]. Understanding here is a low-level understanding, where in this case each individual can be said to understand only by memorizing the existing formula and using the rules of processing or algorithm. Such understanding is the same as instrumental understanding and mechanical understanding. In learning mathematics understanding of the material is necessary. This is because mathematics is not a rote subject, so it takes an understanding. Ruseffendi said understanding is classified into three parts, among them: (a) Understanding of translations is used to convey information in other languages and forms, and concerning the giving of meaning from a variety of information; (b) Understanding of interpretation (explanation) is used to interpret the point of reading, not only with words and phrases, but also includes understanding an information from an idea; (c) Extrapolation (expansion); includes estimations and predictions based on a thought, an overview of an information, also including conclusion [3].

Many teachers teach without using mathematical tools, whereas they are able to visualize mathematics, also make the students interested in learning mathematics because with a real object students explore the concept directly without just the symbols. As in algebra learning, there are many symbols that require students to understand the meaning of the symbols. In learning it, sometimes students are confused to perform the operation. Hence, students are asked actively to solve problems of operation on algebra. One of the solutions is to use instructional media that is also a tool in learning mathematics. Students are asked to explore algebra directly by using mathematical tools dienes Algebraic Experience Models (AEM). Use of this tool is to encourage the ability of students to understand the concept of algebra in order to be more developed. It can be effectively given and instilled through learning in school. Learning algebra aims to make students able to think logically, systematically, analytically, critically, creatively, and cooperation. Based on these description, the authors are interested to conduct research on how to instill the concept of algebra by using Dienes AEM in learning mathematics students in Indonesia.

2. Methods

This research was done by Quasi-Experiment method. The samples taken were high school students in three provinces namely the provinces of DKI Jakarta, West Java, and Banten. From each province was taken one school, which is SMA N 9 Bekasi, West Java, SMA N 94 Jakarta and SMA N 5 Tangerang, Banten. The number of samples in this study was 114 high school students of tenth grade as experimental class and 115 high school students of tenth grade as control class.

IOP Conf. Series: Journal of Physics: Conf. Series **948** (2017) 012044 doi:10.1088/1742-6596/948/1/012044

3. Results and Discussion

Before performing hypothesis testing, it was carried out the normality and homogeneity test. A normality test is assumed that the populations from which the samples are taken are normally distributed [4]. Homogeneity test is an inferential statistic used to assess the equality of variances for a variable calculated for two or more groups [5].

3.1. Test of Normality

Testing of normality carried out by using Lilliefors test with SPSS 19 for windows. The result can be seen in Table 1.

Table 1. The Result of Normality Test

Class	Kolmogorov-Smirnov[a]			Shapiro-Wilk		
	Statistic	df	Sig.	Statistic	df	Sig.
Experiment	.124	114	.162	.955	114	.166
Control	.094	115	.073	.979	115	.068

From Table 1, p-value (Asymp Sig) of the experimental class is $0.162 > 0.05 = \alpha$, and *p-value* (Asymp Sig) control class is $0.073 > 0.05 = \alpha$, hence H_0 is accepted. From the result of the calculation, it can be concluded that the ability of understanding concept for experimental class and control class are normally distributed at the level of significance $\alpha = 0.05$.

3.2. Test of Homogeneity

The homogeneity test is performed by using Cronbach Alpha test. The result can be seen in Table 2 below.

Table 2. The Result of Homogeneity Test

Levene Statistic	Degree of Freedom sample 1	Degree of Freedom sample 1	Sig.
6.782	1	227	.010

Based on Table 2, p-value (Asymp Sig) of the experimental class is $0.162 > 0.05 = \alpha$, and p-value (Asymp Sig) of control class is $0.013 < 0.05 = \alpha$. The value of significance is more than the significance level $\alpha = 0.05$, so it can be concluded that H0 which states the population variance of both groups is accepted. That is, both groups of data are homogeneous.

3.3. Test of Hypothesis

The hypothesis of this research is that there is an influence of learning by using teaching aid Dienes AEM in the instilling of algebra concept against the ability for understanding students' mathematical concepts. The hypothesis proposed was tested by t-test. By using SPSS 19 for Windows program, the following is the t-test results:

Table 3. Test of Hypothesis

		Levene's Test for Equality of Variances		t-test for Equality of Means						
									95% Confidence Interval of the Difference	
		F	Sig.	t	df	Sig. (2-tailed)	Mean Difference	Std. Error Difference	Lower	Upper
Understanding of concept	Equal variances assumed	6.782	.010	6.980	227	.000	3.501	.502	2.512	4.489
	Equal variances not assumed			6.987	216.696	.000	3.501	.501	2.513	4.488

Based on Table 3, p-value (Asymp Sig) of experimental class is $0.355 > 0.05 = \alpha$, and *p-value* (Asymp Sig) of control class is $0.000 < 0.05 = \alpha$, hence H_0 is rejected. This shows that there is an influence of learning by using teaching aid Dienes AEM in the instilling of algebra concept against the ability to understand students' mathematical concepts.

In this study, there are two techniques of learning applied, the first class uses teaching aid Dienes AEM and the second class does not. The first class is known as Experiment class and the second one is Control class.

Table 4. Recapitulation of Test of The Students' Ability Understanding Concept Algebra

Class	Ideal Score	Average Score	Average Grade	Percentage
Experiment	24	16.564	69.1282	69.12%
Control	24	14.775	61.55	61.55%

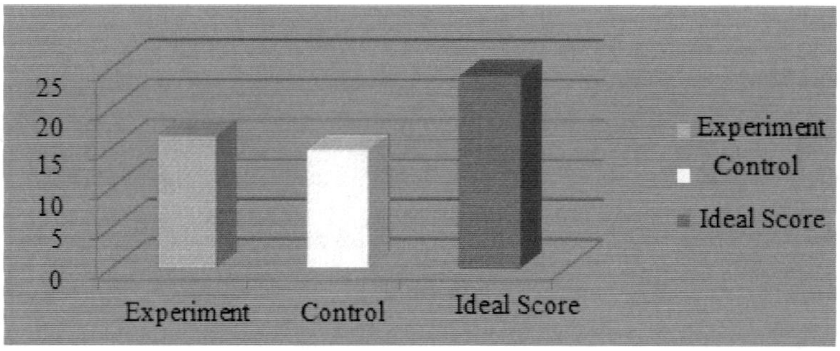

Figure 1. Comparison of ideal score and average score of students' ability to understand the concept of algebra of experimental and control class.

Based on Table 4, the average score of the students' ability to understand the concept of algebraic concepts obtained by the experimental class is higher than the control class and it is depicted in Figure 1. Figure 1 shows the ideal score is 24 and experimental class has score 16.564 or 16.564% of the ideal score, while control class has score 14.775 or 14.775% of the ideal score.

From the Figure 2, it can be seen that score for most item on experimental class is higher than control class. Items 1, 2 and 6 have more than 10% difference and item 4 have less than 10% difference between experimental and control class. In the other hand, items 3 and 5 have a slightly difference result, the control class score is higher than experimental score.

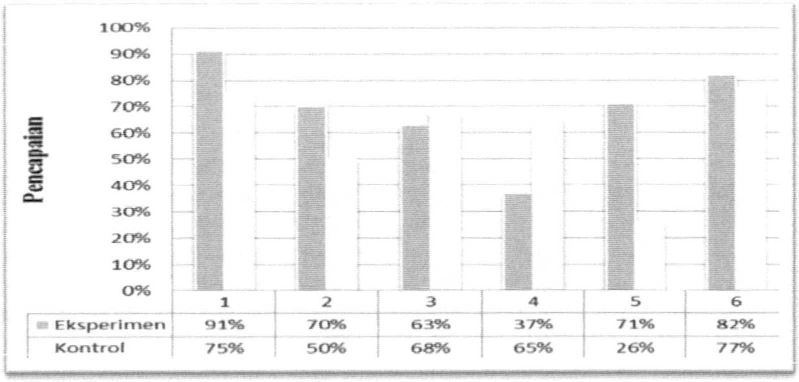

Figure 2. Comparison Score of Experimental and Control Class for Each Item on Students' Ability to Understand the Concept of Algebra

At the meeting, learning is quite good, because students are happy by learning using teaching aid Dienes AEM, which they have not known or seen before. Students pay attention when teacher is explaining how to use teaching aid Dienes AEM, but sometimes the class is crowded, this is because students are unusual to use teaching aid Dienes AEM so they are confused. However, students are interested in learning; they discuss to solve the problem given by using teaching aid Dienes AEM.

At the first meeting, we learn about forming the quadratic equation and determining the root of quadratic equation by factoring, and then the students are given the opportunity to organize the class by forming a group of 4 to 5 people. After that, each student is given activity sheet to be observed and solved before it is taken to group discussion; here is where the thought process occurs. Then each student gathered with the group to discuss their results whether true or not.

Figure 3. Students are discussing to do worksheet

Figure 4. Students present their result of discussion

Then after students work on work sheets, they are given the opportunity to present their findings and other groups were given the opportunity to ask questions and respond to the presentation. Before the learning ends. Learners together with the teacher to draw conclusions from what has been learned at the first meeting. The learners are then instructed to return to their respective seats to prepare for a formative test of material they have learned.

4. Conclusion

From the result, it can be concluded that from the experimental data, it can be concluded that there is an positive influence of learning by using teaching aid Dienes AEM on high school students' ability to understand the concept. Obtained a learning module with Dienes AEM to understand the concept of algebra in high school, and Dienes AEM is appropriate for algebra learning in high school.

References

[1] OECD 2013 *PISA 2012 Result in Focus: What 15-Year-Olds Know and What They Can Do with They Know* OECD Publishing
[2] Anggraeni D and Sumarmo U 2013 *Infinity Journal* **2** 1 p 1-12
[3] Yeni E M 2011 *Pemanfaatan Benda-benda Manipulatif untuk Meningkatkan Pemahaman Konsep Geometri dan Kemampuan Tilikan Ruang Siswa Kelas V Sekolah Dasar* Retrieved from http://jurnal.upi.edu/file/7-Ety_Mukhlesi_Yeni.pdf.
[4] Driscoll P, Lecky F and Crosby M 2000 *J Accid Emerg Med* **17** 3 p 205–211
[5] Levene H 1960 Robust tests for equality of variances In I Olkin et al. (Eds.) *Contributions to Probability and Statistics: Essays in Honor of Harold Hotelling* (Palo Alto: Stanford University Press) p. 278–292

IOP Conf. Series: Journal of Physics: Conf. Series **948** (2018) 012045 doi:10.1088/1742-6596/948/1/012045

Developing learning material of introduction to operation research course based on problem-based learning

Yerizon[1], Jazwinarti[1] and Yarman[1]

[1]Jurusan Matematika, Universitas Negeri Padang, Indonesia

E-mail: yerizon@yahoo.com

Abstract. Students have difficulties experience in the course Introduction to Operational Research (PRO). The purpose of this study is to analyze the requirement of students in the developing lecturing materials PRO based Problem Based Learning which is valid, practice, and effective. Lecture materials are developed based on Plomp's model. The development process of this device consists of 3 phases: front-end analysis/preliminary research, development/prototype phase and assessment phase. Preliminary analysis was obtained by observation and interview. From the research, it is found that students need the student's worksheet (LKM) for several reasons: 1) no LKM available, 2) presentation of subject not yet based on real problem, 3) experiencing difficulties from current learning source.

1. Introduction

Introduction to Research Operations is a compulsory course for students of Mathematics Education Program. One of the materials discussed is the Linear Programming. Linear programming is a mathematical method for allocating limited resources to achieve a goal such as maximizing profits and minimizing costs. Linear programming is widely applied in economic, industrial, military, social and other issues. The basic concept of linear programming has existed at the elementary level of education, which begins the introduction of the symbol of numbers represented by images of objects around us, then addition, subtraction, multiplication and comparing the number of objects. In Junior High School level, its concept is expanded through the Linear Equation System One Variable (SPLSV), then enhanced through the material of Linear Equations System Two Variable (SPLDV), and then in High School (SMA) they have been introduced the system of linear inequality and specific material of linear programming problem that presents everyday problems, and translates the problem into the mathematical model, solves the inequality system which is constraint or limiting, determining the optimum solution, answering the problem.

The method which is used is a graphical method using the point vertices test and the line of sight. At the university level, there is a special course of linear programming that discusses the method of completion of a linear programming whose goal is to maximize profits and minimize the costs. The methods given at the university are graphical methods, simplex methods, dual analytical methods, transportation methods.

However, students have difficulties in understanding the method. The main difficulty of students is to change the real problem into the mathematical model. They also difficult in making verbal representations into mathematical notations or models. This ability is called the ability of mathematical communication. One of the suitable ways to overcome this problem is developing lectures of Introduction to Research Operations course using Problem Based Learning approach for students in improving student's mathematical communication skills. This paper will reveal the difficulties experienced by students in the course of Introduction to Research Operations. This is a preliminary study in order to develop subject matter tools Introduction to Linear Programming based on Problem Based Learning which is valid and effective to improve students' understanding.

Problem Based Learning (PBL) is a learning model which is oriented to problem-solving. According to [1], PBL is a series of learning activities that emphasize the process of solving problems scientifically. Arends [2] states that PBL consists of presenting an authentic and meaningful situational problem to students that can be used as a springboard for investigation and discovery. Furthermore, Wena [3] tells that problem-based learning strategy or PBL is a learning strategy that begins by giving problems to the students, and then they learn through the problems given. In implementing the PBL, the teacher must be able to choose the learning material that contains problems that will be solved by the students. The selection of problem should come from the student's environment. PBLs can also be applied to develop students' thinking skills such as: the ability to analyze situations, apply their knowledge to new situations, recognize differences between facts and opinions, develop their ability to write a hypothesis objectively and problem-solving skills

Savoie and Hughes [3] tell that the PBL learning model has several characteristics: (1) Learning begins with a problem; (2) The given problem must have related to the real world of the students; (3) Organize learning around issues, not around disciplines; (4) Give great responsibility in forming and running their own learning process directly; (5) Using small groups; and (6) Prosecute students to demonstrate what they have learned in the terms of products and performance. In line with that, Sanjaya [1] suggests that there are three main characteristics of PBL are (1) PBL is a series of learning activities, which on the learning process students will carry out activities that are the learning stages of PBL. PBL does not expect students to just listen, record, and then memorize the subject matter, but through this PBL, students will actively think, communicate, search and process data, and finally conclude it; (2) Learning activities are directed to solve the given problem. PBLs place the problem as the focus of learning to determine the solutions; and (3) Problem-solving is done by using scientific thinking approach. The thinking process is done systematically and empirically. Systematic means scientific thinking is done through the stages that have been determined, while empirical means the process of problem-solving is based on facts and clear data. According to Rusman [4] the characteristics of PBL learning are (1) the problem becomes a starting point in learning; (2) the chosen problem is the unstructured real world; (3) problems require multiple perspectives; (4) problems, challenging learners' knowledge, attitudes, and competencies that require the identification of learning needs and new areas of learning; (5) learning self-direction becomes the main thing; (6) the utilization of diverse sources of knowledge, their use, and the evaluation of information resources is an essential process in the learning process; (7) learning is collaborative, communication and cooperative; (8) the development of inquiry skills and problem-solving is just as important as mastering the content of knowledge to find solutions to a problem; (9) the openness of processes in the learning process includes the synthesis and integration of a learning process; and (10) learning process involves evaluating and reviewing learners' experiences and learning processes.

The advantages of PBL Liu [4] are: (1) students are encouraged to have problem-solving skills in real situations, (2) students have the ability to build their own knowledge through learning activities, (3) students accustomed to using the resources (4) studens have the ability to assess their own learning progress, (5) students have the ability to communicate in the discussion activities or presentation of their work results, and (6) Learning difficulties of students individually can be overcome through group work in the form of peer teaching.

From the results of previous research indicating that PBL can improve students' mathematical ability, Rokhmawati [5], Supriyatin [13], and Prawiro [9] found that learning of PBL can improve problem-solving ability and self-efficacy of student. Kazemi [10] found that PBL learning helps reduce student misconceptions about mathematics. The results of Uzel's study [8] Padmavathy [11] show that the achievement and attitude of prospective teachers who study with PBL are significantly better than prospective teachers who learn with traditional learning. Novikasari [12] and Supraptinah [13] reports that prospective elementary school teachers who are studying with PBL have better mathematical content knowledge than prospective elementary teachers who are learning with conventional learning.

2. Methods

The type of this research is the *development research* that is used to develop a student worksheet product. Development research is research used to produce a particular product, and test the effectiveness of these products. The purposes of development research according to [14] are: (1) Better the understanding of the implementation problems of the teachers; (2) Development of prototypical project intervention (training, material, support), Including empiric evidences of Reviews their quality; (3) Generating methodological direction for the design and evaluation of such products or intervention; and (4) Increased (both individual and collective) expertise of the various participant

The first phase is Development and Prototyping Phase. In this phase self-evaluation and expert reviews to examine the validity of the worksheet that has been designed. Self-evaluation is to evaluate its own prototype that has been designed. From the evaluation results, it revised. Once the prototype is believed to be good, and as expected, the next stage is expert reviews. Revision continues until the worksheet is valid. If the worksheet is valid then it followed by one-to-one evaluation. Evaluation of individuals does by asking the three students who provide commentary on the worksheet. Based on the evaluation, revision held to the worksheet. After the worksheet was revised based on feedback on a one-to-one evaluation, it tryout to small group. This tryout is performed on conditions similar to actual conditions. Field trials conducted to look at the practicalities of the worksheet that have been designed. Practicalities of the worksheet are the level of used of the worksheet by the user. Practicalities can be seen during the implementation of learning through observation, interviews and questionnaires.

The second development phase is the assessment phase. In the assessment phase, we tested the effectiveness of the worksheet. The effectiveness of a product means a measure to know whether or not the effect or influence of the developed products. The effectiveness of the materials can be seen from the test results of students' mathematical communication skills after learning. In this paper, we only discuss the first stage.

3. Results and Discussion

The development of metacognition skills assessment instruments based on science literacy on dynamic electrical materials applied Dick and Carey development model with the data collection instrument were in the form of validation sheet, metacognition ability test, and student response questionnaire. This study produced primary and secondary data. The primary data were assessment instrument development process, theoretical validity, empirical validity, reliability, difficulty level, distinguishing factor, and distractor index. The secondary data were in the form of student responses to instrumental assessment of metacognition ability developed.

Based on the facts in the field, we found that the students have difficulty in learning Introduction to Operations Research course. The difficulties experienced of students can be reviewed from several factors, including: 1) during the learning process, 2) lecturers who teach its course, 3) materials discuss, 4) references, and 5) supporting facilities.

In general, difficulties experienced by students is lack of time in solving problems. They are less fluent in doing the steps that must be done. The material in this course must follow certain algorithms. If one stage of error occurs, then the process must be repeated from the beginning. This makes the student takes a lot of time in solving a problem. The other difficulty is in understanding the steps that must be done in solving a problem. Students also experience errors in writing data to be used. If the data is false, then it will result in the result error obtained. The fatal error is the mistake in constructing a mathematical model of a problem. The reason is that students use less appropriate logic in interpreting the language in the given problem.

In general, students do not have significant difficulties during the teaching process given by the lecturer but she/he not use a variety methods in giving lectures. The method used is very monotonous that explains the material, gives examples, and gives exercise. In the presentation, the lecturer gives the material too quickly so difficult for them to follow. This happens because the lecturers are often late, so the lack of time in providing all the material that must be learned.

Difficulties factor of students studying this course is based on content material. Difficult for them to understand the formula. As a result, they have difficulties in applying it to a given problem.

Students have difficulty obtaining an easy-to-understand source book. The books in library are still difficult to understand by the students. This happens because the book is in English and the presentation is still too short. The presentation of matter is still deductive, ie, by defining it followed by giving a theorem and ending with an example. The examples given are not given in detail and are less varied. The book used is not the latest and hard to get the latest book.

Another difficulty experienced by students is during the exam. In general, students lack the time in solving the given problem. It requires high accuracy. If a mistake occurs at the beginning of the work it will be fatal to the final settlement.

4. Conclusion

Based on the result of the research, it can be concluded that the students have difficulties in learning the Introduction to Operation Research course. The difficulties are: 1) lack of time in solving problems or problems given, 2) difficulty in understanding the steps to be done in solving a problem, 3) difficult to understand the contents of the book because the book is in English and the presentation is too short, and 4) less thorough in the algorithm. We recommend that lecturers provide teaching materials that are more easily understood and use a variety of methods.

References

[1] Sanjaya W 2009 *Strategi Pembelajaran Berorientasi Standar Proses Pendidikan* (Jakarta: Kencana Prenada Media Group)
[2] Arends R I 2008 *Classroom Instruction and Management* (New York: The McGraw-Hill Companies, Inc)
[3] Wena M 2013 *Strategi Pembelajaran Inovatif Kontemporer* (Jakarta: Bumi Aksara)
[4] Rusman 2013 *Model-model Pembelajaran* (Jakarta: Raja Grafindo Persada)
[5] Liu M 2004 *Computers in Human Behavior* ***20*** 3 p 357-379
[6] Rokhmawati J D, Djatmika E T and Wardhana L 2016 *IOSR Journal of Research & Method in Education (IOSR-JRME)* ***6*** *3 pp 52-55*
[7] Supriyatin, Kusmayadi T A and Mardiyana 2015 *Jurnal Elektronik Pembelajaran Matematika* ***3*** *9 pp. 985-996*
[8] Uzel D and Ozdemir E 2012 The Effect of Problem Based E-Learning on Prospective Teachers' Achievement and Attitudes Towards Learning Mathematics *Procedia – Social and Behaviour Sciences* ***33*** *1154-1158*
[9] Prawiro Y P, Budiyono & Slamet I 2015 *JMEE* ***5*** *1 pp 51-60*
[10] Kazemi F and Ghoraishi M 2012 Comparison of Problem Based Learning Approach and Traditional Teaching on Attitude, Misconceptions and Mathematics Performance of University Students *Procedia – Social and Behavioral Sciences 46 3852-3856*
[11] Padmavathy R D and Mareesh K 2013 *International Multidisciplinary e-Journal* ***2*** *1 p 45-51*
[12] Novikasari I 2016 *International Journal of Education and Research* ***4*** *6 p 153-162*
[13] Supraptinah U, Budiyono & Subanti S 2015 *Jurnal Elektronik Pembelajaran Matematika* ***3*** *10 pp 1138-1149.*
[14] Plomp T 1997 *Educational and training system design* (Enschede, Netherlands: University of Twente)

IOP Conf. Series: Journal of Physics: Conf. Series **948** (2018) 012046 doi:10.1088/1742-6596/948/1/012046

The effect of Missouri mathematics project learning model on students' mathematical problem solving ability

I Handayani[1], R L Januar [1] and S E Purwanto[1]

[1]Departement of Mathematics Education, Universitas Muhammadiyah Prof. DR. HAMKA, Indonesia.

Email: isnaini-handayani@uhamka.ac.id

Abstract. This research aims to know the influence of Missouri Mathematics Project Learning Model to Mathematical Problem-solving Ability of Students at Junior High School. This research is a quantitative research and uses experimental research method of Quasi Experimental Design. The research population includes all student of grade VII of Junior High School who are enrolled in the even semester of the academic year 2016/2017. The Sample studied are 76 students from experimental and control groups. The sampling technique being used is cluster sampling method. The instrument is consisted of 7 essay questions whose validity, reliability, difficulty level and discriminating power have been tested. Before analyzing the data by using t-test, the data has fulfilled the requirement for normality and homogeneity. The result of data shows that there is the influence of Missouri mathematics project learning model to mathematical problem-solving ability of students at junior high school with medium effect.

1. Introduction

Mathematics is known as queen of science because mathematics can generate other sciences such as physics, chemistry, geography and many other. Therefore, Mathematics science almost can be implemented in many field of life and even becomes one of the measurement of someone IQ.

There is a mathematics exam to measure the mathematics achievement of student aged 15 years, namely PISA *(Program for International Student Assessment)*. PISA is a research program of once three years which aims to periodically do the research concerning the ability of student in reading literacy, mathematics literacy and science literacy. The ability of mathematics literacy in Draft Mathematics Framework PISA 2015 is defined as an individual's capacity to formulate, employ, and interpret mathematics in a variety of contexts. It includes reasoning mathematically and using mathematical concepts, procedures, facts and tools to describe, explain and predict phenomena. It assists individuals to recognize the role that mathematics plays in the world and to make the well-founded judgments and decisions needed by constructive, engaged and reflective citizens [1].

Therefore, PISA students must have the ability to formulate, apply, and interpret mathematics in various contexts and capabilities of mathematical reasoning and the use of concepts, procedures, facts, and mathematical tools to describe, explain, and predict phenomena. That ability is a basic ability that should have to be owned by the student of the participant of PISA to solve mathematics problem.

One of the key aspects assessed in PISA is the problem-solving ability because it is one of the high-order thinking skills assessed in PISA. This is consistent with the PISA mathematical problem which examines three aspects: Content, Context and Competencies Cluster [2]. In this group of competencies PISA develops problem-based mathematical skills.

Indonesia is one of the countries which is included in PISA. However, if it is seen from the result of PISA, then the ability of mathematics literacy of Indonesian student is still low. This can be known from the achievement of Indonesia in following PISA in Indonesian literacy achievement in mathematics from 2003 to 2015.

Table 1. Indonesia Position Based on PISA Study Results

Year	Subject	The average score of Indonesia	International Average Score	Rank Indonesia	Number of participant countries
2003	Mathematics	360	500	38	40
2006	Mathematics	391	500	50	57
2009	Mathematics	371	500	61	65
2012	Mathematics	375	500	64	65
2015	Mathematics	386	500	63	72

The data in Table 1 shows that the students' mathematical literacy skills are still very low when compared to other countries included in PISA. This means that the students mathematical problem-solving ability is still low. Therefore, the need for handling and improving the learning process in Indonesian students in terms of mathematical problem-solving skills is extremely urgent to improve the ability of Indonesian students' mathematical literacy.

To have the ability of mathematics problem-solving is very important for the students. The importance of the ability of mathematics problem-solving for the student is also clearly emphasized by Branca [3], namely: (1) the ability of problem-solving is an important activity in mathematics learning even as the core of mathematics; (2) problem-solving can cover the method, procedure, and strategy or the way which is the core process in mathematics curriculum; and (3) problem-solving is a basic ability in mathematics learning. Similar with the view of Branca, Hartono [4] said that problem-solving is part of very important mathematics curriculum. Relevant to those two experts' opinion, then the ability of mathematics problem-solving is a very important ability that student must have.

The choice of accurate learning model which is in accord with the objective of learning will help much and make the teacher easy in the process of learning in class. One of the learning models that can improve and develop the ability of mathematics problem-solving of the student is learning model known as Missouri Mathematics Project. That statement is relevant with the opinion of Dwiningrat, Suniarsih and Manuaba [5] who stated that one of the learning models that give a chance to the student to train and exercise the ability of problem-solving ability in mathematic problem-solving is learning model known as Missouri Mathematics Project.

Learning model of Missouri Mathematics Project is a learning model which consist of the steps of learning activities in systematic and structured steps along with the activities of student covering cooperative work and independent work and is ended by conclusion and assignment. Krismanto [6] said that learning model the so called Missouri Mathematics Project is one of the models being developed through the research and is a well-structured model. Steps of the Missouri Mathematics Project is review, development, cooperative work, seat work, and assignment.

This learning model is a combination of cooperative working with independent working, so that the ability of student to work together or to cooperate and to work independently can be trained well. In addition, added value of Missouri Mathematics Project is to bridge between teacher centered learning toward student centered learning [7].

Furthermore, Confrey gets the finding that the teacher who plans and implements five steps of mathematics learning will be more successful to be compared with those who use traditional method and approach [8]. The five steps are known as Missouri Mathematics Project which is proven to be more successful. Thus, in accord with the background presented above, the problem can be formulated in a question: Is there an impact of learning model of Missouri Mathematics Project toward the ability of mathematics problem-solving of the student? Based on the formulation of the problem, then the objective of this research is to describe and analyze the impact of the learning model of Missouri Mathematics Project toward the ability of mathematics problem-solving of student.

2. Methods

This research is conducted in Junior High School in the even semester 2016/2017. Sample taking is conducted by using the method of cluster sampling. The sample being taken from this research are two classes namely one as an experiment class and another one is as a controlling class.

This research is using experiment method which is included as Quasi Experimental with the design uses is the static group comparison. The instrument being used in this research is the test ability of mathematics problem-solving of the student. The Instrument being used in this research is the test of ability of mathematical problem-solving. Before the instrument is used to take research data, it should be tested to fulfill validity, reliability, level of difficulty and discriminating power as the condition of good instrument and the data being obtained is valid and not wrong. The result from the try out test of instrument, there are 7 reliable questions to be made as the data for the ability of mathematics problem-solving.

In accord with the precondition of the analysis, before the test of hypothesis (t-test), the test for fulfilling the requirement of analysis whose consists of normality test and homogeneity test should be conducted. The calculation of normality test in this research uses the test of Lilliefors at the level of significance of 5% with the criteria of normal distributed data test if $L_{result} < L_{table}$. Homogeneity test in this research is using the test of Fisher at the level of significance of 5% with the accepted criteria of H_0 if $F_{result} < F_{table}$.

After conducting analysis precondition test, hypothesis test is then conducted. To test formulated hypothesis, t-test statistical analytical is used with the test criteria of declining H_0 if $t_{result} > t_{table}$. To know how big the impact of learning model of Missouri Mathematics Project to the ability of mathematics problem-solving of student is, the data is analyzed by using the formulation of effect size [9].

3. Results and Discussion

Result of analysis of post-test shows average score of the ability of mathematics problem-solving of different students between the experiment group and the controlled one. The different of average score is quite significance, the experiment group got average score higher than that of the control one. Average score of the ability of mathematic problem-solving data of the experiment group is $\overline{X} = 46.39$ whereas average score of the ability of mathematic problem-solving of the controlled group is $\overline{X} = 33.68$. This phenomenon shows that experiment group can do the *post-test* ability of mathematic problem-solving better than the controlled group can do.

Further, hypothesis test is conducted by using t-test. However, before doing t-test, analysis precondition test that consists of normality test and homogeneity test should be conducted.

The result of normality test and homogeneity test is as follow: Result of the calculation of normality test of the experiment group shows that experiment group distributed normally. This can be known from $L_{result} = 0.1020$ and $L_{table} = 0.1437$ so that $L_{result} = 0.1020 < 0.1437 = L_{table}$. So is the result of normality test of controlled group distributed normally. This can be known from $L_{hitung} = 0.1054$ and $L_{tabel} = 0.1437$ so that $L_{result} = 0.1054 < 0.1437 = L_{table}$. Furthermore, based on the calculation of homogeneity test of the experiment group and controlled group, the result obtained is $F_{result} = 1.642$. Whereas for value F_{tabel} at the significance level of 5% is 1.738. Thus, it can be concluded that $F_{result} = 1.642 < 1.738 = F_{table}$ so that both groups can be said homogenous.

After conducting precondition test, the next test that can be conducted is hypothesis test (t-test). It is from the result of hypothesis test using t-test being conducted at the significant level of 5% the result obtained is $t_{result} = 2.753$ and $t_{table} = 1.667$. The following table presents the result of t-test calculation.

Table 2. T-test Result after treatment of t-test

Group	n	dk	t_{result}	t_{table}	Conclusion
Experiment	38	74	2.753	1.667	$t_{result} > t_{table}$,
Control	38				declining H_0

From Table 2 above, it can be known that 2.753 > 1.667 so that, the obtained result is $t_{result} > t_{table}$ then it can be concluded that H_0 is declined. It means that there is a significant difference between experiment group and controlled one, namely the difference between student being taught by using learning model of Missouri Mathematics Project and student being taught not by using learning model of Missouri Mathematics Project. The existence of this difference proves that the implementation of learning model of the Missouri Mathematics Project has an effect on the mathematical problem-solving ability of students in class VII at Junior High School. Furthermore, from effect size calculation, effect size price is obtained (ES) for 0,725 then based on the valid criteria, effect size price is at the medium level.

The experiment group was being taught by using learning model of Missouri Mathematics Project with the following steps: 1) Review, students are reminded on the material lesson concerning rectangular and triangle having been taught at the elementary; 2) Development, students are given the new concepts and new ideas regarding the material that will be studied in the next session; 3) Cooperative Work, students are categorized into several groups and do the work sheet containing the problem of mathematics in group; 4) Seat Work, students are no longer sitting in group, students do the test being given by the teacher individually; 5) students are being given the homework as a routine exercise. Whereas the controlled group is not taught by using the learning model of Missouri Mathematics Project.

In the experiment group, students can learn in group and can discuss to arise their spirit and motivation to solve the problem in the work sheet being given by the teacher. Therefore, students are used to solving the problem and can develop their abilities in solving the problem. The result of post-test answer of the experiment group in a whole has fulfilled some indicators of the ability of mathematic problem-solving correctly, whereas the controlled group in a whole still cannot achieve the indicators of the ability of mathematics problem-solving correctly. Based on the explanation above, it can be concluded that there is an impact of learning model of Missouri Mathematics Project on the ability of mathematic problem-solving of students at 7th grade.

4. Conclusion

Result of post-test data analysis of the ability of mathematic problem-solving of students being obtained shows that average score of the experiment group is higher than that of the controlled one. This phenomenon shows that the average ability of problem-solving of the students being taught by learning model of Missouri Mathematics Project is better than that of the students being taught not by learning model of Missouri Mathematics Project.

Both groups, the experiment group and the controlled one have fulfilled analysis precondition test which means that both data groups have distributed normally and are homogeny. Further, hypothesis test is done by t-test. Based on the calculation of t-test that it is obtained that $t_{result} > t_{table}$ then it can be concluded that there is the influence of Missouri Mathematics Project Learning Model to Mathematical Problem-solving Ability of Students with medium effect.

References

[1] PISA 2013 *Draft Mathematics Frameworks* retrieved from https://www.oecd.org/pisa/pisaproducts/Draft%20PISA%202015%20Mathematics%20Framework%20.pdf
[2] Setiawan H, Dafik and Lestari 2014 *Soal Matematika Dalam PISA Kaitannya Dengan Literasi Matematika Dan Keterampilan Berpikir Tingkat Tinggi* Proceeding of National Conference (Jember: Universitas Jember) retrieved from https://jurnal.unej.ac.id/index.php/psmp/article/download/955/758
[3] Purwosusilo 2014 *Jurnal Pendidikan dan Keguruan* **1** 2 p 30-40
[4] Hartono Y 2014 *Matematika: Strategi Pemecahan Masalah* (Yogyakarta: Graha Ilmu)

[5] Dwiningrat I G A D, Suniarsih N W and Manuaba I B S 2014 *e-Journal MIMBAR PGSD* **2** 1
[6] Krismanto A 2003 *Beberapa Teknik, Model, Dan Strategi Dalam Pembelajaran Matematika* (Yogyakarta: PPPG Matematika)
[7] Suastika I K and Wisulah 2015 *Jurnal Inspirasi Pendidikan* **5** 2 p 722-732
[8] Setiawan 2010 *Strategi Pembelajaran Matematika* (Yogyakarta: P4TK Matematika)
[9] Nurhayati N, Fadilah S and Mutmainah M 2014 *Jurnal Pendidikan Fisika dan Aplikasinya* 4 2 p 4

Improving students' mathematical critical thinking through rigorous teaching and learning model with informal argument

H Hamid

Universitas Khairun, Ternate, Indonesia

E-mail: hasan.hamid66@gmail.com

Abstract. The purpose of this study is to analyze an improvement of students' mathematical critical thinking (CT) ability in Real Analysis course by using Rigorous Teaching and Learning (RTL) model with informal argument. In addition, this research also attempted to understand students' CT on their initial mathematical ability (IMA). This study was conducted at a private university in academic year 2015/2016. The study employed the quasi-experimental method with pretest-posttest control group design. The participants of the study were 83 students in which 43 students were in the experimental group and 40 students were in the control group. The finding of the study showed that students in experimental group outperformed students in control group on mathematical CT ability based on their IMA (high, medium, low) in learning Real Analysis. In addition, based on medium IMA the improvement of mathematical CT ability of students who were exposed to RTL model with informal argument was greater than that of students who were exposed to CI (conventional instruction). There was also no effect of interaction between RTL model and CI model with both (high, medium, and low) IMA increased mathematical CT ability. Finally, based on (high, medium, and low) IMA there was a significant improvement in the achievement of all indicators of mathematical CT ability of students who were exposed to RTL model with informal argument than that of students who were exposed to CI.

1. Introduction

Students' ability to think critically is necessary to develop as every human has potential to think critically. This is in line with Cotton [1] who stated that although many people believed that we were born either with or without creative and critical thinking abilities, research had shown that these skills were teachable and learn-able. Watson and Glaser [2] considered that critical thinking includes: (1) attitudes of inquiry that involve an ability to recognize the existence of problems and an acceptance of the general need for evidence in support of what is asserted to be true; (2) knowledge of the nature of valid inferences, abstractions, and generalizations in which the weight or accuracy of different kinds of evidence are logically determined; and (3) skills in employing and applying the above attitudes and knowledge. Furthermore, to facilitate the interpretation of critical thinking, Watson and Glaser [3] introduce the three factors that are key to critical thinking and abbreviated as RED: (1) Recognizing Assumptions; (2) Evaluating Arguments; and (3) Drawing Conclusions. Assumptions are statements that are assumed to be true in the absence of proof. Identifying assumptions helps in discovering information gaps and enriches views of issues. Assumptions can be stated implicitly and explicitly. The ability to recognize assumptions in presentations, strategies, plans, and ideas is a key element in critical thinking. Being aware of assumptions and directly assessing their appropriateness to the situation helps people evaluate the merits of a proposal, policy, or practice. In addition, arguments are assertions intended to persuade someone to believe or act a certain way. Evaluating arguments is ability to analyze such assertions objectively and accurately. Analyzing arguments aids in determining what weight to put on them and what actions to take. It includes the ability to overcome a confirmation bias as tendency to look for and agree with information that confirms prior beliefs. Emotion plays a key role in evaluating arguments as well. A high level of emotion can cloud objectivity and ability to evaluate arguments accurately. Drawing conclusions consists of arriving at conclusions that logically follow from the available evidence. It includes evaluating all relevant information before drawing a conclusion, judging the plausibility of different conclusions, selecting the most appropriate conclusion, and avoiding overgeneralization beyond the evidence.

Critical thinking in learning mathematics is a cognitive process or mental action in an effort to acquire knowledge of mathematics based on mathematical reasoning. Therefore, critical thinking in mathematics learning includes ability to react to mathematical problems by distinguishing opinion and facts, conclusion and judgments, inductive and deductive arguments, as well as objective and subjective. Thus, critical thinking could be defined as the ability to identify and analyze problems as well as seek and evaluate relevant information to reach the right conclusions.

To facilitate students' critical thinking it is necessary to design precise and meaningful learning situations that give students opportunity to explore cognitive aspects. In consequence, students will be able to interpret and find the right concepts and strategies to facilitate them in learning Real Analysis. The learning model implemented in this study is Rigorous Teaching and Learning (RTL). Rhodes [4] revealed that there are three characteristics of students involved in rigorous learning. Furthermore, Rhodes [4] and Reyes [5] described indicators of rigorous teaching and learning models on academic rigor. It shows the level of challenges that integrate knowledge content, high-level thinking, active engagement, and strategies to design strict learning will enable students to interact with content strictly. This is in line with what Jackson [6] explained on strict learning strategies that describe the way teachers help students learn, manipulate, reorganize, adapt with new situations, and apply learned context. The use of informal arguments in this study is based on the concepts of Strauss and Corbin who acknowledged three activities: Syntactifying, rewarranting and elaborating. These three activities could be drawn in the following scheme [7]:

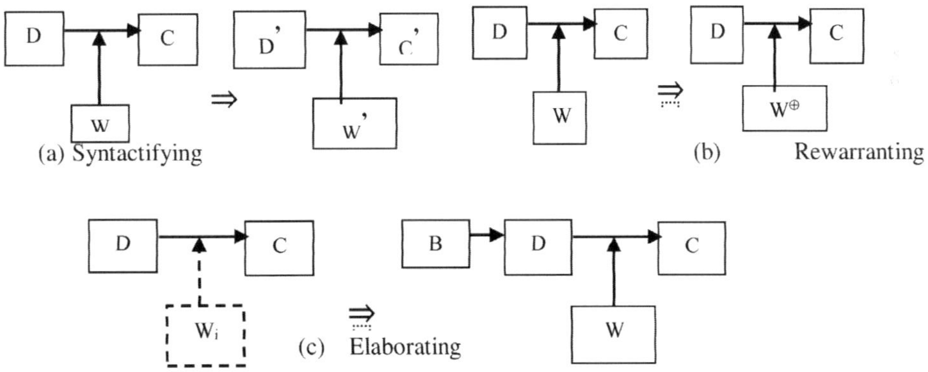

Figure 1. Three Translational Activities

Syntactifying occurs when a student seeks to take a statement in an informal argument given in what is considered to be a strict term and translate it into what is considered a more appropriate representation system for evidence. Such actions include removing references to diagrams used in informal arguments and replacing them with conventional mathematical terms, or introducing logical algebraic notation. In the case of the Toulmin's scheme, we may assume syntactifying as translating data (D), claims (C), and/or warranty (W) from arguments into new data (D'), claims (C'), and/or guarantees (W'). In other representational systems, it is shown without intending to change the meaning of D, C, or W.

Rewarranting occurs when students try to find a deductive reason for the claim that their informal argument is justified by non-deductive means. In the case of the Toulmin's scheme, we may consider rewarranting as a substitute for a reasonable guarantee (W) (i.e., assurance that the student believes it is possible to produce truth) with a valid warranty ($W^{\oplus}$) (i.e., warranty that the student believes is deemed legitimate by Community mathematics).

Elaborating occurs when students try to add in more detail to evidences that do not appear in their informal arguments. This happens in several ways: students will justify the assertion that they are taken for granted in their informal arguments by making explicit assurances that were initially implicit (W_i) in their informal argument, or rather justifying their data (D) (i.e., students trying to Justify the facts taken for granted).

2. Methods

This study employed mixed method. The strategy used is a concurrent embedded strategy [8] that is a research model that combines quantitative and qualitative research methods, and quantitative and qualitative data collected simultaneously. The quantitative method in this study is a primary method, while the qualitative method is a secondary method. The findings of each methods are combined to formulate a comprehensive interpretation of students' mathematical critical thinking (CT) using the Rigorous Teaching and Learning model with utilizing informal arguments. The reason for choosing this method is to make this research more objective, complete and reliable.

The design of this study was quasi-experimental research using experimental group and control group, known as pretest-posttest control group design [9]. The reason for selecting quasi-experimental design is that intact classrooms are used and there are not random assignment. The data collected in this study are quantitative and qualitative data. Quantitative data were obtained from IMA test results, pre-test or post-test mathematical critical thinking (CT) ability. While the qualitative data were obtained from analysis of students works concerning the test of mathematical CT ability, observation and interview.

The data in this study are processed and analyzed descriptively and inferentially. Stages in descriptive data analysis to see the improvement that is by calculating the normalized gain value $\langle g \rangle$ from pretest and posttest mathematical CT, using the following formula [10]:

$$\langle g \rangle = \frac{S_f - S_i}{100 - S_i}$$

Where:
S_f = final test (post-test)
S_i = initial test (pre-test)

Table 1. N-gain of Level Criteria

Interval	Criteria
$\langle g \rangle < 0{,}3$	Low
$0{,}3 \leq \langle g \rangle < 0{,}7$	Medium
$\langle g \rangle \geq 0{,}7$	High

The research hypothesis were tested to understand whether using parametric statistical analysis or nonparametric statistical analysis. Statistic test which is used in study comprise t-test, t'-test, Levene test, Mann-Whitney U test, Kruskal-Wallis test, and Analyzing Estimated Marginal Means through graph.

3. Results and Discussion

Data obtained from both pre-test and post-test results and the normalized gain (N-gain) of mathematical CT capability, this data would be analyzed descriptively or inferentially. This data would be analyzed based on IMA and learning model (RTL and CI). To obtain a deeper picture of the mathematical CT's ability to improve observed aspects, the following explanation is presented descriptively or inferentially.

3.1Descriptive Data Analysis Improvement of Critical Thinking (CT) Ability Mathematical Student

Data analysis of mathematical CT ability were obtained from both pre-test and post-test results including mean, standard deviation and N-gain based on IMA criteria and learning model. The descriptive statistics of students' mathematical CT ability is presented in the following table.

Table 2. Descriptive statistics of students' mathematical CT ability

IMA	Stat.	RTL		CI		Total	
		$\langle g \rangle$	N	$\langle g \rangle$	N	$\langle g \rangle$	N
High	$\bar{x}$	0.81	8	0.79	7	0.80	15
	Std. Dev	0.09		0.74		0.08	
Medium	$\bar{x}$	0.61	26	0.51	27	0.56	53
	Std. Dev	0.11		0.12		0.12	
Low	$\bar{x}$	0.42	9	0.36	6	0.40	15
	Std. Dev	0.11		0.06		0.09	
Total	$\bar{x}$	0.61	43	0.54	40	0.58	83
	Std. Dev	0.16		0.16		0.17	

Description: N = Number of data; $\langle g \rangle$ = Normalized gains

Table 2 shows that in high IMA criteria, mean improvement of students' mathematical CT ability using RTL model with informal arguments is higher than that of the CI students. The normalized gain criteria of the RTL and CI models are included in the high criteria. Furthermore, in medium IMA criteria, the increase in mean of students' mathematical CT ability using RTL model with informal arguments is higher than that of students who received CI, although the criteria of increase in mean for both groups were included in the medium criteria. Whereas, in low medium criteria, the increase in mean of students' mathematical CT ability using the RTL model with informal arguments was higher than that of the CI students, and the criteria of the increase in mean for both groups were categorized as medium criteria. Table 2 generally shows that the increase in mean of students' mathematical CT ability using the RTL model with informal arguments was higher than that of CI students. In addition, the criteria of increase in mean for both groups were included in the medium criterion. To obtain clear description of the increase in mean of mathematical CT ability of RTL students and CI students, the following diagram is shown.

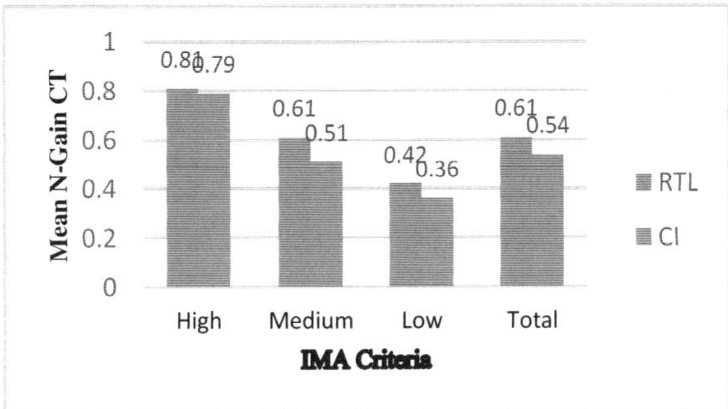

Figure 2. Comparison between increase in mean of student mathematical CT ability across learning model and IMA

IOP Conf. Series: Journal of Physics: Conf. Series **948** (2018) 012047 doi:10.1088/1742-6596/948/1/012047

3.2 Inferential Analysis (Mathematical CT Improvement)

Table 3. Mean Difference of Mathematical CT Ability Score based on IMA Categories

Ability	IMA	Learning model	Sig.	$\alpha = 0.05$	Hypothesis	Test
Mathematical CT	High	RTL = CI	0,293	>	H_0 be accepted	t-test
	Medium	RTL > CI	0,000	<	H_0 rejected	t-test
	Low	RTL = CI	0,100	>	H_0 be accepted	t-test
	Total	RTL > CI	0,000	<	H_0 rejected	ANOVA

Table 3 shows that for high and low category of IMAs there is no significant difference on increase in mean of mathematical CT ability of RTL students and CI students. Furthermore, from Table 3 it appears that the significant difference can be seen for medium IMA.

Table 4. Mean Difference of Mathematical CT Ability Score based on Learning Model

Ability	Learning model	Sig.	$\alpha = 0.05$	Hypothesis Rule	Test
Mathematical CT	RTL > CI	0,02	<	H_0 rejected	t-test

Table 4 shows that the null hypothesis is rejected. This means that there is a significant mean difference of mathematical CT ability scores of RTL students and CI students.

Table 5. Post-hoc Test Result of Mean Difference of N-gain of CT based on IMA category

Learning model	Test Statistics	IMA category	Mean Difference (I-J)	*Sig.*	Testing Rules
RTL >< CI	Bonferroni	Low >< Medium	-0,162	0,000	Different
		Low >< High	-0,405	0,000	Different
		Medium >< High	-0,243	0,000	Different

It can be seen from Table 5 that the ability of CT with IMA category for low and medium, low and high pairs, and medium and high based on the learning model identified different differences. Furthermore, from the table above it can be seen that for the low IMA group (I) and the IMA group being (J), the mean difference value (IJ) is negative, i.e. -0.162, this means that the average CT of mathematical students from the IMA group is better than the average IMA group is low with an average difference of 0.162. For the low IMA group (I) and high IMA groups (J), the mean difference value (IJ) is negative, i.e. -0.405, this means that the mean mathematical CT of students from the high IMA group is better than the average low IMA group with an average difference of 0.405. While in the IMA group being (I) and IMA group high (J), the mean difference value (IJ) is negative, i.e. -0.2243, this means that the average CT of mathematical students from the high IMA group is better than the average IMA group being with the average difference of 0.243.

The interactive graphs between learning models (RTL, and CI) and high, medium, and low category IMAs on improving students' mathematical CT skills are shown in the following figure:

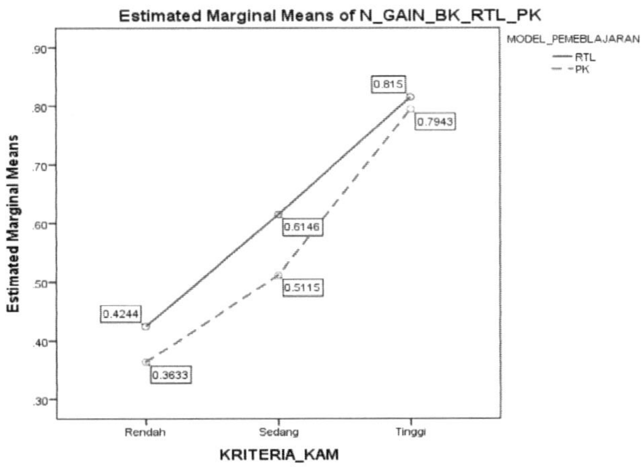

Figure 3. Interaction between Model (RTL, and CI) and IMA on Student Mathematical CT Ability

Based on Figure 4, it can be stated that graph of the average line of increase of mathematical CT ability of RTL students is above the average line increase of the mathematical ability of CI students. This indicates that for each level of the IMA category the RTL model has a greater influence on the improvement of students' mathematical CT ability compared to the CI model. It also appears that the average line graphs for both model groups (RTL and CI) did not change sequence i.e. high-IMA students were better than those with medium and low-IMA categories, as well as for medium IMA categories with low IMA. This indicates that the IMA category has a greater effect on improving mathematical CT ability. Based on the behavior of both graphs that have positive gradient shows that the learning model (RTL and CI) as well as IMA of high, medium and low categories independently give a real influence to the improvement of students' mathematical CT ability. Based on the distance of the graph it appears that for each IMA category tends to be similar and not mutually intersect. Thus, it can be concluded that there is no interaction effect between the model (RTL, and CI) and the IMA categories on improving students' mathematical CT ability.

Table 6. Average Mathematical CT Ability Indicators

No.	Indicators CT	SMI	Stat	Model					
				RTL			CI		
				Pre-test	Post-test	$\langle g \rangle$	Pre-test	Post-test	$\langle g \rangle$
1.	Recognizing Assumptions	14	$\bar{x}$	2.05	10.63	0.72	2.00	10.33	0.69
2.	Evaluating Arguments	14	$\bar{x}$	2.40	10.98	0.74	2.28	10.08	0.67
3.	Drawing Conclusions	21	$\bar{x}$	2.37	10.70	0.45	2.33	8.73	0.34

Table 6 shows that all indicators of students' mathematical critical thinking skills improved in real-time learning after being treated with RTL learning models utilizing both informal and conventional (CI) arguments. For the RTL learning model, the mean of the highest increase of students' mathematical critical thinking ability is on the indicator of evaluating arguments. The average increase of the value of $\langle g \rangle$ is 0.74 and is included in the high category. The high increase was obtained by the students on the indicator of recognizing assumptions with a value of 0.72. For the indicator of drawing conclusions, it shows that the increase in mean of students' mathematical critical thinking ability after the treatment of

RTL learning model is 0.45 and it includes in the medium category. Furthermore, for the CI learning model, the highest increase in mean of students' mathematical critical thinking ability was on the indicator of recognizing assumptions with the value of 0.69 and it includes in the medium category. For the indicator of evaluating arguments and drawing conclusions, the value of increase are 0.67 and 0.34 respectively. This suggests that strong prior knowledge acquisition, rich experience, and familiar with problems that require analysis, synthesis, and evaluation will affect students' mathematical critical thinking ability. The results of this study are consistent with Bloom's opinion [11] in which critical thinking relates to high-level cognitive thinking (analysis, synthesis and evaluation).

4. Conclusion

This study generally indicates that students who were exposed to RTL learning model outperformed students who were exposed to conventional instruction on math mathematical CT ability of in learning of Real Analysis. The increase of mathematical CT ability for both group of students was categorized as medium. In the learning of Real Analysis based on IMA (High, Medium, and Low), the increase in mean of the mathematical CT of RTL students is better than that of CI students. There is no interaction effect between Rigorous Teaching and Learning (RTL) model and conventional learning model (CI) with IMA (High, Medium, and Low) to the increase of students' mathematical CT ability scores. The highest increase in students' mathematical critical thinking ability on the RTL model was on the indicator of evaluate arguments and it felt into high category. The highest increase in students' mathematical critical thinking abilities in the CI model was on the indicator of recognize assumptions and it included in medium category. Therefore, to improve the critical mathematical thinking ability of students with IMA (high, medium and low) in Real Analysis course, the Rigorous Teaching and Learning (RTL) model that utilizes informal arguments could be applied as an alternative model of learning.

References

[1] Cotton K 1991 *Teaching Thinking Skills* [Online] Available at http://www.nwrel.org/scpd/sirs/6cu11.html [Retrieved, December 14, 2014]
[2] Watson G and Glaser E M 2009 *Watson-Glaser II Critical Thinking Appraisal, Technical Manual and User's Guide* (San Antonio, TX: Pearson)
[3] Watson G and Glaser E M 2012 *Watson-Glaser™ Critical Thinking Appraisal, Technical Manual and User's Guide* (San Antonio, TX: Pearson)
[4] Hart P, Natale L and Starr C 2010 Recognizing Rigorous and Engaging Teaching and Learning [Online] Available: http://billmcbride.pbworks.com/f/Recognizing+Rigorous+and+Engaging+Teaching+and+Learning.doc [Retrieved, September 22, 2014]
[5] Reyes N 2010 Transforming Our Teaching and Learning: The Role of Academic Rigor, Inquiry, and Higher Order Thinking *National Center for Teacher Education* (Maricopa County Community College District)
[6] Jackson 2010 *How To Plan Rigorous Instruction* (Virginia, USA: ASCD Alexandria)
[7] Zazkis D, Weber K and Mejia-Ramos J P 2014 Activities that mathematics majors use to bridge the gap between informal arguments and proofs. In P Liljedahl, C Nicol, S Oesterle, D Allan (Eds.) *Proceedings of the 38th Conference for the International Group of the Psychology of Mathematics Education* **4** 417-424 (Vancouver, Canada: PME)
[8] Creswell J W and Clark V L P 2007 *Designing and Conducting Mixed Methods research* (Thousand Oaks: Sage)
[9] Fraenkel J, Wallen N and Hyun H 2012 *How to Design and Evaluate Research in Education 8th* (New York: McGraw-Hill Companies, Inc.)
[10] Hake R R 1998 *Journal of American Association of Physics Teachers* **66** 1 pp 64
[11] Dehghani M, Sani H J, Pakmehr H, and Malekzadeh A 2011 *Procedia Social and Behavioral Sciences* **15** pp 2952

The influence of Missouri mathematics project on seventh grade students' mathematical understanding ability

S Rezeki[1,a), A A Setyawan[2,b) and S Amelia[3,c)

Department of Mathematics Education, Universitas Islam Riau[1,2,3)]

E-mail: sri_rezeki@edu.uir.ac.id[a] E-mail: andokoageng@edu.uir.ac.id[b]
E-mail: sindiamelia88@edu.uir.ac.id[c]

Abstract. Mathematical understanding ability is a primary goal of Indonesian national education goals. However, various sources has shown that Indonesian students' mathematical understanding ability is still relatively low. This study used quasi-experimental research design to examine the effectiveness of the application of Missouri Mathematics Project (MMP) on students' mathematical understanding ability. The participants of the study were seventh grade students in Pekanbaru, Riau Province, Indonesia. They were selected purposively and represented as high, medium, and low-quality schools. The result of this study indicated that there was a significant effect of MMP on the overall students' mathematical understanding ability and in all categories, except for low school level.

1. Introduction

The learning strategy is needed to support all of the competencies that is contained in the Curriculum 2013. The curriculum comprises of what should be learned by students, while learning is how the lesson can be grasped by them. Every learning process must be accordance with the purpose of learning such as to develop students' abilities. Therefore, all mathematics concepts should design to develop students' thinking ability. For example, the concept of arithmetic, geometry, and algebra are taught to develop students' logic and thinking ability [5].

The development of students' mathematical ability is very important in the learning process. The curriculum in Indonesia reveals the importance of developing students' mathematical abilities. One of the abilities that must be developed in learning is mathematical understanding ability [2]. Understanding ability is the lowest level in the cognitive aspects associated with the mastery or understanding. According to Skemp [2], the first ability to be developed is an instrumental ability, while the latter is a relational ability. The relational ability has a higher level than the instrumental understanding. Instrumental understanding and relational understanding need to be improved on mathematics learning. Previous studies indicate that Indonesian students' mathematical ability in terms of instrumental and relational understanding is still low Priatna [8].

There are several factors that cause low achievement in mathematics. Mullis's study [2] explained that 7th grade students in Indonesian who participated in the competition were very weak in solving non-routine problems, both in solving problems about facts and procedures.

Then the results of study conducted by Sri Rezeki et al [8] on the implementation of learning activities in mathematics at Junior High School in Pekanbaru, it was found that many students who were not interested in and not motivated to learn mathematics. It was obviously seen from the following students' activities in the classroom:

1. Students tend to be more passive in classroom, just sitting down to take note what was described by teacher.
2. Students did not ask during learning process even though they didn't understand.
3. Students did not do exercise.
4. Students were not interested in reviewing the previous knowledge that have been discussed.

In response to the above challenges, it should implement learning approaches that can motivate students to participate in learning so that their academic achievement, attitudes, and skills could be improved. Mathematics learning should be focused on developing students' ability so that they will construct and apply their own knowledge actively. Teachers' responsibilities are not limited to prepare the lesson, but teachers should understand students' characteristics to guide them in dealing with learning difficulties, especially in mathematics. To obtain an effective teaching learning process must be in accordance with the method to be used. This is supported by Ruseffendi who argued that in the process of learning mathematics there are ten factors that influence the success of children learn one of them is the model of learning presentation [6]. Not only method of learning also requires the creativity of teachers in presenting an effective learning. This is confirmed by Ruseffendi [2] who argued that one of the abilities that mathematics teachers should have in high school is to use various methods and teaching techniques during teaching.

Based on that case, it is necessary to apply a method that can improve students' mathematical understanding. Teaching model that can be implemented by teachers to improve it might vary, one of which is the Missouri Mathematic Project (MMP) model. Therefore, based on the above description, the study is focused on improving students' mathematical understanding through the model of Missouri Mathematic Project (MMP).

2. Methods

This research was a quasi-experimental research with nonequivalent control group design. The participants of the study was three junior high school students in Pekanbaru in the academic year 2015/2016 which were selected purposively. The selected school represented as high, medium, and low-quality schools.

3. Result and Discussion

The summary of descriptive statistics of students' mathematical understanding ability before and after treatment across school quality is presented in Table 1. From Table 1, it can be seen that not all description of mean value of mathematical understanding of experiment class student is better than control class. The facts in the table above show that at the lower school level, the ability of the two classes looks the same on the pretest and posttest, both from the mean, standard deviation, minimum value, and maximum value. The high school level is also almost balanced for both classes.

Table 1. Descriptive statistics of students' mathematical understanding ability across school quality

Descriptive Analysis	*Pretest*		*Posttest*	
	Experiment	**Control**	**Experiment**	**Control**
Low School Level				
Number of samples (n)	40	40	40	40
Minimum Value	25	25	41	44
Max Value	78	78	94	100
Average ($\bar{x}$)	45.90	41.73	77.50	70.70
Standard Deviation (s)	15.766	14.211	14.825	14.730
Middle School Level				
Number of samples (n)	42	42	42	42
Minimum Value	38	33	50	54
Max Value	83	88	100	92
Average ($\bar{x}$)	65.36	61.71	75.86	70.33
Standard Deviation (s)	12.024	13.793	12.830	10.902
High School Level				
Number of samples (n)	37	37	37	37
Minimum Value	19	26	22	30
Max Value	63	74	93	93
Average ($\bar{x}$)	42.27	44.62	69.81	62.62
Standard Deviation (s)	10.260	9.296	16.060	14.509

To ensure conclusion, this descriptive statistics will be followed by inferential statistics. The data analysis techniques performed in this study were similarity of two average tests for pretest values (2-tailed) and posttest value (1-tailed). To test the data hypothesis before to treatment, we first tested the homogeneity of variance to determine whether the experimental class and control class had homogeneous or non-homogeneous variances. The results are summarized in the Table 2.

Table 2. Homogeneity Test of Pre-test Score of Students' Mathematical Understanding Ability across classrooms and School Categories

School Level	Class	N	sig	Description	Conclusion
Low	Experiment	40	0.241	sig > 0.05	Homogeneous
	Control	40			
Medium	Experiment	42	0.144	sig > 0.05	Homogeneous
	Control	42			
High	Experiment	37	0.273	sig > 0.05	Homogeneous
	Control	37			

Based on the above table, it can be observed that the significance value of all school categories does not reject H0, the conclusion of the two groups at each level has a homogeneous variance. Because the experimental and control classes are homogeneous at each school level, then t-test is conducted to determine the ratio of prior knowledge before treatment between the experimental class and the control class. The results are summarized in Table 3.

Table 3. T-test of Pre-test Score of Students Mathematical Understanding Ability across groups and School Qualities

School Quality	Class	N	S	T_{count}	Sig (2-tailed)	Conclusion
Low	Experiment	40	16.422	1.608	0.116	H_0 is accepted
	Control	40				
Middle	Experiment	42	15.819	1.492	0.143	H_0 is accepted
	Control	42				
High	Experiment	37	13.538	1.056	0.298	H_0 is accepted
	Control	37				

Based on the above table, we get sig (2-tailed) > 0.05 for all school categories, so it is not reasonable to reject H0, it means there is no difference in the average of initial knowledge of the experimental class with the control class. Acceptance of H0 in pre-test data influence the next inferential test stages. The next step is mathematical understanding ability by school qualities based on MMP Learning (Experimental Class) and PK (Control Class) from post-test data. Beginning with the homogeneity test of variance data. The results are summarized in Table 4.

Table 4. Homogeneity Test of Post-test Scores of Students Mathematical Understanding Ability across groups and School Qualities

School Quality	Class	N	Sig	Description	Conclusion
Low	Experiment	40	0.706	sig > 0.05	Homogeneous
	Control	40			
Medium	Experiment	42	0.469	sig > 0.05	Homogeneous
	Control	42			
High	Experiment	37	0.460	sig > 0.05	Homogeneous
	Control	37			

From the above table, it can be observed that all posttest data of low, medium, or high-quality school have homogeneous variance. Therefore, testing the similarity of the two classes will continue on the t-test. The T-test was conducted to determine whether there was any effect of MMP model learning on mathematics comprehension ability at low, middle and high-quality school. In this study the T-test was calculated using the help of SSPS application version 16.0. The results are summarized in Table 5.

Table 5. T-test of Post-test scores of students Mathematical Understanding Ability across groups and School Qualities

School Quality	Class	N	S	T_{count}	Sig (2-tailed)	Conclusion
Low	Experiment	40	21.328	2.016	0.051	H_0 is accepted
	Control	40				
Medium	Experiment	42	13.890	2.577	0.014	H_0 is rejected
	Control	42				
High	Experiment	37	20.218	2.163	0.037	H_0 is rejected
	Control	37				

Based on the significance of the mathematical understanding ability of the experimental class and control class, for the high and medium-quality school accept H0, this means that there is an influence of students' mathematical understanding ability using MMP. While at low-quality school, there is no effect of mathematical ability of students who are exposed to MMP.

This case has been alluded by the descriptive data in Table 1. In low-quality school, the two classes are not much different, so it is natural that in the t-test this school quality data does not reject H0. While at other school quality, high-quality school is also almost not rejecting H0 and this has also been discussed descriptively in Table 1.

This finding is consistent with the research conducted by Anna Fauziah [3] in which the Missouri Mathematics Project (MMP) model had no significant effect on mathematical understanding. According to Fauziah this is possible because the structure of learning on the model of the Missouri Mathematics Project (MMP) is not much different from the usual mathematical teaching structure (SPM). This is in accordance with a theory proposed by Al Krismanto [1] who states that MMP is one of the structured model as well as with SPM.

Another possibility, there are still students who are not actively involved in learnint and get confused when working on students' worksheet, although at subsequent meetings they were faster in working on it. According to Slameto [7], students' activeness in the learning process is one of the factors that can increase learning outcomes. Without students' active participation in learning, learning process will not take place properly. In addition, for meetings 1 and 2, not all stages of the MMP model were implemented. This caused the project provided during the learning process was not performed well.

Nevertheless, researchers found that learning by using the Missouri Mathematics Project model (MMP) could lead students to work collaboratively with group so that it can facilitate students to understand and solve problems by discussing with their peers in group. If they do not understand the subject matter they can ask their group's friends and train students to account for their group's results. With this learning, students can ask and express their opinions without hesitation. This it will certainly have an impact on students' mathematics understanding skills. This is in line with Miftakhul Jannah et al. [4] who claimed that the Missouri Mathematics Project (MMP) learning model encourages students to cooperate, work on worksheets, and to help each other in order to deal with difficulties and exchange ideas. For students who have difficulty in asking teacher and difficulty in understanding the material being studied cooperative learning is very helpful.

4. Conclusion

Based on the results and discussion of the research, it can be concluded that there is significant effect of Missouri Mathematics Project (MMP) on mathematical understanding ability of seventh grade students from high and medium-quality school in Pekanbaru. However, there is significant effect of Missouri Mathematics Project (MMP) on mathematical understanding ability of seventh grade students from low-quality school.

References

[1] Al Krismanto 2003 *Technic, Model, and Strategies of Mathematic Learning* (Yogyakarta: National Education Department)

[2] Andoko Ageng 2014 Application of Learning Connecting Organizing Reflecting Extending Model to Improve Mathematical Understanding and Mathematical Connection of Senior High School Students *Journal of Mathematical Research and Learning (JPPM)* 7 **2**.

[3] Anna Fauziah and Sukasno 2015 Influence of Missouri Mathematic Project Model toward Mathematical Understanding and Mathematical Problem Solving Ability at SMAN 1 Lubuklinggau. *Journal of Science.* 4 **1** pp 18

[4] Miftakhul Jannah et al 2013 Application of Missouri Mathematics Project Model to Increase Mathematical Understanding and Student's Positive attitude. *Journal of Education.* 1 **1** pp 65

[5] Reziyustika, L 2012 *Improved Mathematical Understanding and Connection of Students Using an Open-ended Approach with Co-op Learning* (Thesis: Education University of Indonesia)

[6] Ruseffendi, H. E. T 2005 *Fundamentals of Education Research and Other Non-Exact Fields* (Bandung: Tarsito)

[7] Slameto 1995 *The Models of Junior High Mathematic Learning* (Jakarta: Rineka Cipta)

[8] Sri Rezeki, Andoko Ageng and Mefa Indriati 2015 *Development of Learning Tool with Learning Model of Missouri Mathematics Project at Grade VII of Junior High School* (Proceeding International MSCEIS 2015)

Mobile learning to improve mathematics teachers mathematical competencies

A Hendrayana[1, a)] and Wahyudin[2, b)]

[1]FKIP, Universitas Sultan Ageng Tirtayasa, Jl. Raya Jakarta KM 4 Pakupatan Serang
[2]FMIPA, Universitas Pendidikan Indonesia, Jl. Dr. Setiabudhi No. 229 Bandung

[a]E-mail: aanhendrayana@untirta.ac.id [b]E-mail: prof.wahyudin@gmail.com

Abstract. The role of teachers is crucial to the success of mathematics learning. One of the learning indicator is characterized by the students' improved mathematical proficiency. In order to increase that, it is necessary to improve the teacher's mathematical skills first. For that, it needs an innovative way to get teachers close to easily accessible learning resources through technology. The technology can facilitate teachers to access learning resources anytime and anywhere. The appropriate information technology is mobile learning. Innovations that can make teachers easy to access learning resources are mobile applications that can be accessed anytime and anywhere either online or offline. The research method was research development method. In preliminary analysis, subjects consist of teachers and lecturers in professional teacher education program. The results that the teachers ready to adopt mobile-learning for the improvement of their skills.

1. Introduction

The Indonesian governments' policy about teachers and lecturers is an acknowledgment that teachers are a profession and need to gain fair and solid respect and protection [1]. Therefore, the professional education of teachers should be oriented towards the establishment of academic competence that should be displayed in rhythm with disciplinary content and pedagogical content [2]. In addition, the professional education of teachers also pay attention to the dimensions of improving the professionalism of teachers, namely: support for management competence, empowerment strategies, supervision of the development and research of class action [3]. The hope is that teachers are able to cultivate a high level of motivation to teach and educate teachers [4], because the interest in the teaching profession and achievement motivation has a positive relationship to teaching skills [5].

In mathematics learning, the role of teachers is crucial to the success of learning. The success is characterized by improvement of students' mathematical proficiency [6]. The mathematical proficiency consists of: (1) conceptual understanding; (2) procedural fluency; (3) strategic competence; (4) adaptive reasoning; and (5) productive disposition [6]. Of course, in order to increase students' mathematical proficiency, it is necessary to improve the teacher's mathematical skills first. This is in line with Evans [7] and Samuelsson [8] that teachers 'skills in teaching mathematics can improve students' mathematical skills.

The results of the observations that have been done in the "Education and Training Profession Teachers" program from 2009 to 2016 and "Professional Teacher Education" program from 2013 to 2014, there are many mathematics teachers who do not understand some concepts and its application well. Likewise, there are many teachers who difficult to solve mathematics problems. For example, when teachers are asked about concept of area. Most of them only answer a formula of area, area is length × width. Furthermore, when teachers are asked a follow-up question, "Why is your argumentation about the formula of area?", so most of the teachers didn't know the reason. In addition, when teachers are given a problem to look for an area of triangle. Almost no teacher can explain how formula of triangle area is found. Whereas, if the teacher have good understanding of the concept and also have a good strategy, then they didn't get a difficult to solve it, as shown in Figure 1.

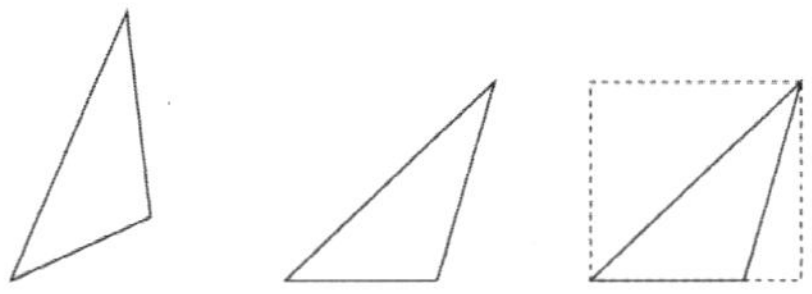

FIGURE 1. An example for finding area of triangle

One way to improve teachers' skills is exploiting technology. The technology can facilitate teachers to access learning resources anytime and anywhere. The technology information that appropriate with that situation is mobile learning. Other studies confirm that this technology can be applied in learning in Indonesia [9, 10]. The mobile learning technology can integrate learning resources, teachers, and experts. One content of mobile learning are video recorded of learning activities. This videos are stored on e-learning that integrated manner with other learning resources, such as documents or tasks. In addition, mathematics experts who are given space will make it easier for teachers to get solutions of the problems they faced.

Based on a study conducted by Wilson [11] shows that applications designed with in-depth thought can present many things, for example, can display dynamically linked recurring presentations, which may not be displayed by silent media such as books or whiteboards. Some applications of computer technology are its ability to display educational processes in an energetic, dynamic visual appearance, especially for mathematics. Mathematics learning with smart phone utilization will make it easier for teachers to deliver course material especially related to graphics, pictures, videos, and interactivity.

The development of mobile learning certainly requires a teacher preparedness study. It can not be arbitrarily created because it will cause a bad effect if the control from the developer can not be done. For that reason, a sensible solution is to pilot the mobile learning that is developed because the researcher has good control over the teacher.

2. Methods

This research is a development study that followed opinion of Borg & Gall [12]. There are four major stages of research, namely: preliminary analysis, product draft development, trial-testing, and effectiveness test. Analysis of preliminary study which examines the phenomenon of research subjects to receive innovation. Subjects consist of teachers and lecturers. Teachers who entangled are students from professional teacher education program. The lecturer is a lecturer who teaches the profession education program.

Development of product draft, namely development of mobile application learning in accordance with the condition of research subjects. Trial is a test activity of product draft on the subject, and conducted to the teacher or lecturer. Effectiveness test is the real test of learning when the product is ready to use. The subjects of the study were lecturers and teachers where teachers' professional education program in the western region of Java. Comparison of mathematical abilities is analyzed both quantitatively and qualitatively. Form of analysis by observing the mathematical ability of teachers before treatment, when treated, and after treatment. Duration of research is three years.

3. Result

We will describe about preparedness of teacher as student and lecturers for applying mobile learning. And also we will give an information about teachers' competency test. There are interesting findings of how the learning interactions use mobile learning.

3.1. Readiness of Learners

Survey results from 2012 to 2017 found that there is an increase in the readiness of teachers in adopting mobile learning technology. The readiness, among others: (1) learners have a strong desire to accept innovation (2) learners have minimal tools, (3) learners have enough prerequisite skills. However, until 2017 there are still obstacles in the teachers, namely (1) no mobile learning application for increasing mathematical competence, (2) learning culture through mobile learning has not been owned, (3) mobile device owned not yet have full readiness for mobile learning, (4) internet networks are sometimes unstable, especially if using economic package (5) quota limitations, (6) interactions can not be made because the network is not smooth yet, (7) learners are often not focused when there is product launching on the application.

3.2. Lecturers Readiness

Lecturers have a better readiness in each year, only the rate of lecturers in accepting the adoption of mobile learning technology is not as high as learners. The findings obtained are, there are two characteristics of lecturers. First, lecturers who are responsive to the development of mobile learning technology. Second, lecturers are not responsive in responding to the development of mobile technology, the age range of lecturers over 50 years. To test the mental readiness of lecturers to adopt mobile learning, 20 lecturers are given the task to conduct online-based learning, the result is good enough that evidenced by the fulfilment of the learner's assignment above 80%.

3.3. Culture of Learning

Prior to the creation of mobile application, researchers created a group in one of the existing mobile application. This is done so that we got a description of how the interaction of lecturers and learners. The first step is recording the teaching activity. Recorded results uploaded to the cloud, in this case YouTube. Links from YouTube are shared on the Learning Management System (LMS) on Google+.

In the LMS is made community "teacher professional education". In the community lecturers interact with learners through video sharing uploaded in the cloud. Another form of interaction is to answer and ask questions. The result, from 18 students who tested, there is only 5 teachers who active interaction using mobile application, 7 teachers who sometimes, and 6 teachers who never appeared in comment in application.

3.4. Competency Test

Learners are given a subject's ability test before being given a learning source based on mobile learning. Once given the test given the source of learning. Furthermore, after completion of learning resources learners are given a test of ability. The results can be seen in Figure 2.

In the interest of answering whether there is an increase in mathematical ability of learners, there are pre-test and post-test learning. Figure 2 shows each learner getting improved mathematical skills after using an application that has few characteristics with mobile learning.

Testing statistics is needed whether the increase in each learner is significant. The test taken is paired t-test, the consideration being that both groups of data are paired and both are normally distributed. The result of t-test yields a significance value of 0,000. This means the two groups of data differ significantly. The average difference of the two groups was 14,444. In other words, on average each prospective teacher increases his ability to about 15.

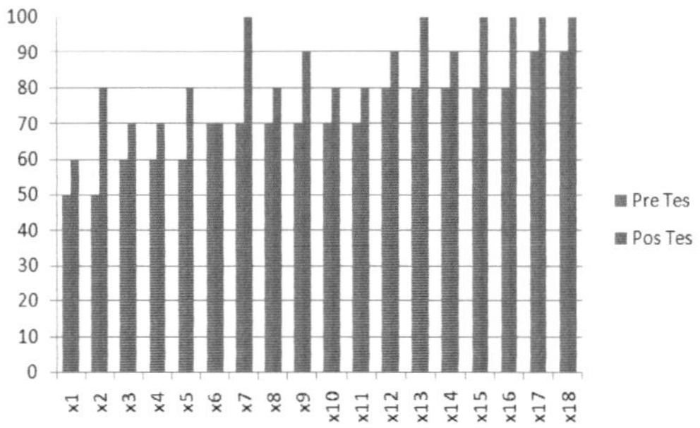

FIGURE 2. Comparison of Scores Before and After Learning

4. Discussion

Considering the findings that learners have the interest, tools, and sufficient skills to adopt mobile learning, it is indication that mobile learning application development will make the ability of teachers to be good. There have been many studies suggesting that interest contributes positively to learning outcomes [13]. Likewise, aids will be a support in learning success [14]. Likewise, the prerequisite skills may be contributed to the success of learners [15]. The absence of an application for learning purposes is just a matter of time, it does not happen if learners want to download paid apps in the play store or App Store as long as they accept to buy the application. One culture of Internet users in Indonesia is looking for free. Not yet formed the culture of learning using mobile learning is to stay conditioned so that learners have no choice other than the mobile learning media then it will be resolved. For example, the option of choosing an online motorcycle taxi has become the choice of many large urban communities in Indonesia because of such a condition. In countries that facilitate e-learning well, it's still a problem [16]. Meanwhile, the readiness of the device in the learner is usually the readiness of battery durability. This can be solved with new technology of battery charging. Simply by educating learners, so this is not a problem. With regard to internet quota, this should have a strong desire from individuals, institutions, and governments. Meaningful learners should calculate the risk of coming to class or college training that requires transportation and accommodation costs. Institutions should provide free hotspot facilities for the academic purpose in certain region to help students who are difficult to pay for internet connection. For the government, the activity of providing free hotspot services can be a policy to educate the citizens. The speed of internet connection in Indonesia at any time there is an increase. However, the increase is not significant enough for the interest of interactive learning. There is the fact that operators are launching a 4G facility that promises internet connectivity to be faster. At the time this study was conducted, 4G facilities have been tested and the results are quite satisfactory. However, until now the range of 4G areas is still limited. As a result, the use of 4G for research purposes can not be utilized properly.

Lecturers are creatures who are accustomed to receive knowledge, for it is not difficult to condition the lecturers to accept new things. Lecturers have three field of jobs, namely: teaching, research, and servicing for community. These three jobs make the lecturer as a creature that is demanded to keep changing in accordance with the condition of society.

The problem of learning culture is strongly related to technological developments and growing problems. When technology makes it easier for learners and becomes a problem for those who do not follow it, then automatically learning culture will change.

Increased competence in learners is good news, this means that mobile apps with not specifically designed have helped learning to improve competence. Especially if there is an application designed specifically for professional education teachers.

5. Conclusion

There are five conclusions generated in this study, namely: (1) In general, learners are ready to implement mobile-based learning, (2) learning habituation will change itself due to advances in technology and community culture (3) Internet connection should be a concern for policy holders so that the benefits of internet can be quickly absorbed, (4) the mathematics competence of learners has improved quite well, although there are many constraints of internet connection, and (5) mobile applications should be developed in accordance with the findings that have been obtained

References

[1] M.F. Gaffar 2007 *Jurnal Administrasi Pendidikan* 5 **1** pp 1-13
[2] N. Dantes 2007 *Jurnal Pendidikan dan pengajaran Undiksha* pp 452-471
[3] I.W. Santyasa 2008 *Jurnal Pendidikan dan pengajaran Undiksha* pp 473-494
[4] B. Nurhayati 2006 *Mimbar Pendidikan* **4** pp 64-70
[5] S. Hidayat 2008 *Jurnal pendidikan dan Kebudayaan* 14 **75** pp 1140-1154
[6] J. Kilpatrick, J. Swafford and B. Findell, B. 2001 *Adding it up: Helping children learn mathematics* pp 115-118.
[7] D. Evans 2007 *Journal of Learning Disabilities* **40** **5** pp 420-426
[8] J. Samuelsson 2010 *International Electronic Journal of Mathematics Education* 5 **2** pp 61-78
[9] I.K. Yusri and R. Goodwin 2013 International Journal of e-education, e-Business, e-Management, and e-Learning 3 **4** pp 293
[10] A. Mutaqin, I. Marethi and S. Syamsuri 2016 *Jurnal Cakrawala Pendidikan* **1** pp 134-141
[11] J. Wilson 2013 *Canadian Journal of Learning and Technology* 39 **1** pp 1-14
[12] W. R. Borg and M.D. Gall 2003 *Educational Reaserch: An Introduction* (Longman, Inc:London)
[13] M.C. Gilbert *Applying contemporary views of mathematical proficiency to the examination of the relationship of motivation and mathematics achievement.*
[14] J.T. Kinard and A. Kozulin 2008 *Rigorous Mathematical Thinking* (Cambridge University Press:Cambridge)
[15] A. Hendrayana 2015 *LPPM Untirta Report*
[16] K.M. Al-Said 2015 T*he Turkish Online Journal of Educational Technology* 14 **2** pp 167 – 181

IOP Conf. Series: Journal of Physics: Conf. Series **948** (2018) 012050 doi:10.1088/1742-6596/948/1/012050

Designing students' worksheet based on open-ended approach to foster students' creative thinking skills

S Romli[1], Abdurrahman[2] and B Riyadi[2]

[1]Postgraduate Physics Education Program, Faculty of Teacher Training and Education, Lampung University, Indonesia
[2]Physics Education Department, Faculty of Teacher Training and Education, Lampung University, Indonesia

[1]E-mail:sunaryoromli@gmail.com

Abstract. This study aimed at designing an open-ended worksheet to enhance students' creative thinking skills. The study was conducted at one private school in Bandar Lampung, Indonesia. The participants of the study were twenty students of tenth grade students and two physics teachers. This study used descriptive data. Data were collected by analyzing qualitative data, literature and focus group discussion to gain information about students' conceptions of physics in the context of open-ended learning. The result showed that students needed innovative learning resources in form of open ended-based worksheet which could give the students an opportunity to develop various solutions related to physics problem. Therefore, students' creative thinking skill could be improved.

1. Introduction

Current educational challenge is that each educational institution should equip their alumni with 21st century competencies. Based on Binkley [3], there are ten 21st century competencies, some of which are creative thinking skill, problem-solving skill, making decision, effective communication skill, multi-representational skill, information literacy, and ICT literacy (Information and Communication Technology). Besides, according to Manik [11] physics is a science that requires more understanding than memorization, then the key to success in learning Physics is the skill of using the three basic things of physics namely concepts, laws or principles, and theories. In the study of physics, the skill to comprehend the concept of physics is an absolute requirement in achieving the success of physics learning [6].

According to Yuliati [8] in the context of physics learning school is something that should be done by students, not done for students. In learning physics students do not only learn about the concept of physics but students are also required to be able to develop skill to ask questions, compile the explanations, and test the explanations and write them again by using their own language. Therefore, in the process of physics learning needs variety of strategies, methods, learning media and appropriate learning resources in order for students to learn actively so that students' thinking skill can be improved.

According to Sumantri [13], thinking is an activity to find the right knowledge. In physics learning is known as skill of mathematical thinking. Correlating to the concept of thinking, mathematical thinking can be seen as a way to improve understanding of physics by compiling data and information which are obtained by conducting research or assessing mathematical objects. The skill of mathematical thinking becomes one measure of the achievement of physics learning goals, especially high-order thinking skill, such as critical, creative, logical, analytical, and reflective thinking.

Creative thinking is the expression of an individual uniqueness in interaction with his/her environment. This creative expression reflects originality of an individual. From the unique personal expression can be expected an emergence of new ideas and innovative products and with characteristics such as: ability to guide themselves to a particular object, describe an idea, analyze the ideas and quality of personal work, and create a new idea in solving problems [13].

According to Manik [11] students are less able to solve more complex problems require process of creative thinking. This is also supported by presentation of merely closed ended problem in daily classrooms. Presented problems are designed in such a way until it has only unique answer or only one problem solving. A student can be said as creative person if able to solve a problem with a lot of ways, so that physics can develop student creative thinking way.

According to Munandar [14] creativity is skill to create new combination, based on data, information or elements that exist. Created results are not always new things, but can also be a combination of available things. In addition, Glegg declared that creativity as an action, idea, or product that replace something old into something new [4].

Limitations of physics learning media will certainly disrupt the process of physics learning in the classrooms, especially if the learning media used is only a textbook that only contains text and image that are less interesting. Based on [12], learning media is anything that can be used to transfer message from sender to receiver so it can stimulate attention and interest and willingness of learners so that learning process runs well in order to achieve the learning objectives effectively.

To improve students' creative thinking skill, one of learning resources that can be used is open ended-based worksheet. Worksheet is one of the instructional instruments to convince effectiveness of teaching and learning process [7]. Learning with an open-ended approach can develop idea, creativity and critical attitude of students. Students may develop variety of methods, ways, or approaches in obtaining the answers, making the process more important than results. This will build mindset of cohesiveness and conceptual understanding [2].

The result of interview with some of teachers and students in senior high school showed that the resource of physics learning used is still rare to use open-ended worksheet. Learning based on open ended can help students think creatively, and support understanding of the concept of Physics. An open-ended approach is a learning approach that presents a problem which could be solve by various method or solutions [1]. The open-ended approach gives students an opportunity to acquire the knowledge or experience, find, recognize, and solve problems in various ways. In open-ended approach, the problem is an open ended problem or incomplete problem. While the basics of disclosure are classified into three types, namely: (1) open process which has many correct solutions, (2) open final result which has many correct answers, and (3) the next way to improve is open in which when students have solved the problem, they can develop a new problem by changing the condition of the previous problem (original).

Learning that uses open-ended problems requires worksheet based on open-ended problems. According to Depdiknas [5] worksheet is a student worksheet is a sheet containing tasks that must be done by students, worksheet usually in form of instructions, steps to complete a task, a task that ordered in activity sheet must be clear about the basic competence that has to be achieved. It is also an instrument in which steps are given to students to learn [17]. The matters are activities doing experiment and observation, writing the result of observation, analyzing the data and drawing the conclusion [12]. The use of worksheet will give opportunity for the students to participate actively in learning. The teacher's role as a facilitator cannot be replaced by worksheet. When using worksheet, teachers are responsible for monitoring students' work during learning process.

Based on the description of the problem above, the researcher designed a learning resource in form of worksheet that could present attractive physics learning using open-ended approach. The author conducted a study aimed at designing an open-ended worksheet to develop students' creative thinking skill.

2. Methods

This research was conducted at one private school at Bandar Lampung involving twenty-four 10th grade students as respondents. Respondents were grouped into three groups based on their achievement: high, medium and low achiever students. FGD respondents involved can be seen in Table 1.

Table 1. Coding of Focus Group Respondents

Focus Group Respondents	Total's Respondents	Allocated Code
High Achiever Students	8	HA
Middle Achiever Students	8	MA
Low Achiever Students	8	LA

The stages of this research were formulating problems, collecting data, and designing worksheet of Physics learning. Data collection was done by analyzing these components: qualitative data obtained from interview, literature studies from various sources, and focus group discussion (FGD). FGD is a method and technique for collecting qualitative data in which a group of people discuss a particular problem or topic guided by a facilitator or moderator. FGD aims at exploring and obtaining a variety of information about a particular problem or topic that may be viewed differently by different explanations. The FGD notes were transcribed and the data were analyzed by content analysis. Subsequently, this resulted in the data being manifest code which identifies the underlying meaning of the topic [15]. In addition, as the informants in the focus groups are all nominated effective guidance teachers, those who have little training in counseling and guidance [16]. The present study demonstrates the possible usefulness of the focus group methodology in investigating perceptions of elements of effective school counseling and guidance [17]. According to [16], when the research experts examined using group focus and / or cognitive interviews as a first step or as an additional source of data for several research instruments.

In the FGD realization process, the questions in each group are open-ended questions and can freely express opinions on the question asked. The researcher engages creative ideas from students, the questions asked consist of four questions: (1) learning resources that help students in learning Physics, (2) the criteria of Physics learning worksheet to be more interesting, (3) learning resources that facilitate open problem solving skills (open ended problem), (4) student's expectations to learn the material of Physics easily. The results of the discussion will be collected and analyzed so as to produce the design of learning resources in form of Student Worksheet.

3. Result and Discussion

Students' opinions about learning resources that can help students in learning physics are: a set of book of physics formulas that have been prepared in detail, modules and student worksheets (that have been designed to help students solve physics problems, as well as practicum activities, and able to understand the concept of physics (HA). A book includes complete material in which there are complex exercises and a power point media with simple and easy-to-understand (MA). Physical learning resources are actually very much like worksheet physics in which facilitate the activity in developing creativity in solving the problem of physics (LA).

Students' opinions about the criteria of physics learning worksheet to be more interesting are: first can be shown from the interesting cover of worksheet, complete content that contains material summary, activity instruction and exercise question (HA). Worksheet uses language which is easy-to-understand, along with interesting picture and interactive and ICT-based (MA). Worksheet contains a complete summary of material, with pictures, sample questions, and exercise questions (LA) can be used independently.

Students' opinions about open-ended learning sources are: if a worksheet is designed based on open-ended problem, it will make students more motivated to find solution so it can develop their creativity in thinking (HA). Worksheet based on open-ended will increase students' interests and challenge students to solve the problem without worrying that the answer will be wrong because it led students to reach correct answer (MA). The worksheet is very scarce until it makes students become curious, because usually open-ended approach is being used in solving math problems. This opinion is supported by Manik [11] which clarifies that learning method that open-ended approach is more effective to improve skill of creative thinking than with conventional learning.

Furthermore, students' opinions about content of the worksheet is that open ended problem part of worksheet was placed at the beginning part of worksheet, so the creativity of students directly accelerated since the beginning of learning (HA). The open-ended content was placed in the middle of the worksheet so the students are not surprised, in the beginning part the students were given closed problem and in the final part is given the conclusion (MA). Meanwhile, according to the group (LA) open-ended content was placed at the end part after the students understand the material and closed problem solving in order to avoid the confusion when solving open ended problem.

Regarding students' opinions about students' expectations to learn physics material becomes easy was from the resource of learning, learning resource should be innovative and interesting. Worksheet should contain brief explanations, practice questions. In addition, the teachers should be more attractive and able to give an inspiration to students (HA). Learning resources used must be complete. In addition to deal with theory, physics lesson must also be enriched with practicum so that students can apply the theory that has been obtained in the class (MA). Teachers should give attractive and detailed materials, and provide practice questions that could develop students' creative thinking. The worksheet used should help students understand the material easily, use simple language with scientific explanation and discussion forum (LA).

Based on the four items of questions that were discussed in the FGD which is followed by 24 students as respondents, result of composition of open ended worksheet based present will be developed as shown in Table 2.

Table 2. Student Worksheet Content Structures

No	Content	Description
1	Summary of Material	Consists of some brief material on the worksheet introduction
2	Open ended problem	Includes various problems which are contained in Physics presented openly (open ended)
3	Constructing the idea	Finding and arranging the solutions of open ended questions so that students are able to improve their creative thinking skills
4	Exploration	After finding the idea and solution, then students discuss with their group to compare the solutions they make.
5	Presentation	Delivering exploration results in form of conclusions.

Based on the FGD result, the module design was found in 5 contents that should be there in open - ended worksheet. The structure of the worksheet consists of a summary part of material, open-ended questions, constructing ideas, exploration, and presentation sections. To improve students' creative thinking skill, the worksheet should be equipped with open-ended problem, construction of ideas, exploration and presentation part. According to Manik [11] application of open-ended approach is effective to improve students' creative thinking skill.

Open-ended problem part comprised of physics questions that are presented in open-ended form. Just as mathematics, Physics can be presented in open ended form because Physics really need mathematical calculation in solving physics questions. Manik [11] stated that the average absorption of students in experimental class that was using open-ended approach in the effective category with the percentage of 77.19% meanwhile in the control class was obtained the average student' absorption in the category less effective with a percentage of 40.54 %. Yuliana [18] stated that one of way to enhance students' creative thinking skill is by presenting open-ended tasks and questions.

The last three parts of the worksheet consisted of construction of ideas, exploration and presentation are important stages to develop students' creative thinking skills. Aspect of constructing ideas and exploration are the most desirable aspects, nearly 80% of students thought that the exploration via discussion lead students to express their opinions about achieved solution. This method is effective in training and developing creative thinking skills, because in discussion there is an exchange of opinions. In the process of exchanging opinions, students can express, consider, reject or accept, combine their own and other and determine its effectiveness. Also discussion also might lead students to draw conclusion that usually find the new solution. These are the things that ultimately develop creative thinking skill of students. At last, it grows the students' creative thinking. It is in line with Noer [9] who stated that students' creative thinking ability in open-ended learning environment are higher than that of in conventional environment.

The FGD results showed that the worksheet structure is in form of an open ended problem statement. With the presence of open-ended part, then the creativity of students in solving the problems of physics will be able to develop. Even, the open-ended approach both develop creative thinking and stimulate other effect such as improve students' confidence [9].

4. Conclusion

Based on the above explanation, the developed open ended-based worksheet was able to develop students' creative thinking skills containing aspect of material summary, open ended problem, exploration, and presentation. The design of worksheet was developed through focus group discussion (FGD) which involved tenth grade students. The FGD results were described and analysed qualitatively so as to develop an open ended-based worksheet that can develop students' creative thinking skill.

References

[1] Becker, J. & Shimada, S 1997 *The Open Ended Approach: A New proposal for Teaching Mathematics* (Reston, VA: NCTM)

[2] Suryapuspitarini, B. K 2011 *Meningkatkan Pemahaman Konsep Matematika dengan Menggunakan LKS Berbasis Open-ended Problem melalui Model Pembelajaran Kooperatif Tipe Team Assisted Individualization (TAI) pada Siswa Kelas VIII SMPN 1 Salaman Magelang.* (Skripsi: Universitas Negeri Yogyakarta).

[3] Binkley M, Erstad O, Herman J, Raizen S, Ripley M, Miller-Ricci M, Rumble M 2012 *Defining twenty-first century skills In Assessment and Teaching of 21st Century Skills* (Netherlands: Springer)

[4] Craft, A 2005 *Creativity in Schools Tensions and Dilemmas* (New York: Routledge)

[5] Depdiknas 2006 *Standar Isi Kurikulum Tingkat Satuan Pendidikan* (Jakarta:Departemen Pendidikan Nasional)

[6] Sakti, Indra dkk. 2012 Pengaruh Model Pembelajaran Langsung (Direct Instruction) Melalui Media Animasi Berbasis Macromedia Flash terhadap Minat Belajar dan Pemahaman Konsep Fisika Siswa di SMA Plus Negeri 7 Kota Bengkulu. *Jurnal Exacta* 10 **1** pp 4.

[7] Kaymakci, S. 2012 A Review of Studies on Worksheets in Turkey US-China *Education Review* 1 pp 57

[8] Lia Yuliati 2010 *Model-Model Pembelajaran Fisika*. (Malang: LP3 UM)

[9] Nenden Faridah, Isrok'atun dan Ani Nur Aeni 2016. Pendekatan open-ended untuk meningkatkan kemampuan berpikir kreatif matematis dan kepercayaan diri siswa. *Jurnal Pena Ilmiah* 1 **1** pp 1061

[10] Noer, Sri Hastuti 2011 Kemampuan berpikir kreatif matematis dan pembelajaran matematika berbasis masalah open-ended. *Jurnal Pendidikan Matematika* 5 **1**.

[11] Olvina Manik, Fakhruddin dan M. Noer 2015 *Efektivitas Pembelajaran Fisika Melalui Pendekatan Open-Ended Pada Materi Listrik Dinamis Terhadap Kemampuan Berpikir Kreatif Siswa di SMAN 10 Pekanbaru.* (Pekanbaru:Universitas Riau)

[12] Sukirman 2012 *Pengembangan Media Pembelajaran* (Yogyakarta: Pedagogi)

[13] Suyanto, S., Paidi., & Wilujeng, I. 2011. Lembar Kerja Siswa (LKS). *Makalah disampikan dalam acara Pembekalan guru daerah terluar, terluar, dan tertinggal di Akademi Angkatan Udara Yogyakarta tanggal 26 Nopember-6 Desember 2011.*

[14] Syaiful Sagala 2007 *Manajemen Strategik dalam Peningkatan Mutu Pendidikan* (Bandung:Alfabeta)

[15] Utami Munandar 1999 *Pengembangan Kreatifitas Anak Berbakat* (Jakarta:Rineka Cipta)

[16] Utami Munandar 2009 *Pengembangan Kreatifitas Anak Berbakat* (Jakarta:Rineka Cipta)

[17] Yildirim, Nagihan, K. U. R. T. Sevil, and A. Y. A. S. Alipaþa 2011 The Effect Of The Worksheets On Students' Achievement In Chemical Equilibrium *Journal of Turkish Science Education* 8 **3**

[18] Yuliana, R 2015 Pengembangan Soal Open-Ended Pada Pembelajaran Matematika Untuk Mengidentifikasi Kemampuan Berfikir Kreatif Siswa (Palembang: *Prosiding Seminar Nasional Pendidikan Matematika (Snaptika) 2015)*

IOP Conf. Series: Journal of Physics: Conf. Series **948** (2018) 012051 doi:10.1088/1742-6596/948/1/012051

Promoting middle school students' abstract-thinking ability through cognitive apprenticeship instruction in mathematics learning

B G P Yusepa[1], Y S Kusumah[2], B G Kartasasmita[3]

[1]Department of Mathematics Education, FKIP Universitas Pasundan Jl. Tamansari No. 6-8 Bandung West Java, Indonesia
[2]Univeritas Pendidikan Indonesia, Jl. Dr. Setiabudhi No. 229 Bandung, Indonesia
[3]Department of Mathematics Education, FKIP Universitas Pasundan Jl. Tamansari No. 6-8 Bandung West Java, Indonesia

[1]E-mail: pyusepa.fkip.pmat@unpas.ac.id [2]E-mail: yskusumaht@upi.edu
[3]E-mail: bana.kartasasmita@gmail.com

Abstract. The aim of this study is to get an in-depth understanding of students' abstract-thinking ability in mathematics learning. This study was an experimental research with pre-test and post-test control group design. The subject of this study was eighth-grade students from two junior high schools in Bandung. In each schools, two parallel groups were selected and assigned into control and experimental groups. The experimental group was exposed to Cognitive Apprenticeship Instruction (CAI) treatment, whereas the control group was exposed to conventional learning. The results showed that abstract-thinking ability of students in experimental group was better than that of those in control group in which it could be observed from the overall and school level. It could be concluded that CAI could be a good alternative learning model to enhance students' abstract-thinking ability.

1. Introduction

Abstract-thinking ability in mathematics learning is very important for students to develop. Mathematics abstraction is a process of the building a continuous mathematical knowledge from concrete to abstract. Abstraction is a fundamental process both in mathematics and in mathematics education [1]. The abstraction is a natural process with full awareness as mind focuses on several aspects of circumstances to make a decision and process of making such concrete situations by finding new meaning in order to establish interconnections within the different elements of the whole [2] [3]. In fact, junior high school students' abstract-thinking ability in Bandung is still relatively low [4] [5].

The indicators of abstract-thinking ability in this study are: (1) the ability to transform problems into symbolic form; (2) the ability to construct an equation; (3) the ability to state relationship between the concept of geometry shapes and linear equations; (4) the ability to make generalizations; and (5) the ability to make equations according to the given situation [4].

According to Yusepa, teachers have an important role in learning process and improving national education and become guidance. Furthermore, Yusepa states that in order for learning to be meaningful, the teacher should be able to teach students with varied models or learning approaches [6][7].

The effective of teacher characteristics are caring, assertive, modeling and enthusiastic, and has high expectations. Teacher's awareness in organizing the classroom or succeeding in all teaching sections

IOP Conf. Series: Journal of Physics: Conf. Series **948** (2018) 012051 doi:10.1088/1742-6596/948/1/012051

should be done earnestly and be able to accommodate students' learning styles. Assertiveness means the ability to improve students' responsibilities and apply those responsibilities to their actions. Modeling and enthusiasm in question are that teacher's belief in teaching and learning is communicated through modeling. By showing enthusiasm, teachers communicate their real-world interest in the topic [8].

Cognitive Apprenticeship Instruction (CAI) is one of learning models that is parallel to the characteristics of student-centered learning and emphasizes on improving students' abstract-thinking abilities. CAI involves real-world problems and contextual problems. To solve problems related to the real world, students need to communicate their ideas through mathematical abstraction. Research shows that if the CAI model is combined with other learning areas (content, method, sequence, and sociology) students will have the ability to apply what they have learned in other learning situations [9]. The components of CAI in this study are (1) modeling; (2) coaching; (3) scaffolding (4) articulation; (5) reflection; (6) exploration.

2. Methodology

This study is an experimental research with a pre-test and post-test control group design. The subject in this study are eighth-grade students from two different types of schools, one of which was categorized as a high-quality school and the rest was a moderate-quality school in Bandung. In each schools, two parallel groups were assigned into experimental and control groups. The experimental group was exposed to Cognitive Apprenticeship Instruction (CAI) treatment while the control group was exposed to conventional learning (PKv).

3. Result and Discussion

The prior ability of students' mathematical abstraction as a whole, based on school level and Mathematical Prior Knowledge, both students who received CAI and PKv did not differ significantly. For more details, the results of statistical data analysis of students' mathematical abstraction abilities are presented in Table below.

Table 1. Normality Test and Compsrison Test Results of Students' Abstract-Thinking Abilities (level of sig = 0,05)

Variable	Learning	Normality Test		Homogeneity		Comparison Test		
		Sig.	Dec.	Sig.	Dec.	Stat Test	Sig.	Conclusion
Prior Abstract-Thinking Abilities	CAI	0.000	Not Normal	-	-	Mann-Whitney	0.120	There is no difference of prior mathematical abstraction abilities
	PKv	0.000	Not Normal					
The Achievement of Abstract-Thinking Abilities	CAI	0.622	Normal	0.604	Homogeneous	t-test	0.000	There is a difference of the achievement mathematical abstraction abilities
	PKv	0.053	Normal					
	PKv	0.037	Not Normal					

Based on Table 1, it is found that: 1) there is no difference in abstract-thinking abilities between students in experimental group and control group, and 2) there is a difference in the achievement of abstract-thinking abilities between students who were exposed to CAI and students were exposed to PKv.

Based on data analysis, this study showed abstract-thinking abilities of students who were exposed to CAI is better than those who were exposed to PKv. It was analyzed by considering all students from high- and moderate-quality schools; and from the high-, moderate- and low- category of MPK. Overall,

IOP Conf. Series: Journal of Physics: Conf. Series **948** (2018) 012051 doi:10.1088/1742-6596/948/1/012051

the average of the achievement of students' abstract-thinking abilities who received CAI is higher than that of students who received PKv. The achievement of students' abstract-thinking abilities who received CAI is significantly better than that of students who received PKv. The average of students' achievement of abstract-thinking abilities at high- and moderate-quality school who received CAI is higher than students who received PKv. The achievement of students' abstract-thinking abilities at high- and moderate-quality school who received CAI is better than that of students who received PKv.

The results of this study also provide an illustration that there is no interaction between students who were exposed to CAI and PKv, school quality (high and middle) to the abstract-thinking abilities. Because based on school quality (high and moderate) who received CAI, the achievement is higher than the achievement of those who received PKv, it can be concluded that CAI can be applied to students who are at high-quality and moderate-quality schools. In other words, the learning factor does not depend on the school quality.

Other factors that influenced the differences in post-test scores of students' abstract-thinking abilities are also seen in each of the MPK categories (high, middle and low). The average of the achievement of students' mathematical abstraction in the high, middle and low category of MPK who received CAI is higher than that of the students who received PKv. The achievement of students' abstract-thinking abilities in the high and middle categories of MPK who received CAI is significantly better than that of the students who received PKv. Whereas, for the low category of MPK, the achievement of mathematical abstraction ability of students who received CAI is similar to that of students who received PKv. The difference in mathematical abstraction achievement of students who received CAI and students who received PKv can be seen from its average. The average achievement of students' abstract-thinking abilities in a higher category of MPK is higher than the middle and low category of MPK, the average of students' achievement of abstract-thinking abilities in middle category of MPK is higher than the low category of MPK. The higher the prior mathematical knowledge, the higher the average of the abstract-thinking abilities.

The results of this study also provide an illustration that there is no interaction between students who received CAI and PKv, MPK categories (high, middle and low) with students' abstract-thinking abilities. This can be seen from the results of performed statistical tests. Because based on MPK categories (high, middle, and low) who received CAI, the achievement is higher than that of students who received PKv, it can be concluded that CAI can be applied to students with high, moderate or low mathematical abilities. In other words, the learning factor does not depend on the category of MPK.

4. Conclusions

Based on the results of this study, it can be concluded that abstract-thinking abilities of students who were exposed to Cognitive Apprenticeship Instruction (CAI) is better than students who were exposed to conventional learning (PKv) viewed from overall, school quality, and prior mathematical knowledge (MPK). The abstract-thinking abilities of students who received CAI in the high, middle and low category of KAM is in middle qualification. Thus it can be concluded that CAI can be a good alternative learning model to enhancement students' mathematical abstraction abilities.

References

[1] M Mitchelmore, P White 2004 Abstraction In Mathematics and Mathematics Learning. *Proceedings of the 28th Conference of the International Group for the Psychology of Mathematics Education* pp 329

[2] E Gray, D Tall 2007 Abstraction as a Natural Process of Mental Compression *Mathematic Education Research Journal* **19** pp 23

[3] J Monahagan, M F Ozmantar 2007 A Dialectical Approach to the Formation of Mathematical Abstractions *Mathematics Education Research Journal* **19** pp 89

[4] B G P Yusepa 2016 Kemampuan Abstraksi Matematis Siswa Sekolah Menengah Pertama (SMP) Kelas VIII *Symmetry: Pasundan Journal of Research in Mathematics Learning and Education* **1** pp 54

[5] Tata 2005 *Peningkatan Kemampuan Pemodelan dan Abstraksi Matematis serta Motivasi Belajar Siswa Sekolah Menengah Pertama Melalui Pembelajaran Kontekstual Kolaboratif* (Disertasi SPS UPI Bandung)
[6] B G P Yusepa 2016c Keserupaan Pembelajaran Matematika Realistik dengan Pembelajaran Matematika Konstektual *Prosiding Seminar Nasional Pendidikan Matematika Universitas Siliwangi* pp 168
[7] B G P Yusepa 2016d Analisis Per-bandingan Kurikulum Pendidikan Indonesia dan Inggris untuk Meningkatkan Kompetensi Pe-dagogik dan Kompetensi Pro-fesional Guru Matematika. *Prosiding Seminar Nasional Matematika dan Pendidikan Matematika:Strategi Mengembangkan Kua-itas Pembelajaran Matematika Berbasis Riset Unswagati* pp 346
[8] A D Jacobsen, P Eggen, D Kauchak 2009 *Methods for Teaching: Promoting Student Learning in K-12 Classrooms* (Yogyakarta: Pustaka Pelajar)
[9] A Collins 2005 Cognitive Apprenticeship. *The Cambridge Handbook of the Learning Sciences* **4** pp 47

IOP Conf. Series: Journal of Physics: Conf. Series **948** (2018) 012052 doi:10.1088/1742-6596/948/1/012052

Developing CORE model-based worksheet with recitation task to facilitate students' mathematical communication skills in linear algebra course

Risnawati[1,a), S Khairinnisa[1, b)] and A H Darwis[1, c)]

[1]State Islamic University of Sultan Syarif Kasim Riau, Indonesia

[a]E-mail: rwati04@gmail.com [b]E-mail: septika.khairinnisa@gmail.com
[c]E-mail: alfina.hadiarti.d@gmail.com

Abstract. The purpose of this study was to develop a CORE model-based worksheet with recitation task that were valid and practical and could facilitate students' communication skills in Linear Algebra course. This study was conducted in mathematics education department of one public university in Riau, Indonesia. Participants of the study were media and subject matter experts as validators as well as students from mathematics education department. The objects of this study are students' worksheet and students' mathematical communication skills. The results of study showed that: (1) based on validation of the experts, the developed students' worksheet was valid and could be applied for students in Linear Algebra courses; (2) based on the group trial, the practicality percentage was 92.14% in small group and 90.19% in large group, so the worksheet was very practical and could attract students to learn; and (3) based on the post test, the average percentage of ideals was 87.83%. In addition, the results showed that the students' worksheet was able to facilitate students' mathematical communication skills in linear algebra course.

1. Introduction

Mathematics is a science whose concepts are arranged continuously, so to learn a concept, one has to master the previous concept called pre-requisite. This also applies to linear algebra courses. Linear algebra is one of the foundations of knowledge in higher mathematics courses such as numerical methods, differential, and other equations. Basic competencies to be achieved after studying linear algebra are mastery of the matrix and its operations, the system of linear equations and solutions, and their vectors and transformations. Kaltz [1] states that *"Algebra: Gateway to a technology Future"*. This shows that algebra has an important position in the development of science and technology.

Basically, algebra is not entirely new material for students, because it has previously been studied at elementary and secondary levels. This is in line with what has been stated by Mason as quoted by Becker and Rivera [2] that every students who starts school has demonstrated the ability to generalize and abstract certain cases and those things are the basic of algebra.

Linear algebra lectures are inseparable from the existence of mathematical communication activities. The communication relates to the delivery of messages/information from one party to another party. Lecturers cannot directly know the contents of the mind or student opinions without any communication. There are two reasons why learning mathematics focuses on communication, namely: (1) mathematics is essentially a language; and (2) mathematics and mathematics learning are, at heart, social activities [3].

Mathematics is a universal language, so by using mathematical form of communication, ideas and opinions possessed by a person can be conveyed clearly, precisely and briefly without having to be hampered by various local languages. In addition, mathematics is a form of social activity which involves various parties such as students and lecturers.

Ability of mathematical communication in the classroom will encourage students to respond based on various interpretations of the problems given by lecturers. Responses submitted by students to be a reference assessment by lecturers based on indicators of mathematical ability. Indicators that can be used to assess the ability of mathematical communication, to wit: (1) stating a situation, image, diagram, or real object into language, symbol, idea, or mathematical model; (2) explaining ideas, situations, and mathematical relations both orally and in writing; (3) listening, discussing, and writing about mathematics; (4) reading with the understanding of a written mathematical representation; and (5) revealing a mathematical description or paragraph in its own language [4]. Mathematical communication is generally divided into two, namely communication indicators in writing and oral. The indicators used in this research are written communication, namely (1) declaring a situation, image, diagram, or real object into language, symbol, idea, or mathematical model; (2) explaining the idea, situation, and mathematical relations in writing; and (3) revealing a mathematical description or paragraph in its own language.

The importance of mathematical communication skills is not yet in line with its implementation in classrooms. Although algebra has been introduced since elementary schools, there are still some students who still have difficulty in following linear algebra lectures. The problems that often occur in the class of linear algebra are, for example giving a story problem in the form of linear equations, asking to create a graph in accordance with the conditions given, and looking for possible settlement. Some students still make mistakes in the use of notation, for example, the notation of a matrix[ ]written as | |, which is the symbol of the determinant of matrix. In addition, there is also the problem of writing matrices and operations, such as writing a trace matrix as A^T which means the transpose matrix and errors in the use of problem solving algorithms. On the story problem, most students have difficulty in declaring the story into mathematical symbols, such as making a mathematical model. Other possible problem is students' difficulties in preparing argument, for example on problems that ask them to write reason in systematic and detailed problem solving. The problem that is seen during the lecture process in the classroom and also during the quiz and mid or end of the semester exam. The score of the students' linear algebra test is still largely low on the story and picture problems, whereas the material tested has been studied previously. These problems indicate the still low mathematical communication ability of students in linear algebra courses.

The problems found in lectures of linear algebra are such as less effective and efficient teaching materials. The material presented in textbooks is largely unstructured according to the curriculum used. In addition, the material contained in the book to be taught and the problems in the book sometimes less appropriate with the level of students' ability. Such teaching materials sometimes make students to have difficulty in seeking reference sources other than those described by the lecturer during the course. Lack of reading resources, references, and examples of problems might result in class conditions that tend to be passive. The students usually only focus on finding the results of a number of problems given by the lecturer. Students are less familiar with different questions. Therefore, various ways to communicate mathematically are less likely to be facilitated. The effort to overcome these things is to develop an appropriate teaching material. The intended teaching materials are those can facilitate students to develop students' mathematical communication skills. One of the teaching materials is students'worksheet.

The use of students' worksheet is to help students shape their own understanding and communicate every ideas in the form of numbers, pictures, graphics, diagrams, words as well as other forms of written communication individually. Students' worksheet developed are teaching materials that contain short material useful as a reference for students to understand the material, but not presented thoroughly. In addition, the students' worksheet also contains various presentation of varied issues, so that students could be facilitated to communicate mathematically in linear algebra lectures. The variation in the given

IOP Conf. Series: Journal of Physics: Conf. Series **948** (2018) 012052 doi:10.1088/1742-6596/948/1/012052

problem provides an opportunity for students from various levels of mathematical ability to be able to communicate mathematically.

The use of students' worksheet in the linear algebra classes should be supported by classroom conditions that are more conducive such as in group-based learning model i.e. the CORE model. The application of the CORE model leads students to perform four frameworks proposed by Miller and Calfee, namely connect, organize, reflect, and extend [5]. Based on these four frameworks, it is found that the stage in the CORE model is connecting old knowledge with knowledge to be learned, organizing knowledge based on relevant and reasonable sources, carrying out reflection on various issues that have been discussed, and deepening and expanding the concept by applying these concepts to further mathematical problems.

CORE model requires students to learn and comprehend the concept thoroughly and continuously so that the lecturer should get students to make preparation before starting the lesson such as giving assignment. The task given before presenting new knowledge is called a recitation task. According to Pasaribu [6], a recitation task is a form of task that is not only aimed at memorizing but also contemplating its contents, representing it in own words, understanding, and interpretation. The assignment aims at reviewing new lessons, memorizing knowledge given, solving problems, collecting materials, giving exercises, assisting lecturers in enhancing more lecturers and in assessing students' readiness and mastery of a lecture material. Therefore, the task should not only be given after lecture, but also can be given beforehand. Giving task before commencing lesson is still rare conducted by most lecturer [7]. It is revealed that the tasks assigned by teachers tend to be carried out at the end of the lesson so that in the process of learning mathematics, most mathematics teachers always use lecture and expository methods [7]. This is not only happening at school but also on classes in higher education. Most lecturers tend to give tasks after teaching a material. Therefore, researchers designed a task given before a lecture so that students have prepared to learn a new material in the next lecture.

Until now, researchers have not found the CORE model-based student' worksheets and recitation assignments aimed at providing student learning aids in linear algebra courses so that students' mathematical communication skills are facilitated effectively and efficiently. Therefore, researchers are interested in conducting and developing the study. The research questions are as follows:

1. How to develop a CORE-based students' worksheet model and provide recitation tasks to facilitate students' mathematical communication skills?
2. What is the student's mathematical communication skill after attending CORE model-based students' worksheet and providing recitation task?

2. Methodology

This research is a sort of development research on linear algebra lecture. This research was conducted to thirty students of mathematics education department of one public university at Riau in the even semester of the academic year 2015/2016. The development procedure of this model uses a model developed by Lee and Owens, namely ADDIE that uses five stages of development i.e. analysis, design, development, implementation, and evaluation [8]. Analysis stage consists of two steps, to wit: needs assessment and front-end analysis (preliminary analysis). The design stage includes activities determining: (1) the schedule, (2) the work team, (3) the design specification, (4) the creation of the material structure, and (5) the evaluation. At this stage, evaluation is carried out by media and material experts. The next stage is development stage which is a trial stage before implementation. The trial consisted of a small group trial consisting of seven people and a large group trial consisting of thirty people. The next stage is implementation which is the application of student worksheets developed. At this stage of implementation, researchers will conduct field trials, which aims to determine whether the resulting product has a good effectiveness in learning. The final step is the evaluation. It is conducted to see the effectiveness of students' worksheet that has been developed.

3. Results and Discussion

The development of a CORE-based students' worksheet model with recitation assignment takes three months (March-June) of 2016. Components contained in students' worksheet developed in linear algebra lectures on matrix materials are as follows:

a. Cover consisting of material title, author's name, course and so on.
b. Preface and table of contents.
c. Lecture materials consisting of pre-requisites for students' worksheet, students' worksheet instructions, material summaries and lecture evaluation questions corresponding to the four stages of the CORE model, as well as recitation tasks.
d. References, about the author.

The students' worksheet developed was validated by media and material experts. Result of the validation showed that the students' worksheet made in the category is very valid. However, there are some things that need to be revised: (1) the use of language should be more practical and concise; (2) the problem should be varied from the easiest to the most difficult; and (3) the material distribution for one lecture should not be too much.

Based on the experts' suggestions, a revision was made to the students' worksheet. After revision, a small group of seven students was tested. The result of the response obtained the average percentage is 92.14% so it can be categorized very practical. Then tested in large groups consisting of thirty students and obtained the average percentage is 90.19% and is categorized very practical. After conducting the two trials, it might be concluded that the students' worksheet that has been made was good. The next stage is implementation stage conducted to see the effectiveness of the lectures. The students' worksheet was used and implemented using the CORE model along with the assignment of recitation tasks. The recitation task was given at the end of each students' worksheet at each meeting as a direction for the next meeting.

The students' activity sheet filled by the observer showed that the students were actively involved in learning process using the developed students' worksheet. Students were able to work on the worksheet in accordance with the guidance provided, address problems correctly, and inquire and maintain actively the results of work either with group members or with other groups. The lecture process using this developed worksheet led students to interact actively with other group member in an effort to understand the material and solve the various mathematical problems given. Reflection and conclusions on the materials of each student might be different even in the same group. The reason is that the worksheet provided an independent evaluation sheet to assess the extent to which students' mathematical ability could be obtained. After that, each group presented the solutions to the existing problems. At the end of the lecture stage, the quiz was given for students to be solved individually.

4. Conclusions

Student activity in linear algebra lectures uses the worksheet developed in line with the score of the final evaluation result (post-test). The result of post-test data analysis of students' mathematical communication skills showed that the average of ideality percentage was 87.83% (high category). The results of this assessment show that lectures of linear algebra using CORE model based student's worksheet with recitation tasks were able to facilitate students' mathematical communication skills. The result of post-test showed that there were minimal mistakes of using symbols / mathematical terms, most of the students have been able to read the table/graph / diagram according to the command in the given problem. Students were getting used to doing mathematical modeling although there were still some errors in the completion algorithm. However, differences of mathematical communication skills between students who use and who do not use students' worksheet-based model CORE along with recitation tasks need to be investigated in further research using quasi-experimental research.

IOP Conf. Series: Journal of Physics: Conf. Series **948** (2018) 012052 doi:10.1088/1742-6596/948/1/012052

References

1. Kaltz. V, J 2007 *Algebra: Gateway to a Technological Future* (University of the District of Columbia:Columbia)
2. Becker, J.R. and F.D. Rivera 2008 Generalization in Algebra: The Foundation of Algebraic Thinkingand Reasoning Across Grades *ZDM Mathematics Education* **40** 1
3. Baroody, AJ. and Niskayuna 1993 *Problem Solving, Reasoning and Communicating, K-8. Helping Children Think Mathematically* (Merril Publishing Company: New York)
4. Sumarmo, Utari 2010 *Berpikir dan Disposisi Matematik: Apa, Mengapa, dan Bagaimana Dikembangkan Pada Siswa* (FPMIPA UPI:Bandung)
5. Miller, R.G. and Calfee, R.C 2004 Making Thinking Visible: A Method to Encourage Science Writing in the Upper Elementary Grades *Science and Children* 42 **3** pp 20
6. Pasaribu, I.L 1999 *Didaktik dan Metodik* (Tarsito:Bandung)
7. Wahyudin 1986 *Kemampuan Guru Matematika, Calon Guru Matematika, dan Mahasiswa Dalam Mata Pelajaran Matematika. Studi Terhadap Tingkat Penguasaan Guru Matematika, Calon Guru Matematika, dan Mahasiswa dalam Mata Pelajaran Matematika, serta Kemampuan Mengajar Para Guru Matematika* (Thesis Universitas Pendidikan Indonesia: Bandung)

IOP Conf. Series: Journal of Physics: Conf. Series **948** (2018) 012053 doi:10.1088/1742-6596/948/1/012053

Indicators that influence prospective mathematics teachers representational and reasoning abilities

Darta[1,a] and J Saputra[1,b]

[1]Universitas Pasundan, Jl. Tamansari No 6-8 Bandung

[a]Email: darta_pmat@unpas.ac.id [b]E-mail: jusepsaputrapmat@unpas.ac.id

Abstract. Representational and mathematical reasoning ability are very important ability as basic in mathematics learning process. The 2013 curriculum suggests that the use of a scientific approach emphasizes higher order thinking skills. Therefore, a scientific approach is required in mathematics learning to improve ability of representation and mathematical reasoning. The objectives of this research are: (1) to analyze representational and reasoning abilities, (2) to analyze indicators affecting the ability of representation and mathematical reasoning, (3) to analyze scientific approaches that can improve the ability of representation and mathematical reasoning. The subject of this research is the students of mathematics prospective teachers in the first semester at Private Higher Education of Bandung City. The research method of this research was descriptive analysis. The research data were collected using reasoning and representation tests on sixty-one students. Data processing was done by descriptive analysis specified based on the indicators of representation ability and mathematical reasoning that influenced it. The results of this first-year study showed that students still had many weaknesses in reasoning and mathematical representation that were influenced by the ability to understand the indicators of both capabilities. After observing the results of the first-year research, then in the second and third year, the development of teaching materials with a scientific approach in accordance with the needs of prospective students was planned.

1. Introduction

The ability of mathematical representation is one of the standard processes in learning mathematics that increasingly gets special attention. NCTM [11] described this ability as the means in which people can understand and use these ideas. In addition to the ability of mathematical representation, mathematical reasoning is an important component that are necessary to be developed in mathematics learning process.

PISA 2012 results showed that Indonesian students are poor at reasoning ability. From the combined results of math, science and reading tests, Indonesia ranks 64th out of 65 participating countries [1]. According to Pranoto [1] this results may indicate that: 1) students are not accustomed to solving non-routine problems; 2) students are weak in modeling real situations into mathematics and interpreting mathematical situations to real situations; and 3) students comprehension and analysis are still weak. Actually, ability to solve non-routine problems and communicate is more required and more complex at this time.

Japan, China, Singapore and South Korea are countries at the top list of PISA and TIMMS result. It might be due to mathematics learning in these countries put greater emphasis on reasoning and problem solving. Therefore, they were able to produce high-achiever students in mathematics tests. Learning to solve problem essentially is about learning to think or learning to reason. Thinking and reasoning are used to apply obtained knowledge to solve unfamiliar problems [9].

The Director General of Teachers and Education Personnel (GTK) reported that based on the results of Teacher Competency Test (UKG) administered at the end of 2015, the average scores was 53.02. The competencies tested are professional competence and pedagogic competency. Based on the data, it is still necessary to improve the quality of teachers in Indonesia.

The improvement of the mathematical representation and reasoning ability of the prospective teachers is a necessity as it might influence students' mathematical ability. This study aims at: (1) analyzing prospective mathematics teachers' representational and reasoning ability; (2) analyzing factors affecting prospective mathematics teachers' representational and reasoning ability, (3) analyzing alternative of the scientific approach to improve prospective mathematics teachers' representational and reasoning ability. In accordance with prior studies, students with high level of comprehension, mastery of material, and problem solving ability of students will determine high level of representational and reasoning ability [13].

Reasoning was described by Copi [6] as a special kind of thinking and drawing conclusions based on premises. It means that reasoning is a thinking process to draw conclusions or construct a new statement by referring to true statements. According to Romberg and Chair [2], the indicators of mathematical reasoning ability are: (1) drawing logical conclusions; (2) providing an explanation using models, facts, traits, and relationships; (3) estimating answers and solutions; (4) using patterns and relationships to analyze mathematical situations, drawing analogies and generalizations; (5) constructing and testing conjectures; (6) giving a counterexamples; (7) following the rules of inference; (8) checking the validity of arguments, compiling valid arguments; and (9) establishing direct, indirect and indirect use of mathematical induction

Darta [7] suggests that representations are classified into: 1) visual representations (e.g. images, charts, or tables); 2) symbolic representations (e.g. mathematical statement, mathematical notation, numerical / algebraic symbol); and 3) verbal representations (e.g. written text or words). The ability of mathematical representation is the ability to express mathematical ideas (e.g. problems, statements, solutions, definitions, etc.) into certain form: 1) drawings, graphs, or tables; 2) mathematical notation, numerical/algebraic symbols; and 3) written text/words, as an interpretation of his thoughts.

In terms of reasoning ability and mathematical representation, researchers conducted interviews with prospective mathematics teachers' attending algebra and trigonometry courses over four years. The courses included difficult subjects required a high level of reasoning, difficult to understand, it is difficult to communicate material because it contains complex concepts. These problems could be observed through the following table.

Table 1. Average Scores of Algebra and Trigonometry Achieved by Mathematics Education Students across Academic Year

Academic year	2012/ 2013	2013/ 2014	2014/ 2015	2015/ 2016
Final Scores	3.1	2.9	2.9	2.9

(Source: Student Data Education Mathematics Study Program of FKIP)

Table 1 shows that the average scores of Algebra and Trigonometry courses achieved by students decreases annually. In general, the value is still under expectations, so it is necessary to make an effort to improve the attained scores thus it might satisfy students, lecturers, and related institutions' needs.

In addition, research conducted by Darta [8] indicated that final year students' self-efficacy are still low which is still below 7 (mean = 6.76) with a scale of 1-9. This situation encourages researchers to conduct further research in order to obtain more comprehensive and in-depth answers about representation and reasoning ability as these abilities contribute positively to self-efficacy.

2. Methods

This study used Research and Development considering four-D model proposed by Thiagarajan [4] with multiphase design through four stages: Define, Design, Develop, and Disseminate. This article describes the results of the first year of research that is defining ability of representation and reasoning of prospective mathematics teacher students as the starting point of the development of the next year. From this analysis, we get a description of facts, hopes, and alternatives to solve basic problems, which facilitate the determination or selection of teaching materials that will be developed.

Analysis of the study of the characteristics of prospective teachers is to get a description of the characteristics of the level of ability of representation and mathematical reasoning based on the Student's Initial Ability (SIA) of each capability indicators under study. Conceptual analysis is conducted to identify the basic concepts of Algebra and Trigonometry to be taught, to compile them in the form of hierarchy, and to break down individual concepts into critical and irrelevant terms. The next analysis is to determine the factors that affect the ability of representation and generalization based on the indicators that make up the two capabilities.

3. Results and Discussion

The results of SIA data were analyzed to determine students' reasoning and mathematical representational ability before the study. The descriptive statistics is presented in Table 2

Tabel 2. Descriptive Statistics

	N	Minimum	Maximum	Mean	Std. Deviation
Reasoning test score	61	.57	3.66	2.3172	.61475
Representation test score	61	.00	3.45	1.9382	.72562

Based on Table 2, the number of participants of the study is 61 students. The scores of Achievement Index (AI) scale range from 0.00 to 4.00. The minimum score of reasoning and representation tests is 0.57 and 0.00, respectively. It indicates that there are several students whose representational abilities is low. While the maximum scores of reasoning and representational ability tests are 3.66 and 3.45 respectively. It indicates that there are several students whose reasoning and representational ability is high. The average of reasoning and representational ability tests is 2.32 and 1.94 respectively. It if indicates that students' reasoning and mathematical representational ability are still low. Furthermore, if criteria of the ability of early mathematics (AEM) is used, the number of students with the category of high, moderate, and low amounted to 4 people, 47 people, and 10 people respectively. Its descriptive statistics of test score across AEM is shown in Table 3.

Table 3. Descriptive Statistics of Test Score across AEM

Test Score	AEM		Statistic
Reasoning score	High	Mean	3.2575
	Moderate	Mean	2.4264
	Low	Mean	1.428
Representation score	High	Mean	2.8625
	Moderate	Mean	2.067
	Low	Mean	0.963

Based on Table 3, the mean of reasoning ability tests from high, moderate, and low achiever are 3.26, 2.43, and 1.43, respectively. While the average score of representation skills of high, moderate, and low achiever respectively are 2.86, 2.07, and 0.96. It might indicate that high, moderate, and low achiever had better in prior mathematical reasoning abilities than in prior mathematical representation ability. It might be seen at the percentage of representation abilities of each indicators shown in Figure 1.

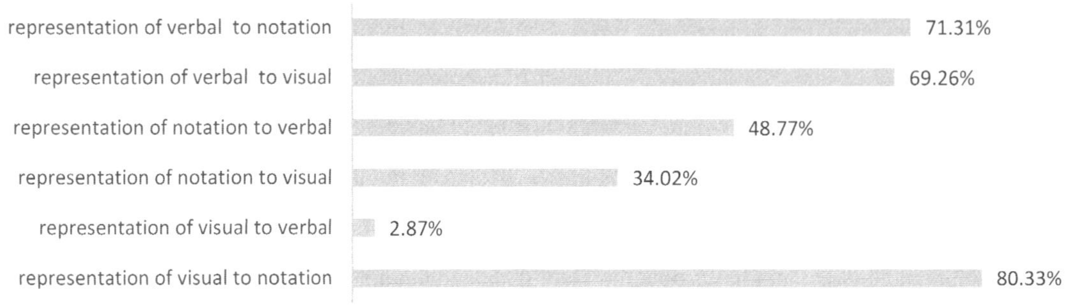

Figure 1. Percentage Ability of Student Mathematical Representation

Finding of this initial study also revealed the fact that students' mathematical reasoning ability is still relatively low. This can be seen at the graph of the percentage of reasoning ability of each indicators in Figure 2

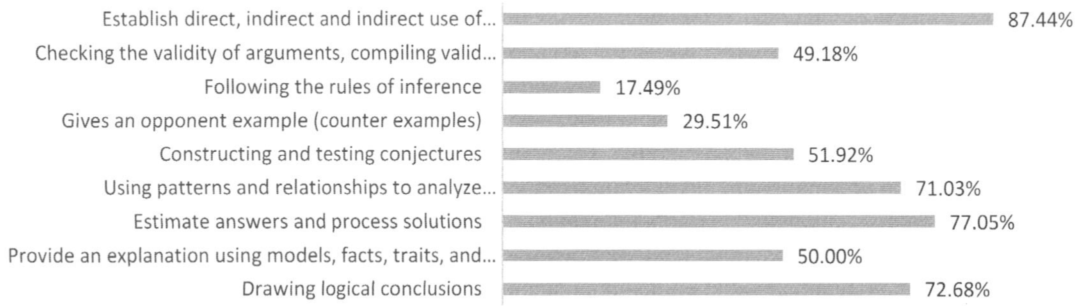

Figure 2. Percentage of Student Mathematical Reasoning Ability

Figures 1 and 2 have shown that mathematical representational and reasoning ability are still low. The possible reason is that mathematics learning has not been direct students to develop their mathematical knowledge. Students depend heavily on lecturers. The dependency might be internalized by students through learning process. However, student are expected to develop reasoning, bring up ideas or ideas that are creative, and represent problems faced. Based on Figure 2, it appears that the indicators follow the rules of inference, give the opponent the example, and check the validity of the argument in a row is an indicator that least controlled by students.

This is in accordance with the research of Peressini and Bassett [12] that without communication especially representations in mathematics we will have little information, data, and facts about students' understanding of mathematical processes and applications. This means that representations that are part of the communication in mathematics which help educators understand the ability of learners to interpret and express their understanding of the concepts and processes of mathematics they learn.

Prabowo [14] revealed several students' weaknesses in effort to master the concept and cultivate a scientific attitude that affect their representational ability. The weaknesses are: a) the inaccuracy of educators in providing visualization of concepts, and lack of explanation of the physical meaning of each mathematical formulation in learning activities; b) problem in construction that spur on the taxonomy of Bloom's lane without the support of process skills; c) educators are not aware and evaluate students' learning progress; d) demonstration and problem-solving activities that do not meet the requirements of teachers with concentration on the fulfillment of material targets; e) the learning environment has not been anticipated by educators in determining teaching and learning strategies. The weaknesses mostly related to the learning process undertaken by educators.

Many studies agree that mathematical representational ability is important to communicate ideas for skeptical, critical, creative, and innovative thinking. This is in line with Darta's research [7] which states that representational ability is a tool for improving students' mathematical competence. In addition, students' representational ability can provide information for teachers about how students think about a context or mathematical idea, and students' strategy in understanding a concept.

In the wake of learning process, we should provide a structured task related to mathematical reasoning. This is done so that the scope of student thinking is not formally constrained, because the more we reduce and formalize the scope of what is considered a mathematical reasoning, the more we limit the potential multidimensionality of learners. In addition, the interaction of teachers and students during the learning process should be minimized so that children can develop their own reasoning. This is in the opinion of [5] stated that we have witnessed in the afterschool program that the more we reduce and formalize the scope of what counts as mathematical reasoning, the more we constrain the multidimensionality of learners' potential. When communicating with students, teachers tend to refine students' work by correcting what students think or how they use language, rather than trying to understand what students is trying to do.

Completion of mathematical representation and reasoning found different answers between students. Here is a description of the answers of high and low achiever students in solving problems whose reasoning indicator is to provide explanations by using models, facts, and relationships, and representational abilities indicators is to change algebraic notation into the image.

Given $f(x) = x^2 - 8x + 16$. (The test adopted from Larson and Palvo [10])

a. Change it into the general form of $f(x) = a(x - h)^2 + k, a \neq 0$? (2 points)
b. From the general form properties, specify the vertex and the cut point? (2 points)
c. Draw the sketch of the function and explain its properties based on the sketch? (4 points).

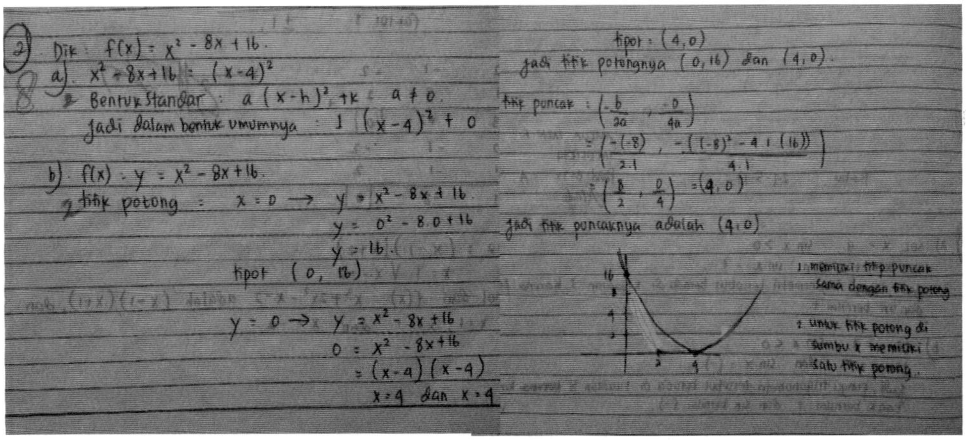

Figure 3. Best Student's Answer

Based on Figure 3, students can solve the problem perfectly and correctly. The student has good reasoning ability. It can be seen at problem 2b (providing an explanation using models, facts, traits, and relationships) in which he applies the formula with discriminant because the student might forgot the peak concept of the standard form $(x - h)^2 + k$ where the peak point (h, k). He told me that he forgot the fastest way, but he reasoned in different ways that produced the same answer. In addition, he has a good representational ability. It could be discerned from the image with the representational abilities indicator namely, providing an explanation by using models, facts, traits, and relationships that he can do. He knows the image properties of the parabola, in this case when $a > 0$, then the parabola opens upward. Furthermore, the answers of the low students are shown in Figure 4.

In Figure 4, the student can only solve the problem with the reasoning indicator namely, providing the explanation using the model, facts, traits, and relationships. The student has a poor reasoning ability as he was not able to solve it. He faced difficulty so there was no response provided. In addition, he has a poor representational ability as well. It can be seen from the image of the 2c (indicators of representation of the form of notation (algebra) into the image) which he can do by means of substitution and producing incorrect answers. Another difficulty is that he does not know the shape of the drawing of the squared function, making the scale on the wrong x and y-axis, resulting in the graph being straight and not curved, the incomplete point retrieval process, consequently, the graph has not cut the x-axis. In accordance with the research result Bergqvist and Lithner [3], most task solutions are based on available algorithms, often without arguments that justify the reasoning, which may lead to rote learning.

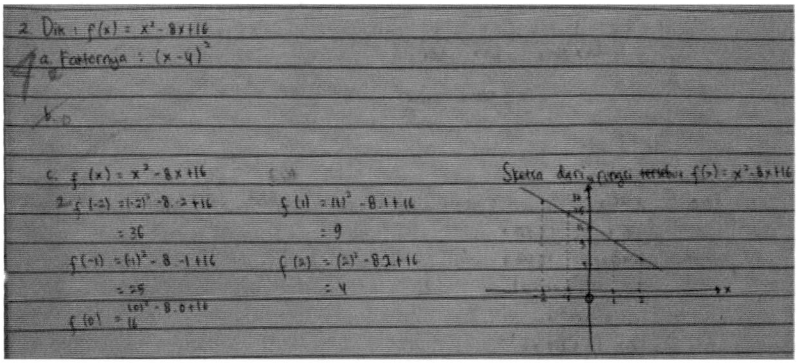

Figure 4. Low Student's Answer

In relation to relevant research results and preliminary analyses, the researcher think that scientific approach encouraged by the government could increase prospective teachers' representational and reasoning ability. Because scientific learning can develop all potential learners covering all aspects. This is in accordance with the implementation of the Curriculum 2013 nationally in which a scientific approach is recommended to use. In addition, studies on scientific approaches, representational and reasoning abilities are necessary to be conducted. In order for the policy to get adequate support from research results in various subjects such as students, teachers, and prospective mathematics teachers as well.

4. Conclusion

This study is an initial sequence of research planned for three years. The results of this first-year study showed that students still had many weaknesses in the representation and mathematical reasoning that was influenced by the ability to understand the indicators of both capabilities. After seeing the results of this first-year study, then the researcher planned following studies in the second and third year for the development of teaching materials with the scientific approach in accordance with the needs of prospective students.

References

[1] Agustiana, R.P. 2014. PISA 2012: Siswa Indonesia Miskin Kemampuan Bernalar. (Tersedia: http://blogs.itb.ac.id/appledore/2014/02/18/32/)

[2] Andriani, S 2012 Pengembangan Bahan Ajar Praktikum Kalkulus Melalui Program Maple untuk Meningkatkan Penalaran dan Representasi Mahasiswa *Jurnal Atikan* 2 **2** pp 295-312

[3] Bergqvist, T., Lithner, J. 2012 Mathematical reasoning in teachers' presentations *Journal of Mathematical Behavior* 31 **2** pp 252–269

IOP Conf. Series: Journal of Physics: Conf. Series **948** (2018) 012053 doi:10.1088/1742-6596/948/1/012053

[4] Buhari, B. 2014. Four-D Model. Model Pengembangan Perangkat Pembelajaran dari Thiagarajan, dkk.
[5] Carlos, Leiva, L., Torres, Z., Lena, Khisty 2013 Acknowledging Spanish and English Resources During Mathematical Reasoning *Cult Stud of Sci Educ* 8 pp 919–934
[6] Copi, I.M 1978 *Introduction to Logic* (New York: Macmillan)
[7] Darta 2013 Peranan Representasi dalam Pembelajaran Matematika di Sekolah *SYMMETRY Jurnal Pendidikan Matematika* 2 **2** pp 261-272
[8] Darta 2016 *Self-Efficacy Mahasiswa Tingkat Akhir untuk Mengajarkan Matematika di Sekolah Menengah* (FKIP Unpas: Laporan Penelitian)
[9] Fisher, D 2013 *Penggunaan Model Core Melalui Pendekatan Keterampilan Metakognitif Dalam Peningkatan Kemampuan Penalaran Matematik Dan Mengembangkan Karakter Siswa SMP* (Tesis:UNPAS)
[10] Larson dan Falvo 2011 *Algebra and Trigonometry* (Belmont: Cengage Learning)
[11] National Council of Teachers of Mathematics 2000 *Principles and standards for school mathematics* (Reston, VA: Author)
[12] Saputra, J., Eliyarti, W 2013 Pengaruh Model Pembelajaran Edutainment dengan Pendekatan Somatic, Auditory, Visualization, And Intelectually (SAVI) terhadap Kemampuan Komunikasi Matematis Siswa SMA *SYMMETRY Jurnal Pendidikan Matematika* **4** 1 pp 217-226.
[13] Saputra, J 2015 Penggunaan Model Problem Based Learning Berbantuan E-Learning dalam Upaya Meningkatkan Kemampuan Pemecahan Masalah Matematis dan Dampaknya terhadap Kemandirian Belajar Mahasiswa *Pasundan Journal of Mathematics Education* **5** 2 pp 77-88.
[14] Sarwanto 2013 Analisis Kemampuan Representasi Mahasiswa Pendidikan Sains *Jurnal Materi dan Pembelajaran Fisika* **3** 2 pp 1-6.

Development of instrument for assessing students' critical and creative thinking ability

R Herpiana[1,a] and U Rosidin[1]

[1]Physics Education of Graduate Program, Lampung University, Jl. Sumantri Brojonegoro No. 01, Lampung, Indonesia.

[a]E-mail: Ria.herpiana@yahoo.co.id

Abstract. The purpose of this study was to develop instruments to measure critical thinking ability and creative students in the topics of physics simple harmonic motion. The research method used was research development with application of procedures including research and data collection, planning, and initial product development. The participants of the study were thirty-four tenth grade students and five physics teachers of physics who were selected randomly from schools in the province of Lampung. The data collected by using test and analyzed in quantitative descriptive. Initial data showed that students' critical and creative thinking ability were still low and instruments to assess students' critical thinking skills and creative students was not yet available. Most of assessment conducted focused on memorization. Thus, the researchers developed a draft of instrument in the form of the test description based on criteria that encouraged students' activity in understanding the concepts, strategies and decision/solution in dealing with problems. The development of the instrument was conducted considering real-world phenomena in the form of pictures and stories, description of the situation, and verbal presentation.

1. Introduction

The primary purpose of education is to improve and develop students' ability. Various factors might contribute to this purpose, one of which is the ability of teachers to perform and utilize assessment, process evaluation, and learning outcomes [1]. Such ability is necessary to understand the application of intended learning objectives that have been set in the curriculum. In addition, the ability lead to improve the learning process performed by the teacher so that learners can understand and apply what they already know. To do that, learners should work to solve a problem, discover by themselves, and try to implement their ideas. In this case, critical and creative thinking ability play important role in reaching the educational goals successfully.

Critical and creative thinking ability is important abilities for the learners to solve the problems encountered [4]. Thus, these abilities are important for all students, especially high school students. However, based on PISA result reported by the Organization for Economic Co-Operation and Development (OECD), Indonesia ranks 69th out of 76 countries [12]. The result indicates that the average of Indonesian students still have problems in terms of the three cognitive aspects, namely knowing, applying, reasoning

According Ministry of National Education and Culture, assessment of learning outcomes performed by educators is the process of collecting information/data about learners' learning achievement in

IOP Conf. Series: Journal of Physics: Conf. Series **948** (2018) 012054 doi:10.1088/1742-6596/948/1/012054

various aspects such as attitude, knowledge, and skill in a planned and systematic ways to monitor the process and learning progress and to improve learning outcomes through assignment and evaluation of learning outcomes [13]. Based on these opinions, it could be concluded that the assessment can provide stimulation to students to develop students' critical and creative thinking skills.

Problems that occur in one of schools in Lampung Province are cognitive assessment instruments used in the form of questions that tend to focus on memorization. Meanwhile problems that encourage students' critical thinking skills and creative learners are not applied. Therefore, students' thinking ability is still low. In addition, it has been exaggerated by teachers' lack of ability to design its instrument designed specifically to improve students' critical and creative thinking skills. Thus, the teachers are in need of guidance to develop the instruments.

The test instrument developed was a test instrument description. The developed test items used image, graphic, and tables which led students to reach taxonomy of educational goals and involve higher-level cognitive processes [11]. The purpose of this study is to develop the design of physics instrument on simple harmonic motion material to measure students' critical and creative thinking ability.

1.1. Assessment Instruments

Assessment is the process of collecting and processing information. In education, assessment means the process of collecting and processing information to determine achievement of students' learning outcomes. To do this, teachers need instruments in the form of questions to assess cognitive, affective, and psychomotor aspects. One way to understand characteristics and quality of learning outcomes are through assessment. Therefore, teachers can understand exactly what students have achieved. Assessment is the process of collecting, interpreting and using evidence to make decisions about student achievement in education [7]. Assessment, in general, can be used to determine the level of achievement of learning outcomes known as the assessment of learning (AoL), and to improve the learning process known as the formative assessment or assessment for learning (AfL) [17]

Assessment of learning is an assessment used to find out what students know, to indicate whether it has met the standards, and to make comparison among students [5]. Assessments is designed to inform teachers to modify their learning activities, differentiate, and understand how students approach learning. In this study, to develop students' critical and creative thinking skills, an assessment of learning is needed that can improve learning in the classroom [6]. Teachers should use assessment to plan lessons, identify student needs in learning and re-teach materials which are not understood by students.

1.2. Critical and Thinking

Critical thinking is a cognitive activity related to the use of mind [4]. Based on Bloom's cognitive dimensions, critical thinking skills occupy the dimensions of analysis (C4), synthesis (C5), and evaluation (C6). It appears that these dimensions are derived from Bloom's old taxonomic system. If matched with Bloom's Bloomfield taxonomy revised by Anderson and Krathwohl [2], then critical thinking skills occupy parts of analysis (C4), and evaluation (C5), because, in the revised version, the synthesis dimension is integrated into the analytical dimension. While creative thinking is a mental activity that fosters original ideas and new understandings. Creative thinking is a habit of the mind trained by paying attention to intuition, animating the imagination, revealing new possibilities, creating stunning perspectives, and generating unexpected ideas [6]. Creative thinking is a mental activity that produces something new that results from development.

2. Methodology

The selected research and development model was developed by the Borg and Gall model which consisted of ten development steps: (1) collecting research and information; (2) planning; (3) developing draft; (4) implementing field trials; (5) revising trial results; (6) re-implementing field trials; (7) results of refined products; (8) field implementation test; (9) refining the final product; and (10) dissemination and implementation [3]. This article only reports three steps: research development from research and information gathering, planning, and early product development. The research steps are presented in Figure 1 below.

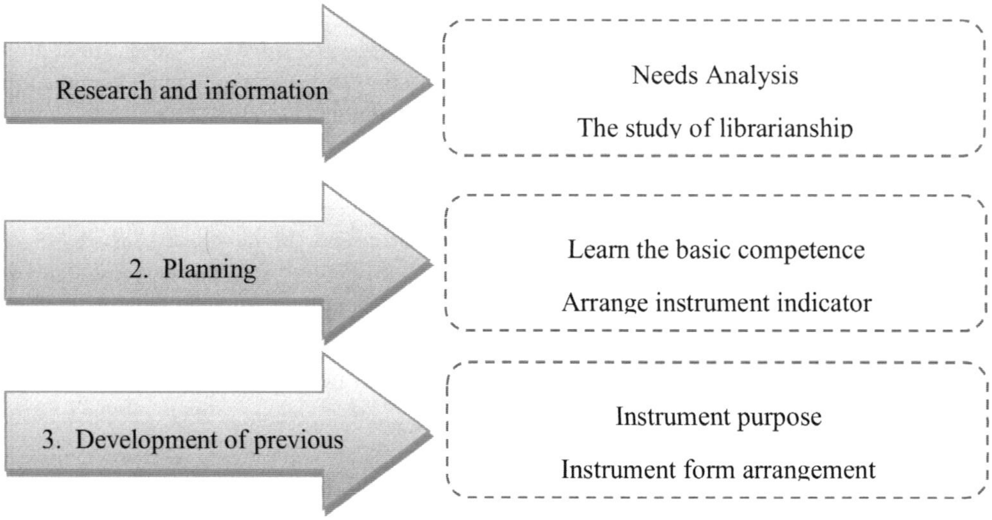

Figure 1. Steps of Development

Research development in this article is

2.1. Research and information gathering

This research and gathering of information is the first step in this research to map the problems that is studied and to bring up the idea of the product to be developed. Form of research and data collection in the form of field needs analysis and literature review. Necessary analysis were performed by providing needs analysis to 34 high school students of class X and five physics teachers who were selected randomly from schools in Lampung Province. The literature review conducted by reviewing several books and research results are relevant to the development of the research.

2.2. Planning

At the planning stage, basic physics competence analysis was conducted such as determining some basic competencies whose phenomenon were found by students easily. Subsequent processes was to develop indicators of assessment instruments adapted to indicators of critical and creative thinking skills. The next steps were to create natural phenomena in the form of pictures and stories, presentation of the problem (description of the situation), and presentation of statements and experimental table data that could be presented in accordance with the basic competence sequence.

2.3. Early product development

The early product development stage determine the purpose of the assessment instrument is to train students' critical and creative thinking skills. The parts planned in the planning phase are designed in such a way that they became design products at this stage. The test instrument should contain some good test instrument parameters. The assessment instrument to be developed was used to help students to practice their critical and creative thinking skills, the following is the assessment instrument design can be seen in the following figure:

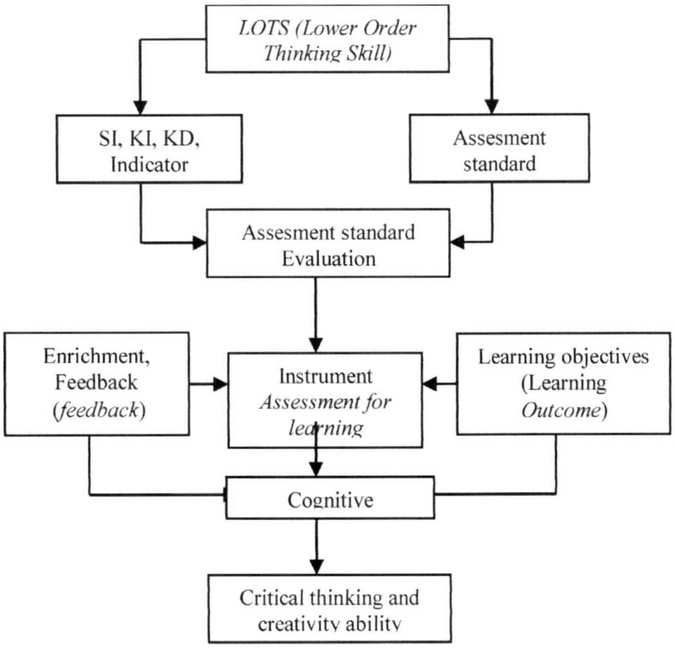

Figure 2. Assessment Instrument Design

3. Results and Discussion

The stages done in this research are that the research and information gathering stage are based on relevant theoretical studies. The information obtained is the need to develop a test instrument used to measure students' critical and creative thinking based on preliminary study conducted in one of the high schools in Lampung. According to preliminary study results, it was found that cognitive assessment instruments used to train students' critical and creative thinking skills have not been widely available. Most questions tend to measure aspects of memory or memorization that cannot be used to train students' thinking skills even higher. Moreover, in solving the problems of simple harmonic motion that requires reasoning and analysis in solving it, so need to develop assessment instruments to measure the ability of critical and creative thinking. So in the next stage, the researcher attempts to develop the instruments that will be developed, such as the question of critical and creative thinking based on indicators of the thinking stage. The type of test is a subjective test in the form of a description test.

The result of the initial product development is the design of the critical and creative thinking criteria of high school students on simple Harmonic Motion materials. The specific purpose of instrument design is that students can:

a. investigate the concept of harmonic vibration on a simple pendulum swing;
b. determining the restoring force on a simple swing;
c. analyze the relationship of style and vibration in everyday life.

The developed instrument is in the form of a description test. The problem developed test adapted to indicators Basic Competence Motion Harmonist Simplified and Indicator Critical thinking and creative thinking. The indicators of critical thinking are: (a) Interpreting, (b) Analyzing, (c) Evaluating and (d) Inferring [4]. The general indicators used in the critical thinking are ability to understand either known or well-asked questions, to identify the relationships between the statements, questions and concepts provided in the questions, and to provide right explanation using right strategy in solving the problem, to complete and correct and students can make the right conclusions.

The test instrument for measuring critical thinking skills is a description test. It contains questions that lead to certain indicators of critical thinking ability. While some creative thinking skills are: (a) Fluency, (b) Flexibility (Dexterity), (c) Originality and (d) Elaboration [6]. Assessment instruments are based on creative thinking skills, test instruments that are created to make students spark a lot of ideas, answers, problem-solving, able to generate varied answers, think something that never thought of others and develop an idea and detailing of an object making it more interesting [15].

The plan for the appraisal of assessment instruments was carried out with the first three meetings of prior learning process in order to develop students' critical and creative thinking skills. A final meeting was conducted to examine the problem as a whole with the objective of knowing the effectiveness of the developed test instrument whether effective in measuring ability think critically and creatively as an assessment of learning students [10]. The first meeting discussed the topic of the concept of harmonic vibration in the pendulum swing, in this meeting the process of teaching and learning with the aim that students can understand the topic discussed to work on the problem of critical thinking and creative skills that have been assembled as an assessment for learning.

After the learning was performed, the students were given some description questions to assess their thinking ability. The second and third meetings were still going through the same process as the first meeting, only the topics discussed was different. The second meeting discussed the restoring style and the third meeting discussed the relationship of style and vibration in everyday life. From these three meetings, the students' value in working on the critical thinking and creative thinking problem was given.

Based on the analysis, it could be said that the test instrument of critical and creative thinking ability was designed to build and train students' critical and creative thinking skills based on the indicators of problem made. The following are examples of problems that could improve students' critical and creative thinking skills in simple harmonic motion materials.

Figure 3. Picture on the left shows a hummingbird flapping its wings while sucking on nectar. Hummingbirds have a body mass of about two grams. This bird flaps its wing 80 times every second. Why does the hummingbird flap its wings? What do you think about it?

Students answered the question in various responses. To measure students' ability to answer the question, it considered analysis of natural phenomena created based on drawings and stories that led students to take decisions in solving the problem.

The results of research conducted by Treagust showed that learning using assessment sheet as an assessment for learning could improve students' ability in science learning. The use of test instruments as an assessment for learning is necessary to develop students' thinking skills [16]. This also agrees with research by Klenowski. The results obtained showed that assessment for learning, a process of identifying development of learning, was able to develop students' thinking ability [8]. Thus, the use of an assessment instrument of critical and creative thinking skills as an assessment for learning is able to develop students' thinking ability.

The use of critical thinking and creative thinking test instruments is one of the alternatives for teachers to determine the level of critical and creative thinking skills of students [14]. By working on the problem of critical thinking and creative students can provide answers in accordance with the ability to think so that teachers know how well the level of thinking ability of students. This, of course, can help teachers to better develop critical and creative thinking test instruments on other topics so that students can develop their thinking ability to be more optimal. The test instrument of critical and creative thinking ability to be developed is considered effective in measuring students' critical and creative thinking ability in accordance with the level of ability they have.

4. Conclusion

Based on the results of research on the literature and analysis, it showed that the assessment instrument is still a rote or memory assessment instrument. Instrument assessment required by students can measure the ability of critical and creative thinking in the form of a description test with the criteria that can stimulate student activity in understanding the concepts, implementing strategies and tactics in solving problems. Development of instrument design is through natural phenomena in form of pictures and stories, presentation of problems (description of situations), presentation of statements and experimental table data so as to strengthen students' conceptual understanding and make correct decisions/solutions in solving problems.

References

[1] Agus dan Jailani 2014 *Pengembangan Instrumen Asesmen Higher Order Thinking Skill (HOTS) pada Mata Pelajaran Matematika SMP Kelas VIII Semester 1* (Universitas Negeri Yogyakarta)

[2] Anderson, L.W. Krathwohl, D.R. 2010 Kerangka landasan untuk pembelajaran, pengajaran, dan asesmen: revisi taksonomi pendidikan Bloom.Terjemahan Agung Prihantoro (Yogyakarta: Pustaka Pelajar)

[3] Borg, W. R. Gall, M.D 1983 *Educational researcher: An introduction* (United States : Pearson education, Inc)

[4] Cottrell, S 2005 *Critical Thinking Skills, Developing Effective Analysis and Argument* (New York: Palgrave Macmillan)

[5] Earl, L 2006 Assessment - a powerful lever for learning *Brock Education* **16** 1

[6] Elaine B. Johnson 2014 *Contextual Teaching & Learning* (Bandung: Kaifa)

[7] Harlen, W 2007 *Assessment of learning* (London: A Sage Publications Ltd)

[8] Klenowski, V 2009 Assesment for Learning revisited: an Asia-Fasific perspective *Assesment in Education: Principles, Policy, Practice* 16 **3** pp 263-268.

[9] Krathwohl, D. R 2002 *A revision of Bloom's Taxonomy: an overview – Theory Into Practice, College of Education* (The Ohio State University Pohl)

[10] Kusuma, Merta 2017 *Pengembangan Perangkat Instrumen Assesmen HOTS Fisika Siswa* (Lampung: Jurusan Pendidikan Fisika Fakultas Pendidikan Matematika dan Ilmu Pengetahuan Alam Universitas Lampung)

[11] Nitko, A. J., Brookhart, S. M 2011 *Educational assessment of student* (Boston: Pearson Education)

[12] OECD 2012 *PISA 2011: Science competencies for tomorrow world volume 1: Analysis. Rosewood* (Drive: OECD)

[13] Permendikbud 2015 *Peraturan Menteri Pendidikan dan Kebudayaan Republik Indonesia nomor 53 tahun 2015, tentang Penilaian hasil belajar Pendidik pada pendidikan dasar dan pendidikan menengah*

[14] Ramos J.L.S., Dolipas, B.B., Villamor, B.B 2013 Higher Order Thinking Skillss and Academic Performance in Physics of College Students: A Regression Analysis *International Journal of Innovative Interdisciplinary Research*, **4** pp 48-60

[15] Schraw, Gregory *et al.* 2011 *Assessment Of Higer Order Thinking Skillss* (America: Information Age Publishing)

[16] Treagust. DF, R. Jacobowitz, JL. Gallagher, and Parker 2001 Using Assesment as a Guide in Teaching for Understanding: A Case Study of a Middle School Science Class Learning about Sound *Science Education* 85 **2** pp 137-157

[17] Weeden, P., Winter, J., & Broadfoot, P 2002 *Assessment: What's in it for schools?* (London: Routledge Falmer)

Developing material for promoting problem-solving ability through bar modeling technique

N Widyasari[1,a) and H Rosiyanti[1, b)

[1]Education Faculty of UMJ, Jakarta

[a]E-mail:widyasari.nurbaiti@gmail.com [b]E-mail: Hastrirosiyanti@yahoo.com

Abstract. This study aimed at developing material for enhancing problem-solving ability through bar modeling technique with thematic learning. Polya's steps of problem-solving were chosen as the basis of the study. The methods of the study were research and development. The subject of this study were five teen students of the fifth grade of Lab-school FIP UMJ elementary school. Expert review and student' response analysis were used to collect the data. Furthermore, the data were analyzed using qualitative descriptive and quantitative. The findings showed that material in theme "Selalu Berhemat Energi" was categorized as valid and practical. The validity was measured by using the aspect of language, contents, and graphics. Based on the expert comments, the materials were easy to implement in the teaching-learning process. In addition, the result of students' response showed that material was both interesting and easy to understand. Thus, students gained more understanding in learning problem-solving.

1. Introduction

One of the objectives of learning Mathematics in school is to solve problems that include ability to understand problems, design mathematical models, solve models, and interpret the solutions obtained [1]. This is because that problem-solving is viewed as a process to find a combination of a number of rules that can be applied in an effort to cope with the new situation. Problem-solving is not merely a form of being able to apply rules that have been mastered through past learning activities, but more than that is the process of obtaining rules at a higher level [2]. According to [3] mentioned that the low ability of problem-solving due to the lessons used by teachers is still dominated by ordinary learning models that give less emphasis on the application of mathematics in everyday life, and students are not accustomed to involved in solving problems. Students tend to solve a problem by mimicking the problem solving that the teacher exhibits when discussing the problems [4].

This is in line with Polya's theory that problem-solving helps students in building their skills, stimulating them to discover new knowledge and actively engaging students in learning activities. According to Polya (1951) problem-solving as a capability includes activities such as [5]; (1) following the step to understand problems that include: identifying known data, identifying the data being asked, identifying necessary data, adequacy of data, and compile a mathematical model of the problem; (2) selecting strategies and executing strategies; (3) implementing calculations and solve mathematical models; (4) interpreting the solution (outcome) to the original problem and checking the correctness of the solution.

Thus, researchers can conclude that solving mathematical problems can help students to improve their abilities, discover new knowledge and play an active role in learning activities. However, the importance of problem-solving skills is not supported by the students' current skills. It is seen that based on TIMSS data in 2015, the score of problem solving ability of grade 4 of elementary school students is only 397 where the benchmark score is 500.

One of the factors that can lead to the condition is the inappropriate use of learning resources used in schools. In general, the source of learning used is only mechanistic less emphasis on mathematical ability, especially problem-solving. Presentation of learning resources is also still abstract in helping students to understand the problems associated with problem-solving abilities. Therefore, it is a necessary to find a source of learning which assist the student in overcoming the problems.

Bar modeling drawing or commonly known as Model Method which is a way of completion developed by the Ministry of Education Singapore since 1980 and has been widely used by other developed countries. This is because the modeling bar gives students a powerful way to model and solve problems related to problem-solving [5]. Furthermore, modeling drawing bars can help students acquire knowledge more meaningfully than using mechanistic means as shown in example 1.

Example : Andi's age is three years older than Sarah's age in this year, but two years younger than Heri's age.

How old is Andi, if the total of their age is 41 years.

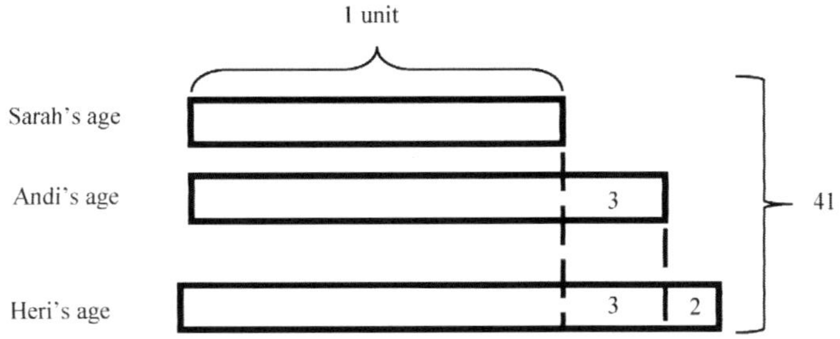

3 units = 41 -2 - 3 - 3
3 units = 33
1 unit = 11 → Sarah's age
So Andi's age is = 11 + 3 = 14 years old

The example shows that the drawing model can be interpreted as a form of translating the story into a diagram or the so-called "model".

Based on the arguments and explanation stated above, the researchers are interested in conducting an experiment by using bar modeling technique as a developed supplementary material to students in elementary school in order to help them improve their mathematical problem-solving ability. Thus, the purposes of this study is material development of the problem- solving ability through bar modeling technique with thematic learning

2. Methods

The research method in this study is RnD research. This research method was chosen because the researchers wanted to develop problem-solving mathematics materials of Bar Modelling Drawing. This corresponds to the opinion of Borg and Gall (1988) [7] who stated that research and development (Research and Development) is a research method used to develop or validate the products used in education and learning. Procedures in this development research apply several stages, namely, (1) observing the location of the research, (2) borrowing thematic books, (3) preparing teaching materials, (4) validating the teaching materials to the bar technics modeling experts, material experts, linguists And design experts, (5) testing into small classes, (6) piloting into large classes.

The material product of bar modeling drawing is validated by experts. There are 3 experts who validate the materials bar modeling drawing, namely (1) bar modeling drawing and material experts, (2)

linguists, (3) design experts. In this research, assessors of the quality of teaching materials bar modeling drawing on thematic learning of mathematics can be seen as follows in table 1.

Table 1. Expert of Material

Number	Full Name	Expert	Institution
1	Irani Syach, S.Si	Bar Modeling Expert dan Material Expert	Sakamoto
2	Wika Sofiana Dewi, M.Hum	Linguist	Muhammadiyah University of Jakarta
3	Khoala Rachman Adzima, M.Si	Design Expert	EsaUnggul University

After being validated by three experts, the product was revised and tested into a small class of three subjects of fifth grade selected from the lowest, medium and high score. The three subjects were given a product and student response sheets to assess the product made by the researchers. After that, the product was revised again and tested in a large class of 12 students and given student response sheets. The data obtained in this research are qualitative and quantitative data. Qualitative data in the form of responses, criticisms, and suggestions are attached in the assessment form. The resulting data relates to the suitability of the development product being made. Quantitative data are percentages obtained from product assessment questionnaires filled by fifth grade of elementary students.

The instrument used in this research is the assessment form used to obtain an expert's assessment of the quality of bar modeling drawing materials. Product quality is reviewed from several aspects, namely (1) material suitability, (2) bar modeling drawing concept, (3) language suitability for elementary students, and (4) suitability of design and visualization of teaching materials. Another instrument is a questionnaire about the student's response to teaching materials that have been made by researchers.

Data analysis was performed to see the value of each aspect on the assessment form and the student response questionnaire. The data provided by experts is collected and analyzed descriptively. The data provided by the students are collected and analyzed by calculating the average score multiplied by 100% and the result will be analyzed descriptively.

3. Result and Discussion

Based on the results of qualitative and quantitative data processing obtained from both the experts and students' responses on teaching materials using bar modeling techniques to improve the ability of mathematical problem-solving, the following data are obtained: (1) Assessment of experts using the assessment scale is the value of 1 to 10, with details that are 1-2 (less once); 3, 4, and 5 (less); 6, 7, and 8 (good); 9 and 10 (very good). Based on the assessment of experts can be summarized as follows as shown as in table 2.

Table 2. Assessment of Expert

No.	Aspect	Skor	Category	Conclusion
1	Language	7.6 = 8	Good	Teaching material worth a trial with a revision
2	Material and bar modeling technique	8.7 = 9	Very Good	Teaching material worth a trial with a revision
3	Design	6.9 = 7	Good	Teaching material worth a trial with a revision

Example of content of material as shown as below.

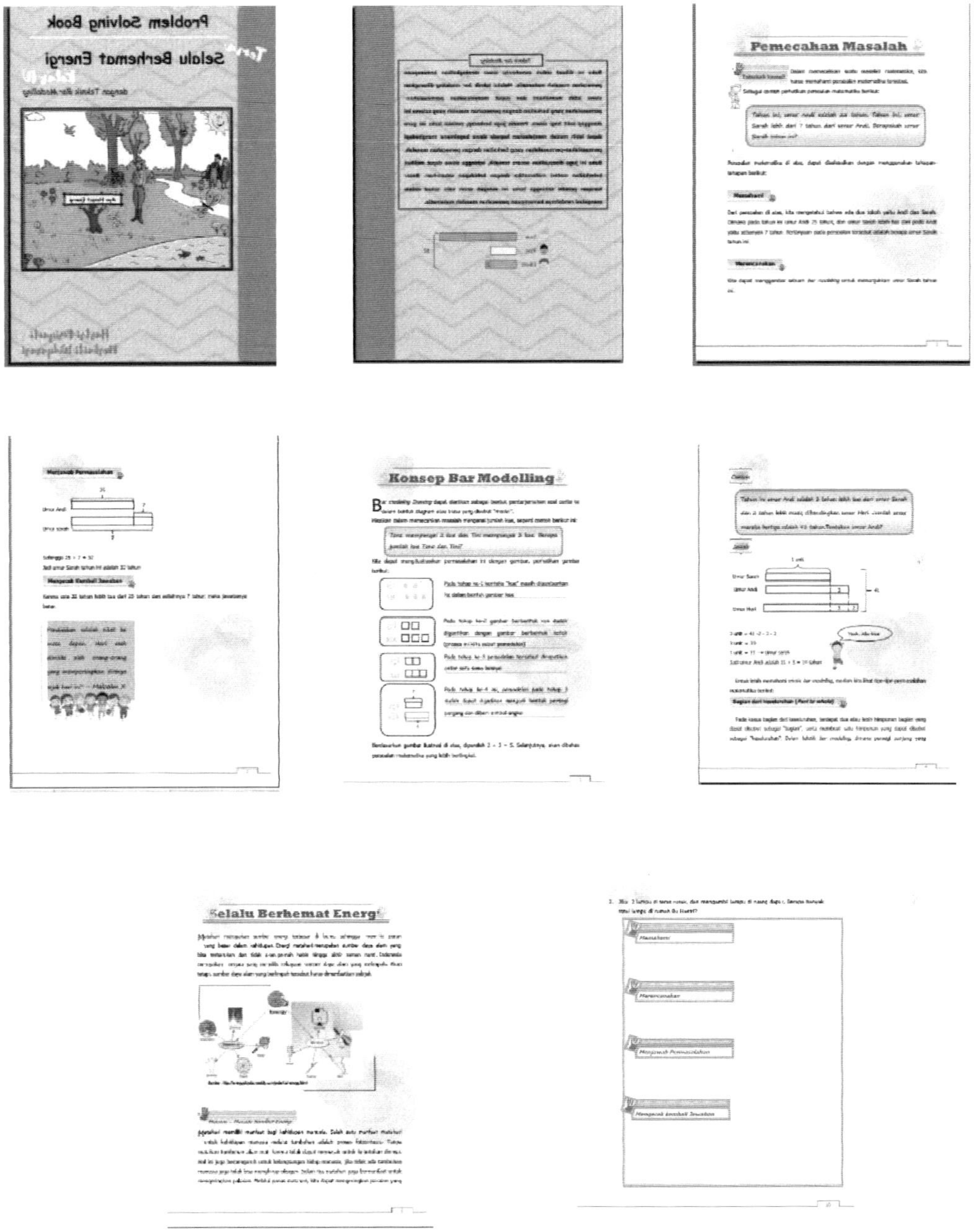

Figure 1. Examples of content materials

In addition, based on all assessments of experts, they stated that the teaching materials are declared as valid and easy to implement in the teaching-learning process. The data based on interview of small class shown that the students of each level (high, medium, and low ability) were easy to understand the material and also it could easy to followed the instruction. Therefore, the materials could be implemented in the big class. The data based on students' responses were obtained as shown as in table 3. Furthermore, the students stated that the teaching materials have an interesting appearance.

IOP Conf. Series: Journal of Physics: Conf. Series **948** (2018) 012055 doi:10.1088/1742-6596/948/1/012055

Table 3. Result of Student Responses

No.	Aspect	%	Conclusion
1	Material	84.8	Agree that the material corresponds to the material, theme, problem-solving, and modeling bar
2	Language	87.3	Agree that the language used is easy to understand
3	Interest	65.5	Agree that teaching materials increase their interest in learning math and problems related to problem-solving

4. Conclusion

The result shows that material in theme "Selalu Berhemat Energi" could be categorized as valid and practical. The validity is measured by using the aspect of language, contents, and design. Based on the expert comments, the materials that we developed can be implemented easily in the teaching-learning process. In addition, the result of the students' response shows that the material is interesting as well as easy to understand. Furthermore, the students have more understanding to learn problem-solving. In conclusion, the material in theme "Selalu Berhemat Energi" deserves to be given to students of the 4th grade.

References

[1] BSNP 2006 *Standar Kompetensi dan Kompetensi Dasar SD/MI* (Balitbang:Jakarta)
[2] I. Hardini, P. Dewi 2012 *Strategi Pembelajaran Terpadu (Teori, Konsep & Impementasi).* (Familia: Yogyakarta)
[3] A. Minarni 2003 *Pengaruh Pembelajaran Berbasis Masalah Terhadap Kemampuan Pemecahan Masalah Matematis*. PROSIDING. 4 **10** pp 92-102
[4] U. Sumarmo 2014 *Kumpulan Makalah Berpikir dan Disposisi Matematika serta Pembelajrannya* (Universitas Pendidikan Indonesia Press: Bandung)
[5] J. Dindyal 2009 *Application & Modelling for The Primary mathematics Classroom* (Prentice Hall: Singapore)
[6] Sugiyono 2009 *Meteode Penelitian Kuantitatif, Kualitatif, dan RnD* (Alfabeta: Bandung)

Higher-order thinking skill problem on data representation in primary school: A case study

R I I Putri[1,a] and Zulkardi Z[1]

[1]Faculty of Teacher Training and Education, University of Sriwijaya, Jalan Srijaya Negara FKIP Bukit Besar Palembang, Indonesia

[a]E-mail: ratu.ilma@yahoo.com

Abstract. This article aimed at reporting research result on a case study of a lesson using a HOTS problem. The task was about data representation using baby growth context. The study used a design research method consisting of three stages: preparing for an experiment, experiment in the classroom (pilot and teaching), and retrospective analysis. Participants were sixth grade students who were learning data representations in a Primary School in Palembang Indonesia. A set of instructional activities were designed using Indonesian version of Realistic Mathematics Education (PMRI) approach. The result showed that students were able to solve the problem and present their solution in front of the classroom. The conclusion indicated that that HOTS problem using the growth of a child as the context could lead students to use their mathematical thinking. During the learning activities along with teacher orchestra's guidance, and discussion, students were able to solve the problem using line graph although some of them used a bar graph. In the future, teachers are necessary to focus on the role of real-world figure in mathematics learning.

1. Introduction

Statistics is the very important topic as statistics is a part of a curriculum in primary and secondary school. Also, quality of teachers is probably the most important factor in education, but teachers need books with well thought out mathematical problems [1]. Data Representation is a part of statistics topic which always uses in daily life. In fact, during students learning data representation, teachers mostly give examples and their solution without students knowing how to represent the data. Hence in teaching data representation, teachers have to use the right theory. Furthermore, teachers can fine right context. It also stated by [2], that data representation was suggested depend on a type of data, sense of researcher, and the use of data. According to [1], at the heart of the process of designing for mathematics education is the choice of suitable problems, for example, graphs of the growth of a sunflower or a child.

The new curriculum aims at developing 21st-century competencies as mathematical thinking, problem-solving by observing, questioning, listening, and communicating the solution to the problem [4]. Curriculum implementation is described, to reach the goal effectively, learning activities need to use principle, one of them is to provide many learning experiences through meaningful and interesting strategy and instructional method [5]. Accordingly with that one of consistent approach is Indonesian version of Realistic Mathematics Education PMRI. PMRI is an instructional theory that fostering students to communicate, argue and justify their solution [6, 2, 7, 17].

Since Indonesia has been joint PISA (Program for International Student Assessment) since 2000, Indonesian students were not able to solve HOT problems [8]. PISA results in mathematics, have been instability. In the year 2000, PISA mathematics, Indonesia was ranked 39 of 41 countries in mathematics. Then in 2003, the rank was 38 of 40 countries. Then the worst result in PISA 2012 that is ranked 64 out of 65 in PISA 2012 [9,15]. This result shows that Indonesian students weak on mathematics literacy, its mean the competence of students in problem-solving, critical and creative thinking in general still low. This article reports a design research that focuses on how grade five students discuss using HOT problem about baby growth on the choice of data representations in a Primary School in Palembang Indonesia.

There are four problem areas where graph types were developed namely (1) spatial organization; (2) discrete comparison; (3) continuous distribution with histogram and line graphs; (4) multivariate distribution and correlation with three-dimensional charts and correlation diagrams [3].

Higher Order Thinking is thinking on the higher level such as mathematical thinking and reasoning, communication, critical attitude, interpretation, reflection, creativity, generalization and mathematizing [10, 18]. According to [11, 19, 20], to assess higher-order thinking are: see how students understand the problems and let students construct their response. Also, give some suggestions for stimulating students to construct mathematical models and have them to make predictions, describe patterns, and provide explanations.

To facilitate interactivity in the classroom, mathematics social norms are used. In Indonesian classrooms, we use the norm to support students learning in both answering and giving reasoning and justification of their answer [4]. The social norms guide the class on how to communicate, to give reasoning and proof of the solution. In mathematical learning, socio-mathematical norms mean use of norms deal with the difference mathematical solution and strategies [2]. The purpose of this study is to develop a local instruction theory based on the Realistic Mathematics Education Theory [13, 4, 16].

2. Method

2.1 The research design

The subject of this research is 30 students of fifth grade primary PMRI pilot school in Palembang Indonesia. Three steps of design research: design, teaching experiments, and retrospective analysis are reported. We observed and watched the interaction in the video and analyzed the samples of students' work. Our analysis focused on the HOT problem of representation data. We gave students of 10 and 11 years old a problem about a girl. The students were asked to draw a graph that would show how the girl had grown [1]. Figure 1 shows the table with the data.

0 year	52 cm
1 year	74 cm
2 year	88 cm
3 year	97 cm
4 year	105 cm
5 year	112 cm
6 year	118 cm

Figure 1. Graphs of the growth of a child problem

Activities during the research, the team, analyzed and designed the lesson, observed the student activities, discussed the observations results, and analyzed the data. Here, the research team focused on the following mini design cycle [14].

1. Design: planning, anticipating, and predicting students' thinking and how the teacher guide the

IOP Conf. Series: Journal of Physics: Conf. Series **948** (2018) 012056 doi:10.1088/1742-6596/948/1/012056

growth and development of students.
2. Teaching experiment where students investigate the problem supported by the teacher.
3. Doing retrospective analysis and revise the lesson

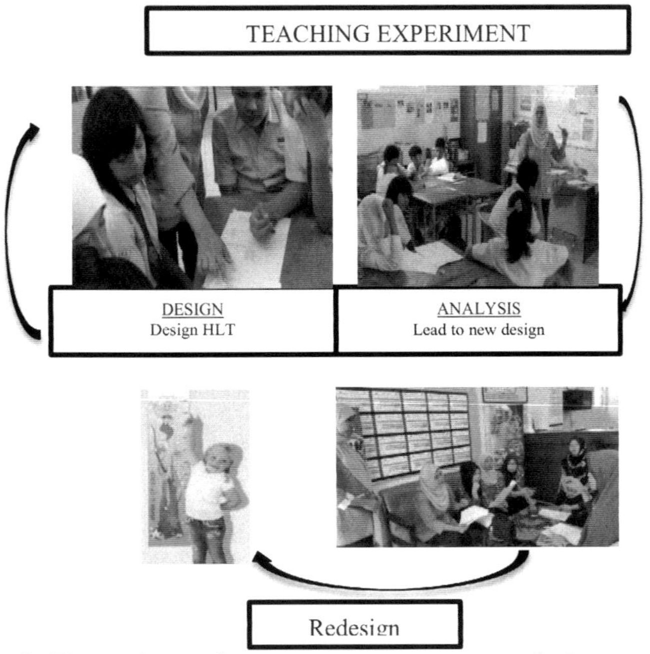

Figure 2. Three phase of design research and its activities

3. Results and Discussion

This part describe three phases of the design research.

3.1 Design Phase

The graphs of the growth of a child problem (see Figure 1) was given as a starting point of the lesson. Then the discussion after the teacher asked for students' to solve the problem. The discussion focused on exploring the students' solutions and strategies.

3.2 Teaching Phase

After Ms. Evi, the teacher, gave the task to all 25 students, they worked in small groups. Students discussed and solved the girl grown problem. The research team observed the class, one research member for each team.

During the discussion, students focused their attention on the graph. One group engaged using bar graph which showed a good understanding of the problem. Others groups also using bar graph although students do not understand the problem. These can be seen in Figure 3.

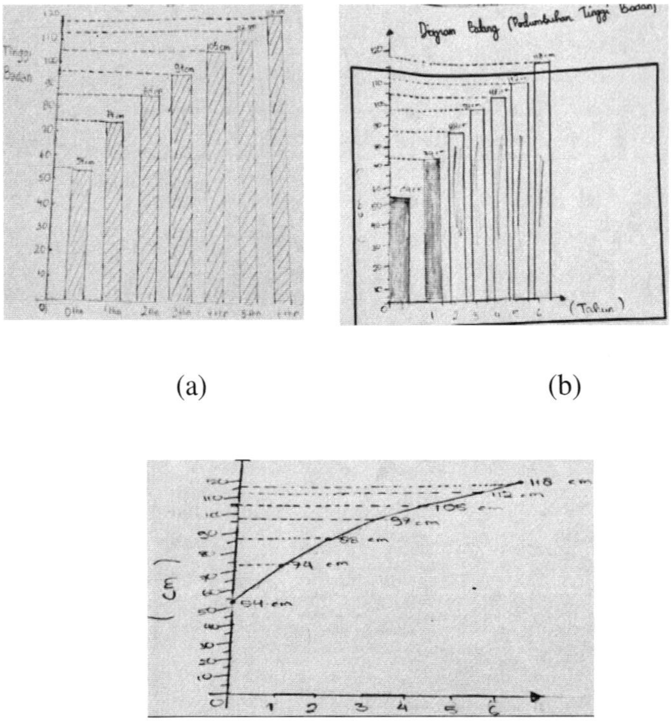

(a) (b)

(c)

Figure 3. Example of students' solution to solve graphs of the growth of a child problem.

One of the interesting of the problems, the students, produced solutions in different strategies that were reflecting the reasoning of the students. But the teacher's reactions make students confused. Group 2 was able to solve the problem correctly. However, they still confuse about zero. This will be discussed below.

T: look, Fina, why zero here and zero there the difference?
S: zero here for year, and zero there for centimetre
T: Look at the position, are you sure?
S: (still confuse…..)

In this fragment, shows that students confuse with the problem due to her unfamiliar with the problem. Teacher moved around and guided students during the group discussion. Many students interested and agreed that the problem is meaningful. However, some students confused with double zero. In another word, they confused on how to put notation zero in the line graph. It came as no surprise that most students did choose some form of a bar graph. One of the solutions was rather surprising because one group choose a line graph. Then the teacher helps students by reintroducing the problem and asking them to draw graphs of the growth of a child. One group drew the representation of the growth of a child as shown in figure 3c. This illustration indicated students understand the problem using the line graph.

3.3 Retrospective Analysis Phase

In this phase, the team analysed and agreed that teachers guided students to solve the problems using their strategies. In the beginning, the students initially attempted to fit on data representation. Then, many of the solutions were not meaningful. Hence, it is important to support students to retreat from the unimportant concept so that students could develop their understanding.

The discussion in groups gave more space and time for students to discuss their strategies constructively. Finally, we noticed that many students actively in communicating, giving their reasoning and justification to their solutions and strategies.

4. Conclusion

The data in this study shows that HOT problem about the growth of a child powerful in bringing out mathematical thinking and communication amongst students. The teacher plays a central role in eliciting the discussion that allows students' thinking to grow.

During the learning activities of this design research study, students can solve the problem using line graph although some of them using the bar graph. In the future, teachers need to focus on the role of the graph in daily life.

References

[1] Galen F 2015 Designing for realistic mathematics education: learning about graphs as an example *Proc. Nas. Conf on Mathematics* (Surabaya: Indonesia)

[2] Widjaya W 2012 Exercising socio-mathematical norms in classroom discourse about data representation: insights from one case study of a grade 6 lesson in Indonesia *The Mathematics Educator* 13 **2** pp 21-38

[3] Bakker, A 2004 *Design research in statistics education: on symbolizing and computer tools.* (Doctoral dissertation). (Utrecht: CD-β Press, Center for Science and Mathematics Education)

[4] Putri R I I, Dolk M, Zulkardi 2015 Professional development of pmri teachers for introducing social norms. *Journal Mathematics Education*, 6 **1** pp 11

[5] Kemendikbud 2013 *Implementasi kurikulum 2013* (Jakarta: Kementerian Pendidikan dan Kebudayaan)

[6] Sembiring R K, Hoogland K, Dolk M 2010 *A decade of PMRI in Indonesia* (Utrecht: APS International)

[7] Putri R I I 2015 Professional development of pmri teachers for introducing social norms *Journal Mathematics Education* 6 **1** 11

[8] Putri R I I 2013 Pengembangan soal tipe pisa siswa sekolah menengah pertama dan implikasinya pada kontes literasi matematika (KLM) 2011 *Proc. Nas. Sem. on Mathematics* (Aceh: Indonesia)

[9] Stacey K, Almuna F, Caraballo M R, Chesne J, Grafunkel S, Gooya Z, …, Zulkardi Z 2015 *PISA's influence on thought and action in Mathematics Education.* Assessing Mathematics Literacy: The PISA Experience ed Stacey & Turner (Switzerland: Springer) pp 275-306

[10] de Lange J 1991 Assessment: No change without problems. *Proc. Nas. Conf. on Assessment in the Mathematical Sciences* (Geelong: Victoria) pp 46-76

[11] Van den Heuvel-Panhuizen M 1996 *Assessment and realistic mathematics education* (Utrecht: Freudenthal Institute, Utrecht University)

[12] Yackel E & Cobb P 1996 Sociomathematical norms, argumentation, and autonomy in mathematics *Journal for Research in Mathematics Education* **27** pp 458

[13] Gravemeijer K & van Eerde D 2009 Design research as a means for building a knowledge base for teachers and teaching in mathematics education *The Elementary School Journal* **109** 5 pp 510

[14] Widjaya W, Dolk M, & Fauzan 2010 The Role of contexts and teacher‟s questioning to enhance students‟ thinking. *Journal of Science and Mathematics The Role of Contexts and and Teacher's Questioning Education in Southeast Asia* 33 **2** pp 168

[15] K. Stacey et al 2015, "PISA's Influence on Thought and Action in Mathematics Education," in *Assessing Mathematics Literacy*, edited by K. Stacey and R. Turner (Springer: Switzerland) pp 275–306

[16] Putri R I I, & Zulkardi 2017 Fraction in shot-put: a learning trajectory *Proc. Int. Conf on Research, Implementation, and Education of Mathematics and Science* AIP Conf. Proc. (Yogyakarta: Indonesia) p 2

IOP Conf. Series: Journal of Physics: Conf. Series **948** (2018) 012056 doi:10.1088/1742-6596/948/1/012056

[17] Sembiring R K, Hadi S and Dolk M 2008 Reforming mathematics learning in Indonesian classrooms through RME *ZDM* **40** pp 927–39

[18] Zulkardi 2002 Developing a learning environment on Realistic Mathematics Education for Indonesian student teacher (Universiteit Twente)

[19] Wijaya A, Doorman L M and Keijze R 2011 Emergent Modelling: From Traditional Indonesian Games to a Standard Unit of Measurement *J. Sci. Math. Educ. Southeast Asia* **34** pp 149–73

[20] D. J. Clarke 2006 "The LPS Research Design," in *Mathematics Classrooms in Twelve Countries: The Insider's Perspective*, edited by D. J. Clarke *et al.* (Sense Publications: Rotterdam) pp 15–36

Acknowledgments

The study were funded by PNBP Hibah Kolaborasi Internasional Unsri, 2015. The authors appreciate and thank to Ms. Evi, the teacher with her students, student teachers and teacher educator for their participations.

IOP Conf. Series: Journal of Physics: Conf. Series **948** (2018) 012057 doi:10.1088/1742-6596/948/1/012057

Arithmetic learning with the use of graphic organiser

F L Sai[1], M Shahrill[2], A Tan[3] and S H Han[4]

[1]Sekolah Menengah Sayyidina Husain, Ministry of Education, Jalan Jerudong, BG 3122, Bandar Seri Begawan, Brunei Darussalam
[2]Sultan Hassanal Bolkiah Institute of Education, Universiti Brunei Darussalam, Jalan Tungku Link, BE 1410, Bandar Seri Begawan, Brunei Darussalam
[3]Faculty of Science, Universiti Brunei Darussalam, Jalan Tungku Link, BE 1410, Bandar Seri Begawan, Brunei Darussalam
[4]Sekolah Menengah Katok, Ministry of Education, Jalan Gadong, BE 1918, Bandar Seri Begawan, Brunei Darussalam

Corresponding author e-mail: masitah.shahrill@ubd.edu.bn

Abstract. For this study, Zollman's four corners-and-a-diamond mathematics graphic organiser embedded with Polya's Problem Solving Model was used to investigate secondary school students' performance in arithmetic word problems. This instructional learning tool was used to help students break down the given information into smaller units for better strategic planning. The participants were Year 7 students, comprised of 21 male and 20 female students, aged between 11-13 years old, from a co-ed secondary school in Brunei Darussalam. This study mainly adopted a quantitative approach to investigate the types of differences found in the arithmetic word problem pre- and post-tests results from the use of the learning tool. Although the findings revealed slight improvements in the overall comparisons of the students' test results, the in-depth analysis of the students' responses in their activity worksheets shows a different outcome. Some students were able to make good attempts in breaking down the key points into smaller information in order to solve the word problems.

1. Introduction

Solving words problems plays a significant role in the current trend of Mathematics education, especially in solving those of real-life situations. It is important for students to build strategic cognitive skills for solving word problems, such as to understand a problem, to apply an appropriate method of solution and skills to relate abstract information [1]. Teachers could provide opportunities that encouraged pupils to model, to describe and to apply Mathematics understanding in appropriate and efficient ways [2, 3].

Problem solving was recommended as means in developing mathematical thinking, and referenced it as a planned action or series of actions performed to aid in the discovery of a solution [4]. Research studies supported that learning by understanding encouraged longer information retention thereby becoming a better problem solver as compared to learning through memorising [5]. Using non-routine based problems in problem solving are the most suitable example to real-life processes, because it does not follow a standard algorithm for extracting and representing the given information [6]. Instead, it provides students with opportunities to use prior knowledge to solve problem and to develop complex skills though the domains of modification over a period of time, in both cognitive and metacognitive processes [7, 8]. Additionally, posing students with open-ended mathematical problems will allow them to use alternative solution methods to solve the problems and that can guide them to a correct solution at a different complexity levels [2]. Activities conducted in the classroom should encourage effective strategies, forming the basis for understanding mathematical processes and concepts [9].

Solving mathematical problems requires students' comprehension and understanding before they are able to apply appropriate strategies to solve the problem [10]. However, students who are weak in Mathematics tend to be lacking in mathematical thinking skills [11, 12]. It is crucial to provide students with the opportunities to solve real-life word problems, and this will require students to apply their transient from algorithmic mathematical thinking to solve problem better [13].

1.1. Integrating the Pólya's problem solving model

George Pólya was a Hungarian mathematician responsible for many well-known conjectures and theorems. Pólya suggested four stages of a problem solving model aimed at helping student learners to formulate logical thinking; to apply reasoning skills and develop understandings on how to connect information from concrete to abstract. The problem solving process was systematised into four stages: understand the problem being posed, devising a plan, carry out the plan and looking back [14]. According to Khalid [15], Pólya argued that for students to understand Mathematics, their experience must be consistent with the way Mathematics is done. There is no particular approach that teachers can use. Different subject requires different kinds of approaches. In this study, Pólya's problem solving model was used as a tool as part of the strategy to examine the development of Mathematics word problems in students. Teachers are encouraged to use instructional techniques in order to improve students' attitudes towards the learning of Mathematics [16–19].

1.2. Graphic organiser embedded with Pólya's problem solving model

Zollman [20–22] introduced a four corners-and-a-diamond mathematics graphic organiser embedded with Pólya's four steps problem solving model (refer to Figure 1) to elementary and middle school students in solving algebraic word problems. The graphic organiser used was a modification from the four-squares writing graphic organiser modelled by Gould and Gould [23]. The findings from Zollman's studies revealed that the students developed positive attitude and higher confidence level towards solving Mathematics word problems with the use of the graphic organiser. The use of visual representation in a graphic organiser as a medium to aid students in understanding the word problems might eventually improve students' conceptual knowledge [24].

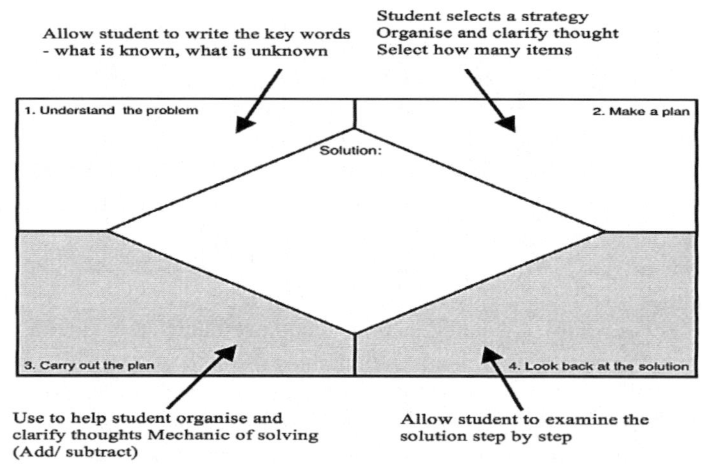

Figure 1. Zollman's graphic organiser embedded with Pólya's Problem Solving Model (taken from Zollman [20–22]).

2. Methodology

This study mainly adopted a quantitative approach to investigate the types of differences found in the arithmetic word problem tests. A diagnostic test consisting of pre- and post-tests were used to compare and analyse the students' performances quantitatively. The Zollman's four corners-and-a-diamond mathematics graphic organiser (Figure 1) embedded with Pólya's Problem Solving Model (hereafter referred to as ZPPSM) was used as the instructional tool in problem solving. The ZPPSM was introduced to provide an alternative tool that would help students improve their Mathematics achievements. We anticipated that the students might be able to break down the information into smaller units for better strategic planning. The research question for this present study is "What differences, from the use of ZPPSM, can be found in the students' arithmetic word problem tests results?"

2.1. Participants

The sample for this study was Year 7 students from a co-ed secondary school in Brunei Darussalam. There were 21 male and 20 female students, aged between 11-13 years old. The same teacher in the sampled school taught the two classes. The permission to conduct the study was approved by the Department of Schools, Ministry of Education. And all participants gave informed consent.

2.2. Procedure

Both classes took part in the pre and post-tests. The study was conducted in a normal classroom setting with students sitting in paired groupings in order to collaborate and solve a mathematical word problem. In addition, it was conducted in the afternoon, for 3 days with a total of 7 lesson periods. Each lesson period was 30 minutes. The pre-test took approximately an hour for the students to answer on Day 1. This artefact was collected as evidence to provide details on the students' problem solving performance, while simultaneously evaluating the common factors in solving arithmetic test items. The written test paper consisted of 17 simple numeral word problem questions.

On Day 2, the students engaged in a group activity and used the ZPPSM to solve the given word problem as seen in Figure 2 below. A total of 30 minutes was allocated for the Mathematics computation work. The first 20-25 minutes was allocated for the briefing of the rules and regulations for the activity task. The artefacts of the word problem was analysed according to the various responses from the students' work. All participants were allowed to use the electronic calculator during the activity. The group activity was done to investigate if the use of ZPPSM showed any significant differences in the students' performances between the arithmetic tests. The activity continued on the final day and a post-test was then administered.

Activity: ***A day out to the Night Market with a budget of $20***

In this activity, each student is given a sum of $20 to purchase any items listed on the piece of worksheet attached. The student must fully make use of the amount given. The student is allowed to choose a minimum of two items, and a maximum of 5. There is no limitation on the chances of having an item being repeatedly selected. However, you are required to show or write the possible mathematical calculation(s) that would match the criteria stated. That is, you will have to show the number of possibilities on how to obtain an answer of $20 in the total sum. The items that were provided to you for the selection, ranged from a price tag of $0.50 to $4 dollars.

Figure 2. A word problem used in the ZPPSM group activity.

3. Results and Discussions

A paired samples t-test conducted on the sample group that used ZPPSM indicated that the scores were not significant for the post-test sub-scale (M=12.73, SD=2.64), in comparison to the pre-test sub-scale (M=12.41, SD=1.96), t (40)=0.94, p=.353). Therefore, we conducted a comparison analysis of the sample that uses ZPPSM with another group that did not practice the use of ZPPSM.

The score difference between sample groups that was taught with ZPPSM and the control group are shown in Table 1. The scores of students between the tests were higher for the group taught with ZPPSM (M=12.73, SD=2.64) than the group without ZPPSM (M=11.40, SD=1.78), $t(69)=2.73$, p=.008, d=0.605, suggesting that the use of the ZPPSM would have an effect on the students' results. In particular, the findings suggested that students who were exposed to ZPPSM tended to perform better. The tests had a Cohen's d value of 0.60, indicating a moderate effect between the tests. A $p <$ 0.05 stated differences between tests, mainly that the students' results in the post-test were slightly higher as compared to the pre-test.

Table 1. Independent t-test between groups (with and without ZPPSM).

	N	Mean	SD	t	df	Sig. (2 tailed)	Cohen's d
Post-test with ZPPSM	41	12.732	2.637	2.728	68.624	0.008	0.605
Post-test without ZPPSM	47	11.404	1.777				

Table 2 shows the results of the Cronbach alpha of each item of the arithmetic pre-test instrument with a reliability coefficient of 0.607, whilst the post-test Cronbach's alpha value of 0.697 was obtained for the group that used the ZPPSM. This increase in value between both groups' reliability tests demonstrated a moderate internal reliability and hence the test was reliable to be carried out for the main study.

Table 2. Instrument item reliability test used in the tests.

	Instrument	No. of Students	No. of Items	Cronbach alpha and reliability coefficient
With ZPPSM	Pre-test	41	18	0.607
	Post-test	41	18	0.697
With & Without ZPPSM	Pre-test	88	18	0.625
	Post-test	88	18	0.647

A cross-tabulation of the paired sample t-test was conducted on the students' attempts for the pre-test items. The results showed that among the 19 items of word problem questions allocated, there were 5 significant items that had a p-value of less than 0.05. The 5 items are listed in Figure 3, indicated as Q3b, Q4b, Q5, Q14 and Q17, whilst the remaining 14 items showed little significant difference before and after the tests.

Compare and find solution: (Q3) Suzi is 12 years older than Fatin, and 7 years younger than Jenny. Jenny is 19 years old. (a) How old is Suzi? (b) How old is Fatin?

Known difference, possible ways: (Q4) Find the solution of a 3-digit number and a 2-digit number whose difference is 23. Write out two possible sets of solutions. (a) Solution answer 1 (b) Solution answer 2

Unknown total (visual/ mental computation): (Q5) Some children stood in a line. Hasya is thirteen from one end and fifth from the other end. How many children are there in the line?

Compare problems, difference unknown: (Q14) Raj has 73 balloons and Matt has 42 balloons. How many more balloons does Raj have than Matt?

Known difference and total sum, but unknown initial value: (Q17) Some money is shared between Ali and Shai, so that Ali gets $5 more than Shai gets. The money to be shared is $47. How much money does Ali and Shai get?

Figure 3. Descriptions of the identified significant question items.

Presented in Table 3 are the descriptive statistics obtained from the students' responses. The results indicated the increase in percentages for all the items with the correct answers in the post-test. Item Q3b showed a higher score in correct answer in the post-test (27.9%) when compared to the pre-test (12.2%). Meanwhile, Q4b showed a 9% increase in students obtaining correct answers and there was a reduction of 18% for 'No attempt'.

Table 3. Descriptive statistics of students' attempts on the arithmetic test items given ($N = 41$).

Item	Pre-test (%)			Post-test (%)		
	Correct Answer	Attempt	No Attempt	Correct Answer	Attempt	No Attempt
Q3a	36.6	61.0	2.4	39.5	55.8	0.0
Q3b	12.2	80.5	7.3	27.9	67.4	0.0
Q4a	9.8	63.4	26.8	18.6	58.1	18.6
Q4b	7.3	56.1	36.6	16.3	60.5	18.6
Q5	4.9	87.8	7.3	16.3	79.1	0.0
Q14	78.0	12.2	9.8	86.0	9.3	0.0
Q17	12.2	70.7	17.1	16.3	79.1	0.0

From Table 3, among the five items identified, item Q4b showed a drastic drop of 18% for 'No attempt' as compared to the pre-test, indicating that most students attempted to answer that question. Comparing both the scales, the students' no attempt decreases and the correct answer increases in the post-test. Thus displaying a significant difference in the students' effort to answer, especially in solving items Q3b, Q5, Q14 and Q17 during the post-test. In summary, the results obtained for the 5 items shown in Table 3 demonstrated the students' tendency to perform better in the post-test. To support the findings further, an analysis of the students' written attempts was made. The results in Table 4 are based on the methods preferred by the students in the tests. The students' algorithmic answers and workings were thematically categorised and analysed.

Table 4. Percentage of the item methods used.

Item	Incorrect Methodology	Incorrect Answer	Incorrect Instruction	Correct Methodology and Answer	No written answer or attempt
Q3a	43.9	4.9	9.8	36.6	4.9
Q3b	19.5	7.3	51.2	12.2	9.8
Q4a	12.2	4.9	46.3	9.8	26.8
Q4b	14.6	0.0	41.5	7.3	36.6
Q5	22.0	31.7	26.8	2.4	12.2
Q14	12.2	0.0	0.0	78.0	9.8
Q17	31.7	7.3	4.9	9.8	17.1

Note: Other methodologies used for Q5 (Attempt use of diagram = 4.9%) and Q17 (Partial correct attempt = 29.3%).

Items Q3, Q4 and Q17 (refer to the descriptions in Figure 3) shared the same patterns of having the initial value as being 'unknown'. Item Q3 is a 'compare and find solution' question requiring the participants to highlight the keywords written on the word problem and be able to solve using the information given. For item Q3a, 43.9% of the population was shown to have used the 'incorrect methodology' to solve the problem. However, 36.6% succeeded in using the 'correct methodology' and were able to obtain the 'Correct answer'. In order to solve Q3b, the answer to item Q3a is required. If a student obtained an incorrect answer in Q3a, it will cause the final response in Q3b to be incorrect. Hence the high percentage (51.2%) observed for item Q3b for the 'incorrect instruction' and 19.5% on the 'incorrect methodology' used.

Item Q4 was a 'known differences' question that had 'multiple possible ways to obtain the correct answer'. Items Q4a and Q4b shared the same type of questioning techniques. Item Q4a showed that 'incorrect instruction attempt' comprised of 46.3% of the population while 'no written answer or attempt' was 26.8%. Item Q4b showed 41.5% of 'incorrect instruction' attempt and 36.6% of 'no written answer or attempt'. Hence, Q4 item showed a high percentage of students that used the methodologies 'incorrect instruction' and 'no written answer or attempt'.

Item Q5 was an item that required participants to find the 'unknown total' either through visual or mental computation and conceptual understanding of Mathematics. For this item, the results showed few participants were able to obtain the 'correct responses' (2.4%) and only few attempts to solve on 'use of diagram' (4.9%). Amongst the participants, 31.7% of the attempts were due to 'incorrect answer', 22% on 'incorrect methodology' and 26.8% from 'incorrect instructions'. Therefore, the findings revealed that a strong relationship of failed attempts were due to the inability to use a pictorial concrete diagram to help the participants to visualise.

Simple comparison problems required the students to find out the differences between two given values, and 78% of the sample was able to obtain 'correct answer' responses with 'correct methodology' for item Q14. Consequently, it shows that the students were able to answer comparative questions when required to find the unknown differences between the initial and final values.

Item Q17 from Figure 3 is a word problem adopted from Stacey and McGregor [25] and Stacey [13]. It is a question that requires students' understanding, before attempting. The item Q17 was provided with information such as, 'total sum' and 'known difference' but not the 'initial unknown' value. The participants were required to find out the 'unknown' value as a pre-requisite task. As shown in Table 4, majority of the participants could either attempt with 'incorrect methodology' (31.7%) or to 'partial attempt with correct responses' (29.3%).

In general, the analysis of the 5 items enabled us to realise that pictorial concrete representation could aid students' understanding in answering a word problem. The students also achieved higher success in comparative questions. Various attempts of problem solving methodology could compromise the lack of mathematical understanding, especially in the case of 'unknown initial'. In addition, the students either use 'incorrect instruction or methodology' leading to incorrect solution and failure to understand. However, the findings did show there were higher rate of attempts by the students to obtain correct answers in the post-test and they scored better in the post-test as compared to the pre-test. Therefore, the use of ZPPSM has improved the students' ability to boost their mathematical knowledge and constructive experiences in benefitting them to solve problems better.

3.1. Analysis from the ZPPSM activity sheet

Most students worked in pairs, except for the Group 1 student who had to work individually due to the odd number of students' attendance that day. From Table 5, there were a total of 7 groups of students with correct responses of zero (3 for Group 1 and 4 for Group 2), and this consisted of 31.8% of the sample. Group 2 had 12 correct answer responses (54.5%) as compared to Group 1 (45.5%).

Table 5. The frequency of students' correct responses using ZPPSM during the activity.

Class	Number of correct responses													Total correct responses
	0	1	2	3	4	5	6	7	8	9	10	11	12	
Group 1	3	0	1	0	1	1	0	1	0	1	1	1	0	10
Group 2	4	3	0	1	0	0	0	1	0	1	0	1	1	12

The group that showed a high success rate in obtaining correct responses had a much more organised key points written on the ZPPSM (refer to Figures 4, 5 and 6). Therefore, the use of the ZPPSM aided the students to break down the information into small pieces (refer to Figures 5 and 6).

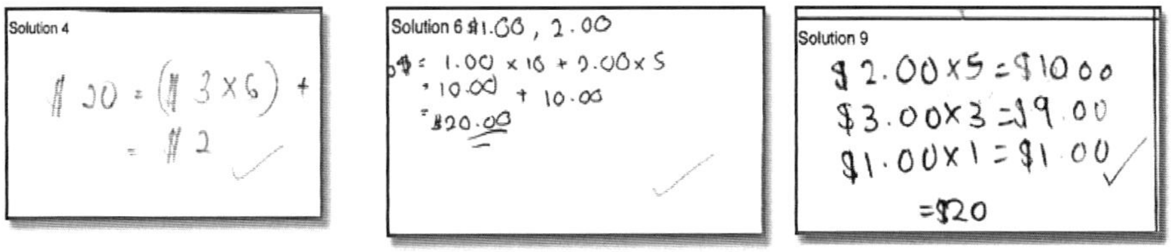

Figure 4. Samples of correct solutions on the activity sheet.

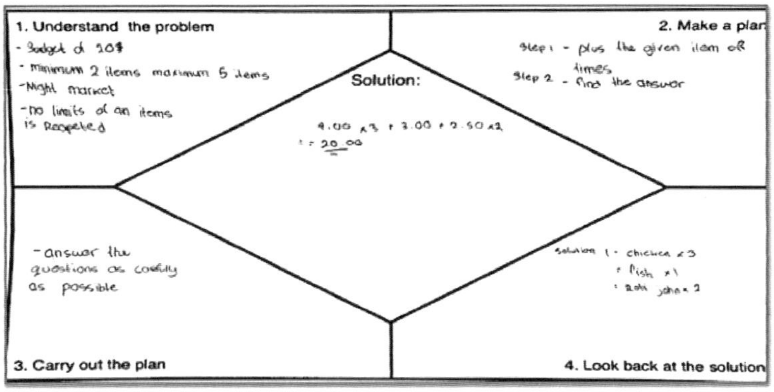

Figure 5. Sampled student 1 artefact with the use of ZPPSM.

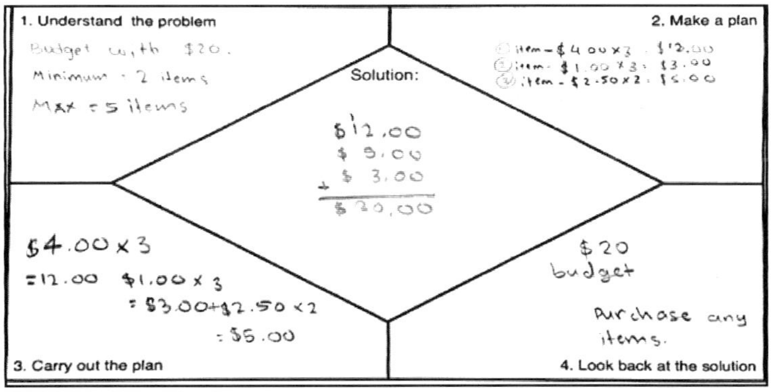

Figure 6. Sampled student 2 artefact with the use of ZPPSM.

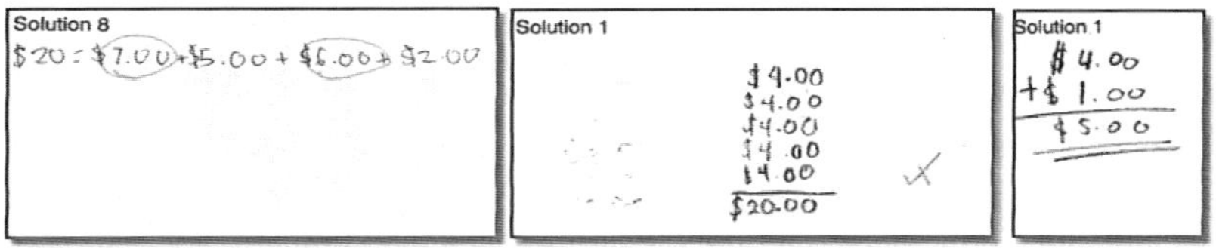

(a) (b) (c)

Figure 7. Samples of incorrect student responses written on the activity sheet.

The three different artefacts shown in Figure 7 demonstrated the students' 'incorrect instruction' attempts. For example, the first artefact Figure 7a), illustrated the 'incorrect instruction' sample in which the $7 and $8 values do not exists on the activity sheet. The artefact Figure 7b) showed that the student obtain the correct answer but using an incorrect instruction. This is because the student was supposed to select at least two different cost items; but only selected the same $4 items to produce a total sum of $20. Thus, it was categorised as the incorrect instruction attempts. The incorrect response shown in Figure 7c), having a total sum of $5 was insufficient to match with the acquired total amount of $20. In reference to the students' written attempts on the ZPPSM worksheets demonstrated that the group did not fully utilise the usage of the ZPPSM, and produced few correct responses. This may also be in part the limitation of this study since the students were only exposed to one cycle on the use of ZPPSM. Nonetheless, the findings above provided a positive grounding in the use of ZPPSM as an effective strategic tool to help students learn arithmetic word problems. The strategic ZPPSM tool would promote transient Mathematics thinking through the active organising and re-organising of concepts and processes [9, 11, 13, 15, 26].

4. Conclusions

Khoo et al. [10] suggested that the Zollman's graphic organiser instructional tool to be an effective tool in helping students to solve problems and develop a positive attitude towards solving Mathematics word problem. Comparing the activity artefacts, it can be concluded that the students who practice Mathematics problem solving with the use of ZPPSM made attempts to break down key points into smaller information. Those students who showed legible and orderly work with ZPPSM tend to have higher outcome success than those who did not use ZPPSM (referring to Table 5 and Figures 4-7). Despite the moderate 'effect size' of using ZPPSM, it did indicate that the arithmetic post-test result was of higher reliability. However, we do realise that the two classes involved in this study, especially from only one secondary school, were insufficient in terms of providing further insights into the students' learning. Yet, the use of ZPPSM has shown to be effective in helping students improve their Mathematics results by breaking down information into smaller chunks [10]. It may be beneficial to introduce the ZPPSM as a strategic tool to student, providing them with opportunities to learn and to develop their understandings.

The analysis using paired sample t-test on the arithmetic tests indicated that the students had difficulties with questions having 'unknown initial' variables. With reference to the above analysis, the students could perform 'Join', 'Separate' and 'Part-whole' kinds of arithmetic questions with ease, except the question items with 'unknown initial' as listed in Figure 3. The students will require more effort to find the solutions for this type of word problems. Moreover, Stacey [13] found that the basic difficulties in problem solving approach were due to the idea of unknown that appears together with the given task, regardless if it is transformative thinking or not. Meanwhile, the results in Table 4 indicated that students with 'no written answer or attempt' had a high tendency towards 'incorrect instructions' leading to 'incorrect methodology' and thus, 'incorrect answer' obtained. Therefore, by providing students with opportunities to learn and willingness to attempt would develop potentials to improve. Nevertheless, strategic tools such as the ZPPSM used in this study will benefit students' brainstorming process with its visual representation and is believed to resolve students' difficulties in solving word problems; and to provide opportunities that encourage students to develop positive attitudes towards solving Mathematics [10].

The findings in this study revealed that the students' performance was slightly better in the post-test when compared to the pre-test. However, there were some students who still faced difficulties in familiarising themselves with the use of the ZPPSM. The findings also indicated that the student groups that obtained the low marks failed to optimise the use of the ZPPSM hence leading to failures to follow instructions given and to obtain correct responses. In addition, the underperforming students would require more time to fill the empty boxes in the graphic organiser; and thus a prolong exposure of the ZPPSM would better benefit and allow the students to acquaint themselves with the four stages of Pólya's problem solving model.

This study has shown that the use of the ZPPSM improved the students' results. Therefore, it is recommended that teachers or educators to consider using the ZPPSM in their lessons. It is advisable that the teacher provides more time and practice for their students to familiarise themselves with the use of ZPPSM as a strategic instructional tool. The teacher should also be well trained on the applications of the Pólya's problem solving model well before implementing them to their lessons.

References

[1] Gasco J, Villarroel J D and Zuazagoitia D 2014 Different procedures for solving Mathematical word problems in high school *International Education Studies* **7** 77–84
[2] Silver E A 1997 Fostering creativity through instruction rich in mathematical problem solving and problem posing *ZDM* **29** 75–80.
[3] Singh P 2006 An analysis of word problems in school Mathematics texts: Operation of addition and subtraction *Journal of Science and Mathematics Education in Southeast Asia* **29** 41–61
[4] Cockcroft W H 1982 *Mathematics Counts* (London: HM Stationery Office)
[5] Hunt E 1994 Problem solving *Thinking and Problem Solving* ed R J Sternberg (New York: Academic Press) pp 215–31
[6] Chong M S F and Shahrill M 2016 The use of an emerging framework to explore students' cognitive competency *Indian Journal of Science and Technology* **9**
[7] Schmidt H G 1989 The rationale behind problem-based learning *New Directions for Medical Education* (New York: Springer) pp 105–11
[8] Schoenfeld A H 1992 Learning to think mathematically: Problem solving, metacognition, and sense making in Mathematics *Handbook of Research on Mathematics Teaching and Learning* ed D Grouws (New York: MacMillan) pp 334–70
[9] Mardjetko E and Macpherson J 2007 Is your classroom mental? Guidelines for enhancing the development of strategies for mental computation *Australian Primary Mathematics Classroom* **12** 4–9
[10] Khoo J S, Shahrill M, Yusof N, Chua G L L and Roslan R 2016 Graphic organizer in action: Solving secondary mathematics problems *Journal on Mathematics Education* **7** 83–90
[11] Botty H M R H, Shahrill M, Jaidin J H, Li H-C and Chong M S F 2016 The implementation of problem-based learning (PBL) in a year 9 mathematics classroom: A study in Brunei Darussalam *International Research in Education* **4** 34–47
[12] Chua G L L, Shahrill M and Tan A 2016 Common misconceptions of algebraic problems: Identifying trends and proposing possible remedial measures *Advanced Science Letters* **22** 1547–50
[13] Stacey K 1999 Priorities for middle school Mathematics: New directions from research for number and algebra *Cross Section* **11** 4–7
[14] Pólya G 1945 *How to Solve It: A New Aspect of Mathematical Method* (New Jersey: Princeton University Press)
[15] Khalid M 2006 *Mathematical thinking in Brunei curriculum: Implementation issues and challenges* Paper presented at the meeting of the APEC-Tsukuba International Conferences Japan
[16] Mundia L 2012 The assessment of math learning difficulties in a primary grade-4 child with high support needs: Mixed methods approach *International Electronic Journal of Elementary Education* **4** 347–66
[17] Yassin N F and Shahrill M 2016 *Using polya's principles as an approach to helping students make sense of problem solving International Journal of Interdisciplinary Educational Studies*, **11** 13–34

[18] Suhaimi Z, Shahrill M, Tengah K A and Abbas N A H 2016 Incorporating the use of writing-to-learn strategy in grade 10 mathematics lessons: The students' perspectives *Journal of Mathematics Education at Teachers College* **7** 11–20
[19] Suhaimi Z, Shahrill M, Abbas N A H, Tengah K A, Roslan R and Yusof N 2017 Exploring the use of journal writing in mathematics classroom *International Journal on Emerging Mathematics Education* **1** 39–50
[20] Zollman A 2009 Mathematical graphic organizers *Teaching Children Mathematics* **16** 222–29
[21] Zollman A 2009 Students use graphic organizers to improve mathematical problem-solving communications *Middle School Journal* **41** 4–12
[22] Zollman A 2012 Write is right: Using graphic organizers to improve student mathematical problem solving *Investigations in Mathematics Learning* **4** 50–60
[23] Gould J and Gould E 1999 *Four Square Writing Method for Grades 1-3* (Illinois: Teaching and Learning Company)
[24] Abdullah N, Zakaria E and Halim L 2012 The effect of a thinking strategy approach through visual representation on achievement and conceptual understanding in solving Mathematical word problems *Asian Social Science* **8** 30–37
[25] Stacey K and MacGregor M 1999 Learning the Algebraic method of solving problems *The Journal of Mathematical Behavior* **18** 149–67
[26] Botty H M R H, Taha H Z H M, Shahrill M and Mahadi M A 2015 Connecting students' achievements with attitudes, the teachings and study habit *Mediterranean Journal of Social Sciences* **6** 113–19

IOP Conf. Series: Journal of Physics: Conf. Series **948** (2018) 012058 doi:10.1088/1742-6596/948/1/012058

Models to support students' understanding of measuring area of circles

S Rejeki[1)] and R I I Putri[2)]

Universitas Muhammadiyah Surakarta, Central Java, Indonesia[1)], Universitas Sriwijaya, South Sumatra, Indonesia[2)]

E-mail: sri.rejeki@ums.ac.id

Abstract. Many studies showed that enormous students got confused about the concepts of measuring area of circles. The main reason is because mathematics classroom practices emphasized on memorizing formulas rather than understanding concepts. Therefore, in this study, a set of learning activities were designed as an innovation in learning area measurement of circles. The activities involved two models namely grid paper and reshaping which are respectively as a means and a strategy to support students' learning of area measurement of circles. Design research was used as the research approach to achieve the aim. Thirty-eight of 8th graders in Indonesia were involved in this study. In this study, together with the contextual problems, the grid paper and reshaping sectors, which used as the models in this learning, helped the students to gradually develop their understanding of the area measurement of circles. The grid papers plays important role in comparing and estimating areas. Whereas, the reshaping sectors might support students' understanding of the circumference and the area measurement of circles. Those two models could be the tool for promoting the informal theory of area measurement. Besides, the whole activities gave important role on distinguishing the area and perimeter of circles.

1. Introduction

Area measurement is an essential part of mathematics contents [1–3]. Cavanagh [2] stated that area measurement is a crucial part in middle-school mathematics curriculum because of two main reasons. First, it is because there are various applications of area concepts in daily lives' activities such as painting, gardening, tiling, and any tasks involving covering a two dimensional surface. Second, it is because area concepts are needed to introduce many other mathematical ideas. Furthermore, area measurement is also the foundation to understand volume measurement in solid geometry [3].

However, it is not only students, but also teachers experience difficulties with this topic. The main difficulty is a confusion between the concepts of area and perimeter [4,5]. For instance, several studies found that some Indonesian students had counted perimeters to answer the questions about areas [6]. In other words, the students used the formula of area and perimeter incorrectly. Previous studies show that the reasons behind students' difficulties is because learning experiences provided in school give more focus on memorizing formulas, rather than understanding concept [7].

Related to the difficulties in learning area measurement, some studies have been conducted to support students' understanding in the concepts [8]. However, these studies talked about the area of all two dimensional figures. There is not yet a particular investigation about area of circles. Moreover, based on an analysis of Indonesian mathematics textbooks for the 8thgrade of secondary school [9–13], the books have provided various contexts on the topic of area of circles. However, in these cases, there is still a gap from the informal level to the formal level, from the concrete situation to the abstract situation [14]. In the case of area measurement of circles, contexts are used only as attributes to introduce the shape of a circle. Afterwards, the lesson is directly presented in formal level. Therefore, an innovation is needed to promote students' learning, particularly in the topics of area and area measurement of circles.

The innovation is not only about an implementation of a way approach of learning mathematics, but also a new way of thinking about the purpose and practices of school mathematics. In this case, Realistic Mathematics Education (RME) is an approach which provides a new way of learning mathematics and a new way of thinking about the purpose and practices of school mathematics [15]. In this study, we designed a set of learning activities based on PMRI approach. This is based on the five tenets of RME, namely phenomenological exploration, using models and symbols for progressive mathematization, using students' own construction, interactivity, and intertwinement [16] .

There were several contexts which aimed at guiding students' thinking to the correct orientation of area. Two models were used to promote students' understanding of area measurement of circles. Models are various vertical instruments used as bridge connected the informal level to the formal level [16]. It could be schemas, diagrams, symbols, strategies, and so forth.

The first model is grid paper. The used of the model is based on the theory which described that the area could be determined by counting or approximating the number of units covering a shape, dependent on the length, width, and configuration, can be calculated using a rule or formula, and determined by the sum of small parts [5]. In this study, grid paper was used as a model to support students understanding of the area of a circle and the way to measure it by estimation. The grids are representation of the units of area.

The second model is reshaping sectors. It is based on the theory about area conservation. The area of a shape equals the sum of the area of the parts, two shapes form a new shape when a part of the boundaries are put together and the area of the new shape is the sum of the original area [5]. In this study, reshaping the sectors of a circle into a rectangle and other two dimensional figures help students to derive the formula to measure the area of a circle.

Considering the issues about students' confusion about the area of circles and the need of an innovation to promote students' learning about area measurement of circles, this study aims to contribute to a local instructional theory which can support the development of students' understanding of the area measurement of circles. However, *this paper will only discuss about how grid papers and reshaping sectors as models used in the learning activities support students' understanding of area measurement of circles.*

2. Methods

Design research was chosen as the research approach used in this study. We first designed a set of learning activities and a Hypothetical Learning Trajectories (HLT) for the learning of area measurement of circles. Afterwards, a pilot experiment was conducted. The results is used to adjust the content of the learning activities and the sequence of the activities which have been established. Then, second cycle was conducted as the actual teaching process in which the sequence of activities was implemented in a natural classroom environment.

Thirty-eight 8th graders of Junior High School were involved in this study. Six students participated in the pilot experiment and thirty-two students contributed in the teaching experiment. Students worked in group with various level of achievement which were seen from the results of pre-test conducted before the experiment.

The data were collected through students' written work, video registration, students' interview, and field notes during the teaching experiment. Those collections of information helped us in interpreting and enable us to make data triangulation. All the data were analyzed qualitatively and the analysis focused on comparing the actual learning and the HLT.

3. Hypothetical Learning Trajectory

The following table describe the refined HLT which is implemented in the teaching experiment. The revision was based on the findings in the pilot experiment and the interview with the teacher. As a result, there were several activities which were conducted in four lessons [17].

Table 1. Refined Hypothetical Learning Trajectory

Activity	Main Goals	Description of Activity	Conjectures of Students' thinking
Comparing and Tiling	Students are able to compare the size of figures and measure the area of plane figures using measurement units in daily life.	- Students are doing an activity of ordering the five biggest islands in Indonesia starting by a contextual problem of covering the miniature of islands by grass.	- Some students might compare and order based on the shape of the islands. - Some students might use ruler and calculate the area of the islands. - Some students might overlap the islands, then do cutting and pasting.
		- Students are continuing a tiling activity with two different size of tiles.	- Some students might continue the tiling activity by drawing the tiles one by one and counting the tiles to determine the number of tiles needed to cover. - Some students might use ruler to continue the tiling activity and determine the number of tiles by using formula.
Measuring Figures Using A Grid Paper	Students are able to measure the area of a plane figure by using a grid paper or by making their own grids.	- Students are doing an activity of comparing two lakes on a grid paper.	- Some students might reshape the figures of the lakes and disregard the grid paper - Some students might compare those two lakes by counting the grids inside.
		- The students are doing an activity of estimating the grass needed to cover a circular garden by making grids.	- Some students might say that the area of the circular garden is ¾ of the area of the square land. - Some student might determine the area as greater than $3r^2$ and less than $4r^2$. - Some students might directly count the grids then draw bigger grids to make their work easier or make smaller grids to make their result more accurate.

Reshaping A Circle into A Rectangle	Students are able to determine the area of a circle which is reshaped into a rectangle.	- The students are doing activities of reshaping the sectors of a circle into a parallelogram, then into a rectangle and determining the area of the rectangle.	- Some students might reshape the sectors of a circle into a parallelogram easily, but they might find difficulty to reshape the parallelogram into a rectangle. - Some students might reshape the parallelogram into a rectangle easily and said that the area of the rectangle could be found by multiplying its length and width. - Some students might say that the area of the rectangle is the radius times half of the circumference of a circle because they relate the parts of the rectangle with the part of the circle.
Reshaping A Circle into Other Plane Figures	Students are able to reshape the sectors of a circle into other plane figures and explain the relations between the formula of the area of a circle and the formulas of the area of those plane figures.	- The students are doing activities of reshaping the sectors of circle into several kinds of plane figures and determining the area.	- Some students might arrange the sectors of circle only into a parallelogram and a rectangle and determine the area. - Some students might reshape the sectors into a rectangle, a parallelogram. A triangle and a trapezoid and determine the area.

4. Results

4.1. Grid Paper

The grid paper was used in the second activity which is measuring figures using a grid paper. There were two activities in this lesson. The first activity was comparing two lakes on a grid paper and the second activity was estimating the grass needed to cover a circular garden on a square land by making grids. The students worked in the group with various level of mathematics performances. The group where students worked was remain the same as the previous lesson.

In the first activity, most students counted the grids to justify the larger lake. As the students did in dealing with the incomplete tiles in the previous lesson, they combined the incomplete grids into one grids and added them with the complete grids. However, some students applied different strategy which was reshaping strategy as shown in this following figure.

IOP Conf. Series: Journal of Physics: Conf. Series **948** (2018) 012058 doi:10.1088/1742-6596/948/1/012058

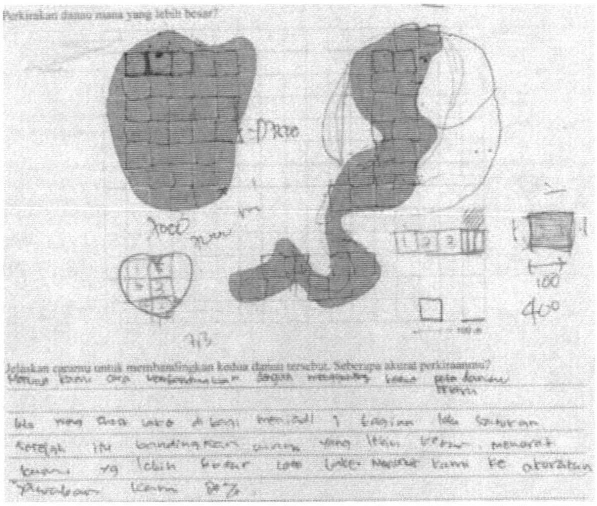

Figure 1. A student's written work on the activity 2-1

The written work shows that the students cut the second figures and reshaped it into more similar figure with the first one. Afterwards, they compared those two figures and justified the larger lake by looking at the remaining parts. However, the class discussion for this activity emphasized the role of grids in a comparing activity.

In the second activity, the students were asked to estimate the amount of grass needed to cover a circular garden on a square land. The expectation was that the students could imitate the strategy they applied in the first activity to solve the problem in this activity. And, as it was expected in the HLT, most students estimate the grass needed to cover the area of the circular land by counting the complete grids and estimate the incomplete grids. None of students estimated by considering the square land, as fore-mentioned in the HLT.

Although the idea was the same which was counting the grids inside the circle, the students' written works showed various way to determine the grass needed to cover the circular garden. Interestingly, all worksheet showed that students also applied formula to determine the area inside the circle. The following figure shows an example of students' written work.

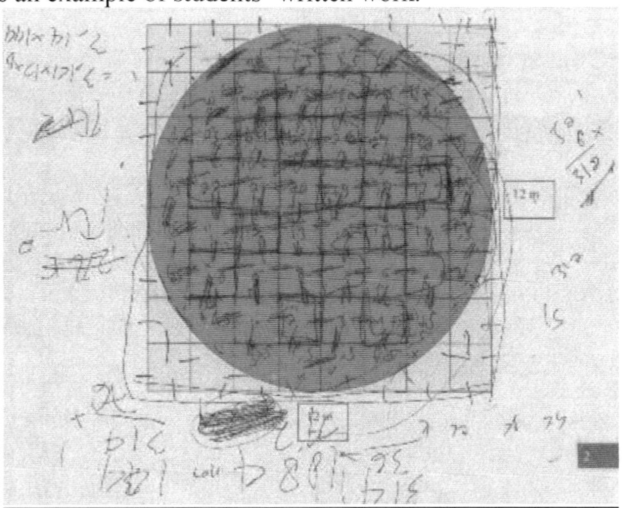

Figure 2. A student's written work on activity 2-2

Generally, they tried to determine the number of grids using their own way which is counting the complete grids and estimate the incomplete grids. However, what they do by applying the formula of the area of a circle indicates that they wanted to make sure about the accuracy of their estimation. Thus, in the class discussion, the students could see that the results of their work and the results of the formula were about the same. Furthermore, it could be a starting point for exploration about the formula of the area of a circle in the next lessons.

4.2. Reshaping Sectors

The mathematical idea in this activity is that the area of a circle equals the radius times half of its circumference. There was only one main activity in this meeting namely reshaping three circles which were cut into different numbers of sectors into rectangles and determining the way to find its area. However, to make the students get the idea of reshaping the sectors of a circle into a rectangle, firstly there was an activity of reshaping a circular garden into a shape resembles parallelogram which was guided by the teacher. In the second problem, the students were given three circle which are divided into 8, 12, and 16 sectors, respectively. The students were asked to reshape the circle into a parallelogram, then into a rectangle. The students worked in the same group as they did in the previous lessons.

After reshaping all three circles, the students could see that the shapes the formed resemble the shape of parallelogram. However, some students did not get the idea how to reshape the parallelogram into a rectangle. Hence, a discussion which lead by the teacher help them to get the shape of rectangles.

As it was expected in the HLT, the students could determine that the area of the rectangle which is length times width. In this activity, they can clearly see that the width of the rectangle is the radius of the circle. However, it was difficult for some students to see the relation between the length of the rectangle and parts of the circle. Hence the teacher and the researcher gave them kinds of guidance as shown in this following transcript.

Transcript 1. Discourse about the problem of circumference

R	: *How many sectors is this circle being cut?*
S1	: *Twelve.*
R+S	: *One-two-three-four-five-six.* *(The researcher is pointing the arcs of the sectors of the circle which are formed a rectangle).*
R	: *How about this one? Which part of a circle is this?* *(The researcher is pointing the initial circle).*
S2	: ***Circumference**.*
R+S	: *One-two-three-four-five-six.* *(The researcher is pointing the arcs of the sectors of the initial circle).*
S2	: ***Half of its circumference**.*

The teacher asked the students to look at the initial circle and to see that the length of all arcs of the twelve sectors is the circumference of the circle. Afterwards, together with the researcher, the students pointed out the arcs of six sectors and a students realized that it is half of the circumference. This fragment indicates that the students could see that the length of the rectangle equals half of the circumference of a circle. However, there were no depth discussion to check whether all students really understand this concept after a student came up with this idea.

In the end of the lesson, the teacher asked a group to share their work in front of class. There were two important conclusions in this lesson. Firstly, the students could conclude that the more number of sectors formed from the initial circle, the more similar the rectangle they formed. Secondly, the area of the rectangle can be determined by multiply the length and the width. The length is eight times the length of arcs and the width is the radius of the circle. In this case, the students who share their work in front of class did not state the length of the rectangle as half of circumference. Hence, the teacher posed a question to the classroom as described in Transcript 2.

Transcript 2. Classroom discussion about the problem of circumference

R : *Your friend said that the length of the rectangle is eight times the length of each arc, do you all agree with his statement?*

S : *Yes (the students answered together, yet some of them just mumbling because the other said yes).*

R : *If you say yes, can you explain how to determine the length of the arcs?*

S : *Mmm, firstly we find the length of an arc, then... (The student get stuck)*

R : *How about the other? Who have another idea?*

S2 : *The length of the rectangle is half of its circumference Miss, because there are eight arcs and the total is sixteen.*

Based on the fragment and the students' written works, we may conclude that some students can determine the area of the rectangle as radius times half of circumference. However, some students still confuse to determine the length of the rectangle and think that it is eight times the length of the arcs without any ideas to find the length of the arcs. However, the discussion help them to understand about this concept.

5. Concluding Discussion

Based on the analysis of several mathematics textbook for junior high school [9–13], the contexts provided in those books are only as attributes. It is because there is still a gap from the informal level to the formal level. In other words, in this case, contexts do not play essential role and can be dropped without affecting the mathematics required [5]. Therefore, the use of models plays important roles in bridging the informal level and the formal level [18]. Furthermore, based on the previous study [19], most of textbook that used in Indonesia contain mainly the set of rules and algorithm which is already formal and lack of applications which are needed by students in order to make the concept be real for them. The learning process only emphasized the procedures to solve the problems and there is insufficient attention for students' reasoning in order to develop their understanding.

In this study, together with the contextual problems, the grid paper and reshaping sectors, which used as the models in this learning, helped the students to gradually develop their understanding of the area of a circle. The grid papers plays important role in not only in estimating areas as expected in the HLT, but also in comparing figures in the beginning activity. This is in line with the theory of calculation using unit of measurement [5]. Moreover, it could be the tool for promoting the informal theory of area measurement [20].

Whereas, the reshaping sectors may support students understanding of the circumference and the area measurement of circles. It is in line with the theory of area conservation and could be the means to support the informal theory of circumference and area measurement [5,20].

However, the students are sometimes coming back to the formal way in determining the area of a circle. Therefore, this study can be done in the very beginning students learn about the area of a circle which is in the sixth grade of elementary school. Yet, we should consider the level of difficulties on the problems designed for the elementary school students.

Moreover, the contexts which used in the problems on each activity were familiar for the students, especially garden and tiling context. However, it is necessary to expand the variation of the context so that the students do not limit the application of the area of a circle just in certain context. Thus, the other suggestion for the next study is give some more contexts as the enrichment material to the students.

6. References

[1] Kordaki M and Potari D 1998 Children's approaches to area measurement through different contexts *J. Math. Behav.* **17** 303–16

[2] Cavanagh M 2008 Area Measurement in Year 7 *Educ. Stud. Math.* **33** 55–8

[3] Battista, M. T 2003 No Title *Learning and Teaching Measurement* ed H Clements, D. and G Bright (Reston: NCTM) pp 122–42
[4] Kidman G and Cooper T J 1997 Area integration rules for grades 4, 6, and 8 students *The 21st international Conference for the Psychology of Mathematics Education* pp 136–43
[5] Oldham E, Valk T V D, Broekman H and Berenson S 1999 Beginning Pre-service Teachers' Approaches to Teaching the Area Concept: identifying tendencies towards realistic, structuralist, mechanist or empiricist mathematics education *Eur. J. Teach. Educ.* **22** 23–43
[6] Abdussakir and Aschadiyah, N. L 2009 Pembelajaran Keliling dan Luas Lingkaran dengan Strategi REACT pada Siswa Kelas VII SMP Negeri Kota Mojokerto *Seminar Nasional Matematika dan Pendidikan Matematika Jurusan Matematika FMIPA UNY* pp 388–401
[7] Baturo A and Nason R 1996 Students Teachers' Subject Matter Knowledge within the Domainof Area Measurement *Educ. Stuies Math.* **31** 235–68
[8] Winarni D W 2010 *Learning the Concepts of Area and Perimeter by Exploring Their Relation* (Sriwijaya University)
[9] Agus N A 2007 *Matematika Untuk SMP dan MTs Kelas VIII* (Jakarta: Pusat Perbukuan Departemen Pendidikan Nasional)
[10] Nuharini D and Wahyuni T 2008 *Matematika konsep dan Aplikasinya* (Jakarta: Pusat Perbukuan Departemen Pendidikan Nasional)
[11] Nugroho H and Meisaroh L 2009 *Matematika Untuk SMP dan MTs Kelas VIII* (Jakarta: Pusat Perbukuan Departemen Pendidikan Nasional)
[12] Dris J and Tasari 2011 *Matematika Untuk SMP dan MTs KElas VIII* (Jakarta: Pusat Kurikulum dan Perbukuan Departemen Pendidikan NAsional)
[13] Marsigit and Al E 2011 *Matematika Untuk SMP dan MTs Kelas VIII* (Jakarta: Pusat Perbukuan Kementrian Pendidikan Nasional)
[14] Murtiyasa B, Rejeki S and Murdaningsih S 2016 An Analysis of Problems on Eight Grade of Mathematics Textbook Based on PIsa's Framework *3rd International Conference on Research, Implementation, and Education of Mathematics and Science (3rd ICRIEMS)* (Yogyakarta: UNY Press) pp 305–8
[15] Sembiring R K, Hadi S and Dolk M 2008 Reforming mathematics learning in Indonesian classrooms through RME *ZDM* **40** 927–39
[16] Treffers A 1987 *Three dimensions. A model of goal and theory description in mathematics instruction the wiskobas project* (Dordrecht: Reidel Publishing Company)
[17] Rejeki S 2014 *Supporting 8th Grade Students' Understanding of The Area Measurement of A Circle* (Utrecht University - Sriwijaya University)
[18] Wijaya A, Doorman L M and Keijze R 2011 Emergent Modelling: From Traditional Indonesian Games to a Standard Unit of Measurement *J. Sci. Math. Educ. Southeast Asia* **34** 149–73
[19] Zulkardi 2002 *Developing a learning environment on Realistic Mathematics Education for Indonesian student teacher* (Universiteit Twente)
[20] Barrett J E, Cullen C, Sarama J, Clements D H, Klanderman D, Miller A L and Rumsey C 2011 Children's unit concepts in measurement: A teaching experiment spanning grades 2 through 5 *ZDM - Int. J. Math. Educ.* **43** 637–50

Acknowledgements

The instructional activities in this study were designed in collaboration with Monica Wijers and Dolly van Eerde, Utrecht University, The Netherlands. The activities were conducted within the International Master Program on Mathematics Education (IMPoME) which was funded jointly by DIKTI, Indonesia and NUFFIC NESO, The Netherlands.

An early version of this paper which discussed about the first cycle of this study was published in Ratu Ilma (Eds). The Second South East Asia Design/ Development Research (SEA-DR) International Conference, April 26th-27th, 2014, Palembang, Indonesia and in Safnil (Eds). The 2015 International Seminar on Education, University of Bengkulu, January 16th-18th, 2015, Bengkulu, Indonesia.

Logical errors on proving theorem

C K Sari[1,a], M Waluyo[1], C M Ainur[1] and E N Darmaningsih[1]

[1]Universitas Muhammadiyah Surakarta, Sukoharjo, Indonesia

[a]E-mail: christina.k.sari@ums.ac.id

Abstract. In tertiary level, students of mathematics education department attend some abstract courses, such as Introduction to Real Analysis which needs an ability to prove mathematical statements almost all the time. In fact, many students have not mastered this ability appropriately. In their Introduction to Real Analysis tests, even though they completed their proof of theorems, they achieved an unsatisfactory score. They thought that they succeeded, but their proof was not valid. In this study, a qualitative research was conducted to describe logical errors that students made in proving the theorem of cluster point. The theorem was given to 54 students. Misconceptions on understanding the definitions seem to occur within cluster point, limit of function, and limit of sequences. The habit of using routine symbol might cause these misconceptions. Suggestions to deal with this condition are described as well.

1. Introduction

When students enroll tertiary level, especially in mathematics field, they are faced with abstract mathematics. They increasingly encounter new cases related to the proof of characteristics and theorems, as in the Introduction to Real Analysis course. The course studies abstract concepts, such as characteristics of real numbers and their proof, sequences of real numbers and characteristics of its convergence. Proving mathematical statements and the proof they face at this level are different from what they used in high school level [1]. The ability to construct proof is crucial for them [2][3]. They need it to explain the truth of mathematical statements [4][3][5][6], to discover new theorems and to arrange mathematics systematically [4][5]. Even though it is important, students often have problems in constructing proof [7][8][9]. In fact, the proof of mathematical statements is not unique. For instance, a theorem could be validated through various proofs. The new proofs of a theorem have evolved to enrich mathematics [10]. Therefore, students have more choices to use effective ways to prove mathematical statements. However, according to Selden and Selden [11], when students face proving problems, they cannot complete it because they get confused in using definitions or implementing mathematical concepts.

Proving involves thinking about a new concept, focusing on important aspects, using existing knowledge, looking at existing relationships, defining new things (if necessary), and making valid arguments [12]. A formal proof is a sequence of formulations that forms, consists of formed axioms or other conclusions, and ends up with a proven theorem [10]. Basically, proofs consist of two aspects, namely the formal-rhetorical part and the problem-centered part. The first part is like proof framework, it consists of statements given by the theorem, definitions, earlier results, or other statements which do not need a deep understanding. The latter is a core statement of the problem which needs deep understanding of concepts and intuitions. Weber [2] categorizes students' proofs into four categories: 1) correct proof, 2) fail to assemble the syntactic knowledge, 3) lack sufficient syntactic knowledge as the basis of proof, and 4) logical error. The first is valid proof, others are incorrect. When students fail to use their syntactic knowledge, they write incoherent statements because they cannot compile their possessed concepts. Meanwhile, students lack sufficient syntactic knowledge if they do not know the definitions, theorems, or other mathematical statements. Therefore, they are not able to complete their proof. The latter, students complete the proof, but their proof is incorrect. Students usually do an incorrect proof by assuming statement that would prove to construct its proof, generating specific example to be generalized into a general statement, doing one way of implication on proving bi-implication statements [13].

Considering that most of the contents of Introduction to Real Analysis are concepts and theorems, reasoning on proving is absolutely necessary. Various students' proofs have to be known, especially logical error on proving, a proof that is usually considered a correct proof. This study aims to identify the proof produced by the students in solving Introduction to Real Analysis problem, especially invalid proofs. Furthermore, the lecturer could give action in the teaching and learning process so that the invalid proofs will be no longer conducted by the students.

2. Methods

A qualitative research was conducted to identify the students' proofs in proving Introduction to Real Analysis case. In this study, the subject of the study was determined by purposive sampling. The subject consisted of 54 students of Introduction to Real Analysis courses. The main data were the result of writing test, while the supporting data were collected by observations of teaching and learning activities and the results of structured tasks. An investigator triangulation was employed to check the validity of data, i.e. checking by some observers or experts [14].

3. Results and Discussion

3.1. General Result

A case of proof related to cluster point was adopted from the book by Bartle and Sherbert [15]. The case asked the students to prove a part of the theorem:

Given a set $A \subseteq R$ *and* $c \in R$. *Show that if there is a sequence of* $(x_n) \subseteq A$, $x_n \neq c$ *for all* $n \in N$ *and* $x_n \to c$, *then* c *is a cluster point of* A.

In the case, R symbolizes the real number system. The case was presented to the 54 students who had attended Introduction to Real Analysis course. As many as thirteen students left the case without responses and the rest tried to prove the case in various ways. The way to prove the theorem is not unique. Proofs of a theorem still appear with wide variations, even more efficient than when the proof is introduced at first [10].

Furthermore, a variety of proof generated by 41 students is discussed further. Figure 1 shows an outline of the proofs.

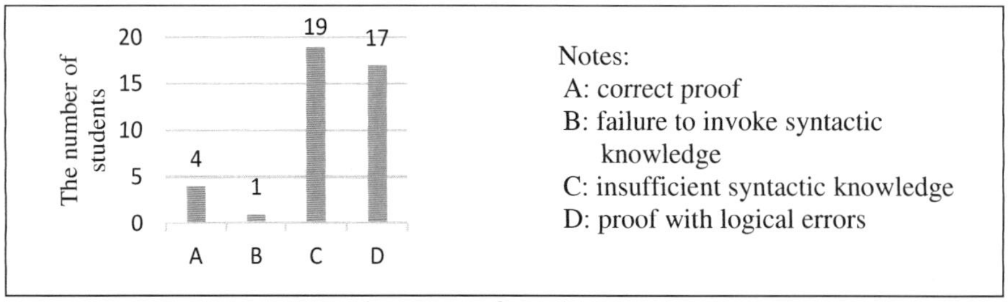

Figure 1. Students' proof on cluster point theorem

Almost a half of number of students who tried to prove the theorem are not able to complete it because they lack syntactic knowledge, such as definitions, lemmas, or other theorems that could be used to prove the theorem.

Several students construct proof but their proof is invalid (41.4%). They make errors in understanding the target of proving, proving by examples, or making errors in understanding concepts related to cluster point. There are five students who are proving by using a specific example. In fact, the theorem is a general statement. More detail about proof by using specific example could be seen in Sari et al [16].

3.2 Logical errors

A total of six students seem to get confused concerning a cluster point of $A \subseteq R$ with the limit of function and or the limit of sequence on R. One of them is Eko. Eko's proof could be seen in Figure 2.

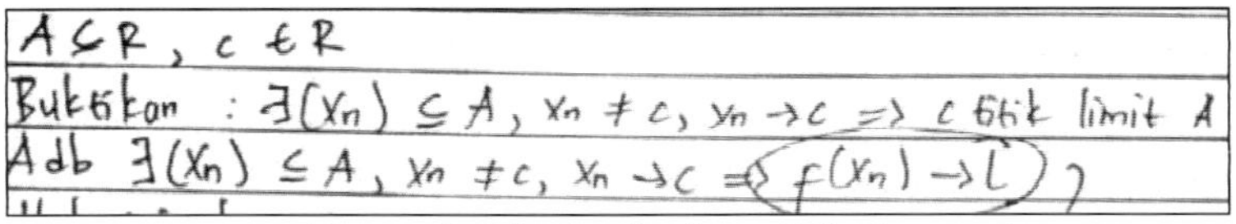

Figure 2. The beginning of Eko's proof

Based on Figure 2, Eko considers that proving a cluster point is same as proving the limit of sequence $f(x_n)$, for a $(x_n) \subseteq A$. Even though both of them discuss the limit, these concepts could not be equated. In the given theorem, function *f* does not appear at all. Moreover, he did not introduce function *f* at first before he used it. After he stated his target of proof, Eko wrote as shown in the Figure 3.

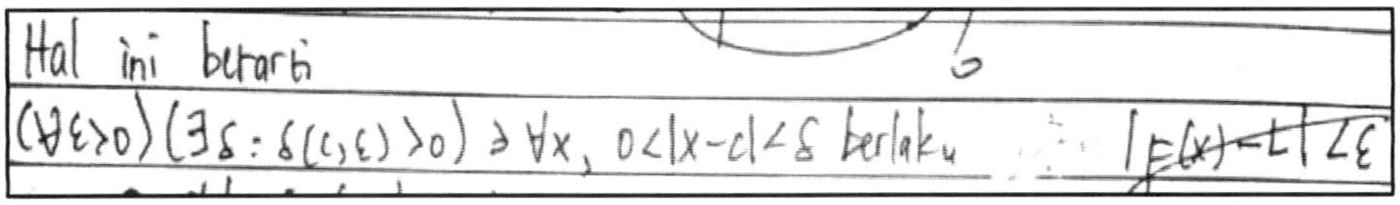

Figure 3. Eko's belief about definition of convergence of $f(x_n)$

The definition of convergence of $f(x_n)$ written by Eko is incorrect. His definition was the definition of the limit of function *f* if *x* tends to *c*. This incorrect definition often occurred when students solved proving problems[13]. As the study of Alcock and Simpson [17], only 29% students used the definition correctly. Furthermore, in carrying the target of proof out, Eko wrote as shown in Figure 4 below.

Figure 4. Eko's first step in carrying the target of proof out.

Eko wrote "dipilih", this means "chosen", instead of "diketahui", this means "given", or other words to state that mathematical object (sequence $(x_n) \subseteq A$ where $x_n \neq c$ for all $n \in N$ and $x_n \to c$) is given on the question. Eko did not write the formal-theoretical part correctly. This leads Eko to failure in constructing proof.

Another logical error was done by Wati. A part of Wati's proof could be seen in Figure 5 below.

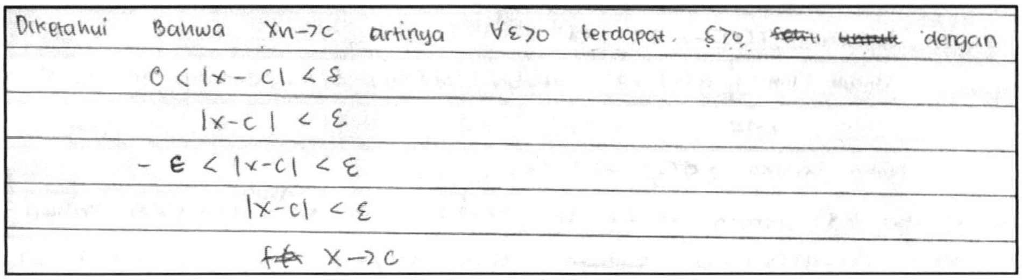

Figure 5. A part of Wati's proof

At a glance, Wati's proof framework led to the proof of limit of function generally, started by taking any $\varepsilon > 0$, showed the existence of $\delta > 0$, then ended up with $|x-c| < \varepsilon$. In her proof, Wati was seen to use the definition of the limit of function on limit of sequence. This part also indicates that Wati do not understand about ε and δ. Even though they are just symbols, mathematicians usually use these symbols in the definition of the limit of function. It is noticeable that Wati played x until the end of her proof without introducing x first. After she wrote $x \to c$, she concluded that c was cluster point of A. The arrow symbol $(\to)$ is associated with convergence. It is like a symbol $x_n \to c$ that states sequence (x_n) converges to c, this means c is the limit of sequence (x_n) [15]. Nevertheless, the statement $x \to c$ that Wati wrote, has no reason to say that c is cluster point of A. According to Stavrou [13], students usually do mistakes on proving because of misconception of understanding the definition.

A complete proof is written by Anto. The last part of Anto's proof could be seen in Figure 6.

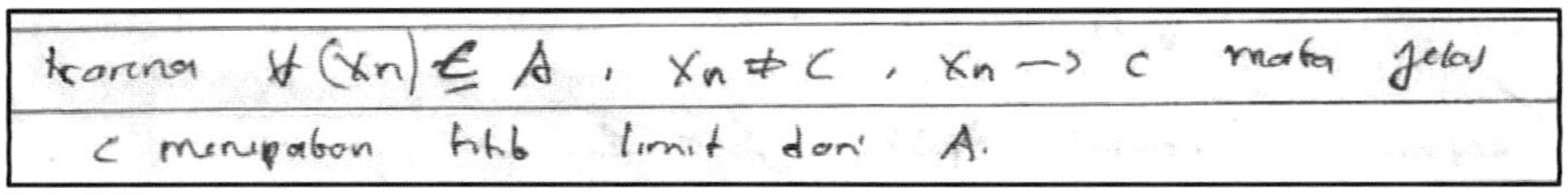

Figure 6. The end of Anto's proof

At the end of his proof, Anto wrote, "Because $\forall(x_n) \subseteq A,\ x_n \neq c,\ x_n \to c$ then it is clear that c is a cluster point of A". In fact, it was what he intended to prove. The theorem given to him to prove, it was not clear. He needed some arguments to ensure it. It is noticeable that his proof framework is incorrect. Moreover, in the beginning of his proof, he wrote as seen in the Figure 7.

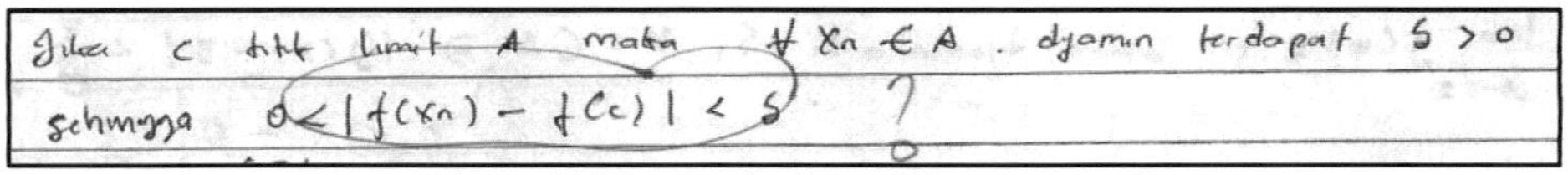

Figure 7. The beginning of Anto's proof

At first, Anto wrote an implication argument. However, it was not a valid argument. In the given theorem, no symbol "f" was given. Moreover, the theorem did not address the function f. He also never declared about f. If other people read his proof, they as if reading a paragraph with incoherent sentences. It was challenging to understand.

It is interesting to note that Anto inscribed an invalid definition as shown in the Figure 8.

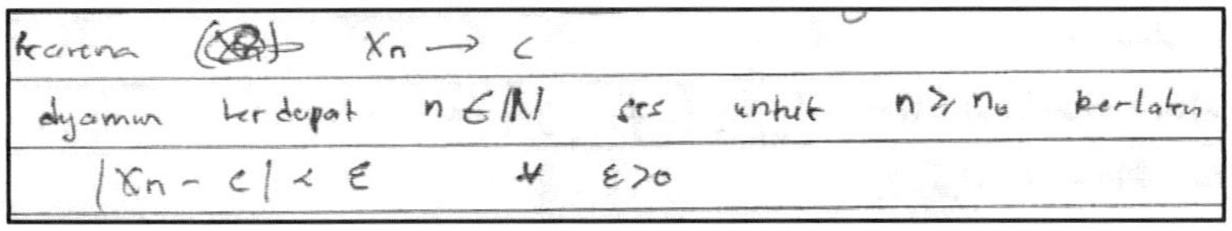

Figure 8. An invalid definition of limit of sequence

At a glance, his definition of the limit of sequence ($x_n \to c$) is true. In point of fact, it is incorrect. Saying "there exists $n_0 \in N$ such that for all $n \geq n_0$ satisfy $|x_n - c| < \varepsilon$, $\forall \varepsilon > 0$" is different from saying "For all $\varepsilon > 0$, there exists $n \in N$ such that for all $n \geq n_0$ satisfies $|x_n - c| < \varepsilon$". On the first statement, the existence of a number $n_0 \in N$ is true for all $\varepsilon > 0$. The latter means that the existence of $n_0 \in N$ depends on the given $\varepsilon > 0$. In a definition or theorem, the order of writing is crucial. Students were not able to reverse the order at will. They often ignore the importance of the order of writing definition or theorem, as is in the Davis and Vinner's paper summary in [18], students neglect the important role of the order when they face the concept of a limit of sequence or series. Moreover, Anto possible lack understanding about ε and δ in his definition. We are not able to conclude this, but his proof indicates this.

4. Conclusion

Based on data analysis, there are some misconceptions in understanding definitions. Students make errors in using the definition of cluster point, the limit of function, and the limit of sequence, but they feel their understanding is correct. Besides, they seem to get confused about each definitions. Moreover, they use some routine symbols, such as symbol "*f*" refers to function, though the theorem does not contain it.

Even though the understanding of definition is not enough to construct a valid proof, it is necessary for students to be encouraged in writing correct proof [11]. A lecturer has to put more emphasis on students' understanding of definitions, how definitions work, and what are the examples and non-examples of the definitions. Besides, for some concepts that have similar "terms" (e.g. convergence of sequence and convergence of function), the lecturer has to explain the similarities and differences. Furthermore, lecturers do not use a routine symbol to state mathematical object. For example, the choice of symbol *f* as a function could be interspersed with other alphabets. By these ways, students might experience many kinds of concepts and mathematical objects and statements. Their understanding could improves when solving proving problems which demand various concepts. In addition, the lecturer should familiarize students with writing a proof framework before they go on next steps of proving. By writing it, students always lead their steps to the target of proof. Therefore, their final proof is in accordance with the formal-rhetorical part. According to Selden and Selden [19], by writing the formal-rhetorical part of proof, students could improve their writing proof, reveal the characteristics of the problem, and focus on the problem. Besides, it enables the lecture to help students become more targeted and effective. A further study which investigates how they prove theorems and why they do their proving steps can be conducted by task-based interview or think aloud method so that we understand every steps that they do.

References

[1] A. Selden 2012 Transitions and Proof adn Proving at Tertiary Level*," in Proof and Proving in Mathematics Education, New ICMI Study Series 15, vol. 15, G. Hanna and M. De Villiers, Eds. Springer Science+Business Media B.V, 2012, pp. 391–420.

[2] K. Weber 2001 Student difficulty in constructing proofs: The need for strategic knowledge *Educ. Stud. Math* 48 **1** pp 101–111

[3] Y. Y. Ko and E. Knuth 2009 Undergraduate mathematics majors' writing performance producing proofs and counterexamples about continuous functions *J. Math. Behav.* 28 pp. 68–77

[4] M. De Villiers 2004 Using dynamic geometry to expand mathematics teachers' understanding of proof *Int. J. Math. Educ. Sci. Technol* 35 **5** pp. 703–724

[5] Y. Lai, K. Weber, J. P. Mejía-Ramos 2012 Mathematicians' perspectives on features of a good pedagogical proof *Cogn. Instr.* 30 **2**, pp 146–169

[6] D. A. Stylianou, M. L. Blanton, and O. Rotou 2015 Undergraduate Students' Understanding of Proof: Relationships between Proof Conceptions, Beliefs, and Classroom Experiences with Learning Proof *Int. J. Res. Undergrad. Math. Educ* 1 pp. 91–134

[7] L. Bills and D. Tall 1998 *Operable Definitions in Advanced Mathematics: The Case of the Least Upper Bound* (Proceedings of PME 22) pp. 104–111

[8] D. R. Thompson, S. L. Senk, G. J. Johnson 2012 Opportunities to Learn Reasoning and Proof in High School Mathematics Textbooks *J. Res. Math. Educ* 43 **3**, pp 253–295

[9] A. Selden, J. Selden 2013 Proof and problem solving at university level *Math. Enthus*. 10 **1** pp 303–334

[10] J. W. Dawson 2006 Why Do Mathematicians Re-prove Theorems? *Philos. Math.* 14 **3** pp 269–286

[11] A. Selden, J. Selden 2007 Overcoming students' difficulties in learning to understand and construct proofs.

[12] D. Tall, O. Yevdokimov, B. Koichu, W. Whiteley, M. Kondratieva, and Y.-H. Cheng 2012 *Cognitive Development of Proof* (The 19th ICMI Study: Proof and Proving in Mathematics Education, G. Hanna and M. De Villiers, Eds. Springer)

[13] S. G. Stravrou 2014 Common Errors and Misconceptions in Mathematical Proving by Education Undergraduates *Issues Undergrad. Math. Prep. Sch. Teach. J.*, 1 pp. 1–8

[14] A. Mackey, M. Gass 2005 *Second Language Research: Methodology and Design* (New Jersey)

[15] R. G. Bartle, D. R. Sherbert 2010 *Introduction to Real Analysis (Fourth Edition)* (John Wiley & Sons, Inc)

[16] K. Sari, M. Waluyo, C. M. Ainur, and E. N. Darmaningsih 2017 Menggunakan Contoh dalam Pembuktian *J. Ilm. Pendidik. Mat.* 2 **1** pp 1–9, 2017

[17] L. Alcock and A. Simpson 2011 Classification and concept consistency *Can. J. Sci. Math. Technol. Educ.* 11 **2** pp. 91–106

[18] J. Mamona-Downs 2001 Letting the intuitive bear on the formal; a didactical approach for the understanding of the limit of a sequence *Educ. Stud. Math.* 48 **2–3**, pp 259–288

[19] A. Selden and J. Selden 2013 Proof and Problem Solving at University Level *Math. Enthus* 10 **1-2** pp 303–334

Students' errors in solving combinatorics problems observed from the characteristics of RME modeling

I Meika[1], D Suryadi[2], Darhim[3]

[1, 2, 3] Indonesia University of Education Bandung
[1] Mathematics Education, FKIP, Mathla'ul Anwar University Banten

[1]E-mail: ikameikamulhat@gmail.com

Abstract. This article was written based on the learning evaluation results of students' errors in solving combinatorics problems observed from the characteristics of Realistic Mathematics Education (RME); that is modeling. Descriptive method was employed by involving 55 students from two international-based pilot state senior high schools in Banten. The findings of the study suggested that the students still committed errors in simplifying the problem as much 46%; errors in making mathematical model (horizontal mathematization) as much 60%; errors in finishing mathematical model (vertical mathematization) as much 65%; and errors in interpretation as well as validation as much 66%.

1. Introduction

Combinatorics is an initial material about probability consisting of basic rules of counting, permutation, and combinations. Probability is very useful concept in everyday life as its role is very important in decision making. At a higher level, permutation and combination materials are the basic materials in learning discrete mathematics [5, 11]. Combinatorics is a branch of contemporary mathematics in which it is widely used in business and industry [11]. Moreover, NCTM states that combinatorics is important as it is an integral part of the mathematics curriculum from the elementary level to the high school level [11].

Using the basic rules of calculation, permutations, and the right combination of problem solving skills are challenging. This is in line with Wraughton & Nolan [14] that understanding the rules of counting is challenging for students especially when they are struggling to determine when and how to apply rules of multiplication, permutation, and combinations in calculation.

According to Mashiach Eizenberg, M., & Zaslavsky, O. [10], combinatorics is one of the mathematical topics that is considered difficult to teach and learn. To overcome such challenges in learning combinatorics, one of the ways is to present the combinatorics problem from daily life (contextual). In studying permutations and combinations, students need real-world context that are relevant to stimulate learning and knowledge [3]. Along with this, Sukoriyanto et al. stated that permutation and combination materials require problems in the form of the real-world issues so that they can retain their knowledge [12].

In order for the real-world problems get mathematical solutions, it takes a simplification of the real situation into a mathematical form called modeling. This is in line with Gravemeijer [6] who stated that modeling is a bridge to convert contextual problems into formal forms. According to Heuvel-Panhuizen [13] and Maaß [9], a model is a form of mathematical representation of a problem. A mathematical model is a representation of a particular phenomenon created in the forms of images, graphs, schemes, equations, or algorithms.

Modeling is also concerned by Cheng [4] in which mathematical modeling is a process of representation of real-world problems in the form of mathematics in an attempt to find a solution of the problem. A model or modeling cannot be separated from the process of mathematization. Some people assume that mathematization is part of the modeling, while some others consider model as part of mathematization. Blum and Leiss [2] consider mathematization as part of the modeling. Mathematical modeling ability can be divided into five stages [1]: (1) simplifying the problem, (2) making mathematical model (horizontal mathematization), (3) finishing mathematical model (vertical mathematization), (4) interpretation, and (5) validation. This study will identify errors in students' answers in solving combinatorics problems observed from the mathematical modeling indicators [1].

Gravemeijer [6] asserts that the development model is formed through four stages, namely situational, referential, general, and formal. The situational stage is the most basic stage of modeling where knowledge and model are still developing in the context of problem situations selected. In the referential stage, the models and strategies developed refer to the context in which the student creates a model to describe the situation from the context. The model at this stage is called 'model of'. In the general stage, it is developed by them has led to the search for solutions mathematically. Such model at this stage is called 'model for'. Finally, at the formal stage, they have worked with symbols and mathematical representations.

2. Methodology

The research method was descriptive study by involving the students from State Senior High School (said 'SMAN') in Banten Province which are RSBI schools in 2015. They were SMAN Cahaya Madani Banten Boarding School (said 'CMBBS') and SMAN 4 Pandeglang. 55 XI science-major students were selected consisting of 30 students from SMAN CMBBS and 25 students from SMAN 4 Pandeglang. To seek the errors in the students' answers in solving combinatorics problems, they had studied combinatorics materials with their respective teachers beforehand.

The instrument used was test. The items were in the form of short essay of 5 questions that considered modeling characteristics in RME whose indicators are: (1) simplifying the problem; (2) making mathematical model (horizontal mathematization); (3) finishing mathematical model (vertical mathematization); (4) interpretation; and (5) validation. The items for piloting study were tested before regarding its validity and reliability by involving 5 experts and 20 students. The test was conducted on May, 16th 2017 at SMAN 4 Pandeglang and May, 23rd 2017 at SMAN CMBBS.

Validity question item is conducted to determine validity of the instrument using content validity. It examined whether the material was really representative. Item 1 and item 5 represented the rules of multiplication and addition, item 2 represented permutation, and item 3 and item 4 represented combination.

3. Findings and Discussion

The percentage of students making errors in solving combinatorics problems observed from the mathematical characteristics of RME is presented in Table 1.

Table 1. The percentage of students' errors

Students' errors	Percentage (%) of students of each question item					Average (%)
	1	2	3	4	5	
Errors in simplifying the problem	69	62	15	44	40	46
Errors in making mathematical model	85	73	38	53	53	60
Errors in finishing mathematical model	89	73	47	58	56	65
Errors in interpretation	89	73	47	58	64	66
Validation	89	73	47	58	64	66

From the table above, it can be seen that the most frequently committed errors by the students are those in interpretation and validation. It is because when they are not able to simplify the problem correctly, it will result in the incorrect creation of mathematical model, and their completion is wrong, so interpretation and validation are also wrong. Table 1 also shows that the error of each question begins

IOP Conf. Series: Journal of Physics: Conf. Series **948** (2018) 012060 doi:10.1088/1742-6596/948/1/012060

to increase sharply on horizontal mathematization or in the stage of making mathematical model. So, the completion of the next step becomes incorrect.

The findings of this study are that the students have difficulty in solving the combinatorics problem, especially in the step of making mathematical model (horizontal mathematization) and its completion. The results of this study are in line with Lockwood [7] in which students face difficulties when solving the problem of combinatorial calculation. In addition, students still make errors in the completion of permutation and combination [12].

The questions are contextual, but they prefer them in the form of numbers because they find it more difficult to model it. It is suggested that permutation and combination problems expressed in the form of context require more mathematical thought, but students will find it more difficult [7].

Question number 1 has got the highest percentage of errors in simplifying the problem compared to other items as much 69%. The text of question number 1 is "In an entry selection test at a favorite high school in Banten, each participants gets a test number consisting of four different digits from 0, 2, 3, 4, 5, 7, and 8. How many test numbers can the admission committee make if the number of the test forms a number in-thousand as the quotient by 5?"

There are three kinds of students' answers. First, some of them did not know the characteristic of the numbers which are divisible by 5. Second, some others knew the characteristic of the numbers divisible by 5, but misplace 0 in the position of thousands. Third, some others knew the characteristic of the numbers divisible by 5 and knew that 0 cannot be placed in the form of thousand, but forgot the context of the question in which 'the mixed numbers are formed by different numbers'. Figure 1 is an example of a student's answer regarding the second stage that is the completion stage of this model or an error in the vertical mathematical stage. It is the most frequently done by the students compared to stage one and stage three. This error resulted in the incorrect stage of interpretation of the solution.

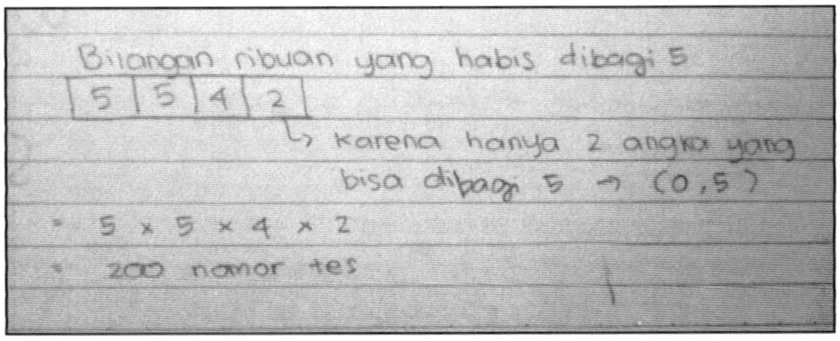

Figure 1. Students' answers on question number 1

For question number 2, this is a case of cyclical permutation. Table 1 shows that there are 62% of the students who committed errors in simplifying the problem (horizontal mathematization) so the errors in the next stage become greater. The text for question number 2 was "A family consisting of father, mother, brother, and sister are having dinner together. They sit around a table. If the father and mother always sit close together, then determine the number of different ways of seating from the family!"

One of the students' answers in Figure 2. below is one of the representatives of students whose answers are toward the expected solution. The students made a model of the given problem, then made formal mathematics (vertical mathematization) that led to the final solution. Here, the students had not finished the model correctly seen in the drawings made by them where AB (representing father and mother) were drawn close together. However, when at the completion stage, they separated A and B by calculating the cyclical permutation of 4, also the factorial number 2 as the solution from the adjacent A and B. So, they committed errors in completing the model as well as its interpretation.

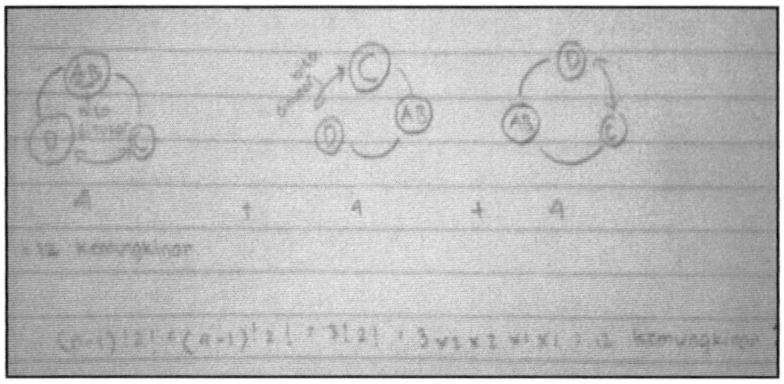

Figure 2. Students' answers on question number 2

For question number 3, Table 1 shows that the students' errors seen from the modeling characteristics were the smallest among other questions. More than half of them had completed each step of the modeling characteristics stage. The question text was "To paint the mosque walls, Mr. Hamid will create mixed colors that are formed from three basic colors. There are six different basic colors. How many mixed colors can Pak Hamid make?"

The incorrect examples of student's answers are shown in Figure 3. There are three kinds of answers that are displayed. Two answers included using a horizontal mathematical process by creating a simplified model of the problem, but the students had not been able to make all the possible answers requested. While, one other answer used formal mathematics. It can be seen that they have not been able to distinguish the use of formal mathematical formulas between permutation and combination.

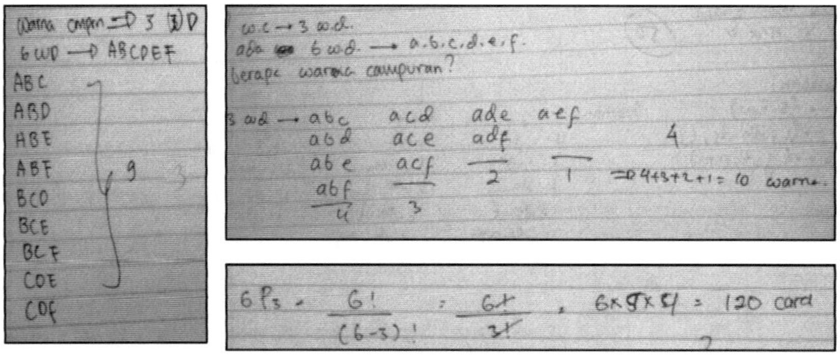

Figure 3. Students' answers on question item 3

Another combination-related case question is on item 4, that is "In a formative test of mathematics, the teacher gives seven multiple choice questions and three short-essay questions. Each student may work on a problem that is considered easier first. Not all questions should be answered. They may choose five multiple choice questions and two short-essay questions. If question number 4 and 7 on multiple choice must be done, how many ways are available to choose the rest of the questions?"

In contrast to the number 3 above, although it is also the case of combination, many students committed errors. It is because the level of difficulty of number 3 which is different from number 4. Their understanding was tested in operating the problem of two combinations in one case. One of the students' answers is presented in Figure 4.

IOP Conf. Series: Journal of Physics: Conf. Series **948** (2018) 012060 doi:10.1088/1742-6596/948/1/012060

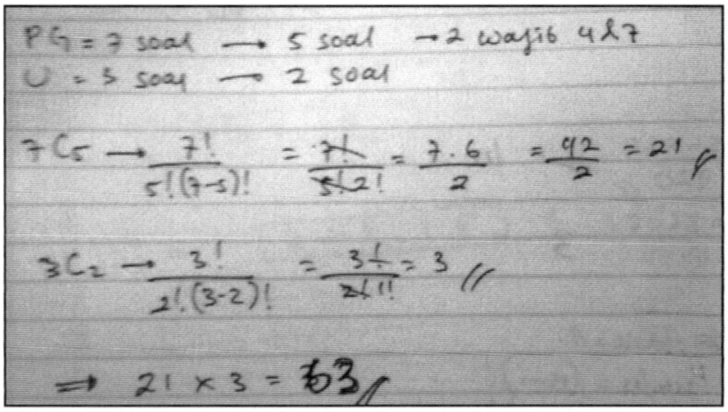

Figure 4 Students' answers on question number 4

Figure 4 shows that the students commit errors in the vertical mathematical stage in which there is information left in the calculation of the remaining multiple choice questions that must be selected. Multiplication operation between two combinations is correct, but because there is an error in the data input, the final answer is also wrong. Such errors often occur where they understand the formal mathematical formulas without understanding the meaning of the symbols they write.

Meanwhile, for question number 5, it is a case on multiplication and addition whose text is " A motor vehicle serial number consists of four-digit numbers. Determine the number of serial numbers that can be created if number 5 appears twice. Draw the position of number 5 in various sequence variations. (Hint: The numbers that can be used are 0, 1,.., 9)"

The answers given by the students vary greatly in which 53% of them committed errors in the horizontal mathematical stage in creating mathematical model. Figure 5 shows that they understand the occurrence of number five is twice, but do not understand the calculation, so that the numbers in the grid box are not clear in meaning.

It suggests that they have not been able to interpret the symbols used in the process of problem solving, so that they had difficulty in the final process of calculation. It is also seen that they were not accustomed to solving challenging contextual problems and training their way of thinking in simplifying (modeling) the given problem.

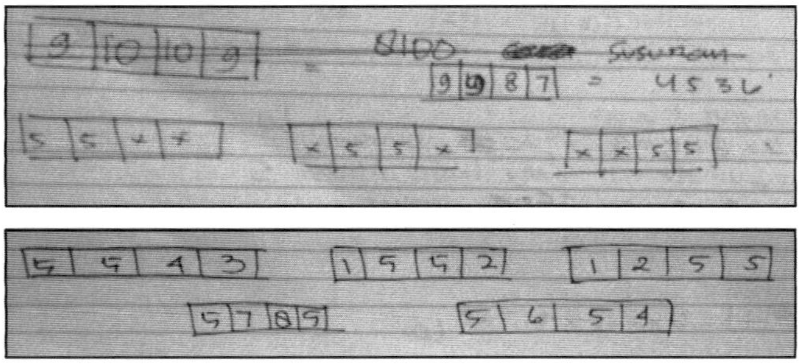

Figure 5. Students' answers on question number 5

4. Conclusions

Based on the finding of the study, students' errors in solving the combinatorics problems observed from the modeling characteristics suggest that their ability in solving the combinatorics problems is still low. It also suggests that the still-low ability is especially regarding horizontal and vertical mathematization because these two stages are the core of modeling-problem completion process under real-world context. Combining the combinatorics problems under real-world contexts that has its own challenges must be continuously trained to the students, so that they are accustomed to finding solutions and being able to create mathematical model to find solutions to the problems they face. For a more effective learning, teachers should make preparation in designing instructional materials in accordance with real-life contexts that train students' thinking skills. Among them is teaching materials that help bring up the modeling characteristics built by students so that they are trained to constrict their own thoughts, and the teachers are guiding them to lead to formal mathematics.

Reference

[1] Blum, W., Leiss, D 2005 Filling up the problem of independence-preserving teacher interventions in lessons with demanding modeling tasks. *CERME 4–Proceedings of the Fourth Congress of the European Society for Research in Mathematics Education* pp 1623

[2] Blum, W. Leiss, D 2007 How do students and teachers deal with mathematical modeling problems? *Mathematical Modeling: Education, Engineering and Economics* pp 222

[3] Busadeea, N., Laosinchai, P 2013 Authentic Problems in High School Probability Lesson: Putting Research into Practice *Procedia-Social and Behavioral Sciences* 93 pp 2043

[4] Cheng, A. K. 2001 Teaching mathematical modeling in Singapore schools. *The Mathematics Educator* 6 **1** pp 63

[5] Garfield, J, Ahlgren, A 1988 Difficulties in learning basic concepts in probability and statistics: Implications for research *Journal for research in Mathematics Education* pp 44

[6] Gravemeijer, K. P. E. 1994 *Developing Realistic Mathematics Education* (Utrecht: Freudenthal Institute)

[7] Lockwood, E. 2013 A model of students' combinatorial thinking *The Journal of Mathematical Behavior* 32 **2** pp 251-26

[8] Lockwood, E., Gibson, B. R. 2016 Combinatorial tasks and outcome listing: Examining productive listing among undergraduate students *Educational Studies in Mathematics* 91 **2** pp 247-270

[9] Maaß, K. 2010 Classification scheme for modeling tasks *Journal für Mathematik-Didaktik* 31 **2** pp 285-311

[10] Mashiach Eizenberg, M., Zaslavsky, O 2004 Students' verification strategies for combinatorial problems *Mathematical Thinking and learning* 6 **1** pp 15-36.

[11] National Council of Teachers of Mathematics 2000 Principles and standards for school mathematics (Reston, VA)

[12] Sukoriyanto, S., Nusantara, T., Subanji, S., Chandra, T. D. 2016 Students' Errors in Solving the Permutation and Combination Problems Based on Problem Solving Steps of Polya *International Education Studies* 9 **2** pp11

[13] Van Den Heuvel-Panhuizen, M 2003 The didactical use of models in realistic mathematics education: An example from a longitudinal trajectory on percentage *Educational studies in Mathematics* 54 **1** pp 9-35

[14] Wraughton, J., Nolan, J. 2012 Pinochle Poker: An Activity for Counting and Probability *Journal of Statistics Education* 20 **2**

IOP Conf. Series: Journal of Physics: Conf. Series **948** (2018) 012061 doi:10.1088/1742-6596/948/1/012061

Differential item functional analysis on pedagogic and content knowledge (PCK) questionnaire for Indonesian teachers using RASCH model

B D Rahmani

Universitas Muhammadiyah Prof. DR. HAMKA

E-mail: bitadwi@uhamka.ac.id

Abstract. The purpose of this paper is to evaluate Indonesian senior high school teacher's pedagogical content knowledge also their perception toward curriculum changing in West Java Indonesia. The data used in this study were derived from a questionnaire survey conducted among teachers in Bandung, West Java. A total of 61 usable responses were collected. The Differential Item Functioning (DIFF) was used to analyze the data whether the item had a difference or not toward gender, education background also on school location. However, the result showed that there was no any significant difference on gender and school location toward the item response but educational background. As a conclusion, the teacher's educational background influence on giving the response to the questionnaire. Therefore, it is suggested in the future to construct the items on the questionnaire which is coped the differences of the participant particularly the educational background.

1. Introduction

Teacher Competency Test or *Ujian Kompetensi Guru (UKG)* is a test to identify teacher's competencies in pedagogic and content knowledge of a subject matter mastered by each teachers in Indonesia. Since 2012, Indonesia Ministry of Education Board has been administering the *UKG* as the prerequisite of being a certified-teacher. However, since 2015, UKG has been conducted to all teachers in all over Indonesia which covers 192 subject matter/skills as a parameter in identifying the level of teacher's competence both in professional and pedagogic competence. Unfortunately, UKG in 2015 did not run successfully since there were many problems faced on the test such as the construct of item test which was not in line with the material, the difficulty level of the test and even the technical problem on the internet connection and many more. Due to these problems, UKG result could not become the most single accurate sources of leveling the teacher's competencies. There should be another instrument to support the result of *UKG*. The teacher self-assessment was seen as a good alternative answer of the problem. The teacher self-assessment test allows the teacher to reflect his/her self on teaching learning process. Moreover, the self-assessment is a powerful technique for self-improvement efficacy in teaching [1]. Thus, this project aims to design a self-assessment questionnaire on pedagogic, content and professional knowledge of the teacher. Despite the differences of sample in filling the questionnaire, it should have been checked whether all the items on the questionnaire were functioning well or not to differentiate the heterogeneity of sample from their gender, educational background also on school location. Thus, the differential item analysis was conducted to analyze the data.

Despite the importance of having a valid and reliable item on the questionnaire, this study has been undertaken to answer some of the following questions: (1) Are there any differences on gender, educational background and/or school location in responding the item on the questionnaire of PCK (Pedagogical and Content Knowledge) and Curriculum Change?; and (2) If yes, how big are the differences on each dimensions?

Due to the limitation of the research, this research only focuses on two aims since those purposes are the basic prerequisite for further analysis. Those two aims are: (1) to investigate whether there are any differences in gender, educational backgrounds and or school location in responding the questionnaire of PCK (Pedagogical and Content Knowledge) and Curriculum Change.

To examine the differences of each item on questionnaire for PCK (Pedagogical and Content Knowledge) and Curriculum toward gender, educational background and or school location. In Indonesia, a ministry of education through *Undang Undang Guru dan Dosen* 2005 or the 2005 Teacher and Lecturer Policy managed that the teachers should have pedagogical, personal, social and professional competences. In order to have all those competencies, the Indonesian Ministry of Education had already ensured the opportunity to improve teachers' quality and competencies through professional development activities. Professional development is an ongoing process of teachers' training, aiming to support teachers' efforts to understand and form their teaching practices [2]. However, the teacher professional program in Indonesia was run based on the teacher competencies level. The teacher's competencies its self was been mapping by the teacher competency test administered by Ministry of Education in 2015. However, the result of UKG was not satisfying. It happened since the curriculum changing also was influenced on teacher pedagogic and content knowledge.

Therefore, there should be another parameter to map the teacher's competencies in Indonesia. Teacher self-assessment is one of the solution in facing the problem. By doing the self-assessment, the teacher could reflect what had already been happened in the classroom and it would be expected to increase the teacher's awareness and efficacy [1]. Moreover, the teacher self-assessment could be as another supporting evidence to mapping the teacher's competencies in Pedagogical and content knowledge aspects.

Many teacher self-assessments has been developed recently. One of them was developed by Schmidt et al [3] Sahin [4]. These self-assessments were more focusing on pedagogical and content knowledge (PCK) of the teacher. It is expected that the teacher would identify by themselves whether they had already implement PCK on their teaching practice. Moreover, due to the happening issue particularly about curriculum change in Indonesia, this questionnaire had already been adjusted to cover that problem also. However, the new construction of the questionnaire really needed a further investigation. The questionnaire had been distributed to several teachers with different gender and educational backgrounds in several schools across Indonesia. Seeing the differences on the sample, the differential item functioning analysis (DIFF) should be conducted on this research. Moreover, DIF can be defined as a statistical phenomenon that occurs when "two individuals with equal ability but from different groups do not have equal probability of success on the item" [5]. In other words, DIF occurs when examinees from different groups show differing probabilities of success on (or endorsing of) the item after matching on the underlying ability that the item is intended to measure. DIF analysis, therefore, is a statistical tool that can be used to improve how a test behaves across groups and to reduce group-based differentials that are not relevant to the construct being measured. In terms of utility, DIF analysis is mainly a tool to assess test fairness, investigate threats to validity, and explore the underlying processes in item responses across groups [6].

2. Methodology

This study gathered a primer data analysis of self-assessment questionnaire that used to investigate the Indonesian ESL teachers based on their Pedagogical and Content Knowledge (PCK). This research is aimed at examining the item response differences toward gender, educational backgrounds also on school locations. One parameter (Rasch model) was used on gathering the data. An instrument was used in this study is in the form of self-assessment questionnaire adapted from Schmidt et al [3] and Sahin [4]. The instrument consists of 25 items for measuring Indonesian ESL teacher knowledge of two categories/ factors which include Pedagogical Content Knowledge (PCK) and Curriculum Changing Knowledge (CC). The 25 questions on the questionnaire used Likert Scale. The Likert Scale can be used to collect attitude data [7] including the human condition [8]. The participants of this study were sixty one Indonesian English Teacher (ESL), most of whom are from West Java, some from East Java and Jakarta, Indonesia. The data collection was held on National Gathering of Senior High School English Teachers 2017 in Bandung, West Java on July 6th 2017. The population was about 200 teachers on the list. However, the questionnaire was collected only from 61 respondents as the sample of this research. In addition, Differential Item Functional Analysis (DIFF) was used to conduct the analysis of the differences of each item response toward the differences of gender, educational background and school location on the questionnaire by using a software Conquest 4.0. The Conquest produced a significant value, estimate and error score. The significant value was counted firstly and it should be less than 0.05 to indicate the items has a difference on each category. Next, when the significant value indicated the difference on each category, the magnitude also (T value) should be counted secondly to see the deviance on each item toward the same category.

3. Results and Discussion

The first running of the data by using Conquest 4.0 in analyzing the differences between item response of Category 1 (PCK) to Gender (Male, Female) showed that there were not any significant differences. It is shown by Chi-square test of parameter equality = 20.41, df = 13, Sig Level = 0.085. In short, male and female teachers were given the same trait in responding the category 1 (PCK) on the questionnaire. Then, the next calculation was counting the differences between items toward the educational background of the teacher. There were two categories of teacher's qualification namely, bachelor degree and master degree. The result showed that there was a significant difference on two categories in responding the item on factor 1 (PCK). It can be seen on the following table below.

Table 1. The Diff of Factor 1 (PCK) on Educational Background

Item	Bachelor Degree	Master Degree	Magnitude	T
1	0.529	-0.529	1.058	2.213953
2	-0.022	0.022	-0.044	-0.09752
3	0.431	-0.431	0.862	1.803806
4	-0.944	0.944	-1.888	-2.24519
5	0.158	-0.158	0.316	0.700344
6	0.038	-0.038	0.076	0.168437
7	0.369	-0.369	0.738	1.565892
8	1.199	-1.199	2.398	3.36734
9	0.37	-0.37	0.74	1.570135
10	-0.655	0.655	-1.31	-2.66
11	0.212	-0.212	0.424	0.915787
12	-0.209	0.209	-0.418	-0.66939
13	-0.533	0.533	-1.066	-1.48474
14	-0.944	0.944	-1.888	-2.65996

Chi-square test of parameter equality = 20.41, df = 13, Sig Level = 0.085

The Table 1 shows that the teacher with bachelor degree could answer easily for six numbers (symbolized by negative value i.e item number 2, 4, 10, 12, 13 and 14). Meanwhile, the most difficult for them was number (1) *I can use teaching methods and techniques that are appropriate for a learning environment* (symbolized with the biggest positive value, 0.529). It happened since the question was asking about kinds of teaching methods and techniques used by them in the classroom.

IOP Conf. Series: Journal of Physics: Conf. Series **948** (2018) 012061 doi:10.1088/1742-6596/948/1/012061

The limitation of the knowledge on kinds of teaching and techniques made the teacher answer unconfidently on their competence. In contrast, the teacher with master degree could answer easily most of the items particularly number 1, 3, 5, 6, 7, 8, and 11. However, the most difficult item for them was item number (14). *I can use appropriate teaching methods and techniques to support students in developing their language skills*. This question was asking not only the appropriate teaching methods and techniques but also how the teaching methods and techniques supported the development of student's language skills. This questions made the teacher think for twice to choose their competence scale whether their competence had already developed the student's skill or not.

In addition, the Table revealed the fact that there was differences between the teacher educational backgrounds. It can be seen on item estimate difference greater than 0.5 logits as the criterion for detecting DIF in 14 items on category 1 [9]. Tennant and Pallant [10] stated that to determine whether there is a difference, three indicators were used, to wit: (1) t value is > 2.0; (2) DIF contrast value is $>$ 0.5; and (3) p (Probability) value is < 0.05.

An item needs to meet those three conditions to be considered bias and be dropped from the instrument. However, if the item meets only one of the conditions, it should not be dropped but it should be separated and fixed. Those are item numbers 1, 4, 8, 10, and 14 which had the magnitude score above 0.5 logits and the T value is greater than 2. In short, there were 5 items on category 1 which has a significant difference toward the educational background of the participant. However, the five items should not be dropped since the probability of the items was significant which is more than 0.05 (p=0.085). In addition, those five items were bias on educational background because of the following reasons below

Item 1: I can use teaching methods and techniques that are appropriate for a learning Environment

The magnitude value of this item was 1.058 (.0.5) and the T value was 2.213953 (>2). The item was biased on teacher's educational background since there was the words "*teaching methods and techniques*" on the questions which asked the teacher's understanding and knowledge to answer the question. In this case, the teacher with master degree were performing well. It is seen from the negative value (-0.529) which means the item was quite easy for them.

Item 4: I can collaborate with school stakeholders (students, parents, teachers, etc.) to support student's learning.

The magnitude value of this item was -1.888 (.0.5) and the T value was -2.24519 (>2). The item was biased on teacher's educational background since there was the words "*collaborate with school stakeholders*" on the questions which asked the teacher's experience engaging the stakeholders involved to support the student's learning. Interestingly, the teacher who graduated from bachelor degree were performing well on this question indicated from the negative value (-0.944) which means the item was quite easy for them. On the other words, the teacher with bachelor degree tended to have a better networking and experience to the stakeholders to support student's learning rather than the teacher with master degree.

Item 8: I can express my ideas and feelings by writing in English

The magnitude value of this item was 2.398 (>0.5) and the T value was 3.36734 (>2). The item was biased on teacher's educational background since the item was asking the teacher's ability on writing. In fact, the teachers who graduated from master degree were doing better in responding this item indicated from the negative value (-1.199) which means the item was quite easy for them. On the other words, the educational background was in line with teacher's competencies.

Item 10: I can understand texts written in English

The magnitude value of this item was -1.31 (>0.5) and the T value was -2.66 (>2). The item was biased on teacher's educational background since the item was asking the teacher's understanding on English text. Interestingly, this item was responded successfully by the teacher with bachelor degree indicated by the negative value (-0.655) which means the item was quite easy for them.

Item 14: I can use appropriate teaching methods and techniques to support students in developing their language skills

The magnitude value of this item was -1.888 (>0.5) and the T value was -2.65996 (>2). The item was biased on teacher's educational background since the item was asking the teacher's knowledge on teaching methods and techniques to support the student's language skills. Interestingly, this item was responded successfully by the teacher with bachelor degree indicated by the negative value (-0.944) which means the item was quite easy for them. It happened since the teacher with master degree were more carefully in responding this item since they had learnt so many teaching methods and techniques on language skills.

The differences on educational background between bachelor and master teacher also happened on category 2 (Curriculum Change perception) of teacher. It can be seen on the following table

Table 2. The Diff of Factor 2 (CC) on Educational Background

Item	Bachelor Degree	Master Degree	magnitude	T
16	0.327	-0.327	0.654	1.418552
17	-0.41	0.41	-0.82	-2.87043
18	-0.492	0.492	-0.984	-3.41075
19	0.406	-0.406	0.812	1.76126
20	0.6	-0.6	1.2	2.274874
21	0.201	-0.201	0.402	0.938142
22	-0.363	0.363	-0.726	-2.55403
23	0.189	-0.189	0.378	0.882133
24	-0.595	0.595	-1.19	-4.24978
25	0.136	-0.136	0.272	0.902972

Chi-square test of parameter equality = 28.15, df = 9, Sig Level = 0.001

The Table 2 shows that the teacher with bachelor degree could answer easily for 4 numbers (symbolized by negative value i.e. item number 17, 18, 22 and 24. Meanwhile, the most difficult for them was number *(20) (I know how to implement the curriculum 2013 in my class* - symbolized with the biggest positive value, 0.6). It happened since the question was asking abstractly how to run the curriculum 2013. The major problem of the question was on which part of the curriculum that the question was asked for. The limitation of teacher's knowledge on curriculum practice made the teacher who graduated from bachelor degree could not be confident enough in responding this item. On the other hand, the teacher with master degree could answer easily on the items particularly number 16, 19, 21, 23, and 25. However, the most difficult item for them was item number *"Curriculum 2013 is easy to implement in my class"*. This item was seen as the unclear statement since there was not any precise word to define in which part of curriculum 2013 to be implemented. In addition, the Table 2 also showed the fact that there was differences between the teacher educational backgrounds on category 2 (curriculum change). It can be seen on item estimate difference greater than 0.5 logits as the criterion for detecting DIF in 10 items on factor 2. The same criteria was used to analyse the item on factor 2. As a result, there are five items (item number 7, 18, 20, 22, and 24) which has the magnitude score above 0.5 logits and the T value is greater than 2. In short, there were 5 items on factor 2 which has a significant difference on educational background toward the educational background of the participant. However, the five items should not be dropped since the probability of the items was quite significant 0.01 (p=0.01). In addition, those five items were bias on educational background because of the following reasons below.

Item 17: I know the reason of curriculum changing

The magnitude value of this item was -0.82 (>-/+ .0.5) and the T value was -2.87043 (>2). The item was biased since the question was asking about the teacher's knowledge on curriculum changing. In this case, the teacher with master degree were performing well indicated from the negative value (-0.327) which means the item was quite easy for them.

IOP Conf. Series: Journal of Physics: Conf. Series **948** (2018) 012061 doi:10.1088/1742-6596/948/1/012061

Item 18: I know the purpose of new curriculum in teaching

The magnitude value of this item was -0.984 (.0.5) and the T value was -3.41075 (>2). The item was biased on teacher's educational background since it was asking about teacher's knowledge on curriculum change. Interestingly, the teacher with bachelor degree performed well than the master degree research in giving the response on this item indicated by negative value (-0.492). This result was contradictory than the previous item number (item 17) which was also asking about the knowledge of the curriculum changing.

Item 20: I know how to implement the curriculum 2013 in my class

The magnitude value of this item was 1.2 (>0.5) and the T value was 2.274874 (>2). The item was was biased since the question was asking about the teacher's knowledge on curriculum changing. In this case, the teacher with master degree were performing well indicated from the negative value (-0.6) which means the item was quite easy for them.

Item 22: Using Curriculum 2013 helps the teaching process

The magnitude value of this item was -0.726 (>0.5) and the T value was -4.24978 (>2). The item was biased on teacher's educational background since it was asking about teacher's knowledge on curriculum change. Interestingly, the teacher with bachelor degree performed well than those with master degree in giving the response on this item indicated by negative value (-0.363) which means the item was quite easy for them.

Item 24: (24). Curriculum 2013 is easy to implement in my class

The magnitude value of this item was -1.19 (>0.5) and the T value was -4.24978 (>2). The item was biased on teacher's educational background since it was asking about the implementation of the the new curriculum in the classroom. Surprisingly, the teacher with bachelor degree performed well than those with master degree in giving the response on this item indicated by negative value (-0.595) which means the item was quite easy for them.

4. Conclusion

Differential Item Functioning (DIFF) had been conducted to analyze the differences on gender, education background, and school location on questionnaire of Pedagogical and Content Knowledge (PCK) and Curriculum Changing (CC) for Indonesian ESL Senior High School Teacher. The questionnaire comprised of 25 item in which 15 items related to PCK and the rest were about Curriculum Changing. In addition, due to there were two categories on the questionnaire, therefore the questionnaire was categorized into two factors namely Pedagogical Content Knowledge (PCK) as factor 1 and Curriculum Change (CC) as factor 2. The analysis on gender, background education and school location were conducted on each category. In short, the result on analyzing the first factor revealed the fact that there was no any difference on gender and school location but educational background among the teachers with bachelor degree and master degree in giving the responses on the questionnaire. It could be observed by a magnitude value which was more than 0.5 and T value which was more than 2.0. In fact, there were 5 items which met those criteria. However, those items were still be maintained since the significant value of all items were more than 0.05 (p = 0.085) which meant no significant to be deleted. On the other hand, the result on analyzing factor 2 showed that the differences also were based on educational background instead of gender and school location. The significant value on educational background of factor 2 was less than 0.05 (p = 0.01) which indicated that there was a significant differences between teacher with bachelor and master degree in responding the item. In fact, it had been also found that there were also five items which had been identified. It meant that there was a different value between teacher with bachelor and master degree. As a conclusion, the questionnaire had different responses where distributed to Indonesian ESL teachers in senior high school particularly on their education background rather than on gender and school location. It is been suggested to do a refinement of the questionnaire which the items could cope all level of education among the teachers' education background.

References

[1] J. A. Ross and C. D. Bruce 2007 Teacher self-assessment: A mechanism for facilitating professional growth *Teach. Teach. Educ.* 23 **2** pp. 146–159

[2] L. Darling-Hammond 1994 *Professional development schools: Schools for developing a profession* ERIC

[3] D. A. Schmidt, E. Baran, A. D. Thompson, P. Mishra, M. J. Koehler, and T. S. Shin 2010 Technological Pedagogical Content Knowledge (TPACK): The development and validation of an assessment instrument for preservice teachers *Journal of Research on Technology in Education* 42 **2**. pp. 123–149

[4] I. Sahin 2011 Development of survey of technological pedagogical and content knowledge (TPACK)," *TOJET Turkish Online J. Educ. Technol* 10 **1**

[5] L. Shepard, G. Camilli, and M. Averill 1981 Comparison of Procedures for Detecting Test-Item Bias with Both Internal and External Ability Criteria *J. Educ. Stat.* 6 **4** pp 317–375

[6] B. D. Zumbo and M. N. Gelin 2005 A Matter of Test Bias in Educational Policy Research: Bringing the Context into Picture by Investigating Sociological/Community Moderated (or Mediated) Test and Item Bias *J. Educ. Res. Policy Stud.* 5 **1** pp. 1–23

[7] R. Likert 1932 A technique for the measurement of attitudes *Archives of Psychology* 22 p. 55

[8] S. Hales 1986 Rethinking the business of psychology *J. Theory Soc. Behav.* 16 **1** pp 57–76

[9] J. D. Scheuneman and R. G. Subhiyah 1998 Evidence for the validity of a Rasch model technique for identifying differential item functioning *J. Outcome Meas.* 2 **1** pp. 33–42

[10] J. F. Pallant and A. Tennant 2007 An introduction to the Rasch measurement model: an example using the Hospital Anxiety and Depression Scale (HADS) *Br. J. Clin. Psychol.* 46 **1** pp 1–18

IOP Conf. Series: Journal of Physics: Conf. Series **948** (2018) 012062 doi:10.1088/1742-6596/948/1/012062

Potential characteristics that relate to teachers mathematics-related beliefs

Y W Purnomo[1;*], T A Aziz[1], P Pramudiani[1], S Darwis[2], and D Suryadi[3]

[1] Universitas Muhammadiyah Prof. DR. HAMKA, Jakarta, Indonesia
[2] Institut Teknologi Bandung, Bandung, Indonesia
[3] Universitas Pendidikan Indonesia, Bandung, Indonesia

*E-mail: yoppy.w.purnomo@uhamka.ac.id

Abstract. A characteristic of the persons was very potential to affect the beliefs they held. This study examines whether there was a significant difference between the beliefs factors with characteristics such as gender, teaching experience, certification status, and grade level assignments. There were 325 primary school teachers in East Jakarta who participated in this research. MANOVA was applied to analyze the data. The findings of this study indicate that only on teaching experience, there was a significant difference between the beliefs held by the teachers, i.e. teachers who have 11-20 years teaching experience were more likely to think absolute than constructivism. Moreover, there was no difference between each belief they held with the other characteristics.

1. Introduction

All this time, research about teachers' beliefs is still very interesting to explore. Some researchers agree that teachers' beliefs might affect decision-making in instructional practices [1–5]. However, several researchers showed that there were complex relationships between teachers' beliefs and practices in mathematics class [1,2,6]. For example, Purnomo [2] found that several beliefs factors about mathematics might be in line with practices in the mathematics classroom, whereas other factors might be inconsistent with its practices. Besides, relationships between corresponding beliefs were not consistent. According to him, the inconsistency could be attributed to either internal or external factors. The possible internal factors were (1) teacher's knowledge about philosophy of mathematics and learning perspective; (2) teachers' resistance to change and innovation; (3) mathematics knowledge for teaching; and (4) experience. However, there have been few studies conducted to discern the effect of teachers' characteristics or internal factor, such as gender, experience, certification status, and grade level assignments on beliefs held.

Gender-related issues seem to be on going focus in mathematics education research [7]. Even though there have not been comprehensive literature that explicitly addresses issues of relationship between gender and teachers beliefs, several researchers consider gender issues to see its dependency on held beliefs [8–11]. Some of these studies found significant differences between male and female on beliefs related to mathematics. For example, Yazici and Ertekein [10] studied the differences of beliefs and anxiety of prospective elementary mathematics teachers based on gender. They found that in learning mathematics, male prospective teachers tend to be more instrumentalist than females according to their mathematical beliefs. In addition, females have more mathematics teaching anxiety than male, especially for subject knowledge in mathematics and self-confidence.

Other possible characteristics affected teachers' beliefs are years of teaching experiences and grade level taught. Perry Wong, and Howard [12] revealed that teachers' beliefs might stem from and limited in cultural communities where teachers live and work, culture and traditions of education system, their own experience as students in school and teacher education program, and their experience as members of the school community. Thus, their years of teaching experience might be a factor that affects beliefs. In addition, Nisbet and Waren [9] found that a static view of mathematics tends to rise along with the grade level assignments, as well as the view of constructivism in mathematics. The reason lies in the fact that teaching complex and challenge mathematical content requires teachers to create classroom environments that consider application of materials, to encourage classroom discussion, and to explore and solve problems in real-world context. Hence, the more complex mathematical content taught by teachers are, the higher possible is that teachers hold beliefs that mathematics is a static and absolute knowledge.

There have not been comprehensive literature about the relationship between teachers' beliefs related to mathematics and their certification status. The reason might be that certified teachers are teachers expected to work professionally as they have multi-year experiences of teaching and involve in professional development program. Consequently, these might affect their beliefs about education.

The above description leads researchers to investigate the relationship between beliefs held by teachers and their characteristics such as gender, years of teaching experience, certification status, and grade level assignments.

2. Methodology

2.1. Participants and Contexts
This study employed survey cross-sectional research design to achieve the purpose. Participants of this study were 325 elementary school teachers in East Jakarta and conveniently selected. East Jakarta is one of six cities in the capital city of Jakarta whose characteristics tend to be similar. There were 69 public schools and six private schools that involved in this study. It consisted of 80.9% female and 17.5% male, while 1.5% were not reported. There were 22.5% aged less than 30 years, 19.4% was located in a range of aged 31-40 years, 19.4% was located in range of 41-50 years, and most of all ware aged over 50 years i.e. amounting to 35.4%, while 3.4% were not reported. Participants came from different ethnic groups, such as Javanese, Sundanese, Betawi, Minangkabau, Malay, Dayak, Bima, and so forth. Detail information of characteristic of the participants is presented in Table 1.

2.2. Data Collection and Analysis
We used the teacher's mathematics-related beliefs questionnaire which has been developed by Purnomo [3,13]. The questionnaire includes three sub-scale, namely, teachers' beliefs about the nature of mathematics (BN-M), teachers' beliefs about teaching mathematics (BT-M), and teachers' beliefs about assessment in mathematics (BA-M). Each of sub-scales consists of two factors. BN-M includes relevant and dynamic factors. BT-M includes relational and instrumental factors. BA-M includes integrated and isolated factors.

The data were analyzed descriptively to describe profiles of the beliefs in each internal factor. Furthermore, the data were analyzed using one way multivariate analyses of variance (MANOVA). Some of the basic assumptions for MANOVA are (1) the independence of observation, (2) multivariate normality of the dependent variable groups of the population, and (3) its homogeneity of variance-matrix group population covariance.

Table 1. Information about participants of the Study

Based on	Frequency	%

Gender	Male	57	17.5
	Female	263	80.9
	Not Reported	5	1.5
	Total	325	100
Certification Status	Yes	184	56.6
	Not yet	137	42.2
	Not Reported	4	1.2
	Total	325	100
Teaching Experience	≤ 3 years	40	12.3
	4-10 years	72	22.2
	11-20 years	71	21.8
	> 20 years	132	40.6
	Not Reported	10	3.1
	Total	325	100
Grade level assignments	Lower grade	140	43.1
	Upper Grade	170	52.3
	Not Reported	15	4.6
	Total	325	100

3. Results and Discussion

Based on analysis of the assumptions, some dimensions do not meet the criteria of multivariate normality, so Pillai's Trace tests were applied to interpret the results. Some researchers argue that Pillai test is very strong and resistant to data that violates the assumptions [14–17]. To answer the research questions. The results of the analysis were explicated based on each characteristic. It is followed by discussion and elaboration of study that related to the findings.

3.1. Gender

Summary of the descriptive data for each of the dimensions of the beliefs related to gender is shown in Table 2. In addition, the results of the MANOVA using the Pillai's trace test is presented in Table 3.

Table 2. Descriptive data for teachers' beliefs and gender

Subscale	Factor	Male		Female	
		M	SD	M	SD
BN-M	Relevant	4.928	0.075	5.036	0.035
	Absolute	4.146	0.151	4.047	0.070
BT-M	Relational	5.041	0.063	5.036	0.030
	Instrumental	4.800	0.100	4.604	0.047
BA-M	Integrated	5.094	0.062	5.132	0.028
	Isolated	4.862	0.084	4.755	0.039

Table 2 indicates that the mean obtained group of female was higher than male's group on two factors, namely the relevant factors (5.036 > 4.928) and integrated (5.132 > 5.094). For others beliefs factors, mean of male's group were higher than female's group. However, the differences are necessary to be statistically verified. MANOVA was conducted to see the difference and its summary is presented in Table 3.

Table 3. MANOVA results for teachers' beliefs and gender

Subscale	Pillai's trace	F	Hypothesis df	Error df	p	η^2

BN-M	0.006	1.006	2.000	315.000	0.367	0.006
BT-M	0.010	1.642	2.000	310.000	0.195	0.010
BA-M	0.005	0.758	2.000	310.000	0.469	0.005

Based on results of the analysis in Table 3, it can be concluded that there is no significant difference between the beliefs and gender. In other words, MANOVA results showed that gender factor has no significant effects on the teachers' beliefs.

3.2. Teaching Experiences

The descriptive analysis of each sub-scales and the characteristics of teaching experience are presented in Table 4. Proceed with the MANOVA's conclusion presented in Table 5.

Table 4. Descriptive data for teachers' beliefs based on teaching experience

Subscale	**Factor**	**≤ 3 year**		**4-10 years**		**11-20 years**		**≥ 20 years**	
		M	SD	M	SD	M	SD	M	SD
BN-M	Relevant	5.108	0.088	5.084	0.066	4.804	0.066	5.071	0.049
	Absolute	3.925	0.179	3.930	0.134	4.042	0.134	4.237	0.099
BT-M	Relational	5.029	0.075	5.047	0.056	5.046	0.057	5.038	0.042
	Instrumental	4.565	0.118	4.506	0.088	4.576	0.090	4.755	0.066
BA-M	Integrated	4.995	0.073	5.180	0.054	5.135	0.055	5.119	0.041
	Isolated	4.645	0.099	4.809	0.074	4.841	0.075	4.771	0.055

Based on Table 4, each sub-scale of the beliefs indicates that the view of constructivist held by participants tends to be more dominant than the traditional view. The mean for the sub-scales of constructivist views is in the range of 4,804 and 5,180. On the other hand, the mean for the sub-scales of traditional views is in the range of 3.925 and 4.841.

Table 5. MANOVA results for teachers' beliefs based on teaching experience

Subscale	**Pillai's trace**	**F**	**Hypothesis df**	**Error df**	**p**	**η^2**
BN-M	0.058	3.059	6.000	618.000	0.006	0.029
BT-M	0.022	1.111	6.000	610.000	0.354	0.011
BA-M	0.022	1.148	6.000	610.000	0.333	0.011

Different from the gender factor, according to Table 5, teaching experience factor has significant effects on teacher beliefs about the nature of mathematics ($p < 0.05$). However, different results are indicated by two other scales, i.e. beliefs about teaching and learning of mathematics and of assessment. The significant value generated by statistical tests Pillai Trace above requires for further analysis. To find out which of the factors that have significant effect on BN-M, MANOVA test was performed as follows.

Table 6. MANOVA results for BN-M

Factor	**df**	**F**	**p**	**η^2**
Relevant	3	4.557	0.004	0.042
Absolute	3	1.527	0.207	0.015

Table 6 shows that there is a statistically significant difference between years of teaching experience on a relevant factor, F = 4.557, p = 0.004. Further analysis was performed to discern the differences across years of teaching experience using the Bonferroni's method of multiple comparisons (see Table 7).

Table 7. Bonferroni method of multiple comparisons

Length of teaching experience (I)	Length of teaching experience (J)	Selisih mean I – J)
3 years or less	4-10 years	0.0242
	11-20 years	0.3047*
	more than 20 years	0.0377
4-10 years	3 years or less	-0.0242
	11-20 years	0.2804*
	more than 20 years	0.0135
11-20 years	3 years or less	-0.3047*
	4-10 years	-0.2804*
	more than 20 years	-0.2669*
more than 20 years	3 years or less	-0.0377
	4-10 years	-0.0135
	11-20 years	0.2669*

*. the significant mean difference on the level of 5%.

Based on Table 7, it indicates that teachers who have11-20 years of teaching experience tend to hold traditional mathematics views. This is verified with a significant mean difference on the level of 5%.

3.3. Certification Status

Descriptive statistics for each sub-scale based on certification status can be seen in Table 8. Whereas, the results of MANOVA can be seen in Table 9.

Table 8. Descriptive data for teachers' beliefs based on certification status

Subscale	Factor	Non-certified		Certified	
		M	SD	M	SD
BN-M	Relevant	4.980	0.049	5.040	0.043
	Absolute	4.039	0.097	4.120	0.086
BT-M	Relational	5.093	0.040	4.990	0.036
	Instrumental	4.581	0.064	4.671	0.057
BA-M	Integrated	5.149	0.039	5.097	0.034
	Isolated	4.780	0.054	4.771	0.047

Table 8 shows that the mean for each of the constructivist view related factors ranges from 4.980 to 5.149, while for the traditional view related factors range from 4.039 to 4.780. The difference between certification status in each sub-scale is presented in Table 9.

Table 9. MANOVA results for teachers' beliefs and certification status

Subscale	Pillai's trace	F	Hypothesis df	Error df	p	η^2
BN-M	0.004	0.659	2.000	305.000	0.518	0.004
BT-M	0.019	2.996	2.000	305.000	0.051	0.019
BA-M	0.003	0.516	2.000	311.000	0.598	0.003

According to Table 9, it indicates that certification status does not have significant effect on beliefs held by the participants ($p > 0.05$).

3.4. Grade Level Assignment

Descriptive statistics for each sub-scale according to grade level assignment is presented in Table 10. Whereas, the MANOVA's result is summarized in Table 11.

Table 10. Descriptive data of teachers' beliefs and grade level assignment

Subscale	Factor	Lower grade		Upper Grade	
		M	SD	M	SD
BN-M	Relevant	4.971	0.048	5.053	0.044
	Absolute	4.000	0.096	4.119	0.087
BT-M	Relational	5.057	0.041	5.022	0.037
	Instrumental	4.663	0.064	4.608	0.058
BA-M	Integrated	5.139	0.038	5.131	0.035
	Isolated	4.798	0.056	4.737	0.051

Means for constructivist view related factors range from 4.971 (SD = 0.048) to 5.139 (SD = 0.038). Meanwhile, means for traditional view related factors range from 4.000 to 4.798. Further analyses to see the difference among factor of beliefs across grade level assignment is summarized as follows.

Table 11. MANOVA results of teachers' beliefs and grade level assignment

Subscale	Pillai's trace	F	Hypothesis df	Error df	p	η^2
BN-M	0.008	1.161	2.000	305.000	0.314	0.008
BT-M	0.002	0.334	2.000	305.000	0.716	0.002
BA-M	0.002	0.331	2.000	305.000	0.719	0.002

Based on Table 11 above, it indicates that there are no significant differences between the beliefs held by teachers and grade level assignment ($p < 0.05$)

Based on the above section, it can be concluded that there are no significant differences of each sub-scales of beliefs on gender, certification status, and grade level assignment. However, teachers who have teaching experience of 11-20 years are more likely to have absolute views

Gender issues often become an attractive factor to be explored. Although there is no rich literature for making a hypothesis related beliefs they held, it remains a compelling issue in the research topics. For example, Golafshani [8] in his doctoral dissertation raised the issue of gender and found that female mathematics teacher in Iran tend to possess more constructivist views than male teachers do. Similar findings were obtained from research conducted by Yazici and Ertekin [10] who studied the difference between beliefs and anxiety of mathematics' prospective teachers of basic mathematics education based on gender. They found that there were differences in the candidates' beliefs about learning math and male are more likely to have instrumentalist beliefs than female teachers do. Nevertheless, the finding of this study indicates that there are no gender differences in the elementary school teachers' beliefs. A plausible reason is that beliefs system from knowledge and experience rather than genetic factors. Further, there are no many comprehensive studies found in the literature investigating relation between beliefs and gender

Besides, it indicates that there is no significant difference between the beliefs held and years of teaching experience. However, there is a difference in beliefs about the nature of mathematics from 11-20 years experienced teachers who lead to hold more absolute mathematical views than any other group. Furthermore, it seems that the higher of years of teaching experience owned by teachers, the more they become absolutist. This finding can be attributed to teacher education that is not the same as the current situation. Teachers who have 11-20 years of teaching experience received education gradually, from the school of teacher education (equivalent to diploma), and lately prosecuted to educated at least Bachelor's degree. Besides, most of them come from non-graduate primary teacher education. In other words, their education has not been embedded with comprehensive mathematics views. Similar findings occur with the certification factor and grade level assignments. A plausible reason is that certification was recognition of the teaching profession using portfolio assessment or a short educational experience, so it is having no effect on changing teachers' beliefs. To change their beliefs, it requires a comprehensive knowledge and encouragement. In the case of grade level assignment, the finding of this study was in line with the findings of Golafshani [8] which indicated that the grade level assignment did not affect significantly because topic and context of mathematics have no relation with system of beliefs.

4. Conclusion

The findings of this research indicate that it is only in the category of years of teaching experiences that have significant difference in beliefs in which 11-20 years experienced teachers tend to have absolutist views. However, there is no significant difference on beliefs based on other characteristics such as gender, certification status, and grade level taught

Acknowledgments

The researchers thank the Kementerian Riset, Teknologi, dan Pendidikan Tinggi Republik Indonesia (Indonesian Ministry of Research, Technology, and Higher Education) which has funded this research. Researchers also give appreciate to the second and third authors in this article who have provided guidance. We thank also to Miss Luluk who had assisted in improving this article in English.

References

[1] Purnomo YW, Suryadi D, Darwis S 2016 Examining pre-service elementary school teacher beliefs and instructional practices in mathematics class *International Electronic Journal of Elementary Education* 8 **4** pp 629

[2] Purnomo YW 2017 The Complex Relationship between Teachers' Mathematics-related Beliefs and their Practices in Mathematics Class *The New Educational Review* 47 **1** pp 200

[3] Purnomo YW 2017 A Scale for Measuring Teachers' Mathematics-Related Beliefs: A Validity and Reliability Study *International Journal of Instruction* 10 **2** pp 23

[4] Ernest P 1989 *The impact of beliefs on the teaching of mathematics. In: Ernest P, editor. Mathematics teaching: The state of the art* (New York: Falmer).

[5] Stipek DJ, Givvin KB, Salmon JM, MacGyvers VL 2001 Teachers' beliefs and practices related to mathematics instruction *Teaching and Teacher Education* 17 **2** pp. 213

[6] Raymond AM 2016 Inconsistency between a beginning elementary school teacher's mathematics beliefs and teaching practice *Journal for Research in Mathematics Education* 28 **5** pp 550

[7] Leder GC 1992 *Mathematics and gender: Changing perspectives. In: Grouws DA, editor. Handbook of research on mathematics teaching and learning* (New York: Macmillan Publishing Co, Inc)

[8] Golafshani N 2005 *Secondary teachers' professed beliefs about mathematics, mathematics teaching and mathematics learning: Iranian perspective* (Toronto: University of Toronto)

[9] Nisbet S, Warren E 2000 Primary school teachers' beliefs relating to mathematics, teaching and assessing mathematics and factors that influence these beliefs *Mathematics Teacher Education and Development* 2 pp 34

[10] Yazici E, Ertekin E 2010 Gender Differences of Elementary Prospective Teachers in Mathematical Beliefs and Mathematics Teaching Anxiety *International Journal of Human and Social Sciences* 5 **9** pp 610

[11] Leedy MG, LaLonde D, Runk K 2003 Gender equity in mathematics: beliefs of students, parents, and teachers *School Science and Mathematics* 103 **6** pp 285

[12] Perry B, Wong N-Y, Howard P 2006 *Comparing primary and secondary mathematics teachers' beliefs about mathematics, mathematics learning and mathematics teaching in Hong Kong and Australia.* In: Leung FKS, Graf K-D, Lopez-Real FJ, editors. *Mathematics Education in Different Cultural Traditions-A Comparative Study of East Asia and the West* (New York: Springer)

IOP Conf. Series: Journal of Physics: Conf. Series **948** (2018) 012062 doi:10.1088/1742-6596/948/1/012062

[13] Purnomo YW 2016 *Eksplorasi keyakinan guru tentang matematika, belajar dan mengajar, penilaian dan hubungannya dengan praktik di kelas matematika*. (Bandung: Universitas Pendidikan Indonesia)
[14] Finch H 2005 Comparison of the Performance of Nonparametric and Parametric MANOVA Test Statistics when Assumptions Are Violated *Methodology* 1 **1** pp 27
[15] Johnson CR, Field CA 1993 Using fixed-effects model multivariate analysis of variance in marine biology and ecology *Oceanography and Marine Biology Annual Review* **31** pp 177
[16] Olson CL 1974 Comparative robustness of six tests in multivariate analysis of variance *Journal of the American Statistical Association* 69 **348** pp 894
[17] Park E, Cho M, Ki C-S 2009 Correct use of repeated measures analysis of variance *The Korean journal of laboratory medicine* 29 **1** pp 1

IOP Conf. Series: Journal of Physics: Conf. Series **948** (2018) 012063 doi:10.1088/1742-6596/948/1/012063

An algorithm for finding a similar subgraph of all Hamiltonian cycles

R Wafdan[1,3], M Ihsan[2,3] and D Suhaimi[2,3]

[1] Graduated Mathematics Program, Syiah Kuala University, Banda Aceh, Indonesia
[2] Department of Mathematics, Syiah Kuala University, Banda Aceh, Indonesia
[3] Multimedia Research Group, Department of Mathematics, Syiah Kuala University, Banda Aceh, Indonesia

E-mail: mahyus@unsyiah.ac.id

Abstract. This paper discusses an algorithm to find a *similar subgraph* called *findSimSubG* algorithm. A *similar subgraph* is a subgraph with a maximum number of edges, contains no isolated vertex and is contained in every Hamiltonian cycle of a Hamiltonian Graph. The algorithm runs only on Hamiltonian graphs with at least two Hamiltonian cycles. The algorithm works by examining whether the initial subgraph of the first Hamiltonian cycle is a subgraph of comparison graphs. If the initial subgraph is not in comparison graphs, the algorithm will remove edges and vertices of the initial subgraph that are not in comparison graphs. There are two main processes in the algorithm, changing Hamiltonian cycle into a cycle graph and removing edges and vertices of the initial subgraph that are not in comparison graphs. The *findSimSubG* algorithm can find the *similar subgraph* without using backtracking method. The *similar subgraph* cannot be found on certain graphs, such as an n-antiprism graph, complete bipartite graph, complete graph, $2n$-crossed prism graph, n-crown graph, n-möbius ladder, prism graph, and wheel graph. The complexity of this algorithm is $O(m|V|)$, where m is the number of Hamiltonian cycles and $|V|$ is the number of vertices of a Hamiltonian graph.

1. Introduction

The problem of finding Hamiltonian cycle is a case to find a cycle that contains all vertices of a Hamiltonian graph. This problem has become a challenge to mathematicians. Even though it is a classic and complicated problem, more than a few mathematicians choose this problem as their research objects. The Hamiltonian cycle problem is not only an important problem in graph theory but also can be used in real-life application such as telecommunication and computer network [1]. Several algorithms have been developed using backtracking or heuristic approach to find Hamiltonian cycles [2]. One of the algorithms that are able to find Hamiltonian cycle is *findSegment* algorithm [3].

The *findSegment* algorithm uses backtracking approach to find the solution. The *findSegment* algorithm works by collecting edges that form Hamiltonian cycle called *segments*. Furthermore, the algorithm will delete edges that are not needed to reduce the search space [3]. By using some modifications, the *findSegment* algorithm can be used to find all Hamiltonian cycles of a Hamiltonian graph.

In addition to finding a Hamiltonian cycle, another thing needed to find a Hamiltonian cycle is finding a similar subgraph. The similar subgraph is a subgraph with a maximum number of edges, contains no isolated vertex and is contained in every Hamiltonian cycle of a Hamiltonian Graph. A graph G' = (V', E') is a subgraph of another graph G = (V, E) if and only if the vertex V' and edge E' sets of G' are, respectively, subsets of the vertex V and edge E sets of G [4]. Finding the similar subgraph is an important case to be analyzed because it is often found in real life such as products distribution using box trucks. Box trucks distribute the products to the same consumers through different routes. However, on some occasions, the trucks use the same routes to distribute products to its destinations. The goal of this research is to find such routes. By finding the routes, the best policy can be made by authorities to avoid a traffic jam or to improve its quality.

In graph theory, if we consider the routes as a Hamiltonian cycle, the routes used by every box trucks can be seen as subgraphs of every Hamiltonian cycle. Applying subgraphs in transportation is not a new problem. Flinsenberg in his book on route planning algorithms for car navigation [5] used subgraphs called cells in his algorithm to divide road graph, that is, a road network represented by a multi-graph. Subgraph is also used by Yousefi and Zamani in their paper [6], they considered that a network graph was composed of several subgraphs. Learning the overall condition of the graph is possible by conducting several searches over each subgraph. This procedure helps discover the shortest path and avoid choosing paths randomly. In addition, Latora and Marchiori, who research about Boston Subway, also used the concept of subgraph [7]. They stated that the MBTA (Boston underground transportation system) is not a closed system, as it can be considered, after all, a subgraph of a wider transportation network, and they concluded that fault tolerance is not a critical issue. In this research, an algorithm to find subgraph that is on every Hamiltonian cycle will be proposed. If it is applied to transportation problem, the algorithm could find routes taken by all trucks of an industry to distribute its products, so the quality of such routes can be improved. This algorithm requires findSegment Algorithm to find Hamiltonian cycles before finding a similar subgraph.

2. Methodology

The algorithm of finding a *similar subgraph* takes the set of solution (every Hamiltonian cycle) of a Hamiltonian graph H = (V, E) as its input. However, the following condition must be met:

The resulting subgraph contains no isolated vertex. Otherwise, the algorithm will not affect anything since every Hamiltonian cycle contains every vertex of a Hamiltonian graph.

The algorithm runs only on Hamiltonian graphs that contain more than one Hamiltonian cycle. On a graph with only one Hamiltonian cycle, the algorithm has no effect since the *similar subgraph* will be the Hamiltonian cycle itself.

For example, the Hamiltonian graph shown in Figure 1 contains four different Hamiltonian cycles, namely (1,3,2,5,7,8,6,4,1), (1,3,2,5,8,7,6,4,1), (1,3,6,7,8,5,2,4,1), and (1,3,6,8,7,5,2,4,1) as shown in Figure 2.

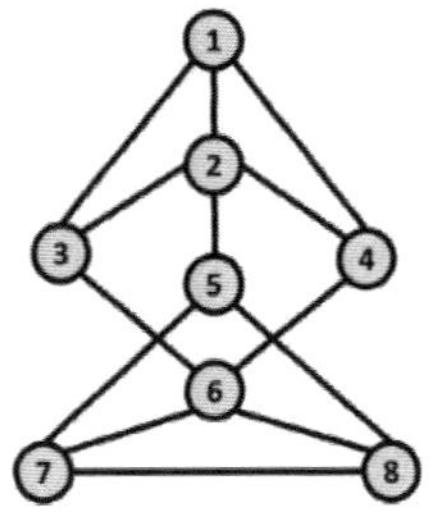

Figure 1. Hamiltonian graph.

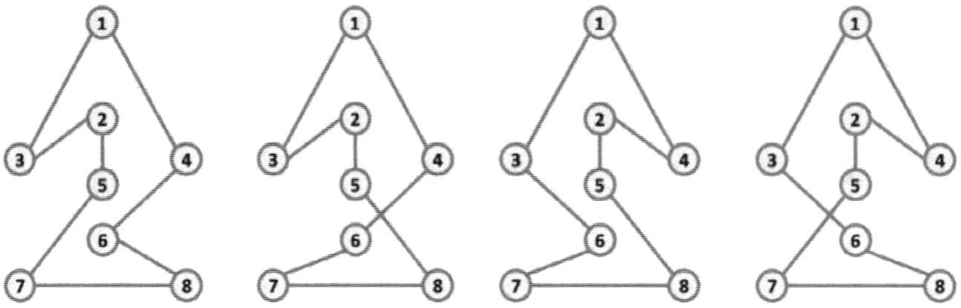

Figure 2. Hamiltonian cycles.

From all four Hamiltonian cycles, a *similar subgraph* is obtained as shown in Figure 3.

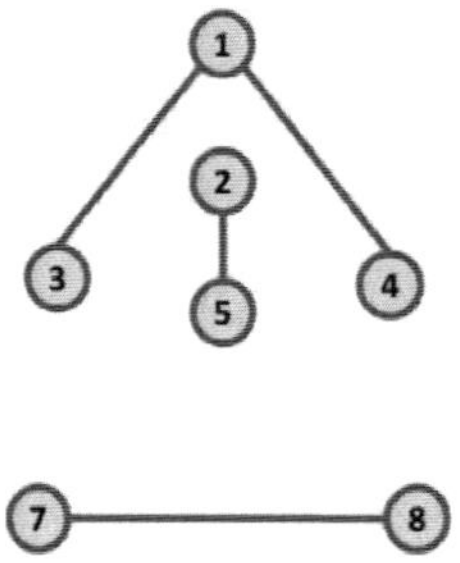

Figure 3. Subgraph of Hamiltonian cycles.

The following is the finding algorithm of the similar subgraph of the graph as shown in Figure 1.

After taking the set of Hamiltonian cycles from findSegment algorithm, set index variable i = 1

Change the i-th Hamiltonian cycle into cycle graph HcG and then set the i-th Hamiltonian cycle as initial subgraph SubG (see Figure 4).

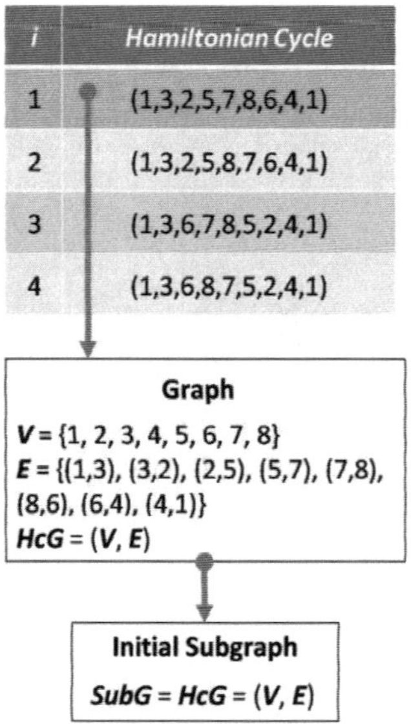

Figure 4. Initial subgraph.

IOP Conf. Series: Journal of Physics: Conf. Series **948** (2018) 012063 doi:10.1088/1742-6596/948/1/012063

- Set $i = i + 1$
- Change the i-th Hamiltonian cycle into cycle graph *HcG*.
- Check whether the initial subgraph *SubG* is a subgraph of the cycle graph *HcG* (see Figure 5).

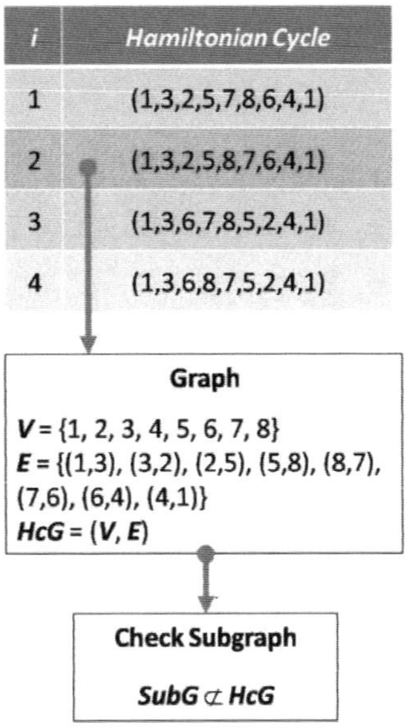

Figure 5. Check Subgraph.

- If not, delete the vertices and edges of the initial subgraph that are not in the cycle graph of the i-th Hamiltonian cycle (see Figure 6).

Initial Subgraph

V_s = {1, 2, 3, 4, 5, 6, 7, 8}
E_s = {(1,3), (3,2), (2,5), (7,8), (6,4), (4,1)}
SubG = (V_s, E_s)

Figure 6. The new initial subgraph.

- Otherwise, go to step 3 until every vertex has been compared.
- The result of the algorithm is an *initial subgraph* that is in every Hamiltonian cycle (see Figure 7).

Initial Subgraph

V_s = {1, 2, 3, 4, 5, 7, 8}
E_s = {(1,3), (2,5), (7,8), (4,1)}
SubG = (V_s, E_s)

Figure 7. The final initial subgraph.

3. Results and Discussion
The first step to finding a *similar subgraph* is finding every Hamiltonian cycle in Hamiltonian graph. The set of Hamiltonian cycles is called *HcList*. The set HcList will then be the input to the algorithm.

IOP Conf. Series: Journal of Physics: Conf. Series **948** (2018) 012063 doi:10.1088/1742-6596/948/1/012063

The finding algorithm is called *findSimSubG*. In general, the subgraph finding algorithm of a Hamiltonian cycle works as follows.

```
findSimSubG(HcList: Set of Hamiltonian cycles): graph
begin
        set subG := convertToGraph(HcList[1])
        for i := 2 to |HcList| do
           set HcG := convertToGraph(HcList[i])
           if subG ⊄ HcG then
                        set subG := removeVE(subG, HcG)
           end
        end
        findSimSubG := subG
end;
```

The *findSimSubG* algorithm consists of two processes, *convertToGraph,* and *removeVE*. *ConvertToGraph* is used to change a Hamiltonian cycle into cycle graph form. The algorithm takes Hamiltonian cycles as the input and outputs a cycle graph. In general, the process *convertToGraph* can be described by the following algorithm.

```
convertToGraph (HamCy : Hamiltonian cycles) : graph
begin
        for i := 1 to |HamCy|-1 do
           set V := V ∪ {HamCy[i]}
        end
        for i := 2 to |HamCy| do
           set E := E ∪ {(HamCy[i-1], HamCy[i])}
        end
        set HcG := (V, E)
        convertToGraph := HcG
end;
```

The *removeVE* process is a process to delete edges that are in subG but not in graph *HcG* and to delete vertices in *subG* with degree 0. In the *removeVE* process, the input of the algorithm is a subgraph *subG* and a cycle graph *HcG*. The algorithm will output a subgraph. In general, the process *removeVE* can be described as the following algorithm.

```
removeVE(subG : Subgraf, HcG : Graf) : graph
begin
        { Let subG = (Vs, Es) and  HcG = (V, E) }
        for each (u, v) ∈ Es  do
           if (u, v) ∉ E then
                        set Es := Es \ {(u, v)}
           else (v, u) ∉ E then
                        set Es := Es \ {(v, u)}
           end
        end
        for each v ∈ Vs  do
           if deg(v) = 0
                        set Vs := Vs \ {(v)}
           end
```

```
        end

        removeVE := subG
end;
```

The *findSimSubG* is tested on eight well-known Hamiltonian graphs, namely *n*-antiprism graph, complete bipartite graph $K_{n,n}$, complete graph K_n, 2*n*-crossed prism graph, *n*-crown graph, *n*-möbius ladder, prism graph Y_n and wheel graph W_n. The aforementioned graphs have no the *similar subgraph*.

The *convertToGraph* algorithm consists of two **for** loops, each **for** loop executes $|V|$ times and there are two processes on the last line of the algorithm. Thus, the total time for the *convertToGraph* algorithm is $O(2|V|+2) = O(|V|)$ where $|V|$ is the number of vertices in the Hamiltonian graph. The *removeVE* algorithm also consists of two **for** loops, each **for** loop executes $|V|$ times and there is a process on the last line of the algorithm. Thus, the total time for the *removeVE* algorithm is $O(2|V|+1) = O(|V|)$. The *findSimSubG* algorithm consists of calling *convertToGraph* process one time outside the **for** loop that is $O(|V|)$, calling *convertToGraph* process one time inside the **for** loop that is $O(m|V|)$ where m is the number of Hamiltonian cycles found in Hamiltonian graph, calling *removeVE* process one time inside the **for** loop that is $O(m|V|)$, and there is a process on the last line of the algorithm. The total time for *findSimSubG* is $O(|V| + m|V| + m|V| + 1) = O(m|V|)$.

4. Conclusion

The *findSimSubG* algorithm is able to find a *similar subgraph* without using backtracking process. Not all Hamiltonian graphs contain the *similar subgraph*. The test using the *findSimSubG* algorithm on eight well-known Hamiltonian graph shows that the graphs have no the *similar subgraph*. If the finding the *similar subgraph* algorithm is applied to transportation problem, then the algorithm could find routes taken by all trucks of an industry to distribute its products, so the quality of such routes can be improved. The limitation of this algorithm is it only works on undirected graphs, while most of the real-life problems will likely be on directed graphs.

Acknowledgments

The authors express gratitude to Directorate of Research and Community Services for grant of Applied Products Research.

References

[1] Serra M and Kent K 2003 *Proc. of Int. Symp. on Circuits and systems* **3** 228-231
[2] Riaz K and Khiyal M S H 2006 *Information Technology Journal* **5** 851-859
[3] Ihsan M, Wafdan R, Hafnani, Halfiani V and Ramli M 2016 *Far East J. Math. Scie.* **100** pp 573-587.
[4] Goodaire E G and Parmenter M M 2002 *Discrete Mathematics with Graph Theory*, (New Jersey: Prentice-Hall)
[5] Flinsenberg I C M 2004 *Route Planning Algorithms for Car Navigation* (Eindhoven: Technische Universiteit)
[6] Yousefi P and Zamani R 2013 *Inter. J. Mach. Learn. Com.* **3** p 44
[7] Latora V and Marchiori M 2002 *Physica A: Statistical Mechanics and its Applications* **314** 1 pp 109-113

IOP Conf. Series: Journal of Physics: Conf. Series **948** (2018) 012064 doi:10.1088/1742-6596/948/1/012064

Interactive graph constructing on graph theory application development

Dairina[1], M Ihsan[2,3] and M Ramli[2]

[1] Mathematics Graduate Study Program, Syiah Kuala University
[2] Department of Mathematics, Syiah Kuala University
[3] Multimedia Research Group, Department of Mathematics, Syiah Kuala University

E-mail: mahyus@unsyiah.ac.id

Abstract. This paper discusses the steps of constructing simple graphs on graph theory application development. Most graphs constructing on application development use an adjacency matrix and an adjacency list before the graph is constructed. However, it is considerably difficult to construct a graph in such way. The steps of constructing a graph in this work were started by directly construct a graph on stage. The graph was constructed by creating the vertices and every edge can be created after creating two vertices as its endpoint. The graphs can be changed by moving the vertex and deleting vertices and edges. Furthermore, the graph can be changed by adding vertices and edges. The adjacency list is then obtained from the given graph and used for further processes. The steps of constructing graph are interactive so that constructing graph can be made easier and more efficient.

1. Introduction

Graph theory is one of the fields in mathematics that accommodate various problems in real life. Many problems can be modeled in the form of a graph, such as shortest-path problem, a scheduling problem, simple mapping graph problem, and traveling salesman problem (TSP). Developing graph theory application is needed to fulfill the need for solving a large complex problem, including accuracy, speed, and interactive use.

A graph $G = (V, E)$ consists of a non-empty set vertex $V(G)$, a set edge $E(G)$ and a relation that map each edge to two vertices [1]. A graph can be represented in many ways, namely sets [2], [3], a graphical form [2], [3], and [4], an adjacency matrix [5], and an adjacency list [5].

In graph theory application development, a graph is formed by inputting an adjacency matrix or an adjacency list first. Afterward, a graph is constructed, and information related to coordinates is added. However, inputting an adjacency matrix or an adjacency list is considered not effective. For adjacency, it is caused by a large number of entries that is n^2. Meanwhile, for adjacency list, the problem arises when determining the pair of vertices that form edges. Another relatively easy way is drawing the graph first. Afterward, an adjacency matrix and an adjacency list can be obtained from the graph.

Inputting a graph using its adjacency matrix can be found in a few of studies, such as finding a spanning tree of a simple graph [2]. Another example is plotting a graph from its adjacency matrix [6], [7], [8], [9], [10], and each vertex's coordinate can be obtained in the following ways: data file, random coordinate and using mouse pointer. Another way to draw a graph is by determining the adjacency list first. For example, a graph is plotted from an adjacency list, and each vertex's coordinate is obtained from data file [11].

Graph application can be used easier if the graph is visually and interactively portrayed. Vertices and edges can be directly inputted by clicking and tapping on stage. Deleting and adding vertices and edges can be done by double-clicking and changing vertex's position can be done by clicking and dragging the vertex. Afterward, an adjacency matrix and an adjacency list can be obtained from the graph and are used for further processes. Therefore, this research will explain the steps needed to draw and construct a simple graph interactively, which can be used in developing a graph theory application.

2. Methodology

In an undirected simple graph, there exists an edge *uv* which connects vertex *u* and vertex *v*. In this context, edge *uv* is identical to edge *vu* which connects vertex *v* and vertex *u*.

The graph is drawn in two-dimensional stage with coordinate (0,0) located in the top-left corner, corresponding to a monitor's coordinate. On stage, the horizontal axis (*x*-axis) is always positive. The same thing applies to the vertical axis (*y*-axis), see figure 1.

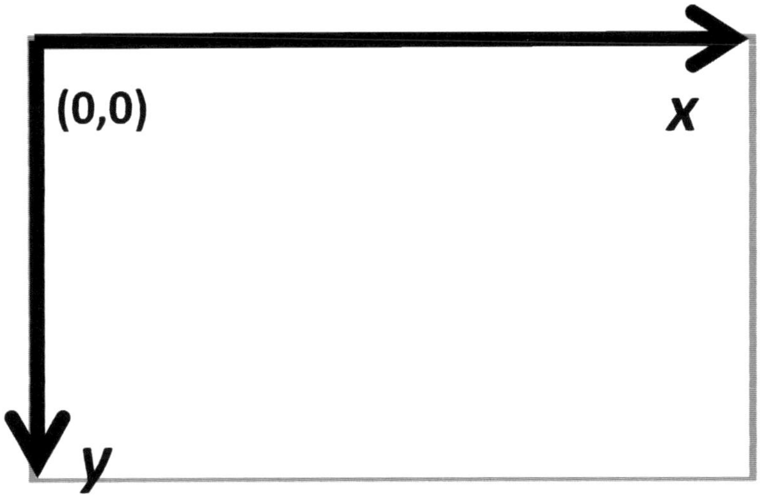

Figure 1. Coordinate of stage.

Constructing a graph is started by drawing or inputting vertices. A vertex can be inputted by clicking directly on stage. The vertex is a small circle with the same size and is marked with a label like as numbers, letters, or indexed variables (such as v_1, v_2, v_3). Its label will be located in the center of the corresponding vertex, see figure 2.

Figure 2. Vertex's label in the centre.

Vertex's label appears in an ascending order such as one, two, three, a, b, c, or v_1, v_2, v_3 and so on. The position or the coordinate of each vertex is obtained from the center where the vertex is located. The tolerance of placing a vertex must meet the minimal distance so the edge between two vertices can be seen. The distance between any two vertices can be calculated using Euclidian formula. The tolerance between any two vertices is $d_i > 2R + k$, and the distance between any two vertices can be computed using equation 1.

$$d = \sqrt{(x_c - x_i)^2 + (y_c - y_i)^2} \tag{1}$$

where d_i is the distance between any two vertices, R is a radius of vertex, k is tolerance distance between any two vertices, (x_c, y_c) is the center point of the added or moved vertex, and (x_i, y_i) is the center point of another vertex on stage.

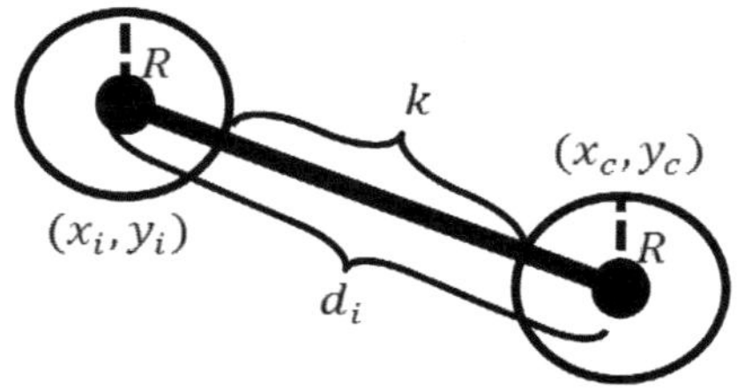

Figure 3. Distance between any two vertices.

An edge can be added after creating two vertices in minimal. The edge can be drawn by clicking two vertices that will be connected. The edge will be a straight line from the center point of one vertex to another. The two vertices can be chosen randomly and do not have to be in ascending order, see figure 4.

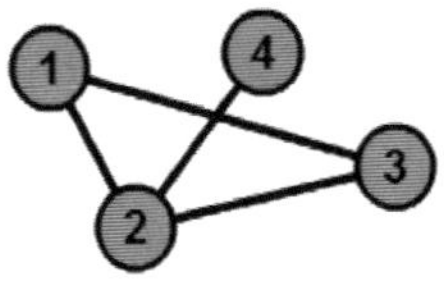

Figure 4. Drawing edge with two random vertices.

For each pair of vertices that is selected, there can only be one edge at maximum (vertex with multiple edges is not allowed). A vertex connected by an edge can be connected with another edge. Thus, that vertex has a degree at least 2. Adding a vertex can be done by creating a small circle with a label on stage. The size of the vertex is the same with other vertices on stage. The label of the vertex is adjusted by the label of the previous vertex.

Deleting a vertex can be done by deleting the small circle, therefore the label and the edges associated with that vertex will also be removed. Changing the position of a vertex can be accomplished by dragging the circle directly on stage. Deleting an edge can be done by choosing two vertices and removing the line that connects the vertices on stage. Adding an edge can be achieved by connecting the center point of the first vertex and the second vertex.

3. Results and discussion

Given a simple graph as shown in figure 4, the graph can be drawn by doing the steps presented in this research.

3.1. Drawing and modifying a graph

Drawing the graph can be done using following ways:

- The first vertex up to the last vertex can be added by clicking the position needed on stage, see figure 5

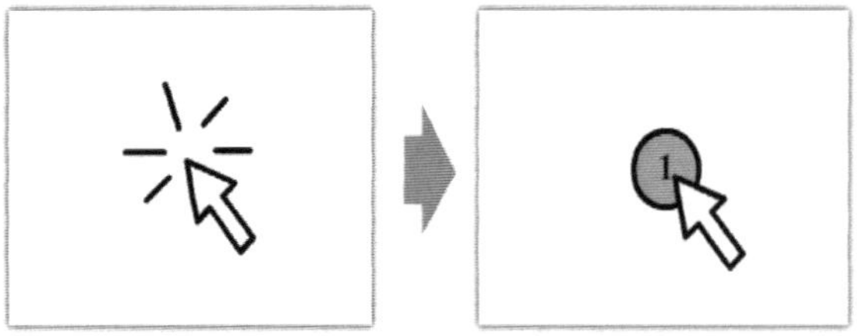

Figure 5. Drawing a vertex.

- The edge can be drawn by clicking two vertices that are required to connect, see figure 6.

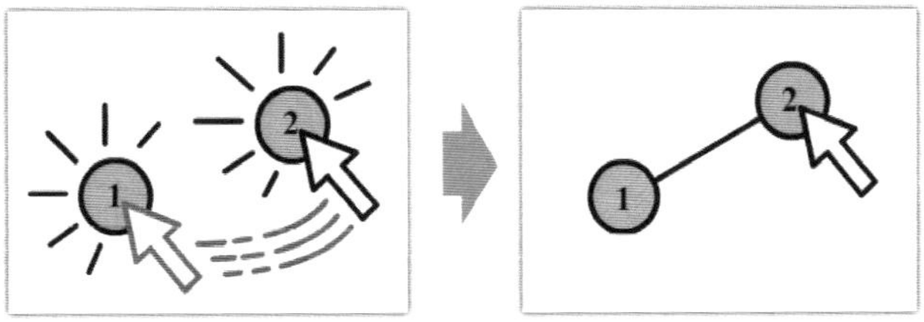

Figure 6. Drawing an edge.

- Adding a vertex and edge can be done using step 1 and 2.
- A vertex cannot be added if the tolerance between the vertices is less than *d* see figure 3.
- The position of a vertex can be changed by clicking and dragging the vertex, see figure 7.

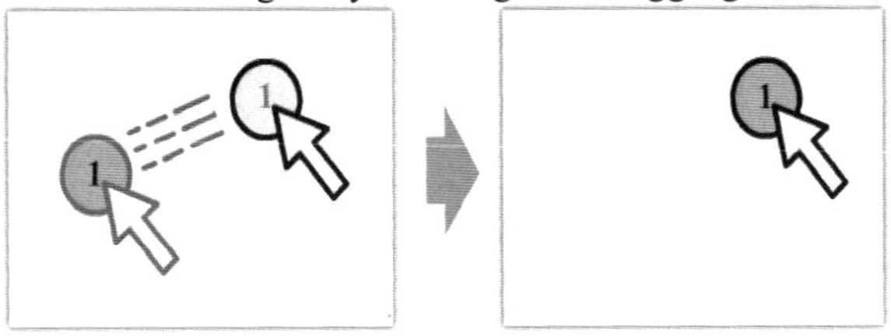

Figure 7. Changing vertex's position.

- Deleting a vertex can be done by double-clicking the vertex needed to remove, see figure 8.

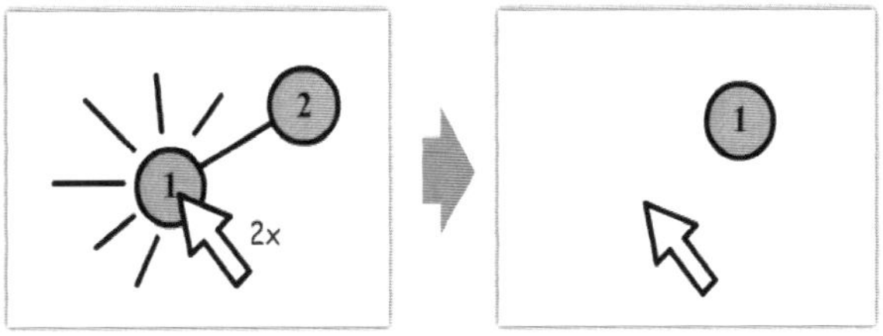

Figure 8. Deleting an edge.

- To draw a complete graph, a feature (clicking a button) can be added after creating all the vertices.

3.2. Storing a Graph

The data of graph shown in figure 4 is needed for further processes. The data of the graph can be presented in the form of adjacency matrix or adjacency list. In this research, the information of the graph which has been created are the number of vertices and edges, the set of edge, and the coordinate for each vertex will be stored in the form of variables, namely adjacency lists and arrays. The process needed to save the information as shown in figure 4 is as follows.

- Each vertex drawn on stage will be counted, and the number will be stored in a variable nV, see figure 10

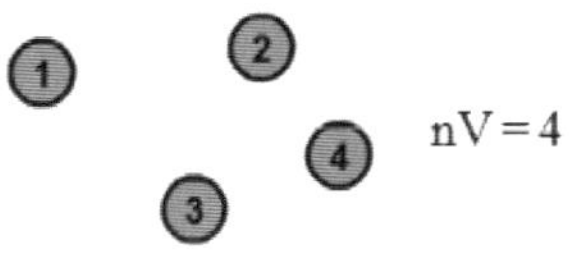

Figure 9. Saving the number of vertices.

- Each edge drawn on stage will be counted, and the number will be stored in a variable *nE*, see figure 11.

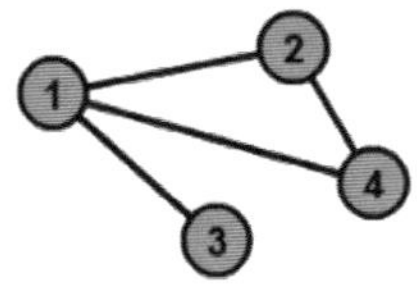

nE = 4

Figure 10. Saving the number of edges.

- Aside from the number of vertices and edges, for each vertex drawn on stage, the position of the vertex will be stored in the form of an array, see figure 12.

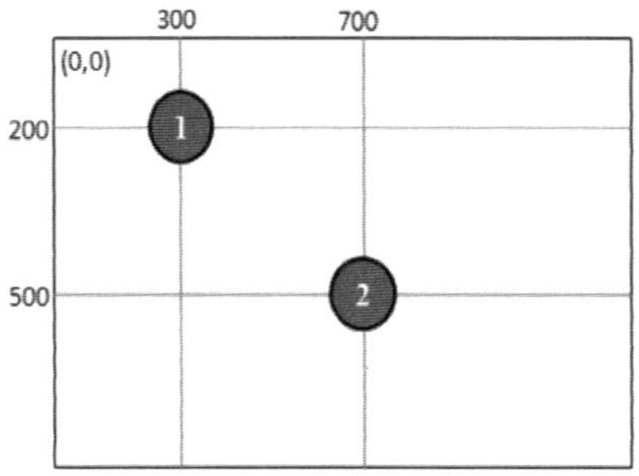

PositionV[1][1]= 300
PositionV[1][2]= 200
PositionV[2][1]= 700
PositionV[2][2]= 500

Figure 11. Saving the position of each vertex.

- Aside from the number of edges, for each edge drawn on stage, two vertices for which the edge connects will be stored in the form of an adjacency list, see figure 13.

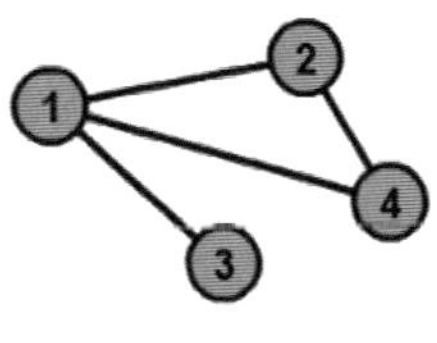

Adj [1][1]= 2
Adj [1][2]= 3
Adj [1][3]= 4
Adj [2][1]= 1
Adj [2][2]= 4
Adj [3][1]= 1
Adj [4][1]= 1
Adj [4][2]= 2

Figure 12. Saving pair of vertices that are connected by an edge.

The information obtained from the graph are stored and used for an algorithm or another related process. The information of the graph are also needed to save the graph in storage so the graph can be displayed if needed. Saving the graph can be done by storing information in a file, such as the number of vertices, the coordinates of each vertex, the number of edges, and the endpoint of each edge.

4. Conclusion

The steps of drawing a graph presented in this research are started by drawing the graph directly on stage. Afterward, an adjacency matrix and an adjacency list are obtained from the graph and used for further processes. Thus, the steps presented in this research are more interactive. Therefore, drawing a graph in graph theory application can be made easier.

Acknowledgments

The authors express gratitude to Directorate of Research and Community Services for grant of Applied Products Research.

References

[1] West D B 2001 *Introduction to Graph Theory* (Pearson Education, Inc.)
[2] Goodaire G E and Parmenter M M 2002 *Discrete Mathematics with Graph Theory* (New Jersey: Prentice-Hall)
[3] Rosen K H 2012 *Discrete Mathematics and Its Applications* 7th *(ed)* (New York: The McGraw-Hill Companies, Inc.)
[4] Wilson R J 1996 *Introduction to Graph Theory* 4th *ed.* (Addison-Wesley)
[5] Cormen T H, Leiserson C E, Rivest R L and Stein C 2009 *Introduction to Algorithms* 3th (London: The MIT Press)
[6] Puspitorini S 2008 *Media Informatika* **6** 39-55.
[7] Nugraha D W 2011 *Jurnal Ilmiah Foristek* **1**
[8] Side S, Sutra and Sulaiman 2013 *Jurnal Sainsmat* **2** 131-139
[9] Lukas S, Anwar T, and Yuliani W 2005 *Seminar Nasional Aplikasi Teknologi Informasi*
[10] Mardlootillah H I, Suyitno A, and Arini F Y 2014 *UNNES Journal of Mathematics* **1**
[11] Ardiansyah, Efendi F S, Syaifullah, Pinto M, Pujianto, and Tempake H S 2010 *J. Infor.* **4**

IOP Conf. Series: Journal of Physics: Conf. Series **948** (2018) 012065 doi:10.1088/1742-6596/948/1/012065

Connectivity algorithm with depth first search (DFS) on simple graphs

O Riansanti[1,2], M Ihsan[1,2] and D Suhaimi[1,2]

[1] Department of Mathematics, Syiah Kuala University, Banda Aceh, Indonesia
[2] Multimedia Research Group, Department of Mathematics, Syiah Kuala University, Banda Aceh, Indonesia

E-mail: mahyus@unsyiah.ac.id

Abstract. This paper discusses an algorithm to detect connectivity of a simple graph using Depth First Search (DFS). The DFS implementation in this paper differs than other research, that is, on counting the number of visited vertices. The algorithm obtains *s* from the number of vertices and visits source vertex, following by its adjacent vertices until the last vertex adjacent to the previous source vertex. Any simple graph is connected if *s* equals 0 and disconnected if *s* is greater than 0. The complexity of the algorithm is $O(n^2)$.

1. Introduction

A graph is connected if any two vertices are joined by a sequence of edges [1]. A connected graph can also be defined as a graph when there is a walk between every pair of vertices [1]. A graph is connected if there exists a path between any pair of vertices [2]. A work of literature also defines a connected graph as a graph that is formed by one component [2].

The connectivity of a graph is often used in the various real-life implementation and problem-solving (algorithm) of a graph, such as a bipartite graph, Hamiltonian graph, Eulerian graph, tree, minimum spanning tree (MST), some shortest path problems (Dijkstra's algorithm), graph coloring, and planar graph. In the real world, the implementation of a connected graph solves problems in telecommunication [3], blood flow [4], computer network [5], and transportation [6]. Determining the connectivity of a graph is so important because of its various benefits.

The connectivity of a graph can be determined using methods or connectivity algorithms. One of the connectivity algorithms is the positive non-diagonal entry of matrix [7]. Other methods are Modified Prüfer's Algorithm Labelling Trees [8], the similarity of virtual vertex set and vertex set [9].

In this research, the algorithm uses the concept of Depth First Search (DFS), which is a well-known concept in computer science. The difference of this algorithm from other algorithms is the way of counting vertices that have been visited [10, 11].

2. Methodology

The connectivity algorithm presented in this article takes input any simple graf G = (V, E). Afterward, the status of each vertex is set to false and is contained in variable Pi, where $1 \leq i \leq |V|$, which means every vertex is an unvisited vertex. Variables are then initialized to |V|, which means no vertex is visited yet.

IOP Conf. Series: Journal of Physics: Conf. Series **948** (2018) 012065 doi:10.1088/1742-6596/948/1/012065

The process of checking connectivity is presented in a recursive function called Visit (source vertex). Let vertex r be source vertex, then the status of vertex r is set to true, which means that vertex r is visited. The algorithm then subtracts 1 from s. Afterward, the algorithm will search for the vertex with the smallest label that is adjacent to r, if there exists such vertex, the recursive function will run on that vertex and then subtract 1 from s. This mechanism follows DFS method. The process will be repeated in the same way as calling the recursive function, but on different source vertex, that is an adjacent vertex to r with next label. The recursive process will stop if there is no adjacent vertex with status false. The last phase is checking the variable s, if s = 0 then the graph is connected and if s > 0 then the graph is disconnected.

3. Results and Discussion

The result of this research shows that any simple graph can be checked its connectivity using connectivity algorithm with DFS. The algorithm shows that if $s = 0$, then the graph is connected, and if $s > 0$, then the graph is disconnected. In general, the algorithm to determine the connectivity of any simple graph is shown as follows.

```
Make_Simple_Graph()
for(i = 1; i <= |V|; i++)
P[i] = 0
s = |V|
Initial VertexStart
Visit(VertexStart)
if(s == 0)
Output "Connected graph"
Else
Output "Disconnected graph"
```

The algorithm consists of three major parts. The first part is inputting any simple graph, initializing the status of each vertex and initializing variable s, and selecting the source vertex. The second part is the process of DFS, which is represented by a recursive function. In general, the second part works as follows.

```
void Visit(int r)
int j
P[r] = 1
s--
for(j = 1; j <=|V|; j++)
if(A[r][j] == 1)
v = j
if(P[v] != 1)
Visit (v)
```

After selecting the source vertex, the algorithm will search for an adjacent vertex of source vertex. This algorithm corresponds to Depth First Search (DFS). After executing the recursive function, the value of variable s will be changed. The third part of the algorithm is checking the value of variable s. If s = 0, then the graph is connected and if s > 0 then the graph is disconnected. The algorithm running on simple graph G = (V, E), where V = {1, 2, 3, 4, 5, 6, 7} and E = {12, 15, 17, 23, 34, 35, 67} is shown in Figure 1 to Figure 4.

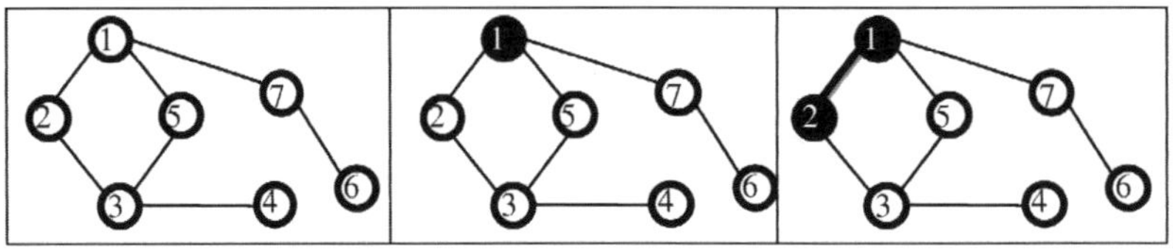

(a) (b) (c)

Figure 1. (a) Inputting simple graph (b) Find source vertex and (c) visited vertex adjacency.

The first step is inputting a simple graph. The graph with $|V| = 7$ is shown in Figure 1 (a). The algorithm then initializes $s = |V|$ where s is a variable to determine the connectivity and for each visited vertex, the algorithm subtracts 1 from s. The algorithm then initializes $P_i = 0$ for each unvisited vertex where $0 \leq i < n$ and $P_i = 1$ for each visited vertex where $0 \leq i < n$. So, $|v| = p_1, p_2, p_3, p_4, p_5, p_6, p_7$ and give a status for each vertices.

As shown in Figure 1 (b), vertex 1 is selected to be the source vertex. Afterward, $s = s - 1$ for each visited vertex so s = 6. Since vertex 1 is a visited vertex, the algorithm sets $P_1 = 1$. The candidates for the adjacent vertex of source vertex are vertex 2, vertex 5, and vertex 7. The algorithm chooses the adjacent vertex with the smallest label. Vertex 2 is therefore chosen to visit. Thus, $s = 5$ and $P_2 = 1$.

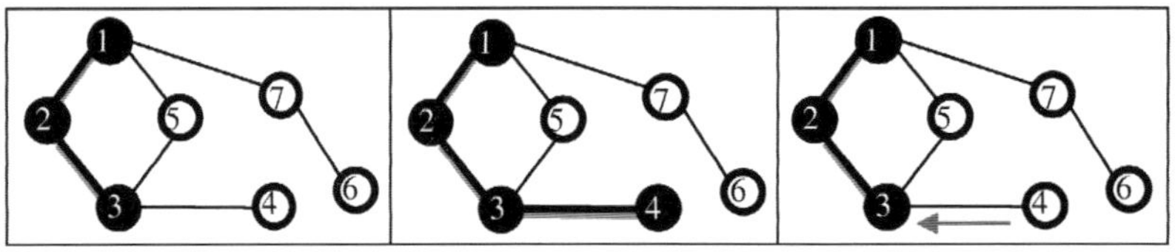

(d) (e) (f)

Figure 2. (d) Visiting adjacent vertex (e) Visiting adjacent vertex (f) Selecting previous source vertex.

Figure 2 (d) shows that vertex 3 is the only vertex adjacent to vertex 2, which is the current source vertex. Therefore, the vertex visited next is vertex 3. Thus, $s = 4$ and $P_3 = 1$. Figure 2 (e) shows that both vertex 4 and vertex 5 are adjacent to vertex 3.The algorithm chooses the one with the smallest label. Vertex 4 is therefore chosen as source vertex. The algorithm then sets s = 3 and $P_4 = 1$. In Figure 2 (f), no vertex is adjacent to vertex 4. The algorithm goes back to previous source vertex, which is vertex 3

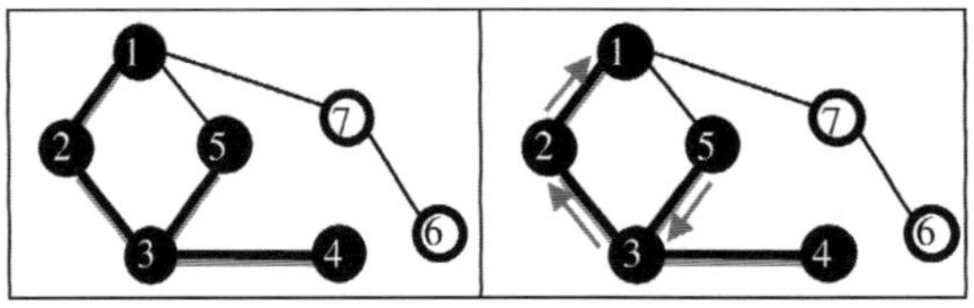

(g) (h)

Figure 3. (g) Visiting adjacent vertex (h) Selecting previous source vertex.

Figure 3 (g) shows that vertex 5 is adjacent to vertex 3, so s = 2 and $P_5 = 1$. In Figure 3 (h), no vertex is adjacent to vertex 5. The algorithm goes back to previous source vertex, which is vertex 3. The process is repeated until vertex 1 is visited as shown in figure 2(h).

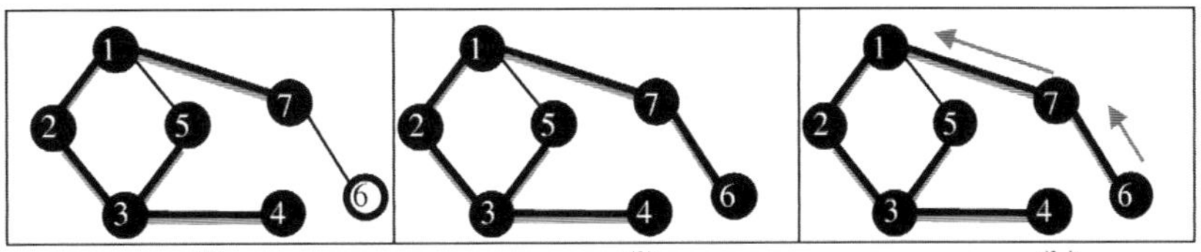

Figure 4. (i) Visiting adjacent vertex (j) Visiting adjacent vertex and (k) Selecting previous source vertex.

Figure 4 (i) shows that vertex 1 is the source vertex. The only vertex adjacent to source vertex is vertex 7. Thus, vertex 7 is visited so $s = 1$ and $P_7 = 1$. Figure 4 (j) shows that the only vertex adjacent to vertex 7 is vertex 6. Thus, vertex 6 is visited so $s = 0$ and $P_6 = 1$. Figure 4 (k) shows no vertex is adjacent to vertex 6, so the algorithm selects the previous vertex and goes back to the first source vertex, which is vertex 1. Vertex 1 has no remaining adjacent vertex. Therefore, the algorithm checks the value of variable s. If $s = 0$, then the graph is connected. However, if $s > 0$, then the graph is disconnected. The graph shown above indicates that $s = 0$, so graph G is a connected graph.

The total time for this algorithm is $O(n^2)$. That is, for each n visited vertices, the algorithm checks its adjacency n times, both adjacent vertices and non-adjacent vertices.

4. Summary

The algorithm can be used to validate an input graph in solving graph problems that need a connected graph, such as verifying a Hamiltonian graph. The proposed algorithm, unfortunately, runs only on a simple graph. Based on this research, it is shown that any simple graph is connected if $s = 0$ and disconnected if $s > 0$ where s is a variable to determine the connectivity of a graph

Acknowledgments

The authors express gratitude to Directorate of Research and Community Services for grant of Applied Products Research.

References

[1] Goodaire E G and Parmenter M M 2002 *Discrete Mathematics with Graph Theory Second Edition* (New Jersey: Prentice-Hall)
[2] Wilson R J 1996 *Introduction to Graph Theory Fourth edition* (London: Longman Group Ltd)
[3] Yang Q and Desmedt Y 2011 Secure communication in multicast graphs. *Advances in Cryptology–ASIACRYPT 2011* pp 538-555
[4] Sci & Educ 2014 Exploring Secondary Students' Epistemological Features Depending on the Evaluation Levels of the Group Model on Blood Circulation, Springer **14** 23 1075–1099
[5] Vegni A M, Campolo C and Molinaro A 2013 *Little TDC Modeling of Intermittent Connectivity in Opportunistic Networks: The Case of Vehicular Ad hoc Networks* (New York: Springer)
[6] Baruah A K 2014 *International J. Com. App.* **86** 11
[7] Aldous J M and Wilson R J 2000 *Graphs and Applications: An Introductory Approach,* (London: Springer-Verlag)
[8] Al A, Mahyus I and Rahma Z 2013 *J. Natural* **13** 19-22
[9] Reza W, Mahyus I and Rahma Z 2014 *J. Natural* **14** 30-33
[10] Hochbaum S D 1993 *The Pseudoflow Algorithm Graph Algorithms and Network Flows, Network* (New Jersey: Prentice-Hall)
[11] Bondy J A and Murty U S R 2008 *Graph Theory* (London: Springer-Verlag)

IOP Conf. Series: Journal of Physics: Conf. Series **948** (2018) 012066 doi:10.1088/1742-6596/948/1/012066

Understanding resistant effect of mosquito on fumigation strategy in dengue control program

D Aldila[1], N Situngkir[2] and K Nareswari[3]

[1]Department of Mathematics, Universitas Indonesia, Depok 16424, Indonesia

E-mail: aldiladipo@sci.ui.ac.id

Abstract. A mathematical model of dengue disease transmission will be introduced in this talk with involving fumigation intervention into mosquito population. Worsening effect of uncontrolled fumigation in the form of resistance of mosquito to fumigation chemicals will also be included into the model to capture the reality in the field. Deterministic approach in a 9 dimensional of ordinary differential equation will be used. Analytical result about the existence and local stability of the equilibrium points followed with the basic reproduction number will be discussed. Some numerical result will be performed for some scenario to give a better interpretation for the analytical results.

1. Introduction

Dengue is an infectious disease, which caused by a DEN virus. This virus has 4 different serotypes, i.e. DEN-1, DEN-2, DEN-3 and DEN-4 [5]. The virus is transmitted from mosquitos' bites, i.e. Aedesaegypti. Actually, there is another mosquito that could spread this virus, i.e. Aedesalbopictus. Unfortunately, majority reported infected case of Dengue in Indonesia is caused by Aedesaegypti bites [4].

There are three types of dengue depend on their infectious stages in human body, i.e. dengue fever, dengue hemorrhagic fever and dengue shock syndrome. The symptoms of this disease are fever, headache, vomiting, pain in the muscle, bleeding under the skin, nosebleeds, and many more [5].

Until today, there is no medicine that can cure people from dengue infection. Vaccines of dengue have been implemented in the last 10 years in many dengue potential countries like Brazil, Thailand, Indonesia, etc. [6]. The most popular policies to control dengue spread that has been implemented in many countries are fumigation and mechanical control. These both policies are to control the mosquito population. Fumigation is given to control adult mosquito population, while mechanical control (promote the clean environment) done in purpose to control mosquitoes breeding site [6].

Unfortunately, implementations of fumigation with chemical material trigger the evolution of mosquito to resist in the effect of this fumigation strategy. According to WHO, resistant define as capability of the organism to be resisting to the toxin that might kill another organism in the same species [7]. The resistant to insecticide is defined as the increase of proportion of insecticide that survives after exposed by an insecticide. This individual enhancement is primarily due to the death of insensitive individuals, providing an opportunity for resistant individuals to continue to breed and pass on resistance genes in off spring. Since first report in 1914 about the resistance to insecticide in San Jose, California [8], the number of cases of insecticide resistant mosquito resistance cases continues to increase like in Indonesia.

Many mathematical model have been introduced to understand how dengue might spread among human and mosquito population with many kind of mathematical approach, such as with deterministic approach [1], [2], [3], statistical approach [9], [10], [11], data analyzing [12], [13], [14]. In this article, mathematical model of dengue with intervention of fumigation to kill adult mosquito will be discussed. Resistant effect of the fumigation strategy in to mosquito population will be involved in to the model to capture previous introduction in the third paragraph.

The paper will be organized as follows. Section 2 introduced the mathematical model along with the assumptions that used to construct the model. Mathematical analysis about the equilibrium points and basic reproduction number will be given in the 3rd section and followed with some numerical experiment in the 4th section. Some conclusions will be given in the last section of the article.

2. The Construction of the Mathematical Model

Let human population divided in to 3-sub population, let call it as susceptible human S_h, infected human I_h and recovered human R_h. In the other hand, mosquito population divided into 6-subpopulation, i.e. normal and resistant mosquito in aquatic stage (E_m and E_r, respectively), normal and resistant susceptible adult mosquito (S_m and S_r, respectively), normal and resistant infected adult mosquito (I_m and I_r, respectively). We assume that recruitment rate of human only coming from newborn with constant rate A_h and A_v for mosquito population. A constant proportion$p$ will separate new born of the mosquito between the natural and the resistant mosquito. This is because we assume that resistant effect of mosquito can be transmitted vertically from newborn process. We also assume that a constant rate of fumigation is given to adult mosquito to control their population with effectiveness rate of the fumigation is ζ_m for natural mosquito and ζ_r for resistant mosquito. Mosquito which not death caused by the fumigation intervention will instantly transferred in to resistant mosquitoes population. Using the transmission dynamic given in figure 1, the mathematical model of dengue spread with fumigation strategy and their resistant effect is given by:

$$\begin{aligned}
\frac{dS_h}{dt} &= A_h - \frac{\beta_h S_h (I_m + I_r)}{S_h + I_h + R_h} - \mu_h S_h + \delta_h R_h \\
\frac{dI_h}{dt} &= \frac{\beta_h S_h (I_m + I_r)}{S_h + I_h + R_h} - (\gamma_h + \mu_h + \alpha_h) I_h \\
\frac{dR_h}{dt} &= \gamma_h I_h - (\mu_h + \delta_h) R_h \\
\frac{dE_m}{dt} &= -\eta_m E_m - \mu_m E_m + p A_v
\end{aligned} \qquad (1)$$

Table 1. Description of parameters for system (1).

Parameter	Description	Value	Dimension
$\boldsymbol{N_h}$	The total of human population	$N_h(t) \geq 0$	human
$\boldsymbol{A_h}$	Human newborn population	$A_h \geq 0$	$\frac{\text{human}}{\text{day}}$
$\boldsymbol{A_v}$	Mosquitoes recruitment rate	$A_v \geq 0$	$\frac{\text{egg}}{\text{day}}$
$\boldsymbol{\beta_h}$	Transmission rate of infection probability from mosquito to human per mosquito	$\beta_h(t) \in [0,1]$	$\frac{\text{human}}{\text{mosquito} \cdot \text{day}}$
$\boldsymbol{\beta_m}$	Transmission rate of infection probability from human to mosquito	$\beta_m(t) \in [0,1]$	$\frac{1}{\text{day}}$
$\boldsymbol{\mu_h}$	Humans natural mortality rate	$\mu_h \in [0,1]$	$\frac{1}{\text{day}}$

μ_m	Natural mortality rate of mosquitoes	$\mu_m \in [0,1]$	$\frac{1}{day}$
δ_h	The rate of transition of recovery human to susceptible human	$\delta_h \in [0,1]$	$\frac{1}{day}$
η_m	The rate of transition of egg to adult mosquito	$\eta_m \in [0,1]$	$\frac{1}{day}$
γ_h	Humans recovery rate	$\gamma_h \in [0,1]$	$\frac{1}{day}$
p	proportion of non-resistant newborn	$\in [0,1]$	–
u	Rate of fumigation	$\in [0,1]$	$\frac{1}{day}$
α_h	Dengue induced mortality in human	$\in [0,1]$	$\frac{1}{day}$
ζ_m	The percentage of normal mosquito mortality caused by the fumigation	$\in [0,1]$	–
ζ_r	The percentage of resist mosquito mortality caused by the fumigation	$\in [0,1]$	–

$$\frac{dS_m}{dt} = \eta_m E_m - \frac{\beta_m S_m (I_h)}{S_h + I_h + R_h} - \mu_m S_m - \zeta_m u S_m - (1-\zeta_m) u S_m$$
$$\frac{dI_m}{dt} = \frac{\beta_m S_m (I_h)}{S_h + I_h + R_h} - \mu_m I_m - \zeta_m u I_m - (1-\zeta_m) u I_m$$
$$\frac{dE_r}{dt} = -\eta_m E_r - \mu_m E_r + (1-p) A_v$$
$$\frac{dS_r}{dt} = \eta_m E_r - \frac{\beta_m S_r I_h}{S_h + I_h + R_h} - \mu_m S_r - \zeta_r u S_r + (1-\zeta_m) u S_m$$
$$\frac{dI_r}{dt} = \frac{\beta_m S_r I_h}{S_h + I_h + R_h} - \mu_m I_r - \zeta_r u I_r + (1-\zeta_m) u I_m,$$

with initial condition of each compartment are non-negative. Please see table 1 for the interpretation of each parameter in system (1).

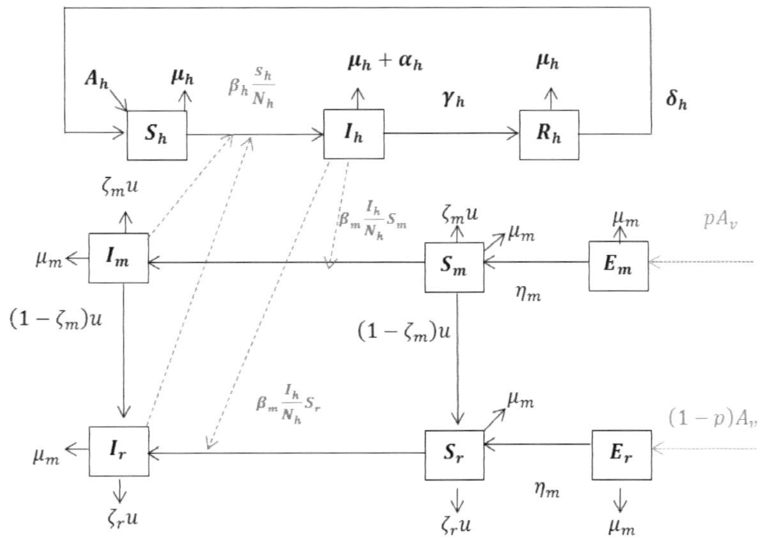

Figure 1. Transmission diagram of system (1) where red arrow present infection process, black arrow for transition process and blue arrow for recruitment rate from newborn.

In the next section, mathematical analysis for the equilibrium points and basic reproduction number will be discussed.

3. Equilibrium Points and Basic Reproduction Number

To find the equilibrium points, first we set all the right hand side of system (1) equal to 0, i.e.

$$\frac{dS_h}{dt} = 0, \frac{dI_h}{dt} = 0, \frac{dR_h}{dt} = 0, \frac{dS_m}{dt} = 0, \frac{dI_m}{dt} = 0, \frac{dS_r}{dt} = 0, \frac{dI_r}{dt} = 0, \frac{dE_m}{dt} = 0, \frac{dE_k}{dt} = 0,$$

and then solve it to each variable. There are 2 different equilibrium points here, i.e. mosquito free equilibrium point (MFE), disease free equilibrium point (DFE) and the endemic equilibrium point (EE).

The MFE is given by

$$\left(S_m = 0, I_m = 0, S_r = 0, I_r = 0, E_m = 0, E_r = 0, I_h = 0, R_h = 0, S_h = \frac{A_h}{\mu_h}\right).$$

This equilibrium represents a situation where no mosquitos exist in the population, which result in to non-existence of the dengue disease. Using the Jacobian matrix of system (1) and evaluate MFE in it, we find that this equilibrium point is always locally stable without any stability criteria.

The DFE is given by

$$\left(\begin{array}{c} E_m = \dfrac{pA_v}{\eta_m + \mu_m}, E_r = -\dfrac{(p-1)A_v}{\eta_m + \mu_m}, I_h = 0, I_m = 0, I_r = 0, R_h = 0, S_h = \dfrac{A_h}{\mu_h}, S_m = \dfrac{\eta_m p A_v}{(u+\mu_m)(\eta_m + \mu_m)}, \\ S_r = -\dfrac{\eta_m A_v (R_1 - 1)}{(u+\mu_m)(\eta_m+\mu_m)(u\zeta_r + \mu_m)} \end{array}\right),$$

where $R_1 = \frac{p(u\zeta_m + \mu_m)}{(u+\mu_m)}$. It can be seen that this equilibrium point will always have a biological meaning (positiveness of each non infected sub-population) if $R_1 < 1$. Using a Jacobian matrix, which evaluated in DFE, the local stability of this equilibrium point is investigated. The characteristic polynomial of the eigenvalues is given by

IOP Conf. Series: Journal of Physics: Conf. Series **948** (2018) 012066 doi:10.1088/1742-6596/948/1/012066

$$\frac{1}{A_h a_{10}}\left((\lambda+\mu_m+\eta_m)^2(\lambda+\mu_m+u)(\lambda+\mu_h+\delta_h)(\lambda+\mu_h)(\lambda+\mu_m+\zeta_r u)\right)(A_h a_{10}\lambda^3 + a_{10}(\mu_h+u\zeta_r+u+\gamma_h+2\mu_m+\alpha_h)\lambda^2+c_{10}\lambda+d_{10})=0,$$

where a_{10}, c_{10}, d_{10} are the function of parameters in table 1 which are not in a simple form to be written in this article. However, according to Routh-Hurwitz criteria [15], the DFE will locally stable if and only if

$$\frac{(\mu_h+u\zeta_r+u+\gamma_h+2\mu_m+\alpha_h)}{A_h}>0, \frac{c_{10}}{A_h a_{10}}>0 \text{ and } \frac{d_{10}}{A_h a_{10}}>0.$$

The endemic equilibrium point (EE) is given by

$$(S_h^*, I_h^*, R_h^*, E_m^*, S_m^*, I_m^*, E_r^*, S_r^*, I_r^*),$$

where

$$S_h^* = -\frac{a_1 I_h - b_1}{c_1},$$
$$R_h^* = \frac{I_h\gamma_h}{\mu_h+\delta_h},$$
$$E_m^* = \frac{pA_v}{\eta_m+\mu_m},$$
$$S_m^* = -\frac{(\mu_m+u)}{(a_3 I_h^2+b_3 I_h)}(a_2 I_h^3+b_2 I_h^2+c_2 I_h+d_2),$$
$$I_m^* = -\frac{1}{(a_5 I_h+b_5)}(a_4 I_h^2+b_4 I_h+c_4),$$
$$E_r^* = \frac{(1-p)A_v}{\eta_m+\mu_m},$$
$$S_r^* = \frac{-1}{(a_7 I_h^2+b_3 I_h)}(a_6 I_h^3+b_6 I_h^2+c_6 I_h+d_6),$$

with $a_1, a_2, a_3, a_4, a_5, a_6, a_7, b_1, b_2, b_3, b_4, b_5, b_6, c_2, c_4, c_6, d_2, d_6$ are a function of parameters in table (1) and I_r, which are not in simple form to be written in this article. It can be seen that all sub population in EE is given as a function of I_h and I_r.

3.1. Basic Reproduction Number

Basic reproduction number ($\mathfrak{R}_0$) is defined as an expected number of secondary cases from one primary infection in a virgin population during one infection period [16]. There is several ways that can be used to find the Basic reproduction number, such as with next-generation matrix approach [17], graph theory approach [16], and many more. In this article, next-generation matrix approach will be used to find the basic reproduction number of system (1), which give us

$$\mathfrak{R}_0 = \frac{\beta_m A_v \eta_m \mu_h \beta_h (pu^2\zeta_m\zeta_r - pu^2\zeta_r^2 + pu^2\zeta_m - pu^2\zeta_r + 2pu\zeta_m\mu_m - 2pu\zeta_r\mu_m - u^2 - 2u\mu_m - \mu_m^2)}{A_h(\gamma_h+\mu_h+\alpha_h)(\eta_m+\mu_m)(u+\mu_m)^2(u\zeta_r+\mu_m)^2}.$$

We find that the endemic equilibrium point will locally stable if $\mathfrak{R}_0 > 1$ and unstable otherwise. Therefore, controlling the value of $\mathfrak{R}_0$ can be use as the scientific reason about strategy to controlling the spread of dengue. In figure 2, sensitivity analysis of $\mathfrak{R}_0$ respect to proportion of non-resistant new born (p) and fumigation rate (u). It can be seen that enlarging fumigation intervention will reduce basic reproduction number significantly. In the other hand, larger value of proportion of non-resistant new born (p) will reduce basic reproduction number.

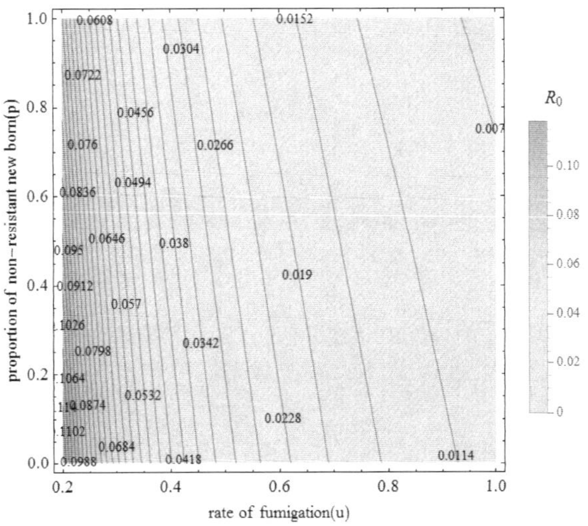

Figure 2. Sensitivity analysis of basic reproduction number respect to rate of fumigation and proportion of non-resistant new born.

In the next section, some numerical simulation will be given to give an interpretation of the analytical results.

4. Numerical Experiments

To perform all simulations in this section, the value of all parameters are $N_h = 10000$; $A_h = \frac{10000}{65\times365}$; $A_v = \frac{1000}{30}$; $\beta_h = 0.65$; $\beta_m = 0.675$; $\mu_h = \frac{1}{65\times365}$; $\mu_m = 0.09$; $\delta_h = \frac{1}{60}$; $\eta_m = 0.15$; $\gamma_h = 0.143$; $p = 0.5$; $u = 0.8$; $\alpha_h = 3.5\times10^{-3}$; $\zeta_m = 0.95$; $\zeta_r = 0.75$, which gave us the magnitude of basic reproduction number is 0.01101066873.

The initial conditions is $S_h = 10000$;$R_h = 40$;$I_h = 0$;$S_m = 30$;$I_m = 80$;$S_r = 180$;$I_r = 300$;$E_r = 50$; $E_m = 60$. The first simulation is given to see how rate of fumigation will affect the reduce of infected human and mosquito as shown in figure 3 where x-axis is time for day while y-axis is for number of human/ mosquitoes. It can be seen that intervention of fumigation could suppress the number of infected human and mosquito in all of the time of simulation. The number of infected people is only slightly different in figure 3(c) since the value of basic reproduction number is less than 1.

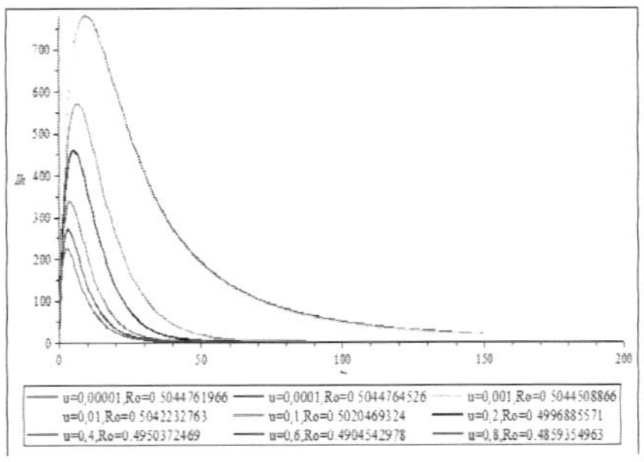

(a) Trajectory of Infected human

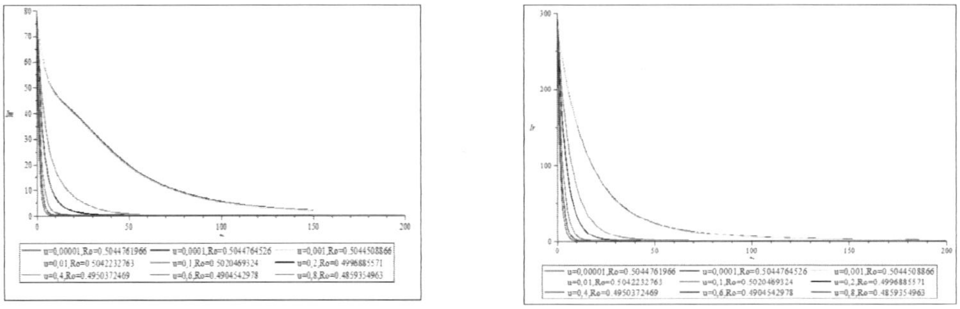

(b) Trajectory of infected natural mosquito (c) Trajectory of infected resistant mosquito

Figure 3. Trajectories of infected human and mosquito for $t \in [0,200]$.

The next simulation is given in figure 4 and 5 which is performed to show how the intervention of fumigation which implemented not in whole of time simulation, but implemented as an impulse function which is given in two different scenario of periodic time. First scenario is when intervention of fumigation is implemented 10 times with rate of u=1 for each day of intervention. In the other hand, second scenario is implemented 5 times with rate of u=2 for each day. Choosing the 1st scenario implemented 10 times while 2nd scenario implemented 5 times, the total rate of fumigation for both scenario are the same, i.e. $\mathbf{u} = \mathbf{10}$.

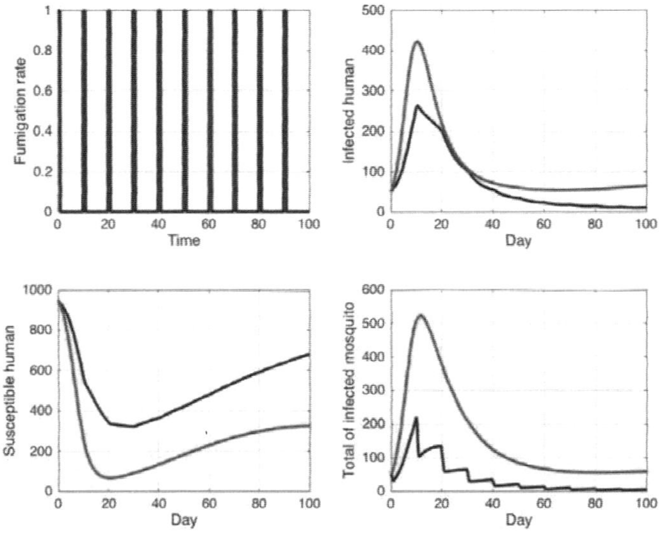

Figure 4. Trajectories of control for 1st scenario (top-left), infected human (top-right), susceptible human (bottom-left) and the total of infected mosquito (bottom-right).

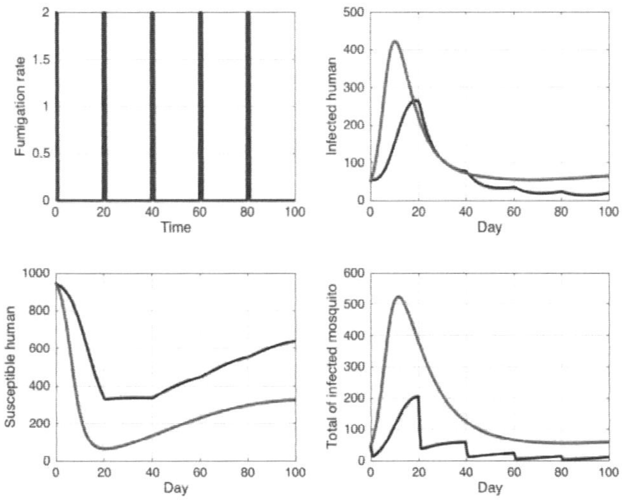

Figure 5. Trajectories of control for 2nd scenario (top-left), infected human (top-right), susceptible human (bottom-left) and the total of infected mosquito (bottom-right).

It can be seen from figure 4 and 5 that both scenario of intervention succeed to reduce number of infected human and mosquito and increase number of susceptible human. However, the successful rate of infection reduction for both scenarios is slightly different as shown in table 2. From this results, we can see that more often the fumigation intervention implemented will give a better result to reduce number of infection in human and mosquito.

Table 2. Numerical result for figure 4 and 5.

Scenario	Number of intervention	Fumigation rate (u)	Number of infected human reduced	Total number of infected mosquito reduced
1st	10 times	1	0.3951	0.7587
2nd	5 times	2	0.3765	0.7476

5. Conclusions

A mathematical model of dengue spread with intervention of fumigation has been intervention in this article. Resistant effect of fumigation is given in to the model to see how the effect of this resistant in mosquito population affects the success of dengue control program. From analytic and numerical result, it can be seen that larger effect of mosquito resistant will make the dengue control program become more difficult to be success. Also we find that more often the intervention is given (periodic-daily intervention) will give a better result in controlling dengue spread. For further research, implementation of fumigation could be reconstruct as an optimal control problem to see how the fumigation change respect to time such that the cost of intervention could be minimized rather than choosing a constant intervention.

Acknowledgments

The authors thank to all reviewers for their constructive comments and suggestion. This research is funded by Universitas Indonesia with PITTA research grand 2017.

References

[1] Aldila D, Gotz T and Soewono E 2012 *Mathematical Biosciences* **242** 9-16
[2] Aldila D, Nuraini N and Soewono E 2015 *AIP Conf., Proc.*, p 1677

[3] Putri Y E, Rozi S, Tasman H and Aldila D 2017 *AIP Conf. Proc.* p 1825
[4] Chandra A 2010 Demam Berdarah Dengue Epidemiologi, Patogenesis, dan Faktor Risiko Penularan, *Aspirator* **2** 110-119
[5] WHO 2016 Dengue and severe dengue Fact sheet No 117 Geneva World Health Organization Available at http://www.who.int/mediacentre/factsheets/fs117/en/
[6] WHO 2016 Background Paper on Dengue Vaccines World Health Organization. Available at http://www.who.int/immunization/sage/meetings/2016/april/1_Background_Paper_Dengue_Vaccines_2016_03_17pdf
[7] WHO 1957 Expert Committee on Insecticides. Seventh Report. World Health Organization Technical Report Series. Available at http://apps.who.int/iris/bitstream//10665/40380/1/WHO_TRS_125.pdf.
[8] Labbé P, Alout H, Djogbenou L, Pasteur N and Weill M 2017 *Genetics and evolution of infectious disease,* 2nd edition (Amsterdam: Elsevier)
[9] Nishiura H 2006 Mathematical and Statistical Analyses of the Spread of Dengue *Dengue Bulletin* **30**
[10] Martinez-Bello D A, Lopez-Quilez A and Torres-Prieto A 2017 Bayesian dynamic modeling of time sries of dengue disease case counts *PLoS Neglected Tropical Diseases* **11**
[11] Earnest A, Tan S B, Wilder-Smith A and Machin D 2012 Comparing Statistical Models to Predict Dengue *Computational and Mathematical Methods in Medicine*
[12] Wu Y, Lee G, Fu X and Hung T 2008 *Proc. Of The World Cong. Eng.*
[13] Kumar A, Devi C, Kartrick P, Ganesan M and Madhan A S 2017 *SSRG Inter. J. Com. Scie. Eng.*
[14] Subitha N and Padmapriya A 2013 *Inter. J. Com. Trends and Tech.* **4** 2646-2651
[15] Edelstein-Keshet L 2005 *Mathematical Models in Biology* (New York: Random House)
[16] Diekmann O, Heesterbeek J A P and Roberts M G 2010 *J. R. Soc. Interface* **7** 873-885
[17] Diekmann O and Heesterbeek J A P 2000 *Mathematical epidemiology of infectious diseases: model building, analysis and interpretation* (Chichester: Wiley)

IOP Conf. Series: Journal of Physics: Conf. Series **948** (2018) 012067 doi:10.1088/1742-6596/948/1/012067

Advection-diffusion model for the simulation of air pollution distribution from a point source emission

S Ulfah[1], **S A Awalludin**[1] **and Wahidin**[1]

[1]Mathematics Education, Universitas Muhammadiyah Prof. DR. HAMKA, Jakarta Indonesia

Email: syafika.ulfah@uhamka.ac.id

Abstract. Advection-diffusion model is one of the mathematical models, which can be used to understand the distribution of air pollutant in the atmosphere. It uses the 2D advection-diffusion model with time-dependent to simulate air pollution distribution in order to find out whether the pollutants are more concentrated at ground level or near the source of emission under particular atmospheric conditions such as stable, unstable, and neutral conditions. Wind profile, eddy diffusivity, and temperature are considered in the model as parameters. The model is solved by using explicit finite difference method, which is then visualized by a computer program developed using Lazarus programming software. The results show that the atmospheric conditions alone influencing the level of concentration of pollutants is not conclusive as the parameters in the model have their own effect on each atmospheric condition.

1. Introduction

Air quality recently has been a global problem, because the air has polluted by human activities and natural processes. [1] reported that in 2012, air pollution exposure was responsible for 7 million premature deaths and this occurs annually. Pollution is able to affect an area far from the source of emission that is caused by a bulk motion of wind. The focus of this article is a source of air pollution from a point source emission such as a chimney. The distribution of air pollution can be studied through mathematical simulation which is meteorological parameters can be observed as they have an effect on the distribution of air pollution. [2] stated that simulation of air pollution provides information about the spread of pollutants in an area, the scale, and level of pollution and estimation.

Modelling and simulation of air pollution distribution are to observe the effects meteorological factors such as the wind, temperature, humidity, pressure, etc. have on the dispersal of pollutants. Reference [3] did a study on air pollution dispersion using a steady state two-dimensional mathematical model. They considered mesoscale wind, which is generated by urban heat island. Heat from the Earth's surface results in the vertical movement and it mixes with the mean wind causing turbulence. Eddy diffusivity plays an important role in turbulence. The strength of turbulence determines the atmospheric stability, whether it is stable, neutral or unstable. The higher the turbulence; the more the atmosphere becomes unstable.

2. Model Development

The advection diffusion model is used to simulate air pollution distribution, which describes the physical processes involved in the dispersal of pollutants. Reference [4] has shown that the dispersion of pollutant can be expressed by the following equation with time dependent:

$$\frac{\partial C}{\partial t}+U\frac{\partial C}{\partial x}+V\frac{\partial C}{\partial y}+W\frac{\partial C}{\partial z}=\frac{\partial}{\partial x}\left(K_x\frac{\partial C}{\partial x}\right)+\frac{\partial}{\partial y}\left(K_y\frac{\partial C}{\partial y}\right)+\frac{\partial}{\partial z}\left(K_z\frac{\partial C}{\partial z}\right)+S \tag{1}$$

where,

$C = C(x, y, z, t)$ is the pollutants concentration at any location (x, y, z) at time t;

U, V and W are the wind profile in downwind (x), crosswind (y) and vertical (z) direction respectively;

K_x, K_y and K_z are eddy diffusivities in the downwind (x), crosswind (y) and vertical (z) direction respectively;

S is the source or sink term of the air pollutants.

In the physical processes, advection carries the pollutants in the x – direction and transports them as far as the wind blows. The mixture of the mean wind and the heat from the Earth's surface together with friction from buildings create a local wind called mesoscale wind. It is considered that the mean wind speed changes with increasing height and the mesoscale wind which is a function that depends on temperature. Diffusion takes place alongside advection and this reduces the concentration of the pollutants by turbulence. It is assumed that the mean concentration of pollutants is constant along the y – direction, thus reducing it to a two-dimensional problem. It is again considered that the wind profile and eddy diffusivity are constant during pollutants emission. Therefore, Eq. (1) becomes:

$$\frac{\partial C}{\partial t}+U\frac{\partial C}{\partial x}+W\frac{\partial C}{\partial z}=K_x\frac{\partial^2 C}{\partial x^2}+K_z\frac{\partial^2 C}{\partial z^2}+S \tag{2}$$

Taking into account the removal mechanism such as precipitation, Eq. (2) can be written as

$$\frac{\partial C}{\partial t}+U\frac{\partial C}{\partial x}+W\frac{\partial C}{\partial z}=K_x\frac{\partial^2 C}{\partial x^2}+K_z\frac{\partial^2 C}{\partial z^2}-\lambda C \tag{3}$$

where λ is the removal mechanism rate.

Temperature would be considered as one of the parameters that influences the distribution of pollutants as it affects turbulence. Therefore, the change in temperature T with time would be as follows:

$$\frac{\partial T}{\partial t}=Q_x\frac{\partial^2 T}{\partial x^2}+Q_z\frac{\partial^2 T}{\partial z^2} \tag{4}$$

where Q_x, Q_z are the heat diffusion rate in the x and z directions respectively.

3. Method of Solution

3.1 The Domain

The objective is to analyze the pollutants concentration in a given two-dimensional domain (x, z) at time t by a developing a program using Lazarus software. Eq. (3) and Eq. (4) would be solved numerically by explicit finite difference scheme to obtain the solution. To apply the finite difference scheme, the domain (x, z) is subdivided into a set of similar rectangles of sides Δx and Δz, by equally spaced grid lines, parallel to the z – axis, determined by $x = i\Delta x, i = 0,1,2,..., M$ and parallel to x – axis, determined by $z = j\Delta z, j = 0,1,2,..., N$. Hence, the domain of interest belongs $M \times N$ grids and the intersection of grid lines (grid points) determine the concentration of pollutants C and also the temperature T. The entire grid would be computed with time difference Δt. The concentration C at a grid point (i, j) with time n is denoted by $C^n_{i,j}$ and the temperature T at a grid point (i, j) with time n is denoted by $T^n_{i,j}$.

3.2 Initial and Boundary Condition
Before the system is run, the initial value is needed to set the system up. At the initial time $t = 0$, temperature T_0 is set in the domain. The boundary for the domain is taken and it is assumed that at the beginning when $t = 0$ there is pollution emission C_0 at some point in the domain. Thus, $C(x, z, 0) = C_0$ at $t = 0$ for $0 < x < M$ and $0 < z < N$ (M – horizontal and N – vertical grid points) and where C_0 is the pollutants concentration at position (x, z) at $t = 0$. As a boundary value problem, the boundary condition is needed to control the system, with $C = C_0$ where the value of the solution at the boundary is known.

3.3 Meteorological Parameters
Meteorological parameters have an influence on air pollutants distribution. The wind can transport the pollutants far away from the source of emission and the diffusivity determines the rate at which pollutants diffuse. These parameters are dependent on the intensity of turbulence, which is influenced by atmospheric stability. In reality, to solve the Eq. (3) the variable wind velocity and the eddy diffusivity are considered to be functions of vertical distance, as suggested by [5]:

$$u = Ur\left(\frac{z}{z_r}\right)^{\alpha}$$

$$K_z = K_r\left(\frac{z}{z_r}\right)^{\beta}$$

where Ur and K_r are the measured wind speed (m/s) and vertical diffusivity at a reference height $z_r\,(m)$, α and β are the constants depending on the atmospheric stability and surface roughness.

The mathematical forms of mesoscale wind in the horizontal and vertical directions as suggested by [6] are:

$$u_e = -ax\left(\frac{z}{z_r}\right)^{\alpha}$$

$$W_e = \frac{az}{\alpha+1}\left(\frac{z}{z_r}\right)^{\alpha}$$

where a is a proportionality constant (s^{-1}).

In the realistic form, temperature affects turbulence, which then affects the mesoscale wind. Therefore, the mathematical form of the mesoscale wind in the vertical direction becomes

$$W = \frac{az}{\alpha+1}\left(\frac{z}{z_r}\right)^{\alpha} + b\left(\frac{T - \min T}{\max T - \min T}\right)$$

where b is the air flow constant which is caused by temperature.

3.4 Numerical Solution
The solution of Eq. (3) and Eq. (4) would be found numerically by explicit finite difference scheme. The explicit finite difference scheme implies that:

$$\frac{\partial C}{\partial t} = \frac{C_{i,j}^{n+1} - C_{i,j}^{n}}{\Delta t} \tag{5}$$

The central difference scheme is used for the advection and diffusion terms; the first derivative is as follows:

$$\frac{\partial C}{\partial x}=\frac{C_{i+1,j}^{n}-C_{i-1,j}^{n}}{2\Delta x} \tag{6}$$

And the second derivative would be

$$\frac{\partial^2 C}{\partial x^2}=\frac{C_{i+1,j}^{n}-2C_{i,j}^{n}+C_{i-1,j}^{n}}{\Delta x^2} \tag{7}$$

The approximations (5), (6) and (7) are applied to Eq. (3) and (4). Substituting the approximations into Eq. (3) results in the equation below,

$$\frac{C_{i,j}^{n+1}-C_{i,j}^{n}}{\Delta t}+U\frac{C_{i+1,j}^{n}-C_{i-1,j}^{n}}{2\Delta x}+W\frac{C_{i,j+1}^{n}-C_{i,j-1}^{n}}{2\Delta z}=K_x\frac{C_{i+1,j}^{n}-2C_{i,j}^{n}+C_{i-1,j}^{n}}{\Delta x^2}+K_z\frac{C_{i,j+1}^{n}-2C_{i,j}^{n}+C_{i,j-1}^{n}}{\Delta z^2}-\lambda\left(C_{i,j}^{n+1}\right) \tag{8}$$

Rearranging Eq. (8) in the finite difference form, $C_{i,j}^{n+1}$ is obtained as:

$$C_{i,j}^{n+1}=A_1C_{i+1,j}^{n}+A_2C_{i-1,j}^{n}+A_3C_{i,j+1}^{n}+A_4C_{i,j-1}^{n}+A_5C_{i,j}^{n}, \tag{9}$$

for $i=1,2,3,\ldots,M$ and $j=1,2,3,\ldots,N$, where,

$$A_1=-\frac{U\Delta t}{2\Delta x}+\frac{K_x\Delta t}{\Delta x^2},$$

$$A_2=\frac{U\Delta t}{2\Delta x}+\frac{K_x\Delta t}{\Delta x^2},$$

$$A_3=-\frac{W\Delta t}{2\Delta z}+\frac{K_z\Delta t}{\Delta z^2},$$

$$A_4=\frac{W\Delta t}{2\Delta z}+\frac{K_z\Delta t}{\Delta z^2},$$

$$A_5=-2\frac{K_x\Delta t}{\Delta x^2}-2\frac{K_z\Delta t}{\Delta z^2}-\lambda\Delta t+1.$$

The explicit scheme will converge and be stable when $0\le\frac{\Delta t}{\Delta x^2}\le\frac{1}{4}$.

The second derivative is applied to the spatial dimensions and then approximations (5) and (7) would be substituted into Eq. (4), it would be:

$$\frac{T_{i,j}^{n}-T_{i,j}^{n-1}}{\Delta t}=Q_x\frac{T_{i+1,j}^{n-1}-2T_{i,j}^{n-1}+T_{i-1,j}^{n-1}}{\Delta x^2}+Q_z\frac{T_{i,j+1}^{n-1}-2T_{i,j}^{n-1}+T_{i,j-1}^{n-1}}{\Delta z^2} \tag{10}$$

Rearranging Eq. (10) gives:

$$T_{i,j}^{n}=\frac{Q_x\Delta t}{\Delta x^2}T_{i+1,j}^{n-1}+\frac{Q_x\Delta t}{\Delta x^2}T_{i-1,j}^{n-1}+\frac{Q_z\Delta t}{\Delta z^2}T_{i,j+1}^{n-1}+\frac{Q_z\Delta t}{\Delta z^2}T_{i,j-1}^{n-1}+\left(-2\frac{Q_x\Delta t}{\Delta x^2}-2\frac{Q_z\Delta t}{\Delta z^2}+1\right)T_{i,j}^{n-1} \tag{11}$$

Eq. (9) and (10) can be computed for each $i=1,2,3,\ldots,M$ and $j=1,2,3,\ldots,N$ with $n=1,2,3,\ldots$, which is true for interior grid points. At the boundary grid points, the boundary conditions are needed for discretization. Therefore, initial concentration value $C_{i,j}^{n}$ is an input data from user, for $i=\left(\frac{M}{2}-1\right),\ldots,\left(\frac{M}{2}+1\right)$ and $j=\left(\frac{N}{2}-1\right),\ldots,\left(\frac{N}{2}+1\right)$ whereby M and N must be even number.

Boundary conditions for the concentration of pollutants are as follows:

$C_{i,0}^{n}=C_{i,N}^{n}=0$, for $i=0,1,2,\ldots,M$ and $j=0,1,2,\ldots,N$,

$C_{0,j}^{n}=C_{M,j}^{n}=0$, for $i=0,1,2,\ldots,M$ and $j=0,1,2,\ldots,N$.

The following is the initial conditions of the temperature:

1) Lower temperature (LowT): $T^n_{i,j}$ are given by input data from user for $i = 0,1,2,...,M$ and $j = 0,1,2,...,N$,
2) Upper temperature (UpperT): $T^n_{i,j}$ are given by input data from user for $i = 0,1,2,...,M$ and $j = N-2, N-1, N$,
3) Temperature between ground level and atmosphere

$$\Delta Temperature = \frac{LowT - UpperT}{N-7},$$

$$T^n_{i,j} = (LowT - \Delta Temperature) * (j-3), \text{for } i = 0,1,2,...,M \text{ and } j = 4,...,N-3.$$

While for the boundary conditions are:

1) $T^n_{i,j} =$ LowT, for $i = 0,1,2,...,M$ and $j = 0,1,2,3$,
2) $T^n_{i,j} =$ UpperT, for $i = 0,1,2,...,M$ and $j = N-2, N-1, N$,
3) $T^n_{i,j} = (LowT - \Delta Temperature) * (j-3)$, for $i = 0$ and $j = 4,...,N-3$,
4) $T^n_{i,j} = (LowT - \Delta Temperature) * (j-3)$, for $i = M$ and $j = 4,...,N-3$.

As stated earlier, the wind component and eddy diffusivity are functions of vertical distance. Therefore, the function of the large-scale wind becomes $U = u + u_e$, hence

$$U[i,j] = Ur\left(\frac{j}{z_r}\right)^{\alpha} + \left(q(i)\left(\frac{j}{z_r}\right)^{\alpha}\right) \tag{12}$$

where q is proportionality constant and the function for the mesoscale wind, which is dependent on the change of temperature, becomes

$$w[i,j] = \frac{a(j)}{\alpha+1}\left(\frac{j}{z_r}\right)^{\alpha} + b\left(\frac{T^n_{i,j} - \min T}{\max T - \min T}\right) \tag{13}$$

For eddy diffusivity,

$$K_z[j] = K_r\left(\frac{j}{z_r}\right)^{\beta} \tag{14}$$

$$K_x[j] = K_r\left(\frac{i}{z_r}\right)^{\beta} \tag{15}$$

All these parameters are counted from $i = 1,2,3,...,M$ and $j = 1,2,3,...,N$, which is the same in the downwind (x) direction with j becoming i.

4. Result and Discussion

This study is to analyze the distribution of air pollution distribution under the influence of large-scale wind, mesoscale wind, and eddy diffusivity. Wind profile was defined in Eq. (12) as large-scale wind and Eq. (13) as mesoscale wind. While eddy diffusivity in vertical and horizontal direction were defined in Eq. (14) and Eq. (15) respectively. There are some unknown parameters, which are $Ur, a, K_r, z_r, \alpha, \beta$ required for the computation of air pollution concentration. The range of values for some of the parameters are as follows: $Ur = -50$ to 50 m/s, $K_r = 0$ to 5 m/s^2, $a = -1$ to 1 s^{-1}, $b = -1$ to 1 s^{-1} $z_r = 10$ m. The values of other unknown parameters α and β are taken according to the stability conditions of the atmosphere and surface roughness as described in some previous studies which is shown in table 1 [7], with these values of α (table 1) corresponding to different stability classes, the value of β can be obtained as $\beta = 1 - \alpha$, based on Schmidt's conjugate power law [8].

Table 1. Variations of the values of α in different atmospheric condition [7].

	Stability	α
Case I	Unstable	0.17
Case II	Neutral	0.27
Case III	Stable	0.61

It is assumed that pollutants are emitted at a constant rate in a uniformly distributed domain. The area extends up to 500 m downwind and 500 m in the vertical direction. The initial concentration and temperature are taken to be $C_0 = 1$ and $T_0 = 27^o C$ and it is assumed that the range of heat diffusion $Q_x = Q_z$ from 0 to 5 m/s^2. The removal of pollutants is assumed to be taking place by either dry deposition or wet deposition processes. The value for the removal parameter $\lambda = 10^{-6}$ s^{-1} is assumed to be constant in all atmospheric conditions.

A numerical method based on explicit scheme requires solutions to be within their range of validity. This analysis is done for $M = 100$ and $N = 100$ independent grids.

To be able to gain insight clearly on air pollution distribution under the effect of large-scale wind, mesoscale wind, and eddy diffusivity for unstable atmospheric condition, the computed concentration values have been analyzed and displayed graphically in figures 2, 3, and 4.

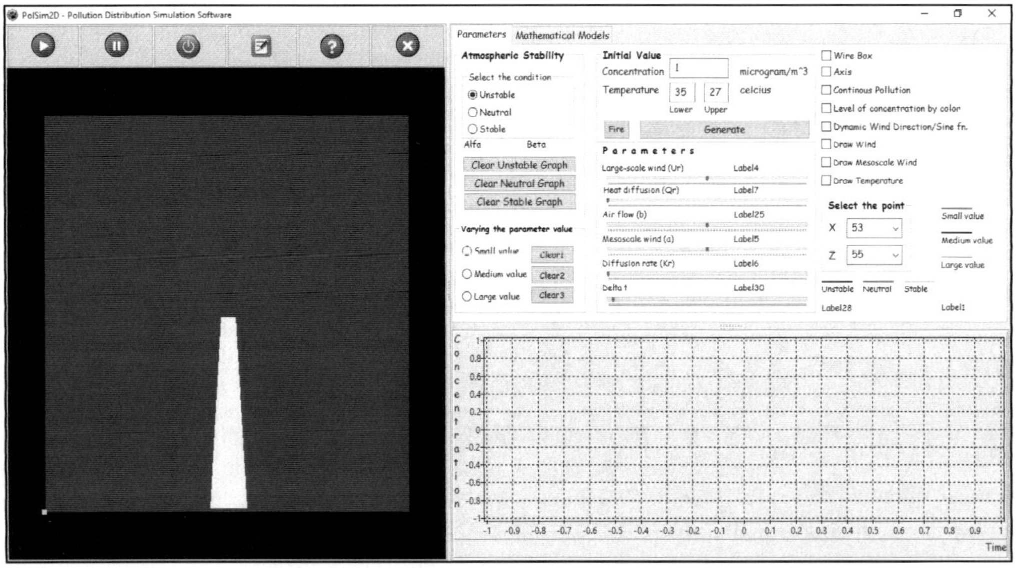

Figure 1. Air pollution simulation interface.

Figure 1. shows the user interface for the program developed using Lazarus programming software. The various buttons are described to show how the program works and the simulation carried out, are done by varying the atmospheric conditions and the parameter values. The various buttons and check boxes are the part where all the parameters are adjusted in order to have a clear understanding of how the pollutants distribute. Arbitrary points are chosen from the plane on the right by changing the X and Z label. When the program is initialized and run, pollutants are emitted from the stack and as the model parameters are adjusted, they then distributed in the plane and the corresponding graph is plotted which shows the decrease or increase in the level of concentration of pollutants by varying any of the parameters in the model.

IOP Conf. Series: Journal of Physics: Conf. Series **948** (2018) 012067 doi:10.1088/1742-6596/948/1/012067

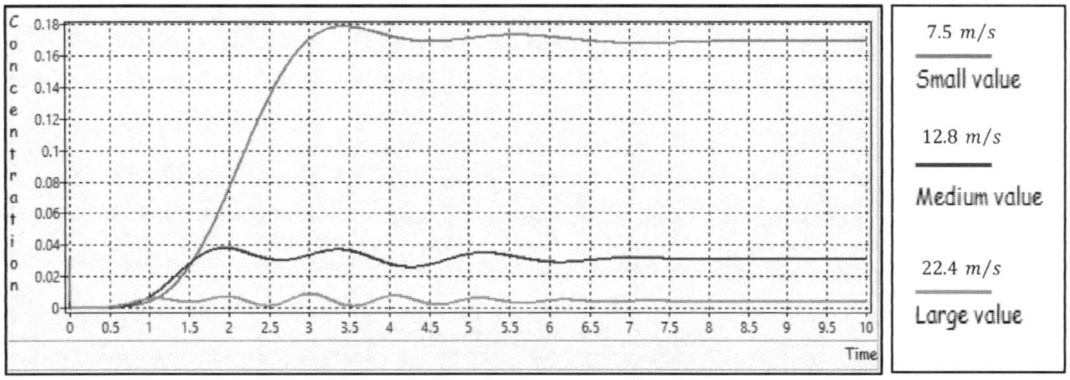

Figure 2. Varying Large-scale wind.

The model parameters are varied and used for the unstable atmospheric condition. This is done to see the effect they have on the distribution of pollutants in the atmosphere. In the simulations, one particular parameter is varied whiles the others are kept constant. The simulations carried out in Figure 1 start with a large-scale wind of $7.5\ m/s$ and this value is varied to $12.8\ m/s$ and $22.4\ m/s$ whiles the other parameters are kept constant. Also the level of constancy in the concentration of pollutants at 8, 7.5 and 6.5 unit of time for the small value, medium value and large value respectively, the concentration of pollutants maintains some level of constancy. This is due to the fact that, the emission is continuous and at the point the change in concentration of the pollutants and the atmosphere is the same so there is not any significance change. Also, the pollutants undergo an advection and diffusion process and therefore there would always be a drop or increase in the concentration levels in the presence of the model parameters. Table 2 shows the model parameters and their values.

Table 2. Varying large-scale wind.

Large-scale wind ($\boldsymbol{U_r}$)	7.5	12.8	22.4
Heat Diffusion ($\boldsymbol{Q_r}$)		0.2	
Air Flow ($\boldsymbol{b}$)		0.16	
Mesoscale Wind ($\boldsymbol{a}$)		0.08	
Diffusion Rate ($\boldsymbol{K_r}$)		1.5	
Delta t		0.00159	

From the figures, it can be seen that as the large-scale wind increases the level of concentration decreases. This is because the bulk transfer of pollutants is as a result of the intensity of the wind in a particular area. As the speed of wind increases, pollutants are carried farther away from the source of emission.

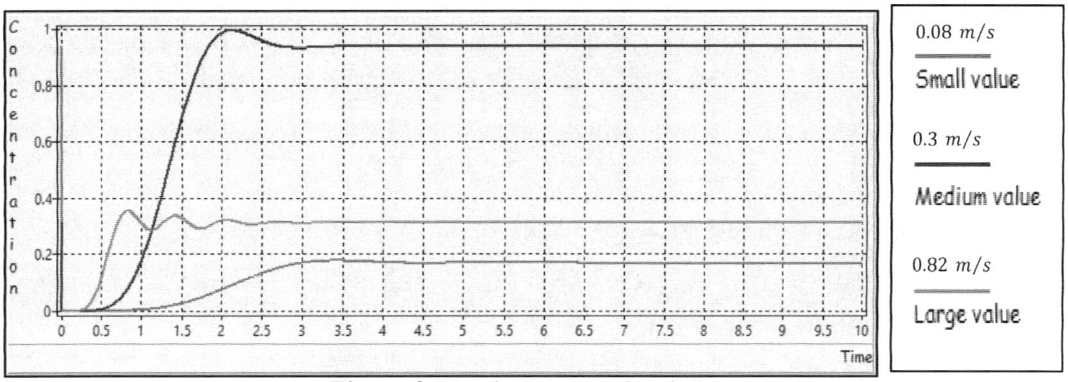

Figure 3. Varying mesoscale wind.

As studied by [3], the mesoscale wind prevents the dispersal of pollutants which results in the increase of pollution concentration in the atmosphere. From Figure 3, the level of concentration of pollutants increase as the mesoscale wind is increased from $0.08\ m/s$ to $0.3\ m/s$ but decreased when it was increased to $0.82\ m/s$. This is because of the position of the arbitrary point that was chosen. The concentration of the pollutants was far from the arbitrary point and hence a decrease in the concentration level. The other model parameters are kept constant as the value of the mesoscale wind is varied. The level of constancy in the concentration of pollutants at 5, 3.5 and 3 for the small value, medium value and large value respectively is as a result of the same reason given under Figure 2. Table 3 shows the values used in each of the simulations carried out.

Table 3. Varying mesoscale wind.

Large-scale wind ($\boldsymbol{U_r}$)		7.5	
Heat Diffusion ($\boldsymbol{Q_r}$)		0.2	
Air Flow ($\boldsymbol{b}$)		0.16	
Mesoscale Wind ($\boldsymbol{a}$)	0.08	0.3	0.82
Diffusion Rate ($\boldsymbol{K_r}$)		1.5	
Delta t		0.00159	

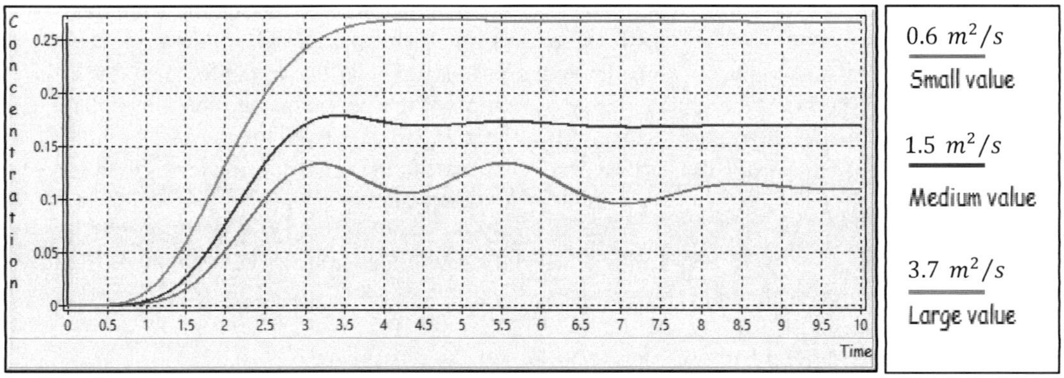

Figure 4. varying diffusion rate.

Diffusion is the movement of molecules from a higher concentration to a lower concentration. In this study, the higher concentration point is the source of emission of pollutants and the lower concentration is the arbitrary point that was chosen. As the diffusivity is increased from $0.6\ m^2/s$ to $1.5\ m^2/s$ and the finally to $3.7\ m^2/s$, the level of concentration of pollutants increase as shown in Figure 35. In this case, the level of constancy in the concentration of pollutants at 9.5, 6.5 and 4.5 unit of time for the small value, medium value and large value respectively can be explained in the same way as that of Figure 2. This due to the fact that the pollutants are diffusing at a faster rate to the arbitrary point which tend to increase the concentration at that point. Table 4 shows the model values that were used to carry out the simulations.

Table 4. Varying diffusivity.

Large-scale wind ($\boldsymbol{U_r}$)		7.5	
Heat Diffusion ($\boldsymbol{Q_r}$)		0.2	
Air Flow ($\boldsymbol{b}$)		0.16	
Mesoscale Wind ($\boldsymbol{a}$)		0.08	
Diffusion Rate ($\boldsymbol{K_r}$)	0.6	1.5	3.7
Delta t		0.00159	

5. Conclusion

Air pollution is one of the major atmospheric problems the world faces nowadays. Mathematical models are used to estimate the impact of the concentration of air pollution, which have on the living organisms as well the entire ecosystem. By using the mathematical model, it becomes easier to learn the basic concept of the physical phenomena that take place during the distribution of pollutants as they undergo advection and diffusion processes under the influence of meteorological parameters. In this study, these parameters represent large-scale wind, mesoscale wind and eddy diffusivity. From the results, various simulations were carried out in order to access how the distribution of air pollution is affected when any of the meteorological parameters are varied.

Acknowledgments

I gratefully acknowledge the assistance and help I received from my supervisor Dr.SompornChuai-Aree and as well as my colleague Collins Bekoe.

References

[1] WHO 2017 *7 Million Premature Deaths Annually Linked to Air Pollution.* http://www.who.int/mediacentre/news/releases/2014/air-pollution/en/ 11 August 2017 (20:10)

[2] Arystanbekova N K 2004 Application of Gaussian plume models for air pollution simulation at instantaneous emissions *Mathematics and Computers in Simulation* **67** pp. 451-8

[3] Agarwal M and Tandon A 2009 Modeling of the urban heat island in the form of mesoscale wind and of its effect on air pollution dispersal *Applied Mathematical Modeling* **34** pp. 2520-30

[4] Lakshminarayanachari K, Sudheer P K L, Siddalinga P M and Pandurangappa C 2013 Advection – diffusion numerical model of air pollutants emitted from an urban area source with removal mechanisms by considering point source on the boundary *International J. App. or Innov. in Eng. and Management* **2** pp 251-268

[5] Lin J S and Hildemann L M 1996 Analytical solution of the atmospheric diffusion equation with multiple sources and height dependent wind speed and eddy diffusivities *Atmos. Environ* **30** pp. 239-254

[6] Dilley J F and Yen K T 1971 Effect of a mesoscale type wind on the pollutant distribution from a line source*Atmos. Environ* **6** pp. 843-851

[7] Erwin J S 1979 Technical note: A theoretic variation of the wind profile power law exponent as a function of surface roughness and stability *Atmos. Environ* **13** pp.191-4

[8] Sharan M and Kumar P 2009 An analytical model for crosswind integrated concentration released from a continuous source in a finite atmospheric boundary layer *Atmos. Environ* **43** pp. 2268-77

Modeling the stock price returns volatility using GARCH(1,1) in some Indonesia stock prices

S A Awalludin[1], S Ulfah[1] and S Soro[1]

[1]Mathematics Education, Universitas Muhammadiyah Prof. DR. HAMKA, Jakarta Indonesia

E-mail: subhanajiz@uhamka.ac.id

Abstract. In the financial field, volatility is one of the key variables to make an appropriate decision. Moreover, modeling volatility is needed in derivative pricing, risk management, and portfolio management. For this reason, this study presented a widely used volatility model so-called GARCH(1,1) for estimating the volatility of daily returns of stock prices of Indonesia from July 2007 to September 2015. The returns can be obtained from stock price by differencing log of the price from one day to the next. Parameters of the model were estimated by Maximum Likelihood Estimation. After obtaining the volatility, natural cubic spline was employed to study the behaviour of the volatility over the period. The result shows that GARCH(1,1) indicate evidence of volatility clustering in the returns of some Indonesia stock prices.

1. Introduction

In the financial field, volatility is one of the key variables to make an appropriate decision. According to [6], the volatility can be defined as a degree of fluctuation in asset price which can be going up or down. In fact, the volatility has taken place in different areas in financial theory and practice, such as risk management, portfolio selection and derivative pricing [2]. In many cases, the volatility is shown by low fluctuation in some period, then following by high fluctuation and vice versa. It indicates that volatility is not constant over time. Estimating the volatility as accurate as possible is needed since return can be obtained from volatility and price can be computed based on the return. We can employ time series model to capture the volatility of returns asset.

The time series model that will be used must agree with heteroscedasticity property. Heteroscedasticity describes the volatility changes over the time horizon. One of the heteroscedasticity models is Generalized Autoregressive Conditional Heteroscedasticity (GARCH) which was proposed by Bollerslev [3]. Estimating volatility using the GARCH has been frequently studied by many researchers. Kamau *et al.*[6] used GARCH(1,1) to estimate the volatility of stock return in Kenyan stock markets. Their finding is that the returns stylized facts including volatility clustering, non-normal distribution and mean. Volatility clustering is the situation that high fluctuations in the returns of an asset are often followed by other high fluctuations, likewise low fluctuations are followed by other low fluctuations.

Their finding is similar to a study by Namugaya *et al.*[8] which showed that Uganda Securities Exchange (USE) returns have a non-normal distribution, positively skewed and stationary. In fact, those returns attributes usually appear in financial time series data. It is well known that the volatility series give important information about the data. Thus, we need to investigate the volatility behavior through the period.

In order to simplify investigation of the volatility behaviour, we need to smooth the volatility series. Numerical method can be employed to do the job. This leads to natural cubic spline function which is a widely used technique for piecewise smooth curve fitting. This function is simply piecewise cubic polynomial which can be constructed so that the connections between adjacent cubic splines are visually smooth [4]. Further, we note that volatility signal is obtained due to the natural cubic spline fitting of the volatility series.

The rest of the paper is organized as follows: (1) theory and methods; (2) results and discussion and (3) conclusions.

2. Methodology

This section describes mathematical and statistical methods which were used for analyzing of the volatility of stock returns in this study. These methods comprise of obtaining the returns from stock price data, estimating volatility series using GARCH(1,1), using Maximum Likelihood Method to Estimate Parameters of GARCH(1,1) and fitting volatility series using natural cubic spline function. The details will be explained as follows.

2.1. Obtaining Returns from Stock Price

We involve data from daily closing prices of the seven companies of Indonesia from July 2007 to September 2015. We can obtain returns series from stock prices data by differencing log of the price from one day to the next. Returns can be defined as the continuously compounded return during day t (between the end of day $t-1$ and the end of day t)[5], as:

$$R_t = \ln \frac{S_t}{S_{t-1}},$$

where S_t is the price at day t. Commonly, continuously compounded return, R_t, is called log return. After obtaining the returns, the information of return fluctuation (volatility series) over time can be estimated by fitting GARCH(1,1) to the returns.

2.2. Estimating Volatility Series Using GARCH(1,1)

Financial data contain non-constant variance over time. It is well known as heteroscedasticity. Capturing heteroscedasticity can be done by GARCH model. We involve the definition of the general process of GARCH which is GARCH (*p,q*).

Definition 1.

Let $(w_t)_{t>0}$ be a sequence of independent and identically distributed (i.i.d) random variables such that $w_t \sim N(0,1)$. The R_t is called the generalized autoregressive conditionally heteroscedasticity or GARCH (*p,q*) process [9] if

$$R_t = \sigma_t w_t, \qquad t \in \mathbb{N},$$

where σ_t is a nonnegative process such that,

$$\sigma_t^2 = \gamma V_L + \alpha_1 R_{t-1}^2 + \ldots + \alpha_q R_{t-q}^2 + \beta_1 \sigma_{t-1}^2 + \ldots + \beta_p \sigma_{t-p}^2, \qquad t \in \mathbb{N},$$

and

$$\gamma > 0,\ \ \alpha_i \geq 0\ \ i = 1,\ldots,q\ \ \beta_i \geq 0\ \ i = 1,\ldots,p,$$

where integers p and q are orders of σ_t^2 and R_t^2, respectively.In particular, GARCH(1,1) is the simplest and frequently useful model [2] which is given by

$$\sigma_t^2 = \gamma V_L + \alpha u_{t-1}^2 + \beta\sigma_{t-1}^2,$$

where γ, α and β are the weight assigned to long-run average variance rate V_L, returns squared R_{t-1}^2, and variance σ_{t-1}^2, respectively. The weights γ, α and β must sum to unity, that is

$$\gamma + \alpha + \beta = 1.$$

Now, we set $\omega = \gamma V_L$, the GARCH(1,1) model can also be written

$$\sigma_t^2 = \omega + \alpha u_{t-1}^2 + \beta\sigma_{t-1}^2, \tag{1}$$

where $\omega > 0$, $\alpha \geq 0$ and $\beta \geq 0$. In order to guarantee the variance to be positive, we set $\alpha + \beta < 1$. The formula (1) is often used for the purpose of estimating the volatility. After that, we estimate the parameters α and β by maximum likelihood method.

2.3. Using Maximum Likelihood Method to Estimate Parameters of GARCH(1,1)

The method gives values of the parameters that maximize the likelihood function of the variable of interest [5]. Now, we have the transformed returns R_t which are approximately normal with mean zero and variance σ_t^2 as required in definition 1. Initially, we determine the probability density function of R_t, $t = 1,2,3,\dots,n$. Since for each t we have

$$f\ r_t = \frac{1}{\sqrt{2\pi\sigma_t^2}} \exp\left(\frac{-r_t^2}{2\sigma_t^2}\right),$$

then the likelihood function $L\ r_t = f\ r_1,\dots,r_n$. For each t , R_t is independence so that

$$L\ r_t = \prod_{i=1}^{n} f\ r_t$$

$$= \prod_{t=1}^{n}\left[\frac{1}{\sqrt{2\pi\sigma_t^2}} \exp\left(\frac{-r_t^2}{2\sigma_t^2}\right)\right], \tag{2}$$

by monotonicity of logarithm function, maximizing likelihood function can be done by maximizing its logarithm [8]. Therefore, we now can maximize (2) by taking natural logarithm. Then we have,

$$l\ r_r = \ln L\ r_r = \ln\left(\prod_{t=1}^{n}\frac{1}{\sqrt{2\pi\sigma_t^2}}\exp\left(\frac{-r_t^2}{2\sigma_t^2}\right)\right)$$
$$= \sum_{t=1}^{n}\left(\frac{1}{2}\ -\ln\ 2\pi\ -\ln\ \sigma_t^2\ +\frac{1}{2}\left(\frac{-r_t^2}{\sigma_t^2}\right)\right).$$

Ignoring constant multiplicative factors of $l\ r_t$ gives

$$\hat{l}\ r_t\ = \sum_{t=1}^{n}\left[-\ln\ \sigma_t^2\ - \frac{r_t^2}{\sigma_t^2}\right], \tag{3}$$

where r_t and σ_t^2 are the returns and the variance at day t, respectively. The parameters that maximize $l(r_t)$, also maximize $\hat{l}(r_t)$. Furthermore, we solve formula (3) numerically by damped Newton's method. In summary, fitting the GARCH (1,1) volatility series of the stock returns. In order to simplify investigation of the volatility in many situations, we need to smooth the volatility series. The volatility series will be smoothed by the natural cubic spline.

2.4. Fitting Volatility Series Using Natural Cubic Spline Function

According to the preceding section, estimating volatility series using the GARCH(1,1), the model gives daily volatility series over the period. In order to study the behavior of volatility, we employ the natural cubic spline to fit volatility series obtaining from GARCH(1,1). It is because the natural cubic spline has such attractive properties as smoothness, continuity of the first and second derivative so that many financial institutions use the method for curve fitting [1]. Therefore, we can get the information on the rate of change and cumulative change of volatility series over the period.

Let $(t_1, y_1), (t_2, y_2), \ldots, (t_n, y_n)$ where $t_1 < t_2 < \ldots < t_n$ and $s(t)$ be a series of knot points and cubic spline function which fits consecutive knot points, respectively. We employed a natural cubic spline which easily to apply in the data. It was improved by McNeil *et al.* [7]. The cubic spline function is defined as:

$$s(t) = a + bt + \sum_{k=1}^{p} c_k (t - t_k)_+^3, \tag{4}$$

where t denotes time, $t_1 < t_2 < \ldots < t_p$ are specified knots and $(t - t_k)_+$ is $t - t_k$ for $t > t_k$ and 0 otherwise. Since the formula (4) is a linear function of the coefficients a, b and c_k, it is fitted to the data using linear regression. However, linearity in the future means that the quadratic and cubic coefficients are 0 for $t > t_p$ by setting $s''(t) = 0$. Therefore the formula (4) can also be written as

$$s(t) = a + bt + \sum_{k=1}^{p-2} c_k \left[(t - t_k)_+^3 - \left(\frac{t_p - t_k}{t_p - t_{p-1}}\right)(t - t_{p-1})_+^3 + \left(\frac{t_{p-1} - t_k}{t_p - t_{p-1}}\right)(t - t_p)_+^3\right].$$

3. Results and Discussion

This study concerns two objectives related to modeling the volatility of stock returns and studying the behaviour of volatility over the period. Firstly, we show the behaviour of stock returns of seven companies. After that, we present the volatility series obtaining from the GARCH(1,1). Finally, we show the result of fitting the natural cubic spline through the volatility series.

3.1 Stock Price Returns

In figure 1, the data are plotted on the y-axis and corresponding quantiles from a standardize normal distribution on the x-axis. It clearly can be seen from all panels that the stock returns are normal in the middle, but have stretched tails on both sides. Points distant from a fitted line indicated non-normality. In other words, the returns distributions contain fat tail (heavy tail). After obtaining the returns, the information of return fluctuation (volatility series) over time can be estimated by fitting GARCH(1,1) to the return.

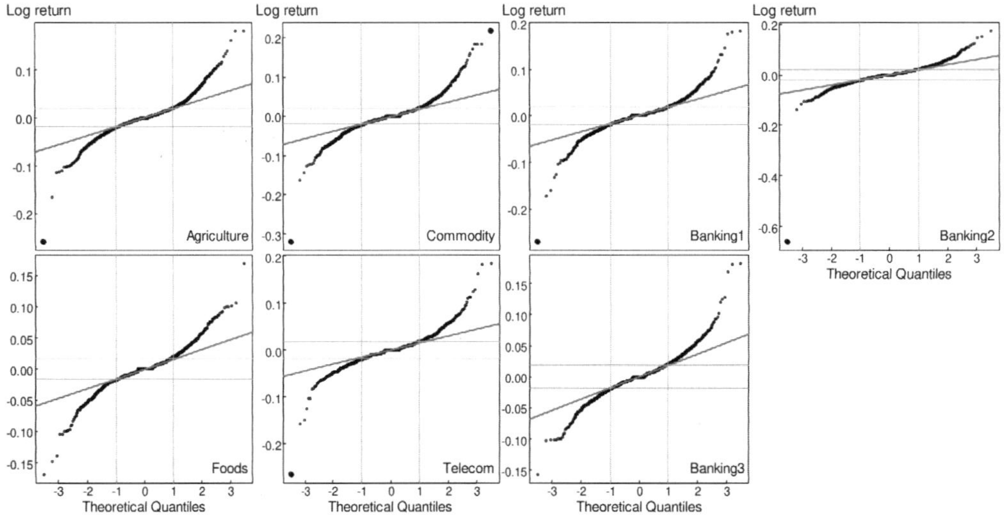

Figure1. Quantile-Quantile plots of stock returns.

3.2 Volatility of Stock Price Returns

Fitting the GARCH(1,1) gives volatility series of the seven stock returns plotted in figure 2 which describes the return fluctuation over the period.

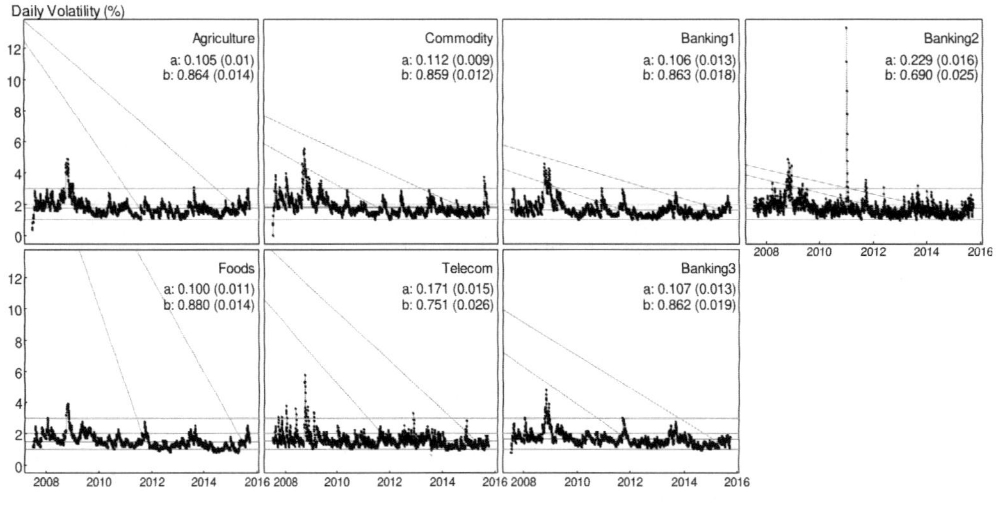

Figure 2. Volatility series of seven stock returns.

It is clear from figure 2 that all stock returns have higher volatility during the end of 2008 but subsequently remained relatively stable. The seven companies are big companies that can possibly reflect the economy of Indonesia. The increasing volatility at the end of 2008 corresponds to the economic crisis in Indonesia at that time. As in figure 2, the volatility series is very fluctuating, we need to smooth the volatility series in order to simplify investigation of their change in many situations. The volatility series will be smoothed using natural cubic spline.

3.3 Fitted Volatility

In summary, we fitted the volatility series of seven stock returns using eight-knot natural cubic spline, the results are graphed in figure 3. It shows the volatility fitted by natural cubic spline which reflects the volatility signals. The lower right panel shows the volatility signals of stock returns for each of the seven stocks on the same axes. It can be seen that the seven volatility signals have the same trends, particularly during the end of 2008. In addition, foods and telecom might simply reflect flat volatility over the period.

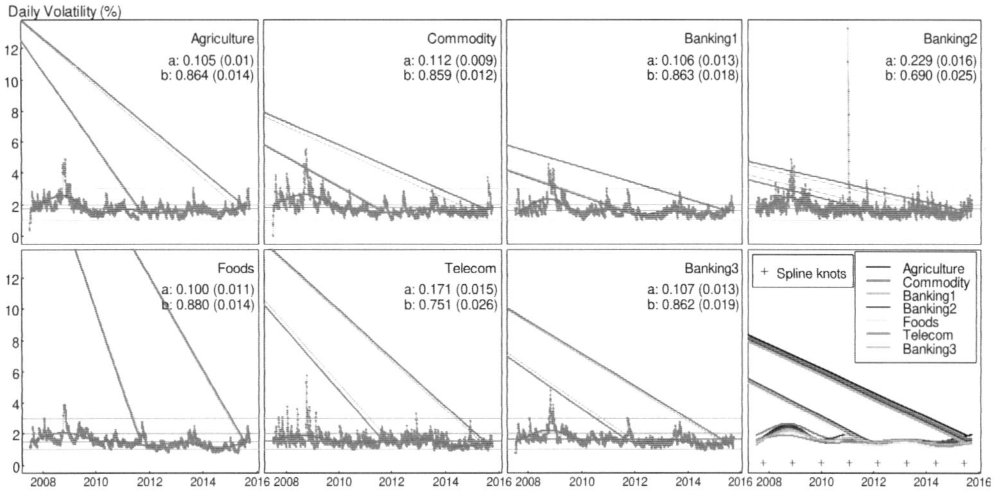

Figure 3. Fitted volatility series.

4. Conclusions

Modelling the volatility is needed in order to predict the volatility of stock price returns. The returns could be obtained from stock price data by differencing log price from one day to the next. GARCH(1,1) was employed to fit the returns over the period. This is done through the estimation of parameters of the model by the Maximum Likelihood Estimation (MLE). After obtaining the volatility, natural cubic spline was used to study the behavior of the volatility. The result shows that GARCH(1,1) indicates evidence of volatility clustering in the returns of some Indonesia stock prices.

Acknowledgments

We would like to address our gratitude to Emeritus Prof. Dr. Don McNeil and Dr. Rattikan Saelim for invaluable assistance that greatly improved the manuscript.

References

[1] Alexander C 2008 *Quantitative Methods in Finance* (West Sussex: John Wiley & Son Ltd) pp 19-21
[2] Arowolo W B 2003 The *International J. Eng. and Sci.***2** 32-37
[3] Bollerslev T 1986 *J. Eco* **31** 307-327
[4] Chapra S C and Canale R P 2010 *Numerical Method for Engineers* 6th ed (New York:

IOP Conf. Series: Journal of Physics: Conf. Series **948** (2018) 012068 doi:10.1088/1742-6596/948/1/012068

McGraw-Hill Companies Inc) pp 528-529
[5] Hull J C 2009 *Option Future and Other Derivative Securities* 7th ed (New Jersey: Pearson Prentice Hall) pp 469-477
[6] Kamau K J, Mwita P N and Nassiuma D K 2015 *Kabarak J. Research and Innov.* **3** 48-54
[7] McNeil N, Odton P and Ueranantasun A 2011 *Songklanakarin J. Sci. and Technol.* **33** 117-120
[8] Namugaya J, Weke P G and Charles W M 2014 *International J. Sci.: Basic and App. Research* **16** 216-223
[9] Posedel P 2005 *Metodoloskizvezki* **2** 243-257

Application of least-squares fitting of ellipse and hyperbola for two dimensional data

M P Lawiyuniarti[1], E Rahmadiantri[1], I M Alamsyah[1] and G Rachmaputri[1,2]

[1]Algebra Research Group, Faculty of Mathematics and Natural Sciences, Institut Teknologi Bandung, Jalan Ganesha No. 10 Bandung 40132

E-mail: gantina@math.itb.ac.id

Abstract. Application of the least-square method of ellipse and hyperbola for two-dimensional data has been applied to analyze the spatial continuity of coal deposits in the mining field, by using the fitting method introduced by Fitzgibbon, Pilu, and Fisher in 1996. This method uses $4a_0a_2 - a_1^2 = 1$ as a constrain function. Meanwhile, in 1994, Gander, Golub and Strebel have introduced ellipse and hyperbola fitting methods using the singular value decomposition approach. This SVD approach can be generalized into a three-dimensional fitting. In this research we, will discuss about those two fitting methods and apply it to four data content of coal that is in the form of ash, calorific value, sulfur and thickness of seam so as to produce form of ellipse or hyperbola. In addition, we compute the error difference resulting from each method and from that calculation, we conclude that although the errors are not much different, the error of the method introduced by Fitzgibbon et al is smaller than the fitting method that introduced by Golub et al.

1. Introduction

As we know, least square fitting method is a way to get curves (functions) through a number of data points (x_1, y_1) to (x_n, y_n) using the polynomial function $f\ (x, y)$. At [2] Fitzgibbon, et al describes the method of elliptical fittings to minimize the distance to the distribution of data by using a constrain of $4a_0a_2 - a_1^2 \ = \ 1$. This fitting technique with constrain has been applied in the field of mining by Fikri et al on [1].

In Golub et al [4] the method of fitting of ellipse and hyperbola to a distribution of data has been introduced. They did not use the same restriction as before but rather using the SVD approach. Recently this approach has been generalized into three dimensional fitting of ellipsoids and hyperboloids by Rahmadiantri et al on [6].

In this research, we will discuss both methods mentioned above and furthermore, we will give the application of the calculation of both methods to the data of coal deposits contained in [1] which consists of the ash, calorific value, sulfur and steam thickness by using the two methods. In addition, we will also compute the difference errors generated from each method used.

2. Methodology

In this section, we will explain the two methods underlying the fitting technique in two-dimensional data with different functions, namely the "Direct Least Square Method" by Fitzgibbon et al in 1996 and the "SVD method" by Gander, et al in 1994.

Initially given a form of presentation of the general equation two-dimensional space as follows

$$f(a,x) = a_0x_1^2 + a_1x_1x_2 + a_2x_2^2 + a_3x_1 + a_4x_2 + a_5 = a \cdot x$$

With:

$\boldsymbol{a} = (a_0 \quad a_1 \quad a_2 \quad a_3 \quad a_4 \quad a_5)^T$

$\boldsymbol{x} = (x_1^2 \quad x_1x_2 \quad x_2^2 \quad x_1 \quad x_2 \quad 1)^T$,

when $f(\boldsymbol{a},\boldsymbol{x}) = 0$ the equation is the general equation of a conic. We will determine $\boldsymbol{a}$ with Direct Least-Square Method and Singular Value Decomposition.

2.1. Direct Least-Square Method

The steps of fitting data in this method are as follow (see [2]):

Step 1. Build Matrix $\boldsymbol{D} = \begin{pmatrix} x_{11}^2 & x_{11}x_{21} & x_{21}^2 & x_{11} & x_{21} & 1 \\ x_{12}^2 & x_{12}x_{22} & x_{22}^2 & x_{12} & x_{22} & 1 \\ \vdots & \vdots & \vdots & \vdots & \vdots & \vdots \\ x_{1n}^2 & x_{1n}x_{2n} & x_{2n}^2 & x_{1n} & x_{2n} & 1 \end{pmatrix}$ and define $\boldsymbol{S} = \boldsymbol{D}^T\boldsymbol{D}$.

Step 2. Use the equation $4a_0a_2 - a_1^2 = 1$ as constrain and build constrain matrix **C** so that $\boldsymbol{a}^T\boldsymbol{C}\boldsymbol{a} = 1$.

Step 3. Use Lagrange method to change the problem minimizing $d(\boldsymbol{a}) = \sum_{1=1}^{N} f^2(a, x_i)$ with constrain $\boldsymbol{a}^T\boldsymbol{C}\boldsymbol{a} = 1$ into finding λ and $\boldsymbol{a}$ satisfying $S\boldsymbol{a} = \lambda\boldsymbol{C}\boldsymbol{a}$ and $\boldsymbol{a}^T\boldsymbol{C}\boldsymbol{a} = 1$.

Step 4. Breaking the equation $\boldsymbol{S}\boldsymbol{a} = \lambda\boldsymbol{C}\boldsymbol{a}$ into :

$$\begin{pmatrix} S_{11} & S_{12} \\ S_{21} & S_{22} \end{pmatrix}\begin{pmatrix} a_1 \\ a_2 \end{pmatrix} = \lambda\begin{pmatrix} C_{11} & 0 \\ \boldsymbol{0} & 0 \end{pmatrix}\begin{pmatrix} a_1 \\ a_2 \end{pmatrix}$$

Step 5. Solve these equations : $\boldsymbol{S}_{11}\boldsymbol{a}_1 + \boldsymbol{S}_{12}\boldsymbol{a}_2 = \lambda\boldsymbol{C}_{11}\boldsymbol{a}_1$ and $\boldsymbol{S}_{21}\boldsymbol{a}_1 + \boldsymbol{S}_{22}\boldsymbol{a}_2 = \boldsymbol{0}$ by solving

$$(\lambda\boldsymbol{I} - \boldsymbol{E})\boldsymbol{a}_1 = 0$$

with $\boldsymbol{I}$ is identity matriks 3×3, $\boldsymbol{E} = \boldsymbol{C}_{11}^{-1}(\boldsymbol{S}_{11} + \boldsymbol{S}_{12}\boldsymbol{S}_{22}^{-1}\boldsymbol{S}_{12}^T)$. Clearly λ is an eigenvalue of $\boldsymbol{E}$, in this case choose positive value. Find λ and corresponding eigenvector, namely $\boldsymbol{v}$.

Step 6. Choose $k = \sqrt{\frac{1}{v^T C_{11} v}}$

Step 7. Compute $\boldsymbol{a}_1 = k\boldsymbol{v}$ and $\boldsymbol{a}_2 = \boldsymbol{S}_{22}^{-1}\boldsymbol{S}_{21}\boldsymbol{a}_1$. Then we get $\boldsymbol{a} = \begin{pmatrix} \boldsymbol{a}_1 \\ \boldsymbol{a}_2 \end{pmatrix}$.

2.2. Singular Value Decomposition Method

The steps of fitting data in this method as follows (see [4]) :

Step 1. From data, build matrix $\boldsymbol{S} = \begin{pmatrix} x_{11} & x_{21} & 1 & {x_{11}}^2 & \sqrt{2}x_{11}x_{21} & x_{21}^2 \\ \vdots & \vdots & \vdots & \vdots & \vdots & \vdots \\ x_{1m} & x_{2m} & 1 & {x_{1m}}^2 & \sqrt{2}x_{1m}x_{2m} & x_{2m}^2 \end{pmatrix}$.

Step 2. Use QR fartorization to get $\boldsymbol{S} = \boldsymbol{Q}\boldsymbol{R}$ with **Q** an orthogonal matrix and **R** an upper triangular matrix.

Step 3. Write $\boldsymbol{R}$ as $= \begin{pmatrix} \boldsymbol{R}_{11} & \boldsymbol{R}_{12} \\ \boldsymbol{0} & \boldsymbol{R}_{22} \end{pmatrix}$ and as $\left\|\boldsymbol{S}\begin{pmatrix} \boldsymbol{v} \\ \boldsymbol{w} \end{pmatrix}\right\| = \left\|\boldsymbol{Q}\boldsymbol{R}\begin{pmatrix} \boldsymbol{v} \\ \boldsymbol{w} \end{pmatrix}\right\| = \left\|\boldsymbol{R}\begin{pmatrix} \boldsymbol{v} \\ \boldsymbol{w} \end{pmatrix}\right\|$, then the problem now is to solve $\min\|\boldsymbol{R}_{22}\boldsymbol{w}\|$ with constrain $\|\boldsymbol{w}\| = 1$.

Step 4. Use SVD on $\boldsymbol{R}_{22}$ to get $\boldsymbol{R}_{22} = \boldsymbol{U}\boldsymbol{\Sigma}\boldsymbol{V}^T$.

Step 5. By Theorem SVD [3] the solution of $\min\|\boldsymbol{R}_{22}\boldsymbol{w}\|$ with constrain $\|\boldsymbol{w}\| = 1$ is $\boldsymbol{w} = \boldsymbol{v}_n$ with $\boldsymbol{v}_n$ is the last column of matriks $\boldsymbol{V}$.

Step 6. Get $\boldsymbol{v} = -{\boldsymbol{R}_{11}}^{-1}\boldsymbol{R}_{12}\boldsymbol{w}$. Then get $\boldsymbol{a} = \begin{pmatrix} \boldsymbol{v} \\ \boldsymbol{w} \end{pmatrix}$.

2.3. Method for Computing Error

In this research will compute error value for each data point (x_i, y_i) by using Lagrange Multiplier (see also [5]). In this case, error is minimum average distance between each data points that belongs to the ellipse or hyperbola curve. The steps to calculate the error value for each method are as follows:

Step 1. Take the equation of ellipse or hyperbola $f(x, y)$ that we get as constrain and define\ $d(x, y) = (x - x_1)^2 + (y - y_1)^2$ with $d^2(x, y)$ is the function that we will minimize.
Step 2. Calculate the first partial derivatives of function $f(x, y)$ and $d^2(x, y)$ on variable x and y.
Step 3. Find x and y that satisfy: $f(x, y) = 0$ and $\nabla d^2(x, y) = \lambda \nabla f(x, y)$.
Step 4. Calculate the value $d(x, y)$ by substituting x and y value that have been obtained.

3. Result and Discussion

The data used to apply this method is the data of coal mining used in [1] consisting of ash content, calorific value, sulfur content and steam thickness for each point drill.
The conic equation is as follows:
$f(a, x) = a_0x^2 + a_1xy + a_2y^2 + a_3x + a_4y + a_5 = \boldsymbol{a} \cdot \boldsymbol{x} = 0$
The ash content data is as follows:

Table 1. Ash Content Data.

Direction	Coordinate	
	X	Y
N-S (North)	0.00	105.00
N-S (South)	0.00	-105.00
E-W (East)	75.00	0.00
E-W (West)	-75.00	0.00
NE – SW (North – East)	68.94	68.94
NE – SW (South – West)	-68.94	-68.94
NW – SE (South – East)	42.43	-42.43
NW – SE (North - West)	-42.43	42.43

Following the previous section :

$$\boldsymbol{S} = \boldsymbol{D}^T\boldsymbol{D} = \begin{pmatrix} 1{,}15 \cdot 10^8 & 3{,}87 \cdot 10^7 & 5{,}17 \cdot 10^7 & 0 & 0 & 24356{,}057 \\ 3{,}87 \cdot 10^7 & 5{,}17 \cdot 10^7 & 3{,}87 \cdot 10^7 & 0 & 0 & 5904{,}8374 \\ 5{,}17 \cdot 10^7 & 3{,}87 \cdot 10^7 & 2{,}95 \cdot 10^8 & 0 & 0 & 35156{,}057 \\ 0 & 0 & 0 & 24356{,}057 & 5904{,}8374 & 0 \\ 0 & 0 & 0 & 5904{,}8374 & 35156{,}057 & 0 \\ 24356{,}057 & 5904{,}8374 & 35156{,}057 & 0 & 0 & 0 \end{pmatrix}$$

By using Maple, we get:

$$E = C_{11}^{-1}(S_{11} + S_{12}S_{22}^{-1}.S_{12}^T) = \begin{pmatrix} -2{,}78 \cdot 10^7 & 6{,}373 \cdot 10^6 & 7{,}01 \cdot 10^7 \\ -2{,}072 \cdot 10^7 & -4{,}73 \cdot 10^7 & -1{,}28 \cdot 10^7 \\ 2{,}04 \cdot 10^7 & 1{,}04 \cdot 10^7 & -2{,}77 \cdot 10^7 \end{pmatrix},$$

the eigenvalue $\lambda = 2{,}92 \cdot 10^6$ and $\boldsymbol{a_1} = \begin{pmatrix} -0.78 \\ 0{,}42 \\ -0{,}38 \end{pmatrix}$, $\boldsymbol{a_2} = \begin{pmatrix} 0 \\ 0 \\ 3718.06 \end{pmatrix}$, so that we get

$$\boldsymbol{a} = \begin{pmatrix} -0{,}78 \\ 0{,}42 \\ -0{,}38 \\ 0 \\ 0 \\ 3718{,}06 \end{pmatrix}.$$

From that calculation we get the equation for ellipse curve as follows:

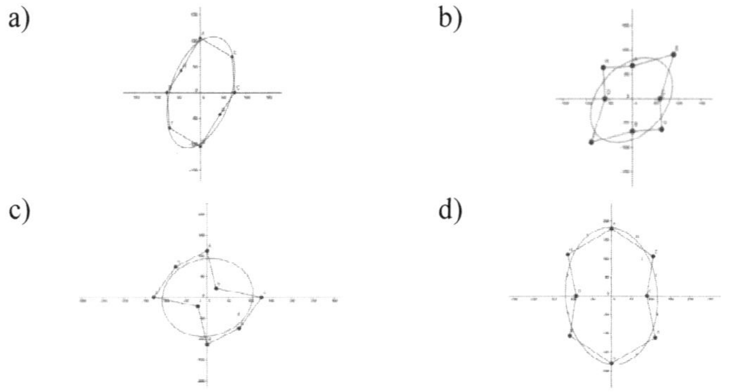

Figure 1. Fitting of ellipse with constrain $4a_0a_2 - a_1^2 = 1$

Figure 1 describes fitting of ellipse with constrain $4a_0a_2 - a_1^2 = 1$: (a) Fitting ash content $f(x) = -0{,}78x^2 + 0{,}42xy - 0{,}38y^2 + 3718{,}06$, (b) Fitting calorific $f(x) = -0{,}52x^2 + 0{,}37xy - 0{,}55y^2 - 0{,}98x - 3{,}13y + 3654{,}63$, (c) Fitting sulfur $f(x) = -0{,}44x^2 + 0{,}87xy - 0{,}57y^2 + 4998{,}12$, and (d) Fitting steam $f(x) = 0{,}78x^2 + 0{,}057xy + 0{,}32y^2 + 10679{,}97$.

Following the previous section and also by using Maple, we get the result as follows:

$$S = \begin{pmatrix} 0 & 105 & 1 & 0 & 0 & 11025 \\ 0 & -105 & 1 & 0 & 0 & 11025 \\ 75 & 0 & 1 & 5625 & 0 & 0 \\ -75 & 0 & 1 & 5625 & 0 & 0 \\ 68.94 & 68.94 & 1 & 4752.72 & 4752.72\sqrt{2} & 4752.72 \\ -68.94 & -68.94 & 1 & 4752.72 & 4752.72\sqrt{2} & 4752.72 \\ 42.43 & -42.43 & 1 & 1800.31 & -1800.31\sqrt{2} & 1800.31 \\ -42.43 & 42.43 & 1 & 1800.31 & -1800.31\sqrt{2} & 1800.31 \end{pmatrix}$$

Then the QR decomposition on the matrix **S** to produces the upper triangular matrix **R** with $\mathbf{R}_{22}$ as follows:

$$\boldsymbol{R}_{22} = \begin{pmatrix} 6386.55 & 3243.89\sqrt{2} & -8670.39 \\ 0 & 8576.45 & 4765.55\sqrt{2} \\ 0 & 0 & 4435.09 \\ 0 & 0 & 0 \\ 0 & 0 & 0 \end{pmatrix}$$

Next, a singular value decomposition is performed on the block matrix $\mathbf{R}_{22}$ and we obtain $\boldsymbol{R}_{22} = \boldsymbol{U\Sigma V^T}$ with

$$\boldsymbol{U} = \begin{pmatrix} -0.79 & -0.57 & 0.24 & 0 & 0 \\ 0.53 & -0.82 & -0.20 & 0 & 0 \\ 0.32 & -0.03 & 0.95 & 0 & 0 \\ 0 & 0 & 0 & 1 & 0 \\ 0 & 0 & 0 & 0 & 1 \end{pmatrix}; \quad \boldsymbol{\Sigma} = \begin{pmatrix} 12856.4 & 0 & 0 \\ 0 & 10345.7 & 0 \\ 0 & 0 & 1826.41 \\ 0 & 0 & 0 \end{pmatrix};$$

$$\boldsymbol{V} = \begin{pmatrix} -0.39 & -0.35 & 0.85 \\ 0.07 & -0.93 & -0.35 \\ 0.92 & -0.07 & 0.39 \end{pmatrix}$$

Next, we get $\boldsymbol{w} = v_3 = \begin{pmatrix} 0.85 \\ -0.35 \\ 0.39 \end{pmatrix}$ and $\boldsymbol{v} = \begin{pmatrix} 0 \\ 0 \\ -3941.81 \end{pmatrix}$. Substituting the values of the vectors $\boldsymbol{v}$ and $\boldsymbol{w}$ into $\boldsymbol{a}$ so we get the equation as follows:

$$f(x,y) = 0.85x^2 - 0.35x + 0.39y^2 - 3941.81$$

$$f(x,y) = -0{,}78x^2 + 0{,}42xy - 0{,}38y^2 + 3718{,}06$$

With the same steps, we get equation of ellipses for all coal deposits that give by figure as follow:

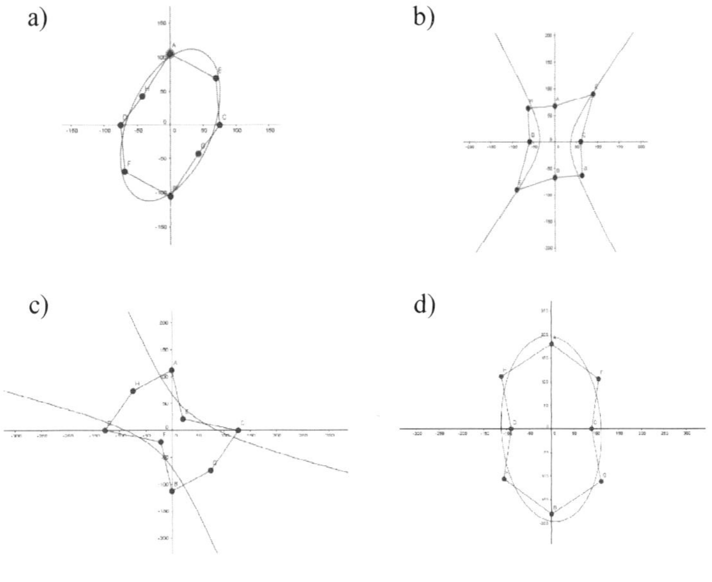

Figure 2. Fitting with SVD approach.

Figure 2 describes fitting with SVD approach: (a) Fitting ash content$f(x) = 0{,}85x^2 - 0{,}35\sqrt{2}xy + 0{,}39y^2 - 3941{,}81$, (b) Fitting calorific value $f(x) = -0{,}86x^2 - 0{,}13\sqrt{2}xy + 0{,}49y^2 + 1126{,}96$, (c) Fitting sulfur $f(x) = 0{,}28x^2 + 0{,}85\sqrt{2}xy + 0{,}45y^2 - 2122{,}91$, and (d) Fitting steam $f(x) = 0{,}95x^2 - 0{,}06\sqrt{2}xy + 0{,}31y^2 - 11859{,}68$.

By the Lagrange Multiplier method explained in previous section, we get the following comparison of error value between the two methods:

Table 2. Comparison Error Value Between Two Method That Used.

Number	**Type of Coal Content**	**Comparison**	
		Least Squares Method with constraining $4a_0a_2 - a_1^2 = 1$	**Least Squares Method With SVD Approach**
1.	Ash	5.42	5.52
2.	Calorie	17.95	21.24
3.	Sulfur	17.64	28.84
4.	Seam Thickness	13.95	16.59

By the above comparison, we conclude that although the errors are not much different, the error of the method introduced by Fitzgibbon et al is smaller than the fitting method that introduced by Golub et al.

4. Conclusion

The results of this research indicates that the application of the least squares method with constraining of $4a_0a_2 - a_1^2 = 1$ to 4 data of coal content of ash, calorific value, sulfur and seam thickness results in fitting results that are all elliptical, while the fittings using the least squares method SVD approach produces ellipse shape for ash data and seam thickness data, whereas for fitting calorific and sulfur data produce a form of hyperbola. Furthermore, the error calculation results for each method in fitting the new data yields that the errors are not much different.

Acknowledgments

This research is supported by Hibah P3MI Institut Teknologi Bandung

References

[1] Fikri H, Heriawan M N, Alamsyah I M and Rachmaputri G 2016 Analysis on The Fitting Method of Ellipsoid Anisotropy Model and Its Influence for Resource Estimation *TPT XXV Association of Indonesian Mining Professionals*

[2] Fitzgibbon A W, Pillu M and Fisher R B 1996 Direct Least Square Fitting of Ellipses *Technical Report DAIRP-794, Departement of Artificial Intelligence, The University of Edinburgh.*

[3] Gander W 2008 The Singular Value Decomposition, *Lecture Notes ETH Zurich.*

[4] Gander W, Golub G H and Strebel R 1994 Least Square Fitting of Circle and Ellipses **34** *558-578.*

[5] Lawiyuniarti M P 2017 Penggunaan Metode Kuadrat Terkecil Elips dan Hiperbola untuk Data Dua Dimensi, Master Program Thesis Institut Teknologi Bandung

[6] Rahmadiantri E, Lawiyuniarti M P, Alamsyah I M and Rachmaputri G 2017 *Proc. 1st Int. Conf. on Applied and Industrial Mathematics and Statistics* (Kuantan: -)

IOP Conf. Series: Journal of Physics: Conf. Series **948** (2018) 012070 doi:10.1088/1742-6596/948/1/012070

Proof of the effect of electric current on function of tara calor using total derivative

Y Soenarto[1], A Kusdiwelirawan[1], A Fitriana[1], I R Ermawaty[1] and V M A Sari [1]

[1] Physics Education Program, Universitas Muhammadiyah Prof. Dr. HAMKA, Jakarta Indonesia

E-mail: imas_re@uhamka.ac.id

Abstract. The research aims to prove the effect of electric currents on taracalor by performing observations, analyze the results using the taracalor formulation which is the literature and by the application of mathematics is the total differential. Steps performed in the research start from the study of theory, subsequent experiments, analysis and comparison of matches against the literature. The result of this research shows that the average values of taracalor for 1 Ampere, 2 Ampere, and 3 Ampere are 0,7725, 0,8894, and 0.9786, respectively. While the average functions for 1 Ampere, 2 Ampere, and 3 Ampere are $(2{,}235\times10^{-3})(dt-90d\Delta T)$, $(2{,}57\times10^{-3})(dt-25{,}71d\Delta T)$, and $(2{,}833\times10^{-3})(dt-12{,}86d\Delta T)$, respectively.

1. Introduction

Physics is a science that studies the motion and behavior of the phenomenon in the sphere of time and space, along with related concepts such as energy and style. The ultimate goal of physics is to understand how the universe works. Mathematical physics is one application of physics, this science is closely related to theoretical physics that seeks to discuss mathematical laws of physics through logical reviewers and the calculation and application of quantitative various laws of physics empirical.

A natural occurrence which is an example of physics is, if there is an electric current flow in a circuit it will arise heat on the electrical circuit that is the place or center of electrical current activity. Therefore, there will be electrical energy that is converted to heat energy. The comparison between the electrical energy given to the heat generated is the taracalor.

In this research empirical approach, concept and mathematical approach. The empirical approach is done by observing the effect of electric current on the heat generated by variant the current by 1 Ampere, 2 Ampere, and 3 Ampere. Determining taracalor has been done by employing total differential.

Electric current is a flow of electric charge. When electric charge moves, we have a current of electricity [1]. In electric circuits this charge is often carried by moving electrons in a wire. It can also be carried by ions in an electrolyte, or by both ions and electrons such as in an ionized gas (plasma) [2].

The SI unit for measuring an electric current is the ampere, which is the flow of electric charge across a surface at the rate of one coulomb per second. Electric current is measured using a device called an ammeter [3]. If there is a current of I ampere in a circuit, and this is such that in t seconds, Q coulombs flow, the rate of flow is Q(t) coulombs per second. This gives us a useful formula:

IOP Conf. Series: Journal of Physics: Conf. Series **948** (2018) 012070 doi:10.1088/1742-6596/948/1/012070

$$I = \frac{Q}{t} \tag{1}$$

Tara calor is defined as the comparison between the electrical energy used and the heat generated. We had seen that when electric current passes through a resistor, the electric energy spented or electric work done in moving the charges is converted into heat energy. In fact work is done in any form mechanical and chemical can be converted into heat energy. Joule conducted a series of experiments and founded that the same work done in different forma produces the same quantity of heat in all cases. This is called Joule's La [6]. Joule showed that there is a definite relation between works done and hear produced by it. Let W is the mechanical work done and Q is the heat produced in calories.

According to Joule's law

$$W \propto Q \tag{2}$$

$$W = JQ \tag{3}$$

$$J = \frac{W}{Q} \tag{4}$$

If a piece of resistance wire immersed in a liquid or clad in solids and inserted as part of the system, the incidence of potential difference V and electric current I in the resistance cord always awaken the flow of energy called workmanship. If this business continues for a time t, then the number of attempts made by W is

$$W = VIt \tag{5}$$

It is the amount of energy added to the system, if the resistivity is no longer part of the system, the energy transfer is called the drainage and during the time t, the amount of energy that is transported is called the quantity Q, where:

$$Q = (m_a c_a + m_k c_k)\Delta T \tag{6}$$

Total Differential

If $y = f(x)$, then dy/dx can be thought of either as the slope of the curve $y = f(x)$ or as the rate of change of y with respect to x. Rates occur frequently in physics; time rates such as velocity, acceleration, and rate of cooling of a hot body are obvious examples. There are also other rates: rate of change of volume of a gas with applied pressure, rate of decrease of the fuel in your automobile tank with distance traveled, and so on. Equations involving rates (differential equations) often need to be solved in applied problems. Derivatives are also used in finding maximum and minimum points of a curve and in finding the power series of a function. All these applications, and more, occur also when we consider a function of several variables.

Let z be a function of two variables x and y; we write $z = f(x, y)$. For a function of two variables, $z = f(x, y)$, we want to do something similar to this. We have said that this equation represents a surface and that the derivatives $\partial f/\partial x$, $\partial f/\partial y$, at a point, are the slopes of the two tangent lines to the surface in the x and y directions at that point. The symbols $\Delta x= dx$ and $\Delta y= dy$ represent changes in the independent variables x and y. The quantity Δz means the corresponding change in z along the surface. We define dz by the equation:

$$dz = \frac{\partial z}{\partial x}dx + \frac{\partial z}{\partial y}dy \tag{7}$$

The differential dz is called the *total differential* of z[7].The total differential of x is the dx whose value is equal to the change of x because y is changed plus the change of x because z is changed.

Everything we have said about functions of two variables works just as well for functions of any number of variables. If $u = f(x, y, z, \cdots)$, then by definition

$$du = \frac{\partial f}{\partial x}dx + \frac{\partial f}{\partial y}dy + \frac{\partial f}{\partial z}dz + \ldots \tag{8}$$

And *du* is a good approximation to Δu if the partial derivatives of *f* are continuous and *dx, dy, dz*, etc., are small.

2. Methodology
Here is the research flow:

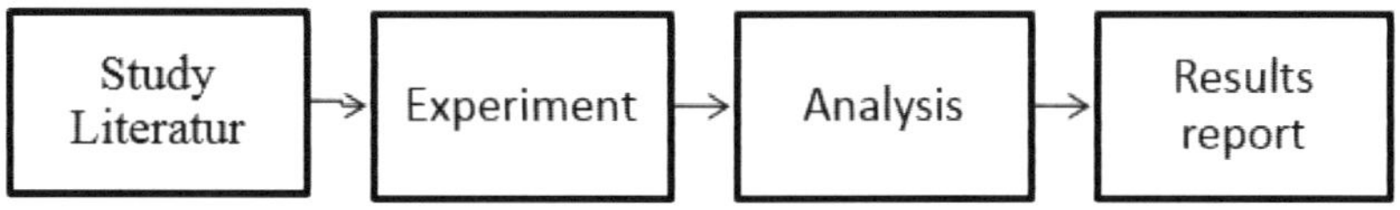

Figure 1. Flow Research.

This research activity begins by studying the theory. Then, after proceeding with experiments or experimental activities as an empirical approach and obtain data to be processed mathematically. The increased of temperature which is the designation of heat which is the main object of the observation of the effect of the current given, so in this study the current is made varies to know the effect of heat generated which later influence on the determination of electric heat calorie. Tools and materials used are amperemeter, calorimeter, resistance, thermometer, power supply, voltmeter, stopwatch, connecting cable. As for a series of experimental steps of preparing tools and materials, filling the calorimeter with water and weighing the mass of water and the mass of calorimeters, assembling tools and connecting with currents, stirring the calorimeter and observing the temperature changes and repeating the steps with different currents. Here's a series of tools used for experiments:

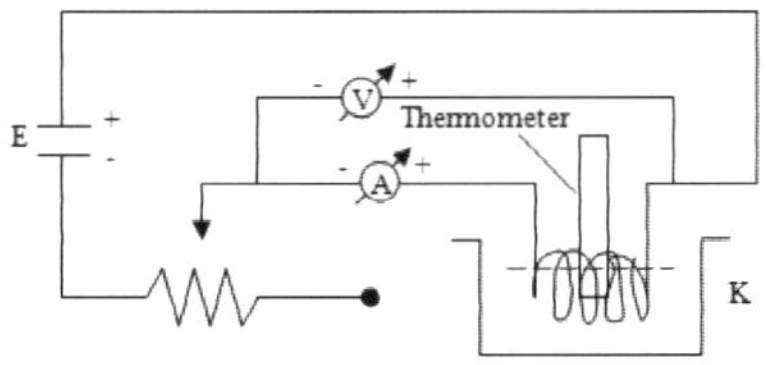

Figure 2. Tool Series.

After obtaining the data, then analyze the calculation by used the existing formula (exact) and by using the total differential. Based on the results of this study then made a report of research results.

3. Results and Discussion
The calculated data based on the exact way can be presented in table 1 and based on how the total differential has obtained the result of the function presented in table 2. If shown in graphical form the effect of the time relation force on the taracalor in each current is presented as figure 1 for the exact calculation and figure 2 in the total differential calculation.

Table 1. Tara Calor Value With Exact Calculations

Current (I)	t(minutes)	ΔT (°C)	Q (Joule)	W (Joule)	J
1	3	2	1584,458	1224	0,7725
	6	4	3168,916	2448	0,7725
	9	6	4753,374	3672	0,7725
	12	8	6337,832	4896	0,7725
Average					0,7725
2	3	7	5545,603	4932	0,8894

	6	14	11091,206	9864	0,8894
	9	21	16636,809	14796	0,8894
	12	28	22182,412	19728	0,8894
Average					0,8894
3	3	14	11091,206	10854	0,9786
	6	28	22182,412	21708	0,9786
	9	42	33273,618	32562	0,9786
	12	56	44364,824	43416	0,9786
Average					0,9786

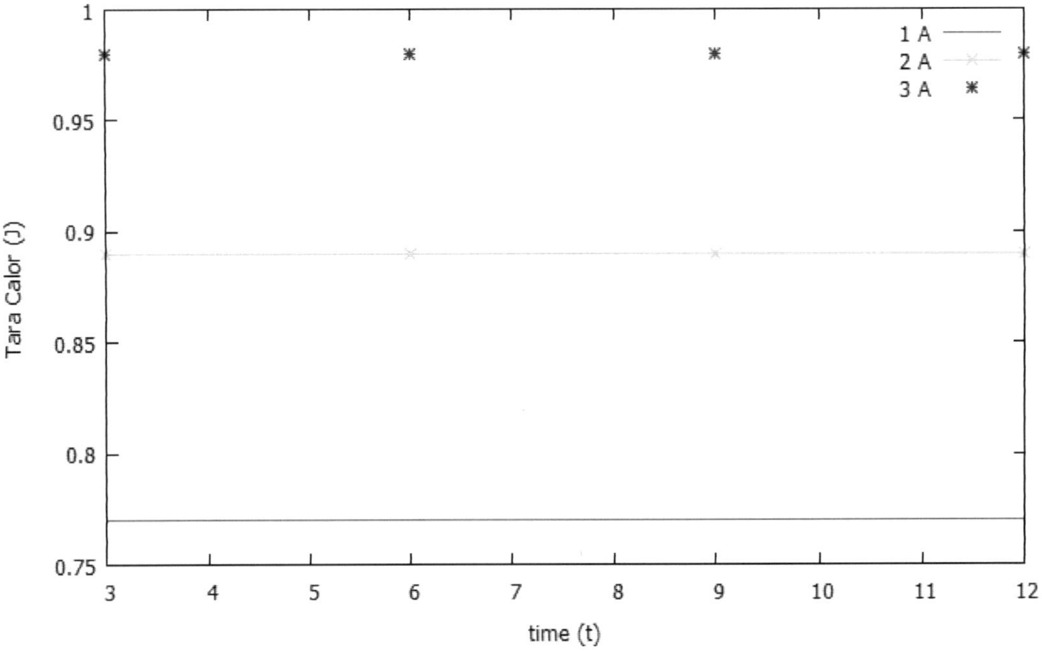

Figure 3. Graph the relationship between time (t) to the taracalor in each current with with exact calculations.

Table 2. Function Of Tara Calor With Total Differential.

Current(I)	T(minutes)	ΔT (°C)	dJ
1	3	2	$(4{,}29\times10^{-3})(dt-90d\Delta T)$
	6	4	$(2{,}15\times10^{-3})(dt-90d\Delta T)$
	9	6	$(1{,}43\times10^{-3})(dt-90d\Delta T)$
	12	8	$(1{,}07\times10^{-3})(dt-90d\Delta T)$
	Average		$(2{,}235\times10^{-3})(dt-90d\Delta T)$
2	3	7	$(4{,}94\times10^{-3})(dt-25{,}71d\Delta T)$
	6	14	$(2{,}47\times10^{-3})(dt-25{,}71d\Delta T)$
	9	21	$(1{,}65\times10^{-3})(dt-25{,}71d\Delta T)$
	12	28	$(1{,}24\times10^{-3})(dt-25{,}71d\Delta T)$
	Average		$(2{,}575\times10^{-3})(dt-25{,}71d\Delta T)$

3	3	14	$(5{,}44\times10^{-3})(dt-12{,}86d\Delta T)$
	6	28	$(2{,}72\times10^{-3})(dt-12{,}86d\Delta T)$
	9	42	$(1{,}81\times10^{-3})(dt-12{,}86d\Delta T)$
	12	56	$(1{,}36\times10^{-3})(dt-12{,}86d\Delta T)$
Average			$(2{,}833\times10^{-3})(dt-12{,}86d\Delta T)$

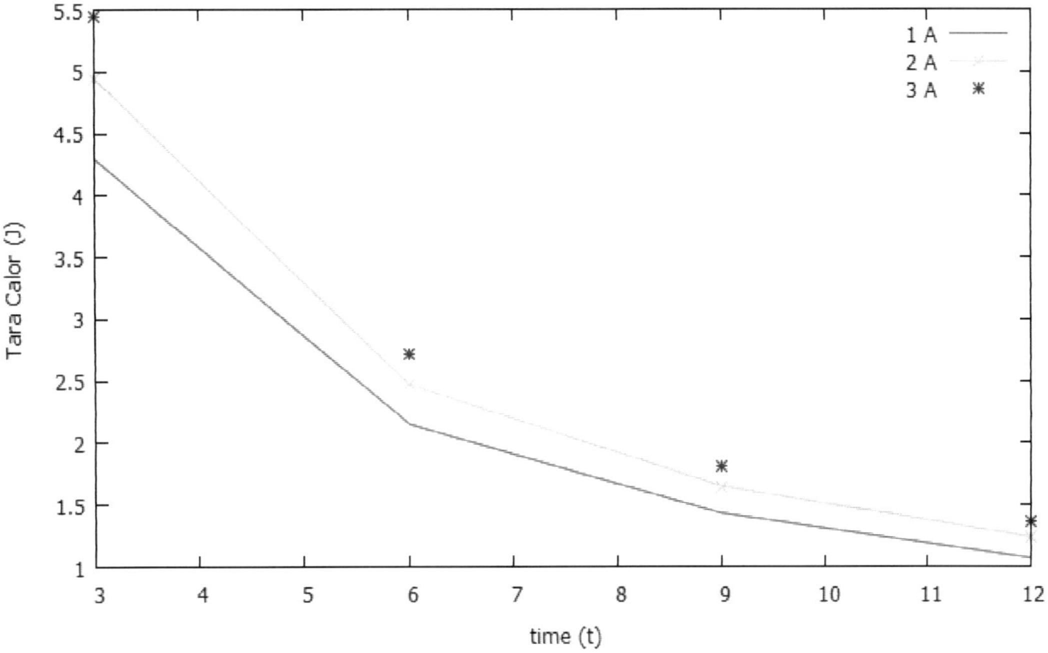

Figure 4. Graph the relationship between time (t) to the taracalor in each current with total differential calculation.

Here is a graph of current to taracalor both exact and total differential calculations.

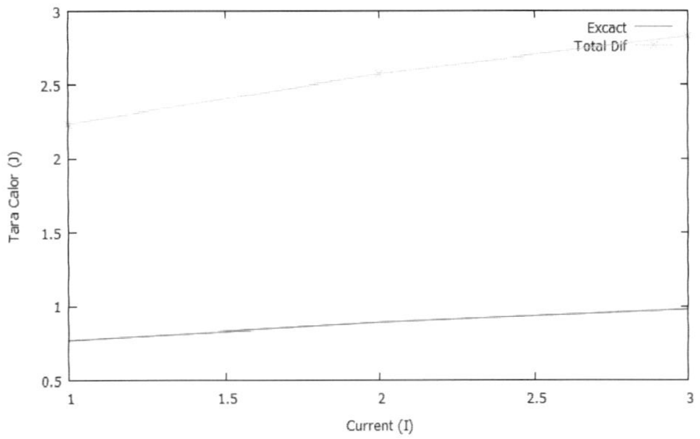

Figure 5. Graph of the current relationship to the average heat rate on each calculation by exact and total differential.

Based on the results of research, it is obtained that the current influences the taracalor, that the greater the current the greater the value of taracalor. In research of taracalor 3 Ampere that is 0.9786

IOP Conf. Series: Journal of Physics: Conf. Series **948** (2018) 012070 doi:10.1088/1742-6596/948/1/012070

bigger than tarakalor with 2 Ampere that is 0.8894, and 1 Ampere that is 0.7725. Similarly, the calculation used the total differential function obtained at current of 3 Ampere is $(dt - 12{,}86 d\Delta T)$ greater than 2 Ampere of. and 1 Ampere of $(dt - 90 d\Delta T)$ this is because the results of (-) on a 3 Ampere is smaller than the (-) of 1 Ampere at a function of temperature $(d\Delta T)$ means the more temperatures rise more rapidly when current of 3 Ampere. The symptom is also felt during the conduct of research (data collection). This is because of the greater flow of electrons when the electrons impinge on the coils. The greater the electron pounding, the heat generated faster so that the required time is less. Beside the similarity between the shape of the resulting graph exact and total differential calculation, resulting in common interpretations proved their influence on the flow of electricity and heat tare can be determined also by the total differential.

References

[1] Anthony C 2004 *The electronics companion* (CRC Press) p 13
[2] Lakatos J, Oenoki K, Judez H, Oenoki K, Hyun K C 1998 *Learn Physics Today!* Lima, Peru: Colegio Dr. Franklin D. Roosevelt. Archived from the original on 2009-02-27. Retrieved 2009-03-10.
[3] John L, Lewis and Philip E H 1969 *Electric Currents* (London: Longman)
[4] Purcell, Edwin J and Dale V 1987 *Calculus with Analytic Geometry Fifth Edition* (New Jersey: Prentice-Hall)
[5] Hamid A A 2007 *Diktat Lecture of Calor and Thermodynamics* (Yogyakarta: UNY)
[6] The IIT Foundation Series 2009 *Physics Class 10 (*New Delhi: Trishna Knowledge Systems)
[7] Boas M L 2006 *Mathematical Methods In The Physical Sciences Third Edition* (New York: Lehigh Press)

IOP Conf. Series: Journal of Physics: Conf. Series **948** (2018) 012071 doi:10.1088/1742-6596/948/1/012071

Some cycle-supermagic labelings of the calendula graphs

T R Pradipta[1] and A N M Salman[2]

[1]Mathematics Education, Universitas Muhammadiyah Prof. DR. HAMKA, Jakarta, Indonesia
[2]Combinatorial Mathematics Research Group Faculty of Mathematics and Natural Sciences, Institut Teknologi Bandung, Indonesia

E-mail: troymath@uhamka.ac.id

Abstract. In this paper, we introduce a calendula graph, denoted by $Cl_{m,n}$. It is a graph constructed from a cycle on m vertices C_m and m copies of C_n which are $C_{n_1}, C_{n_2}, \cdots, C_{n_m}$ and grafting the i-th edge of C_m to an edge of C_{n_i} for each $i \in \{1, 2, \cdots, m\}$. A graph $G = (V, E)$ admits a C_n-covering, if every edge $e \in E(G)$ belongs to a subgraph of G isomorphic to C_n. The graph G is called cycle-magic, if there exists a total labeling $\varphi : V \cup E \rightarrow \{1, 2, \ldots, |V| + |E|\}$ such that for every subgraph $C_n' = (V', E')$ of G isomorphic to C_n has the same weight. In this case, the weight of C_n, denoted by $\varphi(C_n')$, is defined as $\sum_{v \in V(C'_n)} \varphi(v) + \sum_{e \in E(C'_n)} \varphi(e)$. Furthermore, G is called cycle–supermagic, if $\varphi : V \rightarrow \{1, 2, \ldots, |V|\}$. In this paper, we provide some cycle-supermagic labelings of calendula graphs. In order to prove it, we develop a technique, to make a partition of a multiset into m sub-multisets with the same cardinality such that the sum of all elements of each sub-multiset is same. The technique is called an m-balanced multiset.

1. Introduction

The graphs considered here are finite, undirected, and simple. The vertex set and the edge set of a graph G are denoted by $V(G)$ and $E(G)$, respectively. An H-(super)magic labeling was first studied by Gutiérrez and Lladó in 2005 [3]. Lladó and Moragas [5] studied some cycle-(super)magic behavior of several classes of connected graphs. They gave several families of C_r -magic graphs for every $r \geq 3$. Maryati et al. [10] contributed to C_n-supermagic labelings of c copies of C_n. Some other results on C_n-supermagic labelings of several classes of graphs can be found in [1, 2, 6, 7, 8, 11,12, 13, 14].

This paper is organized as follows. In section 2, we define a new class of graph that we call a calendula graph. It is inspired by comb product graph [4, 14]. In section 3, we develop the concept of an m-balanced multiset [8]. It is a technique to partition a multiset to obtain m submultisets such that each submultiset has the same cardinality and the sum of all elements in each submultiset has a same value. This result is used to prove our main result. In the last section, we study C_n-supermagic labelings of calendula graphs. In this paper, we use the notation $[a, b]$ to mean $\{a \leq x \leq b \mid a, b \in \mathbb{Z}^+\}$ and $\sum A$ to mean $\sum_{a \in A} a$. We define $\{a\} \uplus \{a, b\} = \{a, a, b\}$.

2. Calendula Graphs

Let $m \geq 3$ and $n \geq 3$. Let C_n be a cycle on m vertices. A *calendula graph*, denoted by $Cl_{m,n}$, is a graph constructed from C_m and m copies of C_n which are $C_{n_1}, C_{n_2}, \cdots, C_{n_m}$ and grafting the i-th edge of C_m to an edge of C_{n_i} for each $i \in \{1,2,\cdots,m\}$. For illustration, we can see $Cl_{6,4}$ in figure 1.

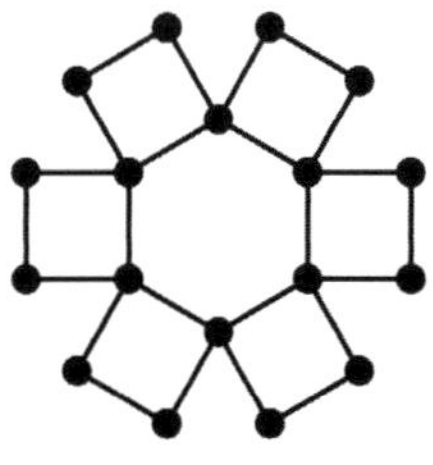

Figure 1. A calendula graph $Cl_{6,4}$.

We can check that the order of $Cl_{m,n}$ is $m(n-1)$ and the measure of $Cl_{m,n}$ is mn. It mean that $\left|V\left(Cl_{m,n}\right)\right| + \left|E\left(Cl_{m,n}\right)\right| = m(2n-1)$. For $m \neq n$, $Cl_{m,n}$ has m subgraph C_n' which isomorphic to C_n. As for $m = n$, $Cl_{m,n}$ has $(m+1)$ subgraph C_n' which isomorphic to C_n. We can also check that $Cl_{m,n}$ contains C_n-covering. Let the vertex set and the edge of $Cl_{m,n}$, respectively, be as follows:

$$V\left(Cl_{m,n}\right) = \left\{v_i^j \mid i \in [1,m] \text{ and } j \in [1,(n-1)]\right\} \text{ and}$$
$$E\left(Cl_{m,n}\right) = \left\{e_i^j \mid i \in [1,m] \text{ and } j \in [1,n]\right\}.$$

3. *m*-Balanced Multiset

A *multiset* is a set which allows the same elements. Let a multiset $V = \{a_1, a_2, ..., a_m\}$ and a multiset $W = \{a_1, a_2, ..., a_n\}$. Define $V \uplus W = \{a_1, a_2, ..., a_m, a_1, a_2, ..., a_n\}$. An m-balanced multiset defined as follows. Let $m \in \mathbb{Z}^+$ and Y is a multiset of positive integers. Y is called *m-balanced*, if there are m submultiset of Y, that is $Y_1, Y_2, \ldots, Y_m$, such that for each $i \in [1,m]$, satisfies $|Y_i| = |Y| m^{-1}$, $\sum Y_i = \left(\sum Y\right) m^{-1} \in \mathbb{Z}^+$, and $\biguplus_{i=1}^m Y_i = Y$. We need the next lemma to prove that $Cl_{m,n}$ is C_n-supermagic.

Lemma 1. Let m and n are positive integers with $m \geq 3$ and $n \geq 3$. If a multiset $X = [1,m] \uplus [1, m(2n-1)]$, then X is m-balanced.

Proof.

For each $i \in [1, m]$ and $j \in [1, 2n]$, define a multiset $X_i = \{a_{i,1}, a_{i,2}, ..., a_{i,2n}\}$ with

$$a_{i,j} = \begin{cases} i, & \text{for } i \in [1,m] \text{ and } j = 1; \\ i+1, & \text{for } i \in [1,m-1] \text{ and } j = 2; \\ 1, & \text{for } i = m \text{ and } j = 2; \\ 2m-i+1, & \text{for } i \in [1,m] \text{ and } j = 3; \\ 3m - & \text{for } i \in [1,m-1] \text{ and } j = 4; \\ 3m, & \text{for } i = m \text{ and } j = 4; \\ (j-1)m-i+1, & \text{for } i \in [1,m] \text{ and } j \in [5,n],\ j \equiv 0 \bmod 2; \\ (j-2)m+i, & \text{for } i \in [1,m] \text{ and } j \in [5,n],\ j \equiv 1 \bmod 2. \end{cases}$$

Next, for every $i\in[1, m-1]$ and $j\in[1, 2n]$, we obtain

$$\sum X_i = 1+(i+1)+(2m-i+1)+(3m-i)+(3m+i)+(5m-i+1)+(5m+1)+(7m-i+1)+\dots$$
$$+\left((2n-3)m+i\right)+\left((2n-1)m-i+1\right)$$
$$=2m+3m+n+\sum_{t=2}^{n-1}4mt$$
$$=2mn^2-2mn+m+n.$$

For every $i=m$ and $j\in[1, 2n]$, we get

$$\sum X_i = m+1+(m+1)+3m+4m+(4m+1)+6m+(6m+1)+\dots+\left((2n-2)m\right)+\left((2n-2)m+1\right)$$
$$=2m+3m+n+\sum_{t=2}^{n-1}4mt$$
$$=2mn^2-2mn+m+n.$$

For $j\in[1, 2n]$, $A_j=\{a_{i,j} \mid 1\le i\le m\}$, let

$$A_j=\begin{cases}[1, m] \text{ for } j\in[1, 2];\\ \left[\left((j-2)m+1\right),\left((j-1)m\right)\right] \text{ for } j\in[3, 2n].\end{cases}$$

It can be checked that $A_1 \uplus A_2 \uplus \dots \uplus A_{2n} = X$ and $\uplus_{i=1}^{m} X_i = X$ Additionally, for each $i\in[1, m]$, we obtain $|X_i| = 2n$ and $\sum X_i = 2mn^2-2mn+m+n$. Therefore, for $m\ge 3$ and $n\ge 3$, we get that X is m-balanced.

4. Calendula Graphs are Some Cycle-Supermagic

In this section we show that a calendula graph $Cl_{m,n}$ for any positive integers m and n with $m\ge 3$ and $n\ge 3$ is C_n-supermagic.

Theorem 2. *Let m and n be two integers with $m\ge 3$ and $n\ge 3$. Let $Cl_{m,n}$ be a calendula graph, then $Cl_{m,n}$ is C_n-supermagic.*

Proof.

Let C_n' be a subgraph of $Cl_{m,n}$ which isomorphic with C_n. Define a total labeling $\varphi: V\left(Cl_{m,n}\right)\cup E\left(Cl_{m,n}\right)\to\{1,2,\dots,m(2n-1)\}$ as follows.

(i) Let $m \ne n$. Let a multiset $X=[1,m]\uplus[1,m(2n-1)]$. Partition X into several submultisets, X_i with $i\in[1,m]$ based on the above Lemma 1. For $i\in[1,m]$, label v_i^j and e_i^j on C_n' by using elements in X_i and the smallest label to label the vertices such that every subgraph C_n' on $Cl_{m,n}$ applies

$$\varphi\left(C_n'\right)=2mn^2-2mn+m+n.$$

Therefore, for $m\ne n$, we obtain φ is an C_n -super magic labeling on $Cl_{m,n}$.

(ii) Let $m = n$. Since $Cl_{m,n}$ has $(n + 1)$ subgraphs C_n', we need a modification labeling (i) such that every subgraph C_n' has the same weight. We divide into two subcases.

(ii.a) For $m = n \equiv 0 \bmod 2$.

First, do the labeling as in (i). Furthermore, re-do the labeling on some edge e_i^j by swapping a pair of edge label e_i^j which are at the same C_n' using the following way:

- exchange thelabel edge e_i^1 with e_i^n, for $i = 1$;
- exchange the label edge e_i^1 with e_i^{n-1}, for $n > 4$, $i \in \left[2, \left(\frac{1}{2}n - 1\right)\right]$ and $i \in \left[\left(\frac{1}{2}n + 1\right), n\right]$;
- exchange the label edge e_i^1 with e_i^{n-2}, for $n > 4$, $i = \frac{1}{2}n$, and for $n = 4$, $i = 2$.

This re-labeling does not change the weight of $n$subgraphC_n' which is obtained on (i). Furthermore, it is obtainedlabelling of a new subgraphC_n' with equal weight such that there are $(n + 1)$ subgraphC_n' which has same weight on $Cl_{m,n}$.

(ii.b) For $m = n \equiv 1 \bmod 2$.

Do labeling as in (i). Furthermore, do re-labeling on some edge e_i^j by swapping a pair of label edge e_i^j which are at the sameC_n' in the following way:

- exchange the label edge e_i^1 with $e_i^{\frac{1}{2}(n+1)}$, for $i = 1, n$;
- exchange the label edge e_i^1 with e_i^n, for $i \in [2, (n - 1)]$.

Similarly to (ii.a), this re-labeling does not change the weight of $n$subgraphC_n' which is obtained in (i). Furthermore, it is obtained labeling a new subgraphC_n' with equal weight such that there are $(n+1)$ subgraphC_n'which has same weight on $Cl_{m,n}$ for $m = n \equiv 1 \bmod 2$.

From (i), (ii.a), and (ii.b), we conclude that $Cl_{m,n}$ is C_n -supermagic for any integers m and n with $m \geq 3$ and $n \geq 3$. □

For illustration, in figure 2, figure 3, and figure 4 we show cycle-supermagic labelings on calendula graphs$Cl_{6,4}$,$Cl_{4,4}$, and$Cl_{5,5}$, respectively.

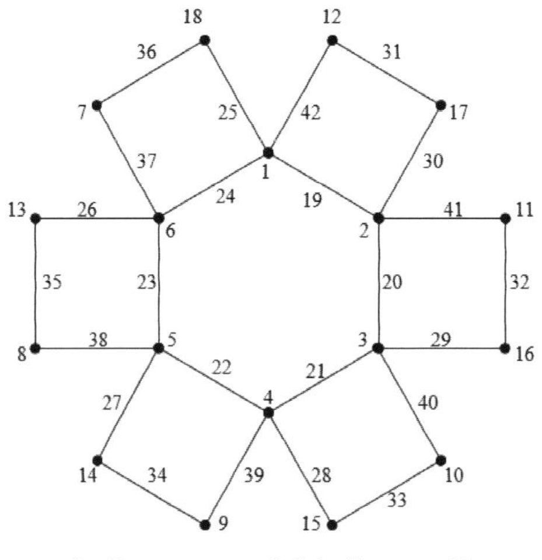

Figure 2. C_4–supermagiclabeling on $Cl_{6,4}$graph.

In figure2, it can be checked that the number of labels of each $C_4$is constant. We obtain the weight of 6 subgraphsC_4as follows.

- $\varphi\left(C_4^1\right) = 1 + 2 + 12 + 17 + 19 + 30 + 31 + 42 = 154$
- $\varphi\left(C_4^2\right) = 2 + 3 + 11 + 16 + 20 + 29 + 32 + 41 = 154$
- $\varphi\left(C_4^3\right) = 3 + 4 + 10 + 15 + 21 + 28 + 33 + 40 = 154$

- $\varphi\left(C_4^4\right)=4+5+9+14+22+27+34+39=154$
- $\varphi\left(C_4^5\right)=5+6+8+13+23+26+35+38=154$
- $\varphi\left(C_4^6\right)=6+1+7+18+24+25+36+37=154$.

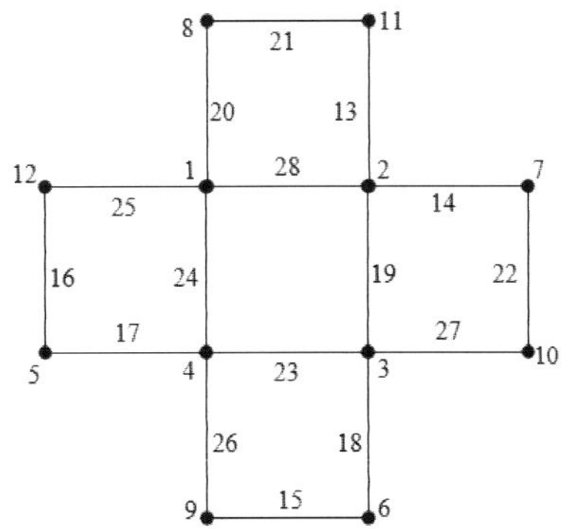

Figure 3. C_4 – supermagic labeling on $Cl_{4,4}$ graph.

In figure3, it can be checked that the number of labelsof each C_4 is constant. We obtain the weight of 5 subgraphs C_4 as follows.

- $\varphi\left(C_4^1\right)=1+2+8+11+13+20+21+28=104$
- $\varphi\left(C_4^2\right)=2+3+7+10+14+19+22+27=104$
- $\varphi\left(C_4^3\right)=3+4+6+9+15+18+23+26=104$
- $\varphi\left(C_4^4\right)=4+1+5+12+16+17+24+25=104$
- $\varphi\left(C_4^5\right)=1+2+3+4+19+23+24+28=104$

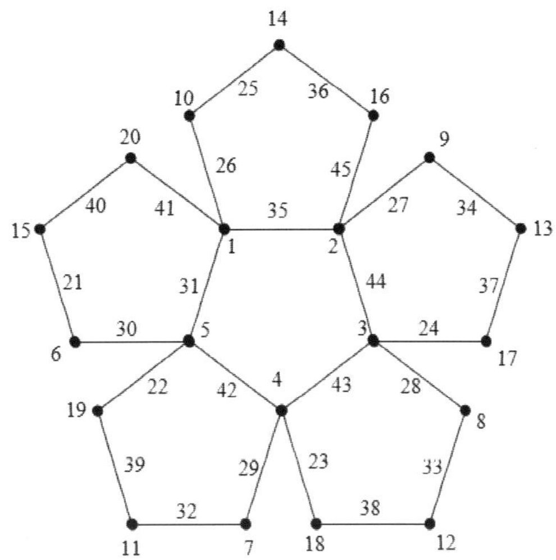

Figure 4. C_5 –supermagiclabeling of $Cl_{5,5}$graph

IOP Conf. Series: Journal of Physics: Conf. Series **948** (2018) 012071 doi:10.1088/1742-6596/948/1/012071

In figure 4, it can be checked that the number of labelsof each$C_5$is constant. We obtain the weight of 6sub graphs C_4 as follows.

- $\varphi\left(C_5^1\right)=1+2+10+14+16+25+26+35+36+45=210$
- $\varphi\left(C_5^2\right)=2+3+9+13+17+24+27+34+37+44=210$
- $\varphi\left(C_5^3\right)=3+4+8+12+18+23+28+33+38+43=210$
- $\varphi\left(C_5^4\right)=4+5+7+11+19+22+29+32+39+42=210$
- $\varphi\left(C_5^5\right)=5+1+6+15+20+21+30+31+40+41=210$
- $\varphi\left(C_5^1\right)=1+2+3+4+5+31+35+42+43+44=210$

5. References

[1] Gallian J A 2015A Dynamic survey of graph labeling *The Electronic J. Comb.* **6**
[2] Hartsfield N and Ringel G 2003 *Pearls in Graph Theory* (New York: Dover Publication,Inc.)
[3] Gutiérrez A and Lladó A 2005 Magic coverings *J. Comb. Math. Comb. Com.* **55** 43-56
[4] Jordan J 2005 Comb graphs and spectral decimation, *Glasg. Math. J.* **51** 71 – 81
[5] Lladó A and Moragas J 2007 Cycle-magic graphs *Discrete Mathematics*, **307** 2925-33
[6] Marbun H T and Salman A N M 2013 Wheel-supermagic labelings for a wheel *k*-multilevel corona with a cycle, *AKCE International J. Graphs and Comb.* **10** 1-9
[7] Maryati T K, Baskoro E Tand Salman A N M 2008 P_h super magic labeling of some trees *J. Comb. Math. Comb. Com.* **65** 197-204
[8] Maryati T K, SalmanA N M,Baskoro E T, Ryan J and Miller M 2010 On *H*-supermagiclabelings for certain shackles and amalgamations of a connected graph, *UtilitasMathematica* **83** 333-342
[9] Maryati T K, Salman A N M andBaskoro E T 2013Supermagic labellings of the disjoint union graphs and amalgamations *Discrete Mathematics* **313** 397-405
[10] Ngurah A A G,Salman A N M and Sudarsana I W 2010 On supermagic coverings of fans and ladders *SUT J. Math.* **46** 67-78
[11] Ngurah A A G, Salman A N M, Susilowati L 2010 *H*-supermagic labelings of graphs, *Discrete Mathematics* **310** 1293-1300
[12] Salman A N M and Purnomo A D 2010 Some cycle-supermagic labelings of the of some complete bipartite graphs. *East-West Journal of Mathematics* 283-291
[13] Salman A N M, NgurahA A G andIzzati N 2010 On (super) edge-magic total labellings of a subdivision of a star S_n *UtilitasMathematica* **81** 275-284
[14] Saputro S W, Mardiana and Purwasih N 2013 The metric dimension of comb product graphs, *Proc. of Graph Theory Conf. in honor of Egawa's 60th birthday* pp 10-14
[15] Wallis W D 2001 *Magic graphs* (Birkhauser Boston, Berlin)

Sample size and power calculation for univariate case in quantile regression

Ferra Yanuar[1]

[1]Mathematics Department, Faculty of Mathematics and Natural Sciences, Andalas University, Kampus Limau Manis, 25163, Padang – Indonesia

E-mail: ferrayanuar@yahoo.co.id

Abstract. The purpose of this study is to calculate the statistical power and sample size in simple linear regression model based on quantile approach. The statistical theoretical framework isthen implemented to generate data using R. For any given covariate and regression coefficient, we generate a random variable and error. There are two conditions for error distributions here; normal and nonnormal distribution. This study resulted that for normal error term, sample size is large if the effect size is small. Meanwhile, the level of statistical power is also affected by effect size, the more effect size the more level of power. For nonnormal error terms, it isn't recommended using small effect size, moderate effect size unless sample size more than 320 and large effect size unless sample size more than 160 because it resulted in lower statistical power.

1. Introduction

The least square estimator has many limitations. This estimator estimates the relationships between the conditional mean of the covariates on the response, or it only measures the mean of the response at given values of covariates [1]. The leastsquare estimator can be applied if the classical assumptions are fulfilled unless the conditional mean is not appropriate anymore. Another limitation of the leastsquare estimator is the response variable must follow normal distribution. If not, the conditional mean cannot appropriate to be applied. Due to these limitations, numerous estimator methods have been proposed. One of more popular estimator methods than others is quantile regression approach. Quantile regression estimates the relationship between covariates at different percentile points of the response variable. Quantile regression also has no assumption about the distribution of error. This estimator is widely used in many disciplines, such as in social science research, health research, and environmental sciences.

In many disciplines of research, the study of statistical power and sample size calculation are important factors to be considered in pilot study which then required for future implementation [2]. The accurate statistical power and reasonable sample size must be allowed in order to achieve the purposes of the study.

The sample size is an estimate of how many objects to be involved in a study. In determining the sample size, it needs predetermined parameters of the corresponding probability distribution, such as mean and standard deviations. Sample size can be determined based on these parameters. There are two criteria that can be used to determine sample size, precision analysis, and power analysis.

In calculating the reasonable sample size, it uses the concept of type I error, denoted by α. Type I error or α is the probability of reaching incorrect conclusion. Sample size calculation in precision analysis is usually referred to as the maximum error of an estimate of the unknown parameter. It allows the maximum half width of the $(1-\alpha)100\%$ confidence interval for this error.

Let $X_1, X_2, \dots, X_n$ are identically independent distribution normal random variables with mean μ and variance σ^2. The $(1-\alpha)100\%$ confidence interval for μ if σ^2 is known, is given by following :

$\bar{X} \pm z_{1-\alpha/2}\frac{\sigma}{\sqrt{n}},$

Then error or ε is defined as :

$$\varepsilon = |\bar{X} - \mu| = z_{1-\alpha/2}\frac{\sigma}{\sqrt{n}}.$$

Hence, the sample size is calculated by :

$$n = \frac{z_{1-\alpha/2}^2 \sigma^2}{\varepsilon^2} \tag{1}$$

Using Chebyshev's inequality to obtain the following nonparametric approach :

$$P(|\bar{X} - \mu| \leq \varepsilon) \geq 1 - \frac{\sigma^2}{n\varepsilon^2},$$

then set $1 - \frac{\sigma^2}{n\varepsilon^2} = 1 - p$, with $p = P\left(|Z| \geq z_{1-\alpha/2}\right)$. Thus, the sample size can be calculated by:

$$n = \frac{\sigma^2}{p\varepsilon^2}\,. \tag{2}$$

Besides controlling the Type I error, or α, the researcher is also suggested to minimize Type II error, β, in hypothesis testing while maintaining Type I error at a certain pre-specifed level. The statistical power here is represented by $1-\beta$, the complement of type II error βwhich is the probability of rejecting the impact or change as it occurs. Meanwhile the $(1-\beta)$ 100% is the power of the test represents the probability that a significant result will be detected when the alternative hypothesis is true. Sample size calculation under assumptions of this power of the test is called power analysis.

For two-sample test with equal sample size and having normal distribution when the variances for both samples are known, the sample size can be calculated by[3]:

$$n = \frac{(\sigma_1^2 + \sigma_2^2)\left(z_{1-\beta} + z_{1-\alpha/2}\right)^2}{\delta^2}, \tag{3}$$

where δ is the difference between two population means which are observed. The problems of this study are how to determine the reasonable power and sample size in simple linear regression based on quantile estimation approach.

2. Data and Methods

In this study, we give brief explanation related to quantile regression approach. This study aims to determine the statistical power and calculate the reasonable sample size for univariate case in the quantile regression. In this research, we use the generated data which the method is then implemented and tested.

2.1. Quantile Regression Approach

Let $\boldsymbol{y}$ is the response variable, $\boldsymbol{x}$ is a $p \times 1$ vector of p indicator variables for the ith observation. Let F_ε be the distribution function of error term which is ε with associated probability density function f_ε. The F_ε is supposed to be strictly increasing and absolutely continuous for every $\varepsilon \in \mathfrak{J}$for some interval$\mathfrak{J} \subset \mathfrak{R}$. Then let $Q_\tau(\boldsymbol{x})$ denote the τ-th ($0 < \tau < 1$) quantile regression function of y given $\boldsymbol{x}$.

The quantile regression model is[4] :

IOP Conf. Series: Journal of Physics: Conf. Series **948** (2018) 012072 doi:10.1088/1742-6596/948/1/012072

$$Q_\tau(\boldsymbol{x}) = \boldsymbol{x}'\beta(\tau) + F_\varepsilon^{-1}(\tau) \tag{4}$$

Normally, we assume $F_\varepsilon^{-1}(\tau)$ is 0 and τ-th quantile of dependent variable can be expresssed as a linear function of indicator variables. For $i = 1,2,\dots,n$ observations, the quantile regression estimation for $\hat{\beta}(\tau)$ are given by :

$$\hat{\beta}(\tau) = min \sum_i \rho_\tau\left(y_i - \boldsymbol{x}_i^T\beta(\tau)\right), \tag{5}$$

where $\rho_\tau(\varepsilon)$ is the loss function defined by:

$$\rho_\tau(\varepsilon) = \varepsilon(\tau - I(\varepsilon < 0)) \tag{6}$$

We also may write (3) as:

$$\rho_\tau(\varepsilon) = \varepsilon(\tau I(\varepsilon > 0) - (1-t)I(\varepsilon < 0)) \text{or } \rho_\tau(\varepsilon) = \frac{|\varepsilon| + (2\tau - 1)\varepsilon}{2}$$

For independent and identic distribution (iid) errors $\varepsilon_i, i = 1,2,\dots,n$, the asymptotic distribution for $\hat{\beta}$ in quantile regression has the following distribution [5][6]:

$$\sqrt{n}\left(\hat{\beta}(\tau) - \beta(\tau)\right) \sim N\left(0, \frac{\tau(1-\tau)}{f(F^{-1}(\tau))^2} E(X'X)^{-1}\right) \tag{7}$$

where $f\left(F^{-1}(\tau)\right) > 0$, and suppose $n^{-1}X'X \equiv Q_n$ converges to a positive definite matrix Q_0, i.e $E(X'X)$.

The errors can be assumed non iid, Koenker [6] showed that the asymptotic distribution of the estimator $\hat{\beta}$ has the following form :

$$\sqrt{n}\left(\hat{\beta}(\tau) - \beta(\tau)\right) \sim N(0, \tau(1-\tau)H^{-1}J_n H^{-1}) \tag{8}$$

where

$J_n = n^{-1}X'X$and $H = \lim_{n\to\infty} n^{-1}X'X\, f_\varepsilon(\tau)$.

2.2. *Statistical Framework for Sample size and Power Calculation*

In univariate case, we have x_i, $i = 1,2,\dots,n_1$ and y_i, $i = 1,2,\dots,n_2$. Both groups of observations are independent and normally distribution with means μ_x and μ_y and variances σ_x^2 and σ_y^2, respectively. We then have the following hypothesis to be tested [7][3] :

H_o: $\mu_x = \mu_y$ vs H_a: $\mu_x \neq \mu_y$

The Type I error of the test is :

$$\alpha = P(Z > c|H_0) \tag{9}$$

Where Z is a test statistic and c is the critical value of the rejecting the null hypothesis H_o.

The Type II error is defined as following:

$$\beta = 1 - P(Z > c|H_a) \tag{10}$$

Thus, we have :

power = 1- $\beta = P(Z > c|H_a)$.

For condition σ_x^2 and σ_y^2 are known, and $n_x = n_y = n$, we can implement Z test:

$$Z = \frac{\mu_x - \mu_y}{\sqrt{\frac{\sigma_x^2 + \sigma_y^2}{n}}} \tag{11}$$

The null hypothesis or H_o is rejected if the absolute value of Z is larger or equal $Z_{1-\alpha/2}$, represented by $|Z| \geq Z_{1-\alpha/2}$.

Under alternative hypothesis we have $\delta = \mu_x - \mu_y$ with $\delta > 0$and $Z_a \sim N(\mu^*, 1)$with $\mu^* = \frac{\delta}{\sqrt{\frac{\sigma_x^2+\sigma_y^2}{n}}} > 0$. Parameter δ is known as predefined constant and as effect size, it plays an important role in the determination of statistical power and reasonable sample size [3].

Therefore, based on equation (10), we have corresponding power here given by :

$$\beta = 1 - P\left(|Z| \geq Z_{1-\frac{\alpha}{2}}|H_a\right) \approx P\left(Z > Z_{1-\frac{\alpha}{2}}\right)$$
$$= P\left(Z - \mu^* > Z_{1-\frac{\alpha}{2}} - \mu^*\right)$$
$$= P\left(Z^* > Z_{1-\frac{\alpha}{2}} - \mu^*\right)$$

With Z^* has standard normal distribution. To achieve desired power of (1-β)100%, set

$$-z_\beta = z_{\frac{\alpha}{2}} - \mu^*.$$

The formula for calculate reasonable sample size then can be obtained by using the same logic. Let β_1 is the coefficient of indicator variable *x*, following is hypothesis test for β_1 :

$$H_0: \beta_1 = 0 \text{ vs } H_a: \beta_1 \neq 0$$

Under H_a, $\delta = \beta_1 + 0 = \beta_1$, with $\mu^* = \frac{\delta}{\sqrt{\frac{\sigma_{\beta_1}^2}{n}}} > 0$.

By modifying the equation (3), the formula to determine sample size is given by:

$$n = \frac{\sigma_{\beta_1}^2 (z_{1-\beta} + z_{1-\alpha/2})^2}{\delta^2} \quad (12)$$

The value of δ is assummed to be known and the unknown variable $\sigma_{\beta_1}^2$ is calculated using (7) for iid errors :

$$\sigma_{\beta_1}^2 = \frac{\tau(1-\tau)}{f(F^{-1}(\tau))^2} E(X'X)^{-1}$$

Test statistics under H_0 to test β_1 is given by :

$$T = \frac{\widehat{\beta_1} - 0}{\sqrt{\frac{s_{\beta_1}^2}{n}}} = \frac{\widehat{\beta_1}}{\sqrt{\frac{s_{\beta_1}^2}{n}}}.$$

Under H_a, statistics T has non-central *t* distribution with $\mu^* = \frac{\delta_1}{\sqrt{\frac{s_{\beta_1}^2}{n}}}$. Sample size is calculated using following approximation:

$$P\left(|T| > t_{n-p,1-\alpha/2}|H_a\right) \approx P\left(T > t_{n-p,1-\alpha/2}|H_a\right) \quad (13)$$

We need *n* large enough to achieve desired power of $(1-\beta)$ x 100%. Since sample size *n*which is involved in degrees of freedom and non-central parameter, it is quite hard to find the closed form of*n*. It can be solved using numerical method.

2.3. Generated Data for Error Distribution

We apply generated data in this present study to implement the theoretical framework. For any given covariate $\boldsymbol{x}_i$ and regression coefficient β, we generate a random variable $\boldsymbol{Y}_i$ and error ε_i. There are two conditions for error disributions here; normal and non normal distribution.

Assuming random variable $\boldsymbol{Y}_i$has normal distribution with mean parameter μ_i and variance σ^2, meanwhileerror term has the same distibution. We set the regresssion coefficient $\beta_0 = 1$, $\beta_1 = 1.5$ and $\tau = 0.1, 0.2, \ldots, 0.90$.

3. Result

The simulation study is done with the design as represented above. Following tables show the result of simulation study. Table 1 and table 2 present the estimated sample size and estimated power respectively for normal error term.

Table 1. Estimated Sample Size for Normal Distribution of Error.

Quantil	**Effect size (δ)**		
(τ)	$\delta = 0.1$	$\delta = 0.25$	$\delta = 0.4$
0.1	22	5	3
0.2	63	11	5
0.3	104	17	7
0.4	147	24	10
0.5	193	32	13
0.6	246	40	16
0.7	310	50	20
0.8	393	64	26
0.9	526	85	34

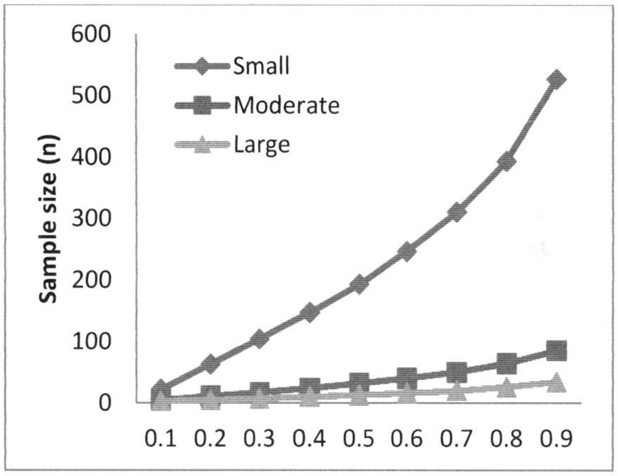

Figure 1. Estimated Sample Size for Normal Distribution of Error.

Table 1 and figure 1 show the estimated sample size under selected points of quantil and three conditions of effect size (**$\boldsymbol{\delta}$**). The sample size range from $n = 3$ to $n = 526$ for the setting that we considered. The table informs us that if effect size is small, it needs large sample size, the opposite sample size is small when effect size is large.

Table 2. Estimated Power for Normal Distribution of Error.

Sample size (n)	**Effect size (δ)**		
	$\delta = 0.1$	$\delta = 0.25$	$\delta = 0.4$
20	0.09	0.19	0.34
40	0.11	0.30	0.55
80	0.16	0.47	0.80
160	0.23	0.72	0.97
320	0.35	0.93	0.99
640	0.56	0.99	1.00

Table 2 shows the empirical power under selected sample size and three conditions of effect sizes, small
($\delta = 0.1$), moderate ($\delta = 0.25$) and large ($\delta = 0.4$). This simulation study observed that the empirical power ranges from 9% to 100%. When effect size is small, the empirical power is far away from 80%. When effect size is moderate, the empirical power is more than 70% for sample size more

than 160. Meanwhile in, large effect size, the empirical power more than 80% when sample size is 80 or more.

The result of simulation study for estimated sample size and estimated power respectively for nonnormal error term are presented in this following Figure 2 and Table 3.

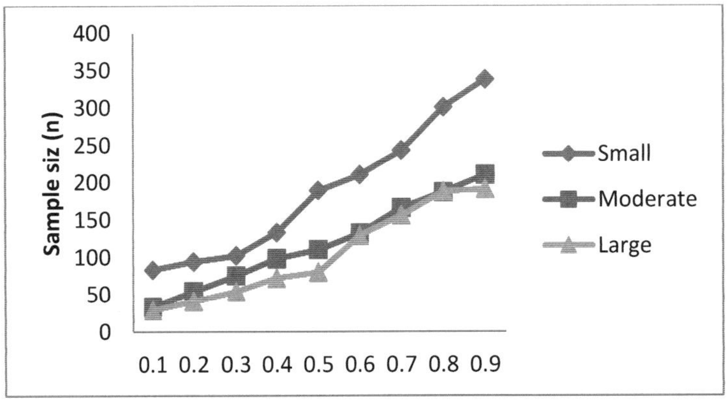

Figure 2. Estimated Sample Size for Non-Normal Distribution of Error.

Figure 2 informs us that sample size is small when effect size is large. Conversely, it needs large sample size if effect size is small. Meanwhile, Table 3 inform us the empirical power for nonnormal error term for selected sample size for three conditions of effect sizes.

Table 3. Estimated Power for Non-Normal Distribution of Error.

Sample size (n)	Effect size (δ)		
	$\delta = 0.1$	$\delta = 0.25$	$\delta = 0.4$
20	-	-	0.27
40	-	0.22	0.43
80	0.10	0.36	0.65
160	0.15	0.65	0.82
320	0.29	0.72	0.89
640	0.52	0.92	0.97

4. Summary

In this present study, we reviewed the statistical power and sample size calculation for univariate case in quantile regression model. The statistical theoretical framework was then implemented to generate data using R software. There are two assumptions for the error term, normal distribution, and nonnormal distribution. This study resulted that for normal error term, sample size is large if the effect size is small. Meanwhile, the level of statistical power is affected by effect size as well, the more effect size the more level of power. For nonnormal error terms, it isn't recommended to use small effect size, moderate effect size unless sample size more than 320 and large effect size unless sample size more than 160.

This study described in how to calculate the statistical power and sample size in univariate case only based on quantile approach. For future work, we will explore the statistical power and sample size calculation for multivariate case then.

References

[1] F Yanuar 2014 The Estimation Process in Bayesian Structural Equation Modeling Approach *J. Phys. Conf. Ser.*

[2] G Shan, S Moonie, and J Shen 2014 Sample size calculation based on efficient unconditional tests for clinical trials with historical controls *J. Biopharm. Stat.*
[3] Q Shao and Y Wang 2009 Statistical power calculation and sample size determination for environmental studies with data below detection limits *Water Resour. Res.* **45** 9
[4] G Tarr 2012 Small sample performance of quantile regression confidence intervals *J. Stat. Comput. Simul.* **82** 1 pp 81-94
[5] R Koenker and G Bassett 1978 Regression Quantiles *Econometrica* **46** 1 pp 33-50
[6] R Koenker 2005 Quantile Regression *J. Stat. Comput. Simul.*
[7] Z Gong 2016 Estimation of Sample Size and Power For Quantile Regression

IOP Conf. Series: Journal of Physics: Conf. Series **948** (2018) 012073 doi:10.1088/1742-6596/948/1/012073

The determination of the tribe of family members in Luhak Limopuluh Koto, West Sumatera Indonesia

S Bahri[1], A Abbas[2] and N N Bakar[3]

[1]Andalas University, Kampus UNAND Limau Manis Padang

E-mail: susilabahri@gmail.com

Abstract. In this paper, some mathematical models that state the tribe of the family members in Nagari Luhak Limopuluh Koto West Sumatera, Indonesia were built. The models were constructed by using the marriage rule and the ethnic data of Minangkabau community which embrace the matrilineal system. The marriage rule prohibits the same inter-tribal marriage while the matrilineal system causes the mother, child, and sibling tribes to be equal. Therefore, the matrices formed by marriage rule, mother-son tribal relation, someone-his/her sibling tribal relation, and the transpose of the matrices, are used in matrix multiplication to obtain the tribal models. The models are consecutively *A*, *AC'*, *(AC')B*, *((AC')B)W'*, *((AC')B)C*, *((AC')B)W*, *((AC')W'* for Denai, Denai's mother, the sister of Denai's mother, the brother of Denai's mother, the husband of Denai's mother's sister, the son of Denai's mother's sister, the wife of Denai's mother's brother, and the Denai's father models.

1. Introduction

The areas in West Sumatera province, Indonesia are grouped into three groups called Luhak Nan Tigo. The three luhak in Minangkabau area are Luhak Tanah Datar, Luhak Agam, and Luhak Limopuluh Koto. The Luhak consists of several Nagaris. The community of each Luhak is classified based on its tribe. Since each Nagari in all Luhak uses a matrilineal system in determining the lineage, then it is applied that

1. A child's tribe equals the mother's tribe
2. Two siblings have the same tribe

Each Nagari has also the marriage rule that prohibits the same inter tribal marriage [3]. It means that a man should not marry a woman from the same tribe.

Luhak Limopuluh Koto as one of Minangkabau's Luhak covers the District of Limopuluh Koto and Payakumbuh city. One of Nagaris in Luhak of Limopuluh Koto is Lubuak Batingkok. People in the Nagari have the various tribes namely Pitopang, Kutianyia, Bodi, Caniago, Sambilan, Tanjung, Koto, Piliang, Payobada, Bendang and Mandahiliang. One of the marriage rules in Luhak Limopuluh Koto is that it is allowed to have a marriage between a man (husband) and a woman (wife) from different tribes such as the marriage between Pitopang tribe (for a man) and Sambilan tribe (for a woman), Kutianyia tribe and Tanjung tribe, Bodi tribe and Piliang tribe, Caniago tribe and Koto Tribe, Sambilan tribe and Pitopang tribe, Tanjung tribe and Kutianyia tribe, Koto tribe and Bodi tribe, Piliang tribe and Mandahiliang tribe, Payobada tribe and Bendang tribe, Bendang tribe and Caniago tribe, and also Mandahiliang tribe and Payobada tribe.

In algebra, matrix can be used to state the presence or the absence of some particular quality or relationship. The binary matrix uses the element of 1 to show that there exists a relation and the element of 0 to express that there is no relation [1]. Therefore, the existence or no existence of the tribal relation among family members and the marriage rule can also be represented by a binary matrix. Next, by applying the concept of matrix multiplication on the binary matrices, we can form some mathematical models of family members who adhere the matrilineal system, especially in Nagari of Luhak Limopuluh Koto.

2. Matrices In Tribe Marriage

A matrix is a rectangular array of numbers. The number is called the element or the entry of the matrix. The size of the matrix represents the total number of its row and columns. If a matrix has m rows and n columns, then the size of the matrix is written as $m \times n$ [2]. For example, a matrix R which has the size of 4 rows and 3 columns is

$$R = \begin{bmatrix} 1 & 2 & 3 \\ 4 & 5 & 6 \\ 7 & 8 & 9 \\ 10 & 11 & 12 \end{bmatrix}$$

Any matrix with one row and n columns is also called as a row matrix. For instance, the following matrix L has the size of 1 row and 4 columns namely,

$$L = \begin{bmatrix} 0 & 0 & 0 & 1 & 0 \end{bmatrix}$$

If some arbitrary matrix has the same number of rows and columns, the matrix is called square matrix. Therefore, the matrix is known as n by n matrix. Furthermore, some square matrix that has ones on the main diagonal and zeros elsewhere is recognized as the identity matrix [2]. For instance, matrix Z is called the identity matrix with size 3×3

$$Z = \begin{bmatrix} 0 & 0 & 1 \\ 1 & 0 & 0 \\ 0 & 1 & 0 \end{bmatrix}$$

Next, any square matrix that has exactly one entry of 1 in each row and each column and 0's elsewhere is said as the permutation matrix. Hence, Z is also permutation matrix. Additionally, any matrix A that is formed by turning all rows into columns and vise versa is called the transpose matrix A'.

Two arbitrary matrices A and B can be multiplied if the column number of A equals to the row number of B. The multiplication will produce a new matrix. Thus, if the size of A is $m \times n$ and the size of B is $n \times p$, then the size of the matrix AB is $n \times p$ [2]. For example, if the matrix R is multiplied with the matrix Z, then the new matrix RZ is

$$RZ = \begin{bmatrix} 1 & 2 & 3 \\ 4 & 5 & 6 \\ 7 & 8 & 9 \\ 10 & 11 & 12 \end{bmatrix} \begin{bmatrix} 0 & 0 & 1 \\ 1 & 0 & 0 \\ 0 & 1 & 0 \end{bmatrix} = \begin{bmatrix} 2 & 3 & 1 \\ 5 & 6 & 4 \\ 8 & 9 & 7 \\ 11 & 12 & 10 \end{bmatrix}$$

As stated in Introduction, any relationship can be denoted by a matrix. Therefore, the marriage rule, the tribal relation between mother and child, and the tribal relation between someone and his/her sibling can also represented by a matrix namely,

2.1. The marriage rule matrix

The marriage rule matrix for Lubuak Batingkok is then

$$W = \begin{array}{c} P_t \\ K_u \\ B_o \\ C \\ S \\ T \\ K_o \\ P_l \\ P_a \\ B_e \\ M \end{array} \begin{bmatrix} 0 & 0 & 0 & 0 & 1 & 0 & 0 & 0 & 0 & 0 & 0 \\ 0 & 0 & 0 & 0 & 0 & 1 & 0 & 0 & 0 & 0 & 0 \\ 0 & 0 & 0 & 0 & 0 & 0 & 0 & 1 & 0 & 0 & 0 \\ 0 & 0 & 0 & 0 & 0 & 0 & 1 & 0 & 0 & 0 & 0 \\ 1 & 0 & 0 & 0 & 0 & 0 & 0 & 0 & 0 & 0 & 0 \\ 0 & 1 & 0 & 0 & 0 & 0 & 0 & 0 & 0 & 0 & 0 \\ 0 & 0 & 1 & 0 & 0 & 0 & 0 & 0 & 0 & 0 & 0 \\ 0 & 0 & 0 & 0 & 0 & 0 & 0 & 0 & 0 & 0 & 1 \\ 0 & 0 & 0 & 0 & 0 & 0 & 0 & 0 & 0 & 1 & 0 \\ 0 & 0 & 0 & 1 & 0 & 0 & 0 & 0 & 0 & 0 & 0 \\ 0 & 0 & 0 & 0 & 0 & 0 & 0 & 0 & 1 & 0 & 0 \end{bmatrix}$$
$$\begin{array}{ccccccccccc} P_t & K_u & B_o & C & S & T & K_o & P_l & P_a & B_e & M \end{array}$$

The symbols P_t, K_u, B_o, C, S, T, K_o,P_l,P_a,B_e, and M are consecutively the symbols for Pitopang, Kutianyia, Bodi, Caniago, Sambilan, Tanjung, Koto, Piliang, Payobada, Bendang and Mandahiliang tribes. The tribes in each rows represents a man (husband) whereas the tribes in each columns represents a woman (wife) tribes. For example, entry 1 in the first row states that a man with Pitopang tribe can marry a woman with Sambilan tribe.

2.2. The tribal relation matrix between mother and child

For Lubuak Batingkok, the relation can be represented by the following matrix C

$$C = \begin{array}{c} P_t \\ K_u \\ B_o \\ C \\ S \\ T \\ K_o \\ P_l \\ P_a \\ B_e \\ M \end{array} \begin{bmatrix} 1 & 0 & 0 & 0 & 0 & 0 & 0 & 0 & 0 & 0 & 0 \\ 0 & 1 & 0 & 0 & 0 & 0 & 0 & 0 & 0 & 0 & 0 \\ 0 & 0 & 1 & 0 & 0 & 0 & 0 & 0 & 0 & 0 & 0 \\ 0 & 0 & 0 & 1 & 0 & 0 & 0 & 0 & 0 & 0 & 0 \\ 0 & 0 & 0 & 0 & 1 & 0 & 0 & 0 & 0 & 0 & 0 \\ 0 & 0 & 0 & 0 & 0 & 1 & 0 & 0 & 0 & 0 & 0 \\ 0 & 0 & 0 & 0 & 0 & 0 & 1 & 0 & 0 & 0 & 0 \\ 0 & 0 & 0 & 0 & 0 & 0 & 0 & 1 & 0 & 0 & 0 \\ 0 & 0 & 0 & 0 & 0 & 0 & 0 & 0 & 1 & 0 & 0 \\ 0 & 0 & 0 & 0 & 0 & 0 & 0 & 0 & 0 & 1 & 0 \\ 0 & 0 & 0 & 0 & 0 & 0 & 0 & 0 & 0 & 0 & 1 \end{bmatrix}$$
$$\begin{array}{ccccccccccc} P_t & K_u & B_o & C & S & T & K_o & P_l & P_a & B_e & M \end{array}$$

The rows in the matrix C shows the mother tribe whereas the column shows the child tribe. The entry 1 means that there is the tribal relation between mother and child whereas the entry 0 means conversely.

2.3. The tribal relation between someone and his/her sibling

The relation can be expressed by using the matrix B namely,

$$B = \begin{array}{c} P_t \\ K_u \\ B_o \\ C \\ S \\ T \\ K_o \\ P_l \\ P_a \\ B_e \\ M \end{array} \begin{bmatrix} 1 & 0 & 0 & 0 & 0 & 0 & 0 & 0 & 0 & 0 & 0 \\ 0 & 1 & 0 & 0 & 0 & 0 & 0 & 0 & 0 & 0 & 0 \\ 0 & 0 & 1 & 0 & 0 & 0 & 0 & 0 & 0 & 0 & 0 \\ 0 & 0 & 0 & 1 & 0 & 0 & 0 & 0 & 0 & 0 & 0 \\ 0 & 0 & 0 & 0 & 1 & 0 & 0 & 0 & 0 & 0 & 0 \\ 0 & 0 & 0 & 0 & 0 & 1 & 0 & 0 & 0 & 0 & 0 \\ 0 & 0 & 0 & 0 & 0 & 0 & 1 & 0 & 0 & 0 & 0 \\ 0 & 0 & 0 & 0 & 0 & 0 & 0 & 1 & 0 & 0 & 0 \\ 0 & 0 & 0 & 0 & 0 & 0 & 0 & 0 & 1 & 0 & 0 \\ 0 & 0 & 0 & 0 & 0 & 0 & 0 & 0 & 0 & 1 & 0 \\ 0 & 0 & 0 & 0 & 0 & 0 & 0 & 0 & 0 & 0 & 1 \end{bmatrix}$$
$$P_t \; K_u \; B_o \; C \; S \; T \; K_o \; P_l \; P_a \; B_e \; M$$

The tribes in rows states the someone tribe whereas the tribes in column states his/her sibling. Because of the lineage in Minangkabau is based on mother, then the sibling must has same tribe. In effect, the entry 1 must be put in the main diagonal of the matrix B whereas the entry 0 is put in another. The entry 1 also means that there is the tribe relation between someane and his/her sibling. Conversely, The entry 0 means there is no the tribe relation between them.

2.4. Transpose of the three matrices (marriage rule, the tribal relation between mother and child, the tribal relation between someone and his/her sibling) are

$$W' = \begin{array}{c} P_t \\ K_u \\ B_o \\ C \\ S \\ T \\ K_o \\ P_l \\ P_a \\ B_e \\ M \end{array} \begin{bmatrix} 0 & 0 & 0 & 0 & 1 & 0 & 0 & 0 & 0 & 0 & 0 \\ 0 & 0 & 0 & 0 & 0 & 1 & 0 & 0 & 0 & 0 & 0 \\ 0 & 0 & 0 & 0 & 0 & 0 & 1 & 0 & 0 & 0 & 0 \\ 0 & 0 & 0 & 0 & 0 & 0 & 0 & 0 & 0 & 1 & 0 \\ 1 & 0 & 0 & 0 & 0 & 0 & 0 & 0 & 0 & 0 & 0 \\ 0 & 1 & 0 & 0 & 0 & 0 & 0 & 0 & 0 & 0 & 0 \\ 0 & 0 & 0 & 1 & 0 & 0 & 0 & 0 & 0 & 0 & 0 \\ 0 & 0 & 1 & 0 & 0 & 0 & 0 & 0 & 0 & 0 & 0 \\ 0 & 0 & 0 & 0 & 0 & 0 & 0 & 0 & 0 & 0 & 1 \\ 0 & 0 & 0 & 0 & 0 & 0 & 0 & 0 & 1 & 0 & 0 \\ 0 & 0 & 0 & 0 & 0 & 0 & 0 & 1 & 0 & 0 & 0 \end{bmatrix}, \quad C' = \begin{array}{c} P_t \\ K_u \\ B_o \\ C \\ S \\ T \\ K_o \\ P_l \\ P_a \\ B_e \\ M \end{array} \begin{bmatrix} 1 & 0 & 0 & 0 & 0 & 0 & 0 & 0 & 0 & 0 & 0 \\ 0 & 1 & 0 & 0 & 0 & 0 & 0 & 0 & 0 & 0 & 0 \\ 0 & 0 & 1 & 0 & 0 & 0 & 0 & 0 & 0 & 0 & 0 \\ 0 & 0 & 0 & 1 & 0 & 0 & 0 & 0 & 0 & 0 & 0 \\ 0 & 0 & 0 & 0 & 1 & 0 & 0 & 0 & 0 & 0 & 0 \\ 0 & 0 & 0 & 0 & 0 & 1 & 0 & 0 & 0 & 0 & 0 \\ 0 & 0 & 0 & 0 & 0 & 0 & 1 & 0 & 0 & 0 & 0 \\ 0 & 0 & 0 & 0 & 0 & 0 & 0 & 1 & 0 & 0 & 0 \\ 0 & 0 & 0 & 0 & 0 & 0 & 0 & 0 & 1 & 0 & 0 \\ 0 & 0 & 0 & 0 & 0 & 0 & 0 & 0 & 0 & 1 & 0 \\ 0 & 0 & 0 & 0 & 0 & 0 & 0 & 0 & 0 & 0 & 1 \end{bmatrix},$$
$$P_t \; K_u \; B_o \; C \; S \; T \; K_o \; P_l \; P_a \; B_e \; M \qquad P_t \; K_u \; B_o \; C \; S \; T \; K_o \; P_l \; P_a \; B_e \; M$$

$$B = \begin{array}{c} P_t \\ K_u \\ B_o \\ C \\ S \\ T \\ K_o \\ P_l \\ P_a \\ B_e \\ M \end{array} \begin{bmatrix} 1 & 0 & 0 & 0 & 0 & 0 & 0 & 0 & 0 & 0 & 0 \\ 0 & 1 & 0 & 0 & 0 & 0 & 0 & 0 & 0 & 0 & 0 \\ 0 & 0 & 1 & 0 & 0 & 0 & 0 & 0 & 0 & 0 & 0 \\ 0 & 0 & 0 & 1 & 0 & 0 & 0 & 0 & 0 & 0 & 0 \\ 0 & 0 & 0 & 0 & 1 & 0 & 0 & 0 & 0 & 0 & 0 \\ 0 & 0 & 0 & 0 & 0 & 1 & 0 & 0 & 0 & 0 & 0 \\ 0 & 0 & 0 & 0 & 0 & 0 & 1 & 0 & 0 & 0 & 0 \\ 0 & 0 & 0 & 0 & 0 & 0 & 0 & 1 & 0 & 0 & 0 \\ 0 & 0 & 0 & 0 & 0 & 0 & 0 & 0 & 1 & 0 & 0 \\ 0 & 0 & 0 & 0 & 0 & 0 & 0 & 0 & 0 & 1 & 0 \\ 0 & 0 & 0 & 0 & 0 & 0 & 0 & 0 & 0 & 0 & 1 \end{bmatrix}$$
$$P_t \; K_u \; B_o \; C \; S \; T \; K_o \; P_l \; P_a \; B_e \; M$$

3. Results and discussion

Some tribal algebra models from some family members in Lubuak Batingkok are as follows:

3.1. Model Denai's tribe

Suppose that Denai's tribe is defined as Bodi. Then, by using row matrix the Denai's tribe can be algebraically expressed by

$$A = \begin{bmatrix} 0 & 0 & 1 & 0 & 0 & 0 & 0 & 0 & 0 & 0 & 0 \end{bmatrix}$$
$$\quad P_t \; K_u \; B_o \; C \; S \; T \; K_o \; P_l \; P_a \; B_e \; M$$

Thus, the model is A.

3.2. The tribal model of Denai's mother

The term of mother child tribal relation refers to the relation between mother and child tribes. Since the system is matrilineal, then the mother's tribe is the same as the child's tribe. The maternal model can be obtained by multiplying the matrices of the child's tribe and mother-child tribal relation as follows,

$$\begin{bmatrix} 0 & 0 & 1 & 0 & 0 & 0 & 0 & 0 & 0 & 0 & 0 \end{bmatrix} \begin{array}{c} P_t \\ K_u \\ B_o \\ C \\ S \\ T \\ K_o \\ P_l \\ P_a \\ B_e \\ M \end{array} \begin{bmatrix} 1 & 0 & 0 & 0 & 0 & 0 & 0 & 0 & 0 & 0 & 0 \\ 0 & 1 & 0 & 0 & 0 & 0 & 0 & 0 & 0 & 0 & 0 \\ 0 & 0 & 1 & 0 & 0 & 0 & 0 & 0 & 0 & 0 & 0 \\ 0 & 0 & 0 & 1 & 0 & 0 & 0 & 0 & 0 & 0 & 0 \\ 0 & 0 & 0 & 0 & 1 & 0 & 0 & 0 & 0 & 0 & 0 \\ 0 & 0 & 0 & 0 & 0 & 1 & 0 & 0 & 0 & 0 & 0 \\ 0 & 0 & 0 & 0 & 0 & 0 & 1 & 0 & 0 & 0 & 0 \\ 0 & 0 & 0 & 0 & 0 & 0 & 0 & 1 & 0 & 0 & 0 \\ 0 & 0 & 0 & 0 & 0 & 0 & 0 & 0 & 1 & 0 & 0 \\ 0 & 0 & 0 & 0 & 0 & 0 & 0 & 0 & 0 & 1 & 0 \\ 0 & 0 & 0 & 0 & 0 & 0 & 0 & 0 & 0 & 0 & 1 \end{bmatrix} = AC'$$
$$P_t \; K_u \; B_o \; C \; S \; T \; K_o \; P_l \; P_a \; B_e \; M$$

$$\begin{bmatrix} 0 & 0 & 1 & 0 & 0 & 0 & 0 & 0 & 0 & 0 & 0 \end{bmatrix} = AC'$$
$$P_t \; K_u \; B_o \; C \; S \; T \; K_o \; P_l \; P_a \; B_e \; M$$

Hence, the model is AC' and the tribe is also Bodi.

3.3. The tribal model of Denai's Aunt (Etek)

Etek means the sister of Denai's mother. Her tribe is also Bodi. It can be yielded by the matrix multiplication of Denai's mothertribe and someone-his/her sibling (the relation between someone and his/her sibling) namely,

$$\begin{bmatrix} 0 & 0 & 1 & 0 & 0 & 0 & 0 & 0 & 0 & 0 & 0 \end{bmatrix} \begin{array}{c} P_t \\ K_u \\ B_o \\ C \\ S \\ T \\ K_o \\ P_l \\ P_a \\ B_e \\ M \end{array} \begin{bmatrix} 1 & 0 & 0 & 0 & 0 & 0 & 0 & 0 & 0 & 0 & 0 \\ 0 & 1 & 0 & 0 & 0 & 0 & 0 & 0 & 0 & 0 & 0 \\ 0 & 0 & 1 & 0 & 0 & 0 & 0 & 0 & 0 & 0 & 0 \\ 0 & 0 & 0 & 1 & 0 & 0 & 0 & 0 & 0 & 0 & 0 \\ 0 & 0 & 0 & 0 & 1 & 0 & 0 & 0 & 0 & 0 & 0 \\ 0 & 0 & 0 & 0 & 0 & 1 & 0 & 0 & 0 & 0 & 0 \\ 0 & 0 & 0 & 0 & 0 & 0 & 1 & 0 & 0 & 0 & 0 \\ 0 & 0 & 0 & 0 & 0 & 0 & 0 & 1 & 0 & 0 & 0 \\ 0 & 0 & 0 & 0 & 0 & 0 & 0 & 0 & 1 & 0 & 0 \\ 0 & 0 & 0 & 0 & 0 & 0 & 0 & 0 & 0 & 1 & 0 \\ 0 & 0 & 0 & 0 & 0 & 0 & 0 & 0 & 0 & 0 & 1 \end{bmatrix} = (AC')B$$
$$P_t \; K_u \; B_o \; C \; S \; T \; K_o \; P_l \; P_a \; B_e \; M$$

$$\begin{bmatrix} 0 & 0 & 1 & 0 & 0 & 0 & 0 & 0 & 0 & 0 & 0 \end{bmatrix} = (AC')B$$
$$P_t \; K_u \; B_o \; C \; S \; T \; K_o \; P_l \; P_a \; B_e \; M$$

So, the Etek's tribe is also Bodi.

3.4. The tribal model of Denai's mother brother (Mamak).

Because Mamak is the brother of Denai's mother, the Mamak tribe is the same as the tribe of Denai's mother. The model can be obtained from the matrices multiplication of Denai's mother tribe and the relation of someone and his/her sibling represented by someone-his/her sibling relation,

$$
\begin{bmatrix} 0 & 0 & 1 & 0 & 0 & 0 & 0 & 0 & 0 & 0 & 0 \end{bmatrix}
\begin{array}{c} P_t \\ K_u \\ B_o \\ C \\ S \\ T \\ K_o \\ P_l \\ P_a \\ B_e \\ M \end{array}
\begin{bmatrix}
1 & 0 & 0 & 0 & 0 & 0 & 0 & 0 & 0 & 0 & 0 \\
0 & 1 & 0 & 0 & 0 & 0 & 0 & 0 & 0 & 0 & 0 \\
0 & 0 & 1 & 0 & 0 & 0 & 0 & 0 & 0 & 0 & 0 \\
0 & 0 & 0 & 1 & 0 & 0 & 0 & 0 & 0 & 0 & 0 \\
0 & 0 & 0 & 0 & 1 & 0 & 0 & 0 & 0 & 0 & 0 \\
0 & 0 & 0 & 0 & 0 & 1 & 0 & 0 & 0 & 0 & 0 \\
0 & 0 & 0 & 0 & 0 & 0 & 1 & 0 & 0 & 0 & 0 \\
0 & 0 & 0 & 0 & 0 & 0 & 0 & 1 & 0 & 0 & 0 \\
0 & 0 & 0 & 0 & 0 & 0 & 0 & 0 & 1 & 0 & 0 \\
0 & 0 & 0 & 0 & 0 & 0 & 0 & 0 & 0 & 1 & 0 \\
0 & 0 & 0 & 0 & 0 & 0 & 0 & 0 & 0 & 0 & 1
\end{bmatrix} = (AC')B
$$

$$P_t \; K_u \; B_o \; C \; S \; T \; K_o \; P_l \; P_a \; B_e \; M$$

$$
\begin{bmatrix} 0 & 0 & 1 & 0 & 0 & 0 & 0 & 0 & 0 & 0 & 0 \end{bmatrix} = (AC')B
$$

$$P_t \; K_u \; B_o \; C \; S \; T \; K_o \; P_l \; P_a \; B_e \; M$$

So, his tribe is also Bodi.

3.5. The tribal model for the husband of Etek (called as Pak Etek)

Since Etek is a wife, her husband's tribe can be determined by multiplying Etek's tribe matrix and the relation matrix of wife and husband indicated by wife-husband relation. The relation matrix is the transpose of the marriage rule matrix. Therefore, the tribal model of Pak Etek is

$$
\begin{bmatrix} 0 & 0 & 1 & 0 & 0 & 0 & 0 & 0 & 0 & 0 & 0 \end{bmatrix}
\begin{array}{c} P_t \\ K_u \\ B_o \\ C \\ S \\ T \\ K_o \\ P_l \\ P_a \\ B_e \\ M \end{array}
\begin{bmatrix}
0 & 0 & 0 & 0 & 1 & 0 & 0 & 0 & 0 & 0 & 0 \\
0 & 0 & 0 & 0 & 0 & 1 & 0 & 0 & 0 & 0 & 0 \\
0 & 0 & 0 & 0 & 0 & 0 & 1 & 0 & 0 & 0 & 0 \\
0 & 0 & 0 & 0 & 0 & 0 & 0 & 0 & 0 & 1 & 0 \\
1 & 0 & 0 & 0 & 0 & 0 & 0 & 0 & 0 & 0 & 0 \\
0 & 1 & 0 & 0 & 0 & 0 & 0 & 0 & 0 & 0 & 0 \\
0 & 0 & 0 & 1 & 0 & 0 & 0 & 0 & 0 & 0 & 0 \\
0 & 0 & 1 & 0 & 0 & 0 & 0 & 0 & 0 & 0 & 0 \\
0 & 0 & 0 & 0 & 0 & 0 & 0 & 0 & 0 & 0 & 1 \\
0 & 0 & 0 & 0 & 0 & 0 & 0 & 0 & 1 & 0 & 0 \\
0 & 0 & 0 & 0 & 0 & 0 & 0 & 1 & 0 & 0 & 0
\end{bmatrix} = ((AC')B)W'
$$

$$P_t \; K_u \; B_o \; C \; S \; T \; K_o \; P_l \; P_a \; B_e \; M$$

$$
\begin{bmatrix} 0 & 0 & 0 & 0 & 0 & 0 & 1 & 0 & 0 & 0 & 0 \end{bmatrix} = ((AC')B)W'
$$

$$P_t \; K_u \; B_o \; C \; S \; T \; K_o \; P_l \; P_a \; B_e \; M$$

Then the model is Koto.

3.6. The tribal model of Etek's child

This model is generated by multiplying the Etek's matrix and the relation matrix of mother and child tribe indicated by mother-child tribal relation namely,

$$\begin{bmatrix}0&0&1&0&0&0&0&0&0&0&0\end{bmatrix}\begin{array}{c}P_t\\K_u\\B_o\\C\\S\\T\\K_o\\P_l\\P_a\\B_e\\M\end{array}\begin{bmatrix}1&0&0&0&0&0&0&0&0&0&0\\0&1&0&0&0&0&0&0&0&0&0\\0&0&1&0&0&0&0&0&0&0&0\\0&0&0&1&0&0&0&0&0&0&0\\0&0&0&0&1&0&0&0&0&0&0\\0&0&0&0&0&1&0&0&0&0&0\\0&0&0&0&0&0&1&0&0&0&0\\0&0&0&0&0&0&0&1&0&0&0\\0&0&0&0&0&0&0&0&1&0&0\\0&0&0&0&0&0&0&0&0&1&0\\0&0&0&0&0&0&0&0&0&0&1\end{bmatrix}=((AC')B)C$$

$$\quad P_t\ K_u\ B_o\ C\ S\ T\ K_o\ P_l\ P_a\ B_e\ M$$

$$\begin{bmatrix}0&0&1&0&0&0&0&0&0&0&0\end{bmatrix}=((AC')B)C$$

$$P_t\ K_u\ B_o\ C\ S\ T\ K_o\ P_l\ P_a\ B_e\ M$$

So the Etek's child tribe is Bodi.

3.7. The tribal model for the wife of Mamak (Mintuo)

The mathematical model o f the Mintuo tribe can be determined by multiplying Mamak tribe and the transpose of the marriage rule matrix

$$\begin{bmatrix}0&0&1&0&0&0&0&0&0&0&0\end{bmatrix}\begin{array}{c}P_t\\K_u\\B_o\\C\\S\\T\\K_o\\P_l\\P_a\\B_e\\M\end{array}\begin{bmatrix}0&0&0&0&1&0&0&0&0&0&0\\0&0&0&0&0&1&0&0&0&0&0\\0&0&0&0&0&0&0&1&0&0&0\\0&0&0&0&0&0&1&0&0&0&0\\1&0&0&0&0&0&0&0&0&0&0\\0&1&0&0&0&0&0&0&0&0&0\\0&0&1&0&0&0&0&0&0&0&0\\0&0&0&0&0&0&0&0&0&0&1\\0&0&0&0&0&0&0&0&0&1&0\\0&0&0&1&0&0&0&0&0&0&0\\0&0&0&0&0&0&0&0&1&0&0\end{bmatrix}=((AC'')B)W$$

$$\quad P_t\ K_u\ B_o\ C\ S\ T\ K_o\ P_l\ P_a\ B_e\ M$$

$$\begin{bmatrix}0&0&0&0&0&0&0&1&0&0&0\end{bmatrix}=((AC'')B)W$$

$$P_t\ K_u\ B_o\ C\ S\ T\ K_o\ P_l\ P_a\ B_e\ M$$

Hence, the Mintuo tribe is Piliang.

3.8. Model Denai's father

Since the tribal system in Minangkabau is matrilineal, the matrix of child-his/her father tribal relation is obtained by multiplying the mother tribe and the transpose of marriage rule matrix,

$$
\begin{bmatrix} 0 & 0 & 1 & 0 & 0 & 0 & 0 & 0 & 0 & 0 & 0 \end{bmatrix}
\begin{array}{c} P_t \\ K_u \\ B_o \\ C \\ S \\ T \\ K_o \\ P_l \\ P_a \\ B_e \\ M \end{array}
\begin{bmatrix}
0 & 0 & 0 & 0 & 1 & 0 & 0 & 0 & 0 & 0 & 0 \\
0 & 0 & 0 & 0 & 0 & 1 & 0 & 0 & 0 & 0 & 0 \\
0 & 0 & 0 & 0 & 0 & 0 & 1 & 0 & 0 & 0 & 0 \\
0 & 0 & 0 & 0 & 0 & 0 & 0 & 0 & 0 & 1 & 0 \\
1 & 0 & 0 & 0 & 0 & 0 & 0 & 0 & 0 & 0 & 0 \\
0 & 1 & 0 & 0 & 0 & 0 & 0 & 0 & 0 & 0 & 0 \\
0 & 0 & 0 & 1 & 0 & 0 & 0 & 0 & 0 & 0 & 0 \\
0 & 0 & 1 & 0 & 0 & 0 & 0 & 0 & 0 & 0 & 0 \\
0 & 0 & 0 & 0 & 0 & 0 & 0 & 0 & 0 & 0 & 1 \\
0 & 0 & 0 & 0 & 0 & 0 & 0 & 0 & 1 & 0 & 0 \\
0 & 0 & 0 & 0 & 0 & 0 & 0 & 1 & 0 & 0 & 0
\end{bmatrix}
= (AC'')W''
$$

$$
\quad P_t \; K_u \; B_o \; C \; S \; T \; K_o \; P_l \; P_a \; B_e \; M
$$

$$
\begin{bmatrix} 0 & 0 & 0 & 0 & 0 & 0 & 1 & 0 & 0 & 0 & 0 \end{bmatrix} = (AC')W'
$$

$$
P_t \; K_u \; B_o \; C \; S \; T \; K_o \; P_l \; P_a \; B_e \; M
$$

Thus, the father's tribe is Koto.

4. Summary

Every tribal model of a family member can be constructed by using one of three fundamental matrices. The matrices are the matrices of the matrices of the marriage rule mother-child tribal relation, someone-his/her sibling tribal relation, and the transpose of the last three matrices. Furthermore, according to the matrilineal system, if the model constructed shows mother-child tribal relation or two sibling tribal relation, then the resulting tribe must be the same as the matrilineal system is adhered although the models of both family members are different.

Acknowledgments

This research was supported by DIKTI through the research scheme "Penelitian Unggulan Perguruan Tinggi (PUPT)" in 2017.

References

[1] Bradley, I and Meek, R L and Matrices and Society 1986 *Matrix Algebra and Its Application in The Social Sciences* (Princeton University Press,Princeton,New Jersey, pp 61-78.

[2] Harville D A 2008 *Matrix Algebra From A Statistician's Perspective* (New York: Springer Science Business Media) pp 1-9

[3] Sukmasari F 2014 Perkawinan Adat Minangkabau available at http://adat-budaya-minang.blogspot.co.id/2008/01/2-perkawinan-adat-minangkabau.html

IOP Conf. Series: Journal of Physics: Conf. Series **948** (2018) 012074 doi:10.1088/1742-6596/948/1/012074

Linear systems on balancing chemical reaction problem

R A Kafi[1] and B Abdillah[2]

[1]Department of Mathematics, Universitas Indonesia, Depok, Indonesia
[2]Faculty of Computer Science, Universitas Indonesia, Depok, Indonesia

E-mail: rahmat.alkafi@sci.ui.ac.id

Abstract. The concept of linear systems appears in a variety of applications. This paper presents a small sample of the wide variety of real-world problems regarding our study of linear systems. We show that the problem in balancing chemical reaction can be described by homogeneous linear systems. The solution of the systems is obtained by performing elementary row operations. The obtained solution represents the finding coefficients of chemical reaction. In addition, we present a computational calculation to show that mathematical software such as Matlab can be used to simplify completion of the systems, instead of manually using row operations.

1. Introduction

A chemical reaction is an expression showing a symbolic representation of the reactants and products [1]. The molecules to the left of the arrow are called reactants and those to the right the products. If the number of atoms for each type of element on the reactants is the same as the number of atoms of the corresponding type of the products, then the chemical reaction is said to be balanced [2], otherwise, it is not. The most important part of balancing chemical reaction is to determine the stoichiometric coefficients. There are two methods for balancing chemical reaction: by trial and algebraic [3]. The balancing by trial involves making successive intelligent guesses at making the coefficients that will balance a reaction equal and continue until the reaction attain the balance condition [4]. For simple chemical reactions, this method is very practical. However, for more complicated chemical reactions can take highly refined inspection. In this case, the algebraic is a systematic method that can overcome the looping provided in the trial method and can handle complex chemical reactions.

The algebraic method involves putting unknown coefficients in front of each molecule in the chemical reaction and solving for the unknown [3]. This then followed by write down the balance conditions on each element. There are various approaches that can be used. In this paper, we use systems of linear equations to balance chemical reaction. This proposed approach, we write down the systems in matrix form, obtain a homogeneous system of equations. We then perform elementary row operations on the matrix to obtain the solution of the systems. Furthermore, we present a computational calculation using Matlab to examine the obtained solution.

2. Methodology

2.1. Theoretical Basis

With referring to [2] we introduce the operation on the augmented matrix correspond to the equations in the associated system in order to solve the system. The operations below are called elementary row operations on a matrix:

IOP Conf. Series: Journal of Physics: Conf. Series **948** (2018) 012074 doi:10.1088/1742-6596/948/1/012074

1. Multiply a row through by a nonzero constant
2. Interchange two rows
3. Add constant time one row to another

Moreover, the augmented matrix can be simplified to a form from which the solution of the system can be ascertained by performing certain elementary row operations on the augmented matrix, by following below definition.

Definition 1. An $m \times n$ matrix A is said to be in reduced row echelon form when the following properties are satisfied:

1. If a row does not consist entirely of zeros, then the first non-zero number in the row is a 1. We call this a *leading 1.*
2. If there are any rows that consist entirely of zeros, then they are grouped together at the bottom of the matrix.
3. In any two successive rows that do not consist entirely of zeros, the *leading 1* in lower row occurs farther to the right than the *leading 1* in the higher row.

Each column that contains a *leading 1* has zeros everywhere else in that column.

2.2. Numerical Examples

In this section, the use of linear systems to solve the problem of balancing chemical reaction is discussed. The objective is to find the stoichiometric coefficients that fulfill the reaction. All of the stoichiometric coefficients must be the smallest integer [5].

Example 1. Propane (C3H8) burns in oxygen to form carbon dioxide and steam (propane combustion reaction).

$$\mathrm{C3H8} + \mathrm{O2} \rightarrow \mathrm{CO2} + \mathrm{H2O}$$

Let the stoichiometric coefficients be a, b, c, and d such that

$$a\mathrm{C3H8} + b\mathrm{O2} \rightarrow c\mathrm{CO2} + d\mathrm{H2O}$$

We compare the number of Carbon (C), Hydrogen (H), and Oxygen (O) atoms of the reactants with the number of atoms of the products. We obtain the following set of equations in the table 1.

Table 1. Stoichiometric coefficient.

	Left Side		Right Side
Carbon	$3a$	=	c
Hydrogen	$8a$	=	$2d$
Oxygen	$2b$	=	$2c + d$

From which we obtain the homogeneous linear system

- $$\begin{bmatrix} 3 & 0 & -1 & 0 \\ 8 & 0 & 0 & -2 \\ 0 & 2 & -2 & -1 \end{bmatrix} \begin{bmatrix} a \\ b \\ c \\ d \end{bmatrix} = \begin{bmatrix} 0 \\ 0 \\ 0 \end{bmatrix}$$

with augmented matrix for this system is

- $$\begin{bmatrix} 3 & 0 & -1 & 0 & 0 \\ 8 & 0 & 0 & -2 & 0 \\ 0 & 2 & -2 & -1 & 0 \end{bmatrix}$$

In the first step of elementary row operation, replace row two by three times row two minus eight times row one, i.e. $R_2 \leftrightarrow 3R_2 - 8R_1$ to yield

- $$\begin{bmatrix} 3 & 0 & -1 & 0 & 0 \\ 0 & 0 & 8 & -6 & 0 \\ 0 & 2 & -2 & -1 & 0 \end{bmatrix}$$

Exchange a half row two with row three i.e. $\frac{1}{2}R_2 \leftrightarrow R_3$, yield

- $$\begin{bmatrix} 3 & 0 & -1 & 0 & 0 \\ 0 & 2 & -2 & -1 & 0 \\ 0 & 0 & 4 & -3 & 0 \end{bmatrix}$$

Replace row one by four times row one plus row three $R_1 \leftrightarrow 4R_1 + R_3$ and replace row two by two times row two plus row three $R_2 \leftrightarrow 2R_2 + R_3$ to yield

- $$\begin{bmatrix} 12 & 0 & 0 & -3 & 0 \\ 0 & 4 & 0 & -5 & 0 \\ 0 & 0 & 4 & -3 & 0 \end{bmatrix}$$

Replace row one with a twelfth-row one $R_1 \leftrightarrow \frac{1}{12}R_1$, row two with a fourth row two $R_2 \leftrightarrow \frac{1}{4}R_2$ and row three with a fourth row three $R_3 \leftrightarrow \frac{1}{4}R_3$ to obtain the reduced row echelon form

- $$\begin{bmatrix} 1 & 0 & 0 & -\frac{1}{4} & 0 \\ 0 & 1 & 0 & -\frac{5}{4} & 0 \\ 0 & 0 & 1 & -\frac{3}{4} & 0 \end{bmatrix}$$

from which we conclude that the general solution of the system is

- $a = \frac{1}{4}d,\ \ b = \frac{5}{4}d,\ \ c = \frac{3}{4}d$

where d is arbitrary. Let $d = 4$ so that $a = 1$, $b = 5$ and $c = 3$. Hence, the chemical reaction can be balanced as

$$C_3H_8 + 5O_2 \rightarrow 3CO_2 + 4H_2O$$

Example 2. Consider the following reaction of sugar fermentation.

$$C_6H_{12}O_6 \rightarrow CO_2 + C_2H_5OH$$

Let a, b, and c be the stoichiometric coefficients such that

$$aC_6H_{12}O_6 \rightarrow bCO_2 + cC_2H_5OH$$

As in the previous example, for each of the atoms in the reaction, the number of atoms on the reactant must be equal to the number of atoms on the product. Then we obtain the following set of equations in the table2.

Table 2. Reaction of sugar fermentation.

	Left Side		Right Side
Carbon	$6a$	=	$b + 2c$
Hydrogen	$12a$	=	$6c$
Oxygen	$6a$	=	$2b + c$

From which we obtain the homogeneous linear system

- $$\begin{bmatrix} 6 & -1 & -2 \\ 12 & 0 & -6 \\ 6 & -2 & -1 \end{bmatrix}\begin{bmatrix} a \\ b \\ c \end{bmatrix} = \begin{bmatrix} 0 \\ 0 \\ 0 \end{bmatrix}$$

with augmented matrix for this system is

- $$\begin{bmatrix} 6 & -1 & -2 & 0 \\ 12 & 0 & -6 & 0 \\ 6 & -2 & -1 & 0 \end{bmatrix}$$

The following elementary row operations $R_2 \leftrightarrow R_2 - 2R_1$ and $R_3 \leftrightarrow R_3 - R_1$ reduces the above matrix to

- $$\begin{bmatrix} 6 & -1 & -2 & 0 \\ 0 & 2 & -2 & 0 \\ 0 & -1 & 1 & 0 \end{bmatrix}$$

In the same vein, the following row operations $R_2 \leftrightarrow \frac{1}{2}R_2$ and $R_3 \leftrightarrow 2R_3 + R_2$ reduces the above matrix to

- $$\begin{bmatrix} 6 & -1 & -2 & 0 \\ 0 & 1 & -1 & 0 \\ 0 & 0 & 0 & 0 \end{bmatrix}$$

Finally, $R_1 \leftrightarrow \frac{1}{6}R_1 + \frac{1}{6}R_2$ reduces the matrix to the reduce row echelon form

- $$\begin{bmatrix} 1 & 0 & -\frac{1}{2} & 0 \\ 0 & 1 & -1 & 0 \\ 0 & 0 & 0 & 0 \end{bmatrix}$$

from which we conclude that the general solution of the system is

$$a = \frac{1}{2}c, \quad b = c$$

where c is arbitrary. Let $c = 2$ so that $a = 1$ and $b = 2$. Hence, the chemical reaction can be balanced as

$$C_6H_{12}C_6 \rightarrow 2CO_2 + 2C_2H_5OH$$

Example 3. Consider the following reaction of photosynthesis.

$$CH_3COF + H_2O \rightarrow CH_3COOH + HF$$

Let a, b, c and d be the stoichiometric coefficients such that

$$aCH_3COF + bH_2O \rightarrow cCH_3COOH + dHF$$

For each of the atoms in the reaction, the number of atoms on the reactant must be equal to the number of atoms on the product. Then we obtain the following set of equations in the table 3.

Table 3. Reaction of photosynthesis.

	Left Side		Right Side
Carbon	$2a$	=	$2c$
Hydrogen	$3a + 2b$	=	$4c + d$
Oxygen	$a + b$	=	$2c$
Fluor	a	=	d

From which we obtain the homogeneous linear system

- $$\begin{bmatrix} 2 & 0 & -2 & 0 \\ 3 & 2 & -4 & -1 \\ 1 & 1 & -2 & 0 \\ 1 & 0 & 0 & -1 \end{bmatrix} \begin{bmatrix} a \\ b \\ c \\ d \end{bmatrix} = \begin{bmatrix} 0 \\ 0 \\ 0 \\ 0 \end{bmatrix}$$

with augmented matrix for this system is

- $\begin{bmatrix} 2 & 0 & -2 & 0 & 0 \\ 3 & 2 & -4 & -1 & 0 \\ 1 & 1 & -2 & 0 & 0 \\ 1 & 0 & 0 & -1 & 0 \end{bmatrix}$

The following elementary row operations $R_1 \leftrightarrow R_1 - 2R_4$, $R_2 \leftrightarrow R_2 - 3R_4$ and $R_3 \leftrightarrow R_3 - R_4$ reduces the above matrix to

- $\begin{bmatrix} 0 & 0 & -2 & 2 & 0 \\ 0 & 2 & -4 & 2 & 0 \\ 0 & 1 & -2 & 1 & 0 \\ 1 & 0 & 0 & -1 & 0 \end{bmatrix}$

In the same vein, the following row operations $-\frac{1}{2}R_1 \leftrightarrow R_4$ and $R_2 \leftrightarrow R_2 - 2R_3$ reduces the above matrix to

- $\begin{bmatrix} 1 & 0 & 0 & -1 & 0 \\ 0 & 0 & 0 & 0 & 0 \\ 0 & 1 & -2 & 1 & 0 \\ 0 & 0 & 1 & -1 & 0 \end{bmatrix}$

The following elementary row operations $R_3 \leftrightarrow R_2$ and $R_3 \leftrightarrow R_3 + 2R_4$ reduces the above matrix to

- $\begin{bmatrix} 1 & 0 & 0 & -1 & 0 \\ 0 & 1 & 0 & -1 & 0 \\ 0 & 0 & 0 & 0 & 0 \\ 0 & 0 & 1 & -1 & 0 \end{bmatrix}$

Finally, $R_4 \leftrightarrow R_3$ reduces the matrix to the reduce row echelon form

- $\begin{bmatrix} 1 & 0 & 0 & -1 & 0 \\ 0 & 1 & 0 & -1 & 0 \\ 0 & 0 & 1 & -1 & 0 \\ 0 & 0 & 0 & 0 & 0 \end{bmatrix}$

from which we conclude that the general solution of the system is

$$a = d, \quad b = d, \quad c = d$$

where d is arbitrary. Let $d = 1$ so that $a = 1, b = 1$ and $c = 1$. Hence, the chemical reaction can be balanced as

$$CH_3COF + H_2O \rightarrow CH_3COOH + HF$$

2.3. Computational Results

In this section, we use Matlab to solve linear systems correspond to the chemical reactions given in the previous section. Remark that we constructed our Matlab program.

Example 4. Solving linear system given in example 1.

Type the augmented matrix [3 0 -1 0 0; 8 0 0 -2 0; 0 2 -2 -1 0] on the terminal. If we run the program, then this gives the same reduced row echelon matrix as in example 1 (see figure 1).

```
1.0000         0         0   -0.2500         0
     0    1.0000         0   -1.2500         0
     0         0    1.0000   -0.7500         0
```

Figure 1. Reduced row echelon matrix on example 1 using Matlab.

Hence, we obtain the solution as given in figure 2.

```
ans =

     1     5     3     4
```

Figure 2. The solution of the system of example 1 using Matlab.

Example 5. Solving linear system given in example 2.
Type the augmented matrix [6 -1 -2 0; 12 0 -6 0; 6 -2 -1 0] on the terminal. If we run the program, then this gives the same reduced row echelon matrix as in example 2 (see figure 3).

```
1.0000         0   -0.5000         0
     0    1.0000   -1.0000         0
     0         0         0         0
```

Figure 3. The solution of the system of example 2 with Matlab.

Example 6. Solving linear system given in example 3.
Type the augmented matrix [2 0 -2 0 0; 3 2 -4 -1 0; 1 1 -2 0 0; 1 0 0 -1 0] on the terminal. If we run the program, then this gives the same reduced row echelon matrix as in example 3 (see figure 5).

```
1     0     0    -1     0
0     1     0    -1     0
0     0     1    -1     0
0     0     0     0     0
```

Figure 4. The solution of the system of example 3 using Matlab.

3. Conclusions

In this paper, we have shown how to balance chemical reactions using linear systems. In all the examples presented, the linear systems aresuitable approach to balance chemical reactions problem. The elementary row operations that are applied to augmented matrix correspond to the equations in the associated system are easy to perform. The correctness of our manual calculation on each given examples have been confirmed by our Matlab program and it shows an excellent agreement.

References

[1] Risteski I B 2009 *Chin. Chem. Soc.* **56** 65-79
[2] Anton H and Rorres C 2013 *Elementary Linear Algebra: Application version 11th Edition* (New Jersey: John Wiley & Sons, Inc.)
[3] Tuckerman M E 2011 available at http://www.nyu.edu/classes/tuckerman/adv. chem/lectures/ lecture_2/node3.html
[4] Gabriel C I and Onwuka G I 2015 *Nat. Sci. Resea.* **3** 29-36
[5] Hayes R E and Mmbaga J P 2012 *Introduction to Chemical Reactor Analysis 2nd Edition* (Florida: CRC Press)

IOP Conf. Series: Journal of Physics: Conf. Series **948** (2018) 012075 doi:10.1088/1742-6596/948/1/012075

Application of multi response optimization with grey relational analysis and fuzzy logic method

Sri Winarni[1], and Sapto Wahyu Indratno[2]

[1]Faculty of Mathematics and Natural Sciences, Padjadjaran University
[2]Faculty of Mathematics and Natural Sciences, Bandung Institute of Technology

Email: sri.winarni@unpad.ac.id

Abstract. Multi-response optimization is an optimization process by considering multiple responses simultaneously. The purpose of this research is to get the optimum point on multi-response optimization process using grey relational analysis and fuzzy logic method. The optimum point is determined from the Fuzzy-GRG (Grey Relational Grade) variable which is the conversion of the Signal to Noise Ratio of the responses involved. The case study used in this research are case optimization of electrical process parameters in electrical disharge machining. It was found that the combination of treatments resulting to optimum MRR and SR was a 70 V gap voltage factor, peak current 9 A and duty factor 0.8.

1. Introduction

Optimization of quality is an effort to get factors that produce optimum quality. There are two optimization approaches, namely single response optimization and multi-response optimization. The single response optimization approach is used if the quality of the product generated only considers a single product characteristic. When product quality not only considers a product characteristic, the optimization approach used is multi-response optimization. On the single response optimization approach, the optimum point obtained is not necessarily the optimum point simultaneously, [1]. The quality of a product often not only considers a single quality characteristic, but some characteristics are considered simultaneously. The optimization process by considering more than one response (product characteristics) is called multi responses optimization. It can be illustrated in Fig. 1.

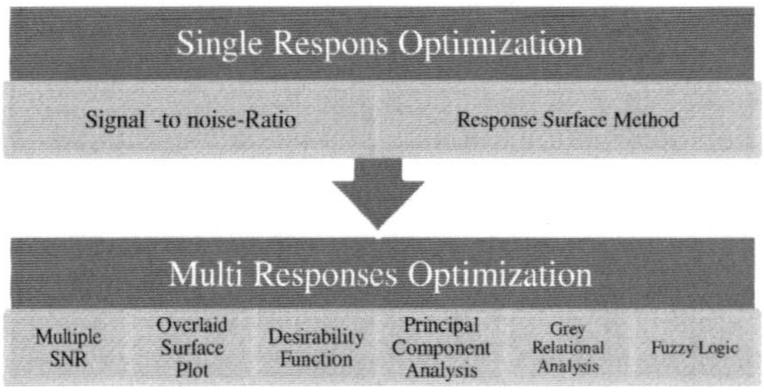

Figure 1. Single and multi responses optimization

In the scheme of Figure 1, The method analysis that be used in single response approach are the Signal to Noise Ratio (SNR) and response surface method. The analysis on the SNR method is relatively easy, so this method is the most commonly used for single response optimization. In the multi responses optimization approach there are several analytical methods used. Multiple SNR is a development of the SNR method. Overlaid surface plot and desirability function method is the development of response surface method. Other methods that can be used in multi-response optimization are principal component analysis, grey relational analysis, and fuzzy logic.

Fuzzy Logic and grey relational analysis is a decision-making technique used to optimize multirespon problem. This research will be carried out the development of multi-response optimization analysis using grey relational analysis and fuzzy logic method [3]. The purpose of this research is to get the optimum point in the case of taguchi design multi responses optimization using grey relational analysis, and fuzzy logic method.

2. Methods and material

The case studies used in this research were cases of optimization on Electrical Discharge Machining (EDM), "Taguchi-grey relational based multi response optimization of electrical process parameters in electrical discharge machining [4] In this study grey relational analysis will be combined with fuzzy logic method into fuzzy-grey relational analysis.

2.1. The Case Study

Electrical Discharge Machining (EDM) is a thermal erosion process, in which the material removal from the work piece is obtained by high thermal energy due to the ionization of dielectric medium between tool and electrode. Figure 2 show the set up of electrical parameters are mainly contributed factor to produce thermal energy, its needed to find optimal combination of electrical parameters on machinability.

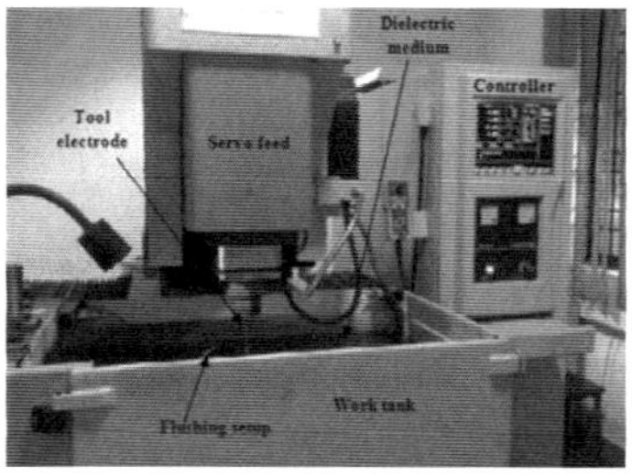

Factors	Level 1	Level 2	Level 3
A : Gap Voltage (V)	40	60	70
B : Peak current (A)	9	12	15
C : Duty factor	0.4	0.6	0.8

Respon :
Material Removal Rate (mm^3/min)
Surface Roughness (mm)

Figure 2. Case of EDM optimization

Since the discharge energy is mainly determined by gap voltage, peak current and duty factor, these factors have been taken as the input parameters. Duty factor is defined as ratio of pulse duration to total pulse cycle. The process variables used in the present study 40 V, 60 V and 70 V have been considered as gap voltage variables. 0.4, 0.6 and 0.8 have been chosen as duty factor with the discharge current value of 9 A, 12 A, and 15 A.

Engine performance is measured by response of Material Removal Rate (MRR) and Surface Roughness (SR). Both of these responses will be considered simultaneously to obtain the treatment composition which yields the optimum MRR and SR. The experimental design used in this case is the taguchi L9 design. Experimental data coding are given in Table 1.

Table 1. Data experiment for case of EDM optimization

Ru	Factors			Responses	
n	A	B	C	MRR	SR
1	1	1	1	4.74	3.26
2	1	2	2	8.60	6.98
3	1	3	3	9.51	11.78
4	2	1	2	7.46	4.58

Run	Factors			Responses	
	A	B	C	MRR	SR
5	2	2	3	12.8	9.62
6	2	3	1	8.38	6.99
7	3	1	3	15.51	6.53
8	3	2	1	10.90	4.82
9	3	3	2	13.36	8.98

2.2. Fuzzy-Grey Relational Analysis

The method of analysis used in this research is fuzzy-grey relational analysis method. The fuzzy-grey relational analysis method is a combination of fuzzy logic and grey relational analysis methods. Stages of analysis begins with the orthogonal array design that is the input of fuzzy logic. The analysis step is given in Fig. 3.

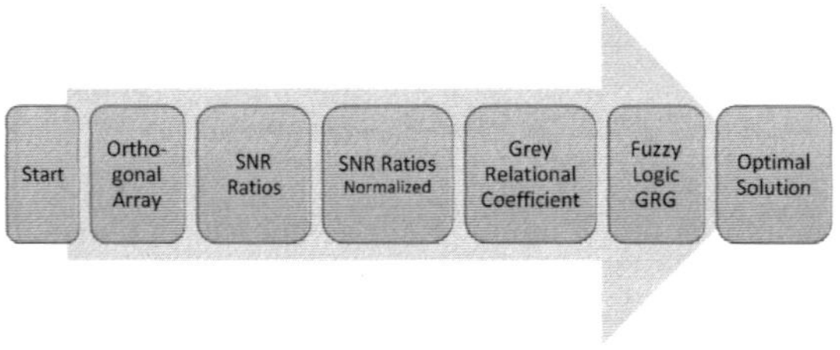

Figure 3. Stages of fuzzy-grey relational analysis method

The fuzzy logic GRG (Grey Relational Grade) analysis stage consists of three stages: fuzzifier, fuzzy inference engine, and defuzzifier. The output of fuzzy logic GRG analysis is Fuzzy-GRG. The optimum solution is determined from the influence of the factor with the highest Fuzzy-GRG value. The complete analysis phase can be explained as follows, [3]:

(a) Step 1: Calculation of S/N Ratios

There are three types of S/N ratio: larger-the-better, smaller-the-better and nominal the best. SNR larger-the-better is used when the goal of response optimization is maximize response. The material removal rate is a larger-the-better performance characteristic, since the maximization of the quality characteristic of interest is sought and can be expressed as Equation 1, [5].

$$\eta_{ij} = -10log\left(\frac{1}{n}\sum_{j=1}^{n} y_{ij}^2\right) \quad (1)$$

Where n = number of replicas, y_{ij} = observed value, and $i = 1,2,\dots n$, $j = 1,2,\dots k$

SNR smaller-the-better is used when the goal of response optimization is minimize response. The surface roughness is the smaller-the-better performance characteristic and the loss function for the same can be expressed as Equation 2.

$$\eta_{ij} = -10log\left(\frac{1}{n}\sum_{j=1}^{n} \frac{1}{y_{ij}^2}\right) \quad (2)$$

SNR nominal-the-best is used when the goal of response optimization is target response. In the case study used in this research there was no response with nominal-the-best purposes.

(b) Step 2: Normalization of S/N Ratios

The beginning step in the GRA is normalization of the S/N ratio, which is performed to prepare raw data for the analysis where the original sequence is transferred to a comparable sequence. Linear normalization is usually demanded since the range and unit in one data sequence may differ from the others. A linear normalization of the S/N ratio in the range between zero and unity is also called as the grey relational generation, [3]. Formula to normalize SNR with greater the better criterion given in Equation 3.

$$\eta_{ij}^{*} = \frac{\eta_{ij} - \min \eta_{ij}}{\max\eta_{ij} - \min \eta_{ij}} \tag{3}$$

With η_{ij}^{*} is the normalized value of the SNR of the *i*-th response and the *j*-th treatment. min η_{ij} is the minimum value of η_{ij} and max η_{ij} is the maximum value of η_{ij}. SNR normalization with smaller the better criteria is given in the Equation 4.

$$\eta_{ij}^{*} = \frac{\max \eta_{ij} - \eta_{ij}}{\max\eta_{ij} - \min \eta_{ij}} \tag{4}$$

Normalized SNR values range from 0 to 1. Value 1 indicates that the response is close to the target.

Step 3: Determination of Grey Relational Coefficient (GRC)

After the SNR normalization process, the next step is to determine the GRC value. GRC for all the sequences expresses the relationship between the ideal (best) and actual normalized S/N ratio. If the two sequences agree at all points, then their GRC is 1, [3]. The formulation for GRC is given in the Equation 5.

$$\gamma(x_{0j}, x_{ij}) = \frac{\Delta Min + \zeta \Delta Max}{\Delta_{ij} + \zeta \Delta Max} \tag{5}$$

$$\text{with } i = 1,2,\dots,m \; j = 1,2,\dots n$$

$$\Delta Min = Min\{\Delta_{ij}, i = 1,2,\dots,m; j = 1,2,\dots,n\}$$

$$\Delta Max = Max\{\Delta_{ij}, i = 1,2,\dots,m; j = 1,2,\dots,n\}$$

$\Delta_{ij} = x_{0j} - x_{ij}$ is an absolute difference between values $x_0(j)$ and $x_i(j)$. ζ is the identification coefficient and is used to offset the influence Δ_{max} when it has a value too large. In this case the value is taken $\zeta = 0,5$.

Step 4: Determination of Fuzzy-Grey Relational Grade (Fuzzy-GRG)

Fuzzy logic GRG analysis is divided into fuzzifier, fuzzy engine inference, defuzzifier. At the fuzzifier stage, the conversion of linguistic variables into fuzzy variables. in the fuzzy engine inference, fuzzy rules are established based on the "IF and THEN" rules. For two input variables and one output variable, the fuzzy rule can be formed as follows. [6]:

Rule 1 : if x_1 is A_1 and x_2 is B_2 then y is C_1 else,
Rule 2 : if x_1 is A_2 and x_2 is B_2 then y is C_2 else,
Rule 3 : if x_1 is A_3 and x_2 is B_3 then y is C_3 else,

.

.

.

Rule n : if x_1 is A_n and x_2 is B_n then y is C_n else.

The variables Ai, Bi, and Ci are fuzzy subsets that connect the MRR normalization input variables, SR normalization and Fuzzy-GRG output variables. In this research input variables are formed in three subset, namely low (L), medium (M), and Hight (H). While the output variables are formed in five subsets, namely very small (VS), small (S), medium (M), large (L) and very large (VL). With fuzzy sets of input and output variables can be formed fuzzy rule given in Table 2.

Table 2. Fuzzy Rule

SR Normalized	MRR Normalized: Low	Medium	Hight
Low	Very Small	Small	Medium
Medium	Small	Medium	Large
Hight	Medium	Large	Very Large

If normalization of MRR low and normalization of SR low then fuzzy-GRG very small, if normalization of MRR medium and normalization of SR low then fuzzy-GRG small, and so on. This rule can be simulated with mamdani operators as follows, [7]:

$$\mu_{C_0}(y) = \left(\mu_{A_1}(x_1) \wedge \mu_{B_1}(x_2)\right) \vee \left(\mu_{A_2}(x_1) \wedge \mu_{B_2}(x_2)\right) \vee \ldots \vee \left(\mu_{A_n}(x_1) \wedge \mu_{B_n}(x_2)\right) \quad (6)$$

Next is the defuzzifier stage, which transforms the fuzzy output variable into a non fuzzy variable. The transformation uses Equation 7.

$$y_0 = \frac{\sum y \mu_{C_0}(y)}{\sum \mu_{C_0}(y)} \quad (7)$$

The fuzzy-GRG variable is used to determine the optimum point.

Step 5: Determination of The Optimum Point

The optimum point is determined from the influence of the largest fuzzy-GRG main factor. Determination of the influence of the main factors using the main influence plot, [8]..

3. Results and discussion

Fuzzy-grey relational analysis is done with the initial phase of data preparation. The response value is converted to the SNR value which is then normalized. Through the grey relational analysis process, the normalized SNR are converted to GRC value. The obtained GRC values will be used as inputs to the fuzzy process and produce the MPCI output used to determine the optimum solution.

3.1. *Results Determination of SNR, SNR Normalized and GRC Values*

In the case study of EDM optimization considers MRR response with larger the better criteria and SR response with smaller the better criteria. The value of SNR of response, its normalization, and GRC results are given in Table 3.

TABLE 3. SNR, Normalized of SNR, GRC for MRR and SR Response

Run	Response		SNR		Normalized of SNR		GRC	
	MRR	SR	MRR	SR	MRR	SR	MRR	SR
1	4.74	3.26	13.516	-10.264	0.000	1.000	0.333	1.000
2	8.60	6.98	18.690	-16.877	0.502	0.408	0.501	0.458
3	9.51	11.78	19.564	-21.423	0.587	0.000	0.548	0.333
4	7.46	4.58	17.455	-13.217	0.383	0.736	0.448	0.654
5	12.80	9.62	22.144	-19.663	0.838	0.159	0.755	0.373
6	8.38	6.99	18.469	-16.889	0.481	0.407	0.491	0.457
7	15.51	6.53	23.812	-16.298	1.000	0.460	1.000	0.481

Run	Response		SNR		Normalized of SNR		GRC	
	MRR	SR	MRR	SR	MRR	SR	MRR	SR
8	10.90	4.82	20.748	-13.661	0.702	0.696	0.627	0.622
9	13.36	8.98	22.516	-19.065	0.874	0.213	0.799	0.389

SNR value of MRR and SR is obtained from Equation 1 and Equation 2. Normalization of SNR is done by Equation 3 and 4. Normalization value is in the interval value 0 to 1. The normalized SNR close to 1 indicates that the response value is close to the target. In MRR with larger the better criteria, a large response value will result the SNR normalized close to 1. As for the SR response with smaller the better criteria, the normalized SNR that are worth close to 1 are for small value responses. In contrast to the SNR normalized close to 0 are for MRR of small value and SR of great value. Determination of GRC value is done by first determining the value of Δ_{ij} which is the absolute difference between the value of $x_0(j)$ with $x_i(j)$. Value $x_0(j)$ is the largest normalized SNR value in each response. The GRC value is determined by Equation 5.

3.2. Results Analysis of Fuzzy-Grey Relational Analysis

In this research, fuzzy-grey relational analysis is done with matlab software. The initial stage is defining the input and output variables given in Fig. 4.

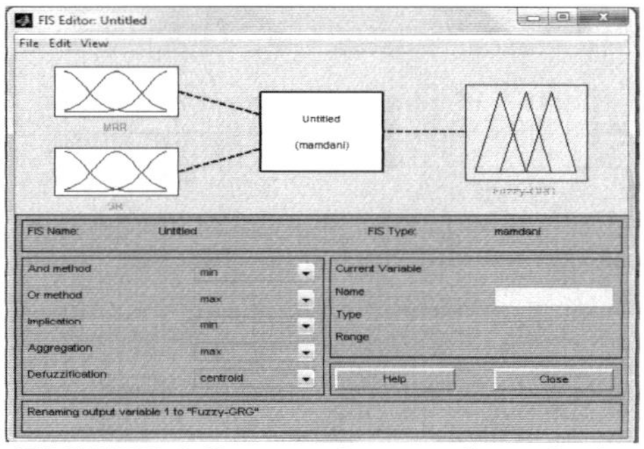

FIGURE 4. Input and output fuzzy logic

The input variable used is GRC value for MRR and SR from Table 3, while its output is fuzzy-GRG. The fuzzy variable definition is given in Fig. 5.

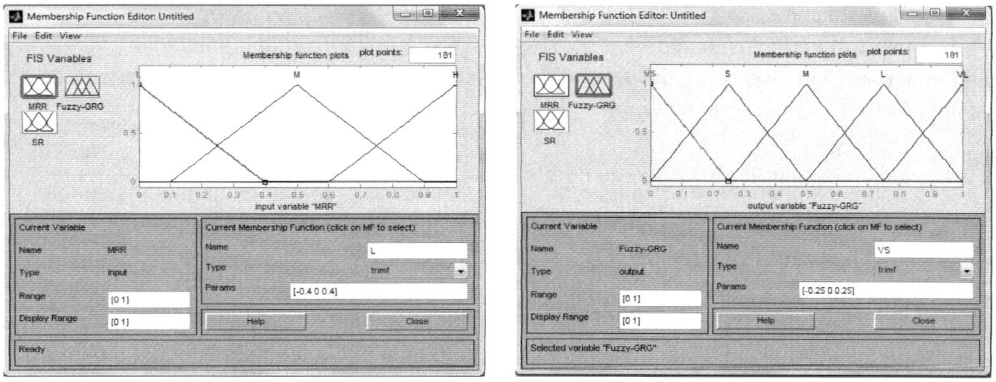

(a) (b)

FIGURE 5. Fuzzy variable for input (a) and output (b)

Fuzzy variables for GRC of MRR and GRC of SR are divided into three subsets; Low (L), Medium (M) and Hight (H), figure 5(a). As for the fuzzy variables MPCI output is divided into five subsets; Very Small (VS), Small (S), Medium (M), Large (L) and Very Large (VL), figure 5(b). Fuzzy rule at the fuzzy engine inference stage and the defuzzifier stage is given in Fig. 6.

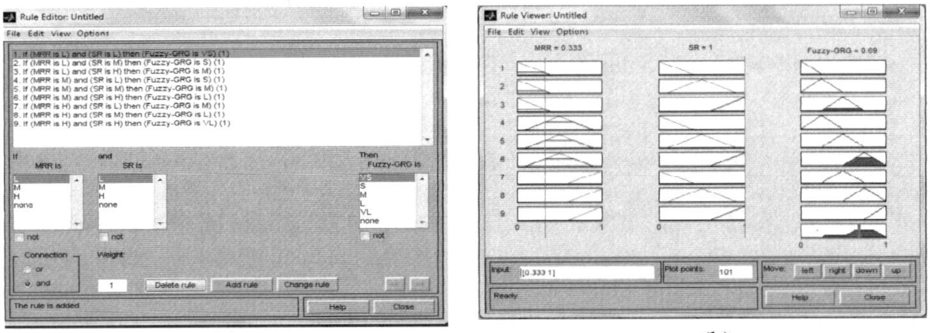

(a) (b)

FIGURE 6. Fuzzy rule (a) and defuzzifier stage (b)

At the fuzzy engine inference stage there are nine fuzzy rules that are formed (Equation 6). The rule transforms the input fuzzy variable into fuzzy output. Next is the defuzzifier stage, transforming fuzzy output into fuzzy-GRG output (Equation 7). In Figure 6 if the GRC of MRR is 0.333 and GRC of SR is 1 then fuzzy-GRG will be 0.690. fuzzy-GRG results are given in Table 4.

TABLE 4. Output results of fuzzy-GRG

Factors			**Fuzzy-GRG**
A	**B**	**C**	
1	1	1	0,690
1	2	2	0,500
1	3	3	0,440
2	1	2	0,549
2	2	3	0,597
2	3	1	0,500
3	1	3	0,750
3	2	1	0,527
3	3	2	0,649

Fuzzy-GRG value close to 1 indicates that the response value is close to the target value, otherwise fuzzy-GRG value close to 0 indicates that the response value is far from its target value. Fuzzy-GRG is a conversion variable of some of the responses considered in the optimization process. Furthermore fuzzy-GRG is used to determine the optimum point.

3.3. Result of Determination of Optimum Solution

The optimum solution is determined from the influence of the factors of each experimental factor. The determination of the optimum solution visually done with the main effect plot given in Fig. 7.

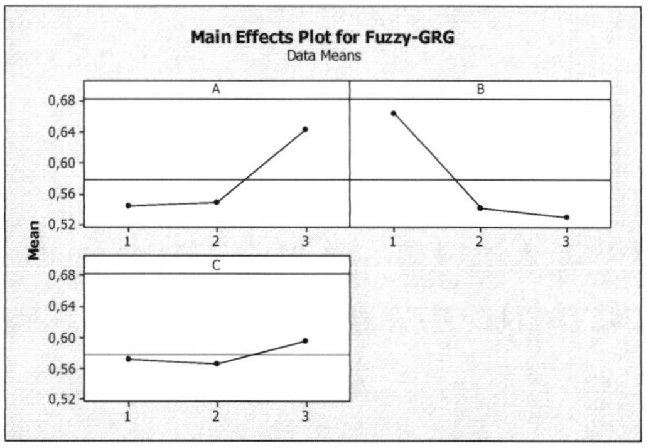

FIGURE 7. Main effect plot for fuzzy-GRG

It is found that for factor A the greatest influence of fuzzy-GRG is given by level 3. In factor B, the greatest influence of fuzzy-GRG is given by level 1. Whereas in the C factor, the largest effect of fuzzy-GRG is given by level 3. Thus the optimum solution is A3B1C3. This means that the optimum MRR and SR can be generated by the treatment of A3B1C3.

4. Conclusion

The conclusion of this research is that fuzzy-grey relational analysis method can be used as an alternative method on multi-response optimization. In the case of EDM optimization used in this study it was found that the combination of treatments resulting in optimum MRR and SR was a 70 V gap voltage factor, peak current 9 A and duty factor 0.8.

References

[1] D C Montgomery 2013 *Design and Analysis of Experiments* (Internatio: John Wiley & Sons)
[2] P V Ramaiah, N Rajesh, and K. D. Reddy 2013 *Determination of Optimum Influential Parameters in Turning of Al6061 Using Fuzzy Logic* **2** 10 pp 5555–5560
[3] G Rajyalakshmi and P V Ramaiah 2015 *Indian J. Sci. Technol*
[4] T Muthuramalingam and B Mohan 2013 *Indian J. Eng. Mater. Sci.*
[5] B O P Soepangkat, A Soesanti, and B Pramujati The Use of Taguchi-Grey-Fuzzy to Optimize Performance Characteristics in Turning of AISI D2
[6] N Senthilkumar and Sudha A grey-fuzzy approach for optimizing machining parameters and the approach angle in turning AISI 1045 steel
[7] B Das, S Roy, R N Rai, and S C Saha 2104 *J. Eng. Sci. Technol. Rev.* **7** 217–222
[8] K. Anand Babu and G. Vijaya Kumar 2015 *Int. J. Eng. Technol.* **7** 1200–1211

Ethnomathematics: The use of multiple linier regression $Y = b_1X_1 + b_2X_2 + e$ in traditional house construction Saka Roras in Songan Village

J B Darmayasa [1], Wahyudin[1] and T Mulyana[1]

[1]Departement of Mathematics Education, Universitas Pendidikan Indonesia
Jl. Dr. Setiabudhi No. 229, Bandung 40154, Indonesia

E-mail: jeromath@student.upi.edu

Abstract: Ethnomathematics may be the connecting bridge between culture and technology and arts. Therefore, the exploration of mathematics values that intersects with cultural anthropology should be significantly conducted. One case containing such issue is the construction of Traditional House of Saka Roras in Bali. Thus, this research aimed to explore the mathematic concept adopted in the construction of such traditional Bale (house) located in Songan Village, Kintamani, Bali. Specifically, this research also aimed to investigate the selection of linear regression coefficient for the saka (pillar) in the Bale. This research applied Embedded Mix-Method Design. Meanwhile, the data collection was conducted by interview, observation and measurement of pillars of 32 Bale Saka Roras. The result of this research revealed that the connection between the width and height of pillars was stated in the formula $Y = 26{,}3 + 18{,}2X$, where X acted as stimulus variable. The coefficient value amounted to 18.2 showed that most preceding architects in Songan Village were more likely to use 19 as the coefficient towards the pillar width than the other coefficients such as 17, 20 and 21 as mentioned in book/palm-leaf manuscript entitled Kosala-Kosali. The last but not least, the researchers also figured out that the pillar width depended on the length of the house-owner candidate's index finger.

1. Introduction

Indonesia is an archipelagic country that becomes home to various tribes. Such variety is apparently followed by cultural diversity. However, the influence of globalization causes several regions' turmoil due to the confusion of keeping up with the world's development or preserving the inherited culture. Thus, today's youth generation may have to face the challenge of how to combine the inherited culture with technology in the purpose of maintaining self-existence in society.

Similar challenge is apparently faced by the Balinese society's successive generation. Balinese society surely understand that tourist attractiveness is not entirely located in natural beauty, since culture and customs are also the integral part of it. As the consequence, culture and noble values preservation arethe absolute obligation. Thus, to embody related obligation, the researchers believe that formal education plays important role. Because by such education, the noble values of culture may be integrated in students' courses.

One of courses that allows the integration of culture and its noble values is Mathematics. Specifically, the presence of ethnomathematics becomes the solution upon the challenge of cultural preservation and technology and arts mastery. D'Ambrosio [1] states that Ethnomathematics is a form stating the connection between culture and mathematics.

Therefore, further step to perform should be the exploration of Ethonamthematics in Bali as well as its mapping with school's mathematic concept. Since Ethnomatematics is a new concept or program in education world, there are only a few of works relating to Balinese culture containing such field. This is in line with Powell [2] who states that Ethnomathematics is a relatively new study in the form of coherent one.

A research conducted by Puspadewi [3] revealed that Ethnomathematics found in Balinese woven handicraft shows the use of tessellation in woven handicraft pattern. On the other side, Suryanatha [4] delivers that there are several mathematical concepts contained in Balinese culture, such as Least Common Multiple (LCM) in the calculation of *Otonan*, circular concept in the construction of *Tamas*, symmetrical concept in Balinese carving and food paper wrap, and modulo concept in newborn's name.

The other interesting concept as the result of exploration may be found in the concept of multiple linier regression applied by the architecture of Balinese traditional house. Bale *Saka Roras* is Balinese traditional building which applies the concept of multiple linier regression in its construction.

Several villages where its residents remain living in Bale *Saka Roras* are Songan, Pinggan, Belandingan and Terunyan. The four villages are geographically located in the Kaldera of Ancient Batur Mountain, in which most of the residents are the decendants of *Bali Mula*. Such villages are believed as remotely developed ones which start their development in the last few years.

According to the interview conducted by the researchers with the aging residents of Songan Village, the village's residents had to face the difficulty of obtaining equipment for house construction. On the other hand, most of Bale *Saka Roras* were constructed approximately 30-100 years ago. Based on such fact, the researchers aimed to investigate the accuracy of measurement by the architects of Bale *Saka Roras* in Songan Village.

2. Method

This research was conducted in March to April 2017 in Songan Village, Kintamani Sub-District, Bangli Regency, Bali by using qualitative as well as quantitative methods (mixed method). More importantly, Embedded Mixed Method was also used in this research [5].

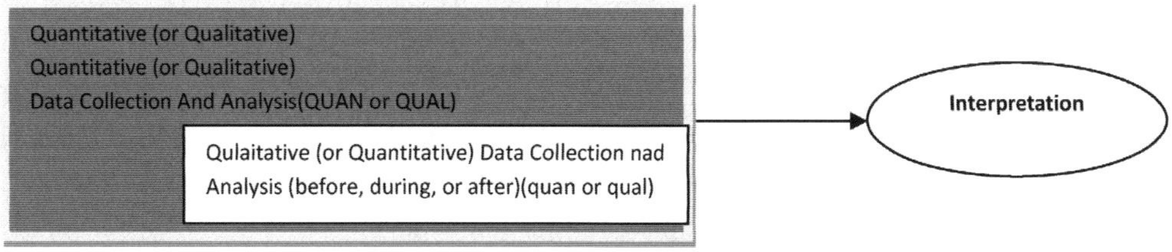

Figure 1. Design of Embedded Mixed Method.

Referring to such type, the qualitative or quantitative data was collected before collectively analyzed and interpreted by the researchers. In the stage of Qualitative, the researchers conducted an interview with two architects of traditional house and one figure who had particular knowledge in traditional house. The researchers then paired the information obtained from the interviewees with the proposed theories in several related literatures. The researchers also performed the same activity with the chosen residents of Bale Saka Roras or specific people who had comprehension in the history of such traditional building. Subsequently, the researchers quantitatively collected the data on the pillars' width and length by measurement. There were 32 Bale Saka Roras' pillars measured.

3. Result

Bale *Saka Roras* is Balinese traditional house. In its construction, the architects greatly considerd the aspect of harmony. Therefore, the division of spaces (*Mandala*) in its construction was inspired by the *Tri Hita Karana* that teaches about three factors of harmony. *Tri Hita Karana* is a commandment regulating the relationship between human and God, among humans, and between human and environment.

The dimension of time (*kala*) also needed to be considered in the construction of Balinese traditional house. Widana [6] writes that according to Balinese society's perspective, each time has the aspect of merit and demerit due to natural movement (Macrocosms). Such movement, in case of measuring traditional building in Bali, always refers to the measurement of the house owner's body part size, in the hope of the space compliance between the building and its owner [6].

As the foothold of the implementation of time dimension and the harmony of concept, the architects referred to the book/palm-leaf manuscript entitled AstaKosala-Kosali. AstaKosali is a name of palm-leaf manuscript describing the measurement of house construction, while *Asta Kosala* explains the measurement of tower or high building, media, bade, coffin [7]. By referring to regulations contained in AstaKosali, the measurement of Bale *Saka Roras* may be considered to such an extent, which depended on the body posture of man (head of family) who would live in the house. Bale *Saka Roras* is a house (*Bale*) with 12 (*Roras*) pillars (*saka*). The form of Bale *Saka Roras* may be described in the following figure :

Figure 2. Form of *Bale Saka Roras.*

One of interesting things on Bale *Saka Roras* wasthe relationship between its pillar measurement and Mathematics. The pillars measurement is extensively specified in *Asta Kosali*. Theoretically, the followings are several pillars measurement that may be applied in the construction of house or worship place [7].

1. Length of *21 rahi maurip anyari kacing*, namely *Bhatara Asih* (Main)
2. Length of *21 rahi maurip aguli madu*, namely *Bhatara Asih* (Medium)
3. Length of *20 rahi maurip aguli linjong*, namely *Prabu Nyakra Negara* (Medium)
4. Length of *19 rahi maurip anyari linjong*, namely *Mitra Asih* (Main)
5. Length of *19 rahi maurip anyari linjong*, namely *Istri Asih* (Medium)
6. Length of *25 rahi maurip anyari kacing*, namely *Prabu Murti Jinem*
7. Length of *23 rahi maurip anyari kacing*, namely *Tri Gegana*
8. Length of *23 rahi maurip aguli*, namely *Sang Hyang Gana Tunggal.*
9. Length of *20 rahi maurip aguli tujuh*, namely *Kusuma Mahadewi.*
10. Length of *19 rahi maurip acaping (atelek),* namely *Sang Hyang Nawagana.*

Several Balinese terms used are *rahi* (*a rai*), *a guli, a nyari* that are the measurements of house-owner candidate's body part, which is more popular as Anthropometry. The followings are the measurement of *rai, nyari,* and *guli* [8].

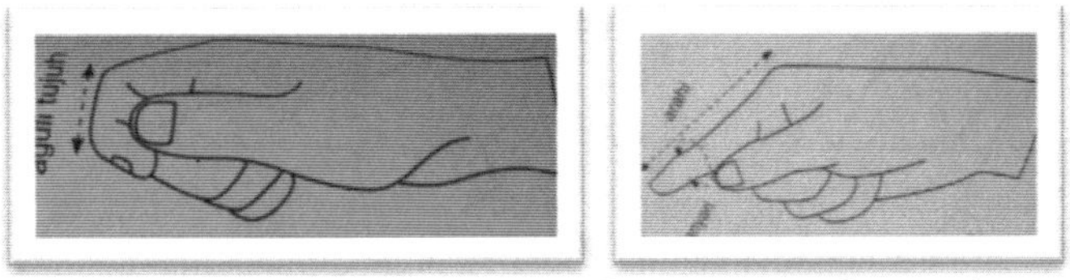

Figure 3. Measurement of *a nyari, a rahi, a guli.*

The other attraction that may be explored from the above mentioned measurement was the concept of Linear Regression. If there is one variable that acts as stimulus, such condition may be called as the general form of $Y = A + BX$ and as multiple linear regression when the stimulus variables are two or more. Two stimulus variables may be formed in $Y = A + B_1X_1 + B_2X_2$.

In relation with the specification of pillar measurement and *Bale Saka Roras* construction, the measurement of the house-owner candidate's body part acted as the variable. For example, "The Length of *19 Rai Maurip a nyari Linjong"* may be written in the following mathematic model:

$$Y = 0 + 19X_1 + 1X_2 \text{ atau } Y = 19X_1 + X_2 \quad (1)$$

with

Y = Length/Height of Pillar

X_1 = *a Rai* (Length of house-owner candidate's index finger)

X_2 =*a nyari linjong* (Width of house-owner candidate's middle finger)

Such function was only an example. There were actually more than 10 other functions with different coefficients and stimulus variables. Each function or specification had its own strength and weakness as well as relationship with the house-owner candidate's social stratification. Therefore, in spite of the same location, the researchers thought that it was possible that one's house had different measurement to another, in this case, the case study of the Bale *Saka Roras* measurement in Songan Village, Kintamanai Sub-District, Bangli Regency, Bali.

Songan Village is one of villages located in the inner of Ancient Batur Mountain Caldera. The residents in Songan Village are the mix of *Bali Mu*la and *Bali Arya* society. Such cultural assimilation allowed the existence of different perception in the measurement of traditional house construction. Besides, considering the sanctity of the measurement and its impact towards the life of the house-owner candidate or architect in-charge, the measurement must be performed under strict specification. This issue became the other factor causing the difference of measurement between one house to another.

The limitation of construction equipment such as saw, ax, measuring tool and pencil experienced by the architects in Songan Village in 1930s became the factor that inhibited the accuracy of measurement amounted up to 1 millimeter, eventhough the architects were accustomed to be highly accurate and persistent in constructing ideal houses.

In order to figure out the most-applied measurement by preceding architects in Songan Village in constructing Bale *Saka Roras*, the researchers measured 32 Bale *Saka Roras*. Specifically, the researchers measured the width and height of pillars. Since it was difficult to figure out the measurement of *pangurip* due to its unification with pillar's length/height, the researchers subtracted the measured pillar with 3 cm (average measurement of *a guli)* as the most-applied *pangurip.*

The combination of measurement method and interview with the resident or specific people having the knowledge related to the history of traditional house construction provided an advantageous information to the researchers. Based on the interview, the researchers found out that there were 4 house (*Bale*), in which in its construction process, did not fulfill the house-owner candidate's expectation and material availability. Thus, the researchers decided to exclude those houses from the data analysis. In other words, there were only 28 houses to analyze. From 28 houses, the researcher figured out the relationship between the pillar's width and height in the form of the following regression equation:

$$Y = 26,3 + 18,2X \quad (2)$$

With *Y* stated the pillar's height and *X* stated the pillar's width. By referring to the equation (2), the researchers conclude that most of preceding architects presumably used 19 as the coefficient towards the measurement result of house-owner candidate's body part.

4. Discussion

The researchers' inquisitiveness on the applied coefficients (17, 19, 20, 21, 23, or 25) was based on the numerous *Bali Mula* society's distinctive cultures. The relationship between the amount of coefficient and social stratification of the house owners also became the other significant reason. Therefore, the researchers believed that it was necessary to figure out the coefficient used by most of *Bale Saka Roras* in Songan Village. The finding on coefficient of 19 tended to refer to the geographic condition of Songan Village (in the mountain). This kind of location then caused them to construct shorter house' height since this may prevent them from the cold air. The use of *Saka Roras* as the main house by most of *Bali Mula* society in Songan Village indicated that they are not the King's descendants. This was in line with the genealogical information stating that their ancestors was *Mpu Kamareka* (a poet). Furthermore, the size of *a rai* in the pillars' width may be used to figure out the body anatomy of the *Bali Mula* society's ancestors, such as the body's height. Such height may be figured out by firstly finding the relationship between index finger's length (*a rai*) and body's height. The researchers may also predict the other body's anatomies such as head diameter and lips circumference. The former was predicted by measuring the width of front door, while the latter was by measuring the width of traditional stove. Thus, by figuring out the sizes of parts of inherited *Bale Saka Roras*, the researchers were possibly able to reveal the characteristics of the tribe's ancestors, at least the four generations above. This would be critical information, since the photographs of such traditional building inheritance are not available.

References

[1] D'Ambrosio and Ubiratan 2001 *Teaching Children Mathematics* **7** p 308
[2] Powell A B and Frankenstein M 1997 *Ethnomathematics: Challenging Eurocentrism in mathematics education* (New York: State University of New York Press)
[3] Puspadewi K R and Putra I 2014 *J. Math* **4** 80-89
[4] Suryanatha I P A and Apsari R A 2013 *Etnomatematika: Ketika Matematika Bernapas dalam Budaya* https://p4mriundiksha.wordpress.com/2013/11/10/etnomatematika/
[5] Cresswell J W 2014 *Research Design (Qualitative, quantitative, and mixed methods approaches) 4th Edition* (USA: SAGE Publications)
[6] Widana, I B G 2011 *Dharmaning Hasta Kosali : Arsitektur Tradisional Bali.* (Denpasar: Dharma Pura)
[7] Tonjaya I N G and Bandesa K 1982 *Lintasan Asta Kosali* (Denpasar: Ria)
[8] Pulasari J M 2007 *Cakepan Asta Kosala-Kosali, Uperenggeniya Lan Dharmaning Bhagawan Siswa Karma* (Surabaya: Paramita)

Data mining for dengue hemorrhagic fever (DHF) prediction with naive Bayes method

Ria Arafiyah[1] and Fariani Hermin[1]

[1]Computer Science Program, Faculty of Mathematics and Natural Sciences Jakarta State University, Jl. Rawamangun Muka East Jakarta Indonesia.

E-mail: riaarafiah@unj.ac.id

Abstract. Handling of infectious diseases is determined by the accuracy and speed of diagnosis. Government through the Regulation of the Minister of Health of the Republic of Indonesia No. 82 of 2014 on the Control of Communicable Diseases establishes Dengue Hemorrhagic Fever (DHF) has made DHF prevention a national priority. Various attempts were made to overcome this misdiagnosis. The treatment and diagnosis of DHF using ANFIS has result an application program that can decide whether a patient has dengue fever or not [1]. An expert system of dengue prevention by using ANFIS has predict the weather and the number of sufferers [2]. The large number of data on DHF often cannot affect a person in making decisions. The use of data mining method, able to build data base support in decision makers diagnose DHF disease [3]. This study predicts DHF with the method of Naive Bayes. Parameter of The input variable is the patient's medical data (temperature, spotting, bleeding, and tornuine test) and the output variable suffers from DBD or not while the system output is diagnosis of the patient suffering from DHF or not. Result of model test by using tools of Orange 3.4.5 obtained level of precision model is 77,3%.

1. Introduction

One of the problems of prevention and eradication of infectious diseases which until now is still a public health problem is Dengue Hemorrhagic Fever disease or better known as Dengue Hemorrhagic Fever (DBD). This dengue fever vector is Aides aegypti mosquito through repetitive bites to susceptiblepeople. Malaysia and Singapore have succeeded in establishing dengue-free due to the Government's attention to the issue of Environmental Health, by legitimizing environmental health issues in the form of regulations and sanctions for houses with mosquito larvae. While in Indonesia, since 1968 the spread of dengue disease has been increasingly wide spread and even it has always been an epidemic in some areas every year. Efforts to prevent the spread of dengue fever through mosquito nets eradication (PSN) and case control has not run optimally. It is because in the eradication, not only Aides aegypti mosquitoes are eradicated, but also it eradicates dengue virus carried by mosquitoes. Thus the emphasis of eradication is also directed at efforts to reduce the number of mosquitoes that can bring dengue virus by killing the larva. Meanwhile, to remove larvae has received less attention from the community because it is considered an effort that doesn't have clear result compared to fumigation [4].Dengue hemorrhagic fever (DHF) is still a public health problem and has both social and economic impacts. The survey results pointed out that public knowledge about BDB is lacking [5].Data mining techniques with various methods are widely used to predictdisease [6]. Diagnosing heart disease by using Naive Bayes algorithm [7][8] offers more than 86% of accuracy. [9] [10]It offers more than 79% of accuracy ina work to predict diabetes disease.This study aims to predict dengue fever with Naive Bayes method. It is used because it has advantages: It enhances the classification performance by eliminating the un-related features, It has a good performance and takes less computational time [11]. Parameter of the input variable is the patient's medical data (temperature, spot, bleeding, and tornuine test) and the output variable suffers from DHF or not.

2. Research Method

This research was conducted from April 2017 until November 2017. Field research was conducted in DKI Jakarta. Laboratory research was conducted in Computer Laboratory, Faculty of Mathematics and Natural Sciences of State University of Jakarta and Computer Science Department of Bogor Agricultural University. The research data is secondary data from research of M. Syafii in 2006. The sample of research was taken from medical record (medical record) hospital Dr. Sardjito Yogyakarta on 13-16 December 2005. The sample consists of data of patients suffering from dengue fever and dengue fever (DD) as many as 212 patients. According to international classification of diseases tenth revision (ICD 10) DHF disease is coded A.91 and DD disease with code A.90. Samples were taken from all hospitalized patients from January to November 2005.The requirements of the medical records taken as samples were in the medical record that there were 4 (four) notes on the clinical criteria: fever (heat), patches, spontaneous bleeding signs (nosebleeds, bleeding gums, bloody vomitus and black colored stools) and results tornikuet test also noted criteria laboratory results of blood platelet examination. DHF was used as a comparison in this study due to consideration of 4 clinical criteria and 1 laboratory criteria. Processing and data analysis were using software Orange 3.4.5.

3. Research Result

By using patient medical data input, that is; fever (temperature), presence of spotting, the presence of bleeding, and tornikuet test and output data, diagnose of the patient whether or not having dengue fever is made. Data processing using Naïve bayes method with tools Orange, obtained in figure 1.

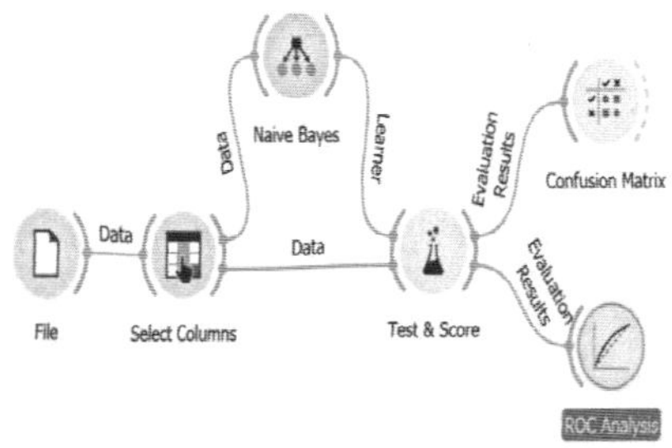

Figure 1. Widget data processing with Orange.

Figure 2. The results of the Naïve Bayes prediction test.

Classification Accuracy (CA) is the accurate percentage of data records classified correctly after testing on the classification results [11]. While precision or confidence is the proportion of positive predicted cases that are also positively correct on the actual data. Recall or sensitivity is the proportion of true positive cases that are correctly predicted positively [11]. The F1 score can be interpreted as a weighted average of the precision and recall, where an F1 score reaches its best value at 1 and the worst score at 0. The relative contribution of precision and recall to the F1 score are equal. Based on figure 2 Classification accuracy 0,693, precision 0,773 and Recall 0,708. The area under curve (AUC) is calculated to measure the performance difference of the method used. AUC is worth 0.677 or has an accuracy rate between 0.60-0.70, meaning poor classification [11]. F1 score is 0.739.The proportion between the predicted and actual class shows in confusion matrix.

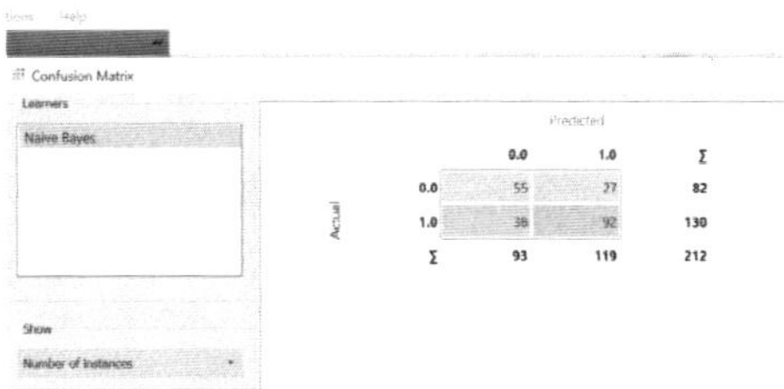

Figure 3. Confusion matrix.

From Figure 3.show model is able to predict false of 55 and true equal to 92. In other words accuracy prediction is equal to 69%. The performance of the classification Naïve Bayes algorithm in graphical form can be seen in the graph of Receiver Operating Characteristic (ROC) in figure 4:

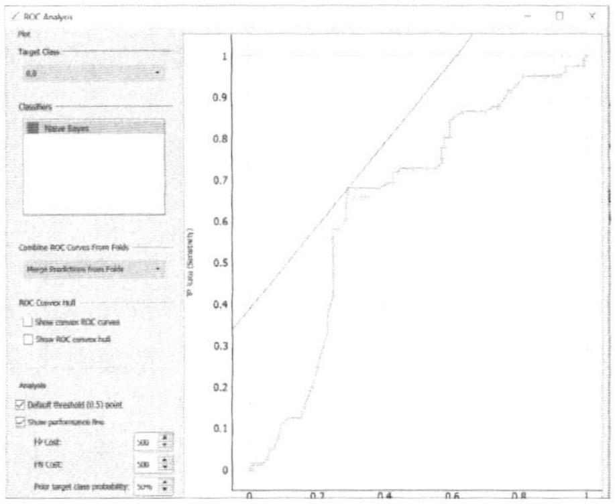

Figure 4. Performance of classification Naïve Bayes algorithm.

From figure 4, performance classification of Naïve Bayes algorithm can be seen from the shape of the green curve. The performance of the algorithm is bad, if the resulting curve approaches the baseline line or the line across from the 0.0 point (dotted line), and is good, if the curve approaches the point 0.1.

4. Conclusion and Recommendation

This research using patient medical input data, that is; fever (temperature), presence of spotting, the presence of bleeding, and tornikuet test and output data diagnose of the patient either having DHF or not. Precision model prediction uses Naïve bayes method with 77,3%. The Naïve Bayes classification method is included in the category of poor classification, so another method needed [12]. Naïve Bayes method can be optimized by using Particle Swarm Optimization (PSO) to improve performance in this case [13].

Acknowledgments
The authors would like to thank the directorate of research and community service of the ministry of research technology and higher education Directorate General of Higher Education for funding and supporting the Institutes of research and community service, Jakarta State University, Jakarta.

Reference

[1] Syafii M 2006 Adaptive Neuro Fuzzy Inference System (ANFIS) untuk Diagnosa dan Tatalaksana Penyakit Demam Berdarah Dengue. Tesis Ilmu Komputer IPB, Bogor.
[2] Arafiyah R 2015 The Expert SystemsPrevention Of Epidemic Dengue Hemorrhagic Fever Based Community Development *Proc. Inter. Sem. Universitas Mercubuana*, Yogyakarta.
[3] Lukito H 2007 Perumusan Pola Penyebaran Demam Berdarah Melalui Data Mining Pada Database Dinas Kesehatan DKI Jakarta Institut Pertanian Bogor
[4] Sumantri A 2008 Model Pencegahan Berbasis Lingkungan Terhadap Penyakit Demam Berdarah Dengue Propinsi DKI Jakarta, Disertasi Program Doktor Studi Pengelolaan Sumber Daya Alam dan Lingkungan Sekolah Pascasarjana IPB Bogor
[5] Soeparmanto P 2000 Peningkatan Penanggulangan Demam Berdarah Dengue (DBD) Berbasis Masyarakat dengan Pendekatan Pendidikan Kesehatan,Center for Research and Development of Health Services and Technology, NIHRD
[6] Fatima, Meherwar and Pasha M 2017 Survey of Machine Learning Algorthms for Disease Diagnostic *J. Intel. Learn. Systems and Applications*
[7] Vembandasamy K, Sasipriya R and Deepa E 2015 Heart Diseases DetectionUsing Naive Bayes Algorithm *Inter. J. Innov. Sci. Eng. & Tech.* **2** 441-4
[8] Chaurasia V and Pal S 2013 Data Mining Approach to Detect Heart Disease *Inter. J. Adv. Com. Sci. and Infor. Tech.* **2** 56-66
[9] Iyer A, Jeyalatha S and Sumbaly R 2015 Diagnosis of Diabetes Using Classification Mining Techniques *Inter. J. Data Mining & Knowledge Management Process* 5 1-14
[10] Sarwar A and Sharma V 2012 Intelligent Naïve Bayes Approach to DiagnoseDiabetes Type-2 *Inter. J. Com. App.* **3** 14-16
[11] Han J and Kamber M 2006 Data *Mining Concepts And Techniques 2nd Edition* (San Fransisco:: Elsevier)
[12] Gorunescu F 2011 *Data Mining Concepts, Models and Techniques* (Berlin: Springer)
[13] Hu X, Eberhart R C and Shi Y 2003 Swarm intelligence for permutation optimization: a case study of n-queens problem. IEEE swarm intelligence symposium (Indianapolis: IEEE)

IOP Conf. Series: Journal of Physics: Conf. Series **948** (2018) 012078 doi:10.1088/1742-6596/948/1/012078

New method in muon-hadron absorption on $Th_x DUO_2$ nano material structure at 561 MHz quantum gyro-magnetic

M Hardiyanto[1,2] and I R Ermawaty[3]

[1]Industrial Engineering Department, Institut Teknologi Indonesia
[2]Betha Group LHC– CERN, Lyon, France
[3]Physics Educational Department, Universitas Muhammadiyah Prof. Dr. HAMKA, Jakarta Indonesia

E-mail: moh_hardiyanto_iti@yahoo.com

Abstract. We present an experimental of muan-hadron tunneling chain investigation with new methods of Th_x DUO_2nano structure based on Josephson's tunneling and Abrikosov-Balseiro-Russel (ABR) formulation with quantum quadrupole interacting with a strongly localized high gyro-magnetic optical field as encountered in high-resolution near-field optical microscopy for 1.2 nano meter lambda-function. The strong gradients of these localized gyro-magnetic fields suggest that higher-order multipolar interactions will affect the standard magnetic quadrupole transition rates in 1.8 x 10^3 currie/mm fuel energy in nuclear moderator pool and selection rules with quatum dot. For muan-hadron absorption in Josephson's tunnelling quantum quadrupole in the strong confinement limit we calculated the inter band of gyro-magnetic quadrupole absorption rate and the associated selection rules. Founded that the magnetic quadrupole absorption rate is comparable with the absorption rate calculated in the gyro-magneticdipole approximation of Th_xDUO_2nano material structure. This implies that near-field optical techniques can extend the range of spectroscopic measurements for 545 MHz at quantum gyro-magnetic field until 561 MHz deployment quantum field at $\vec{B}$ around 455-485 tesla beyond the standard dipole approximation. However, we also show that spatial resolution could be improved by the selective excitation of ABR formulation in quantum quadrupole transitions.

1. Introduction

Quantum entanglement between physically separated systems is a resource that is essential for implementing many quantum information processing. The vast majority of entangled muon-hadron sources that are in use in various laboratories around the world today rely on spontaneous parametric down conversion in χ crystals to find out of new fuel energy on nuclear ordered chain reaction. However, our research recently demonstrated new methods of source with Abrikosov-Balseiro-Russell (ABR) formalism in non-Abellian system and Josephson's tunneling [1] such as

$$\Big| i \cong \sum_{\alpha} x_{\alpha}^{i} \, | \alpha >,$$

which muon-hadron pairs were generated through non degenerate four-muon scattering. In ABR formalism, the parametric χ has resulted by a few equations consists of

IOP Conf. Series: Journal of Physics: Conf. Series **948** (2018) 012078 doi:10.1088/1742-6596/948/1/012078

- $\sum_{j}^{occ} \sum_{\beta\gamma\delta} \chi_{\beta}^{i*} \chi_{\gamma}^{j} (\mu\beta|V|\gamma\delta) < \chi|CV|\gamma \cong \sum_{\beta\delta} (\mu\chi\beta|V|\gamma\delta)\rho_{\delta\beta}$
- $< \Phi|H\chi|\Phi^{\uparrow} > = \chi^{**}$

For fast neutron floating at 1.8 x 10^3currie/mm, the microstructure in Dirac's condition was symbolized by $\Phi$and if not at Abellian-system was denoted by $\Phi^{\uparrow}$. This condition including of Schwinger transmutation by χ^{*} and the invariant denoted by $\chi^{\downarrow}$ for non-Abellian system in quantum quadrupole for 455 tesla gyro-magnetic on ThxDUO2 nano material structure [3]. The design of the experiment with Catch-Nuc and Interstellar Nuclear Beam equipments [2] to produce correlated photon-pairs in dispersion-shifted Thx nano structure near 1.1 nano Lambda-function.

Nano structures interacting with optical near fields do not necessarily behave in the same way as nanostructures interacting with far-field radiation the response of a quantum well when it is excited by the diffracted field of an aperture causes the enhancement of quadrupole transitions, giving rise to a modified absorption spectrum of the quantum well.

In this paper we focus on the interaction of a spherical of Thx DUO2 nano structure with quantum quadrupole a highly confined gyro-magnetic optical near field. It was shown that such fields could be generated near laser-illuminated sharply pointed tips.

1.1. Quantum Quadrupole Wave Functions (Strong Confinement)

We assume that a spherical quantum dot is made from a direct band gap conductor for which the bulk gyro-magnetic dipole transitions are allowed between the valence band and the conduction band. In a generic manner, we assume that the valence band has a p-like character and that the conduction band has an s-like character. The latter assumption is commonly encountered for several conductors such as $Th_xO_{1.7}$. Adopt this geometry and approximate the fields near the tip by an oscillating magnetic dipole oriented along the tip axis. In the research reported it was demonstrated that this is a reasonable approximation and that the dipole moment can be related to the computationally determined field enhancement factor using by Volkov's detector and Multi Channel Spectroscopy Nuclear Beam at LHC-CERN nuclear reactor [4].

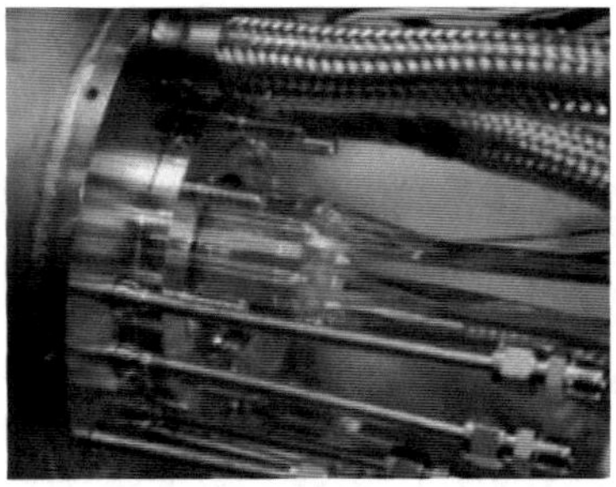

Figure 1. Volkov's detector

(Courtesy of Betha Group LHC-CERN nuclear reactor, Lyon, France, 2016)

Figure 2. Multi Channel Spectroscopy Nuclear Beam (Courtesy of Betha Group LHC-CERN nuclear reactor, Lyon, France, 2016)

The two fundamental differences between the work reported here and the earlier experiments are the wavelength of operation at 749 nm versus 1537 nm, and the use of a linear configuration as opposed to a Sagnac-loop configuration. After floating of 1.8 x 10^3currie/mm muon-hadron scattering in nuclear chamber, the barrier from Th_x has resulted to be hold of anti-neutrino particle existing. This is a good condition for finding out of new fuel energy in nuclear reactor before we continued to Josephson's tunneling condition after the muon-hadron floating, and based on a few equations in ABR formalism, the quantum-correlations approaching will be an adjustment the critical value of pump wavelength by Th_x DUO_2nano-structure [5].

The quantum theory of four-wave interactions in Th_x nano-structure has developed by ABR formalism with very simplicity equations such as:

$$\frac{\partial A_1}{\partial z} = -\frac{\alpha}{2} A_1 + i\chi\left(\left|A_1\right|^2 A_1\right)$$

$$\frac{\partial A_2}{\partial z} = -\frac{\alpha}{2} A_2 + i\chi\left[2\left|A_1\right|^2 A_2 + A_3^{i\Delta k z}\right]$$

In these equations, the field amplitudes of the pump signal and idler waves, respectively, and α is the attenuation coefficient of the Th_xnano-structure according to $P(z)=P(0)e^{-\alpha z}$ where P is power and z is propagation distance.

1.2 ABR Formalin Model

The analysis described above applies equally well to four muon-hadron interactions in any type of Th_xnano material matrix of nuclear structure. The advantages of using quantum correlations for demonstrating nonlinear-Th_x nuclear structure effects arise from several novel properties: the nonlinear coefficient is enhanced in small-core a few μm^2 which support a single transverse mode over an extremely broad wavelength range (370 nm – 1600 nm). These four properties combine to allow efficient interactions to occur FPS which are either much less efficient or not possible at all in standard Th_xnano-structure nuclear.

Table 1. Various properties of the $Th_x DUO_2$nano material structure in FPS

Th_x properties	Value
Length	1.36μm
Core diameter	1.89μm
Attenuation	92.4 dB/nm
Cutoff wavelength	< 571μm
λ_0	Mode A : 749 ± 4.1μm Mode B : 743 ± 4.1μm
D_{slope}	Mode A : 0.64 ps Mode B : 0.68 ps
γ	97.5 (W nm)$^{-1}$

Source : courtesy of Betha Group LHC-CERN nuclear reactor, Lyon, France, 2016

Strong confinement is achieved if the Bohr radii of electron b_e and b_h hole are much larger than the radius of the quantum quadrupole. We assume the mentioned conditions in gyro-magnetic for 545 MHz to 561 MHz quantum gyro-magneticcan express the wave function band as

- $$\Psi^E(r) = \frac{1}{\sqrt{V_0}} u_{c,0}(r)\xi^e(r)$$

Here $u_{c,0}(r)$ is the conduction band Bloch function (with lattice periodicity) that has the corresponding eigenvalue $k = 0$, and V_o is the volume of the unit cell.

2. Methodology

2.1. *Muon-Hadron Quadropole in Josephson's Tunneling*

The gyro-magnetic quadrupole interaction on muon-hadron for Josephson's tunneling $\hat{H}^Q$ can be represented as

- $\hat{H}^Q = \hat{\Psi}^+(r)H^Q(r)\hat{\Psi}(r)d^3r$

The interband terms are found by substitution of Eq. (6) and its adjoint into Eq. (7), thus (Levy, V., et. all, 2015) The group delay is then plotted as a function of wavelength and fitted to a low-order polynomial up to χ^2 or χ^3. The first derivative of the group-delay curve, normalized to the length of the Th_x microstructure under test, the gives the parameter GVD, *D*. The Dirac's condition will be supporting to existing of twin-photons at Abellian-system.

2.2. *A subsectionAbsorption Rates in Th_xDUO Nano Structure*

To compare the magnetic dipole and the magnetic quadrupole absorption rates in strongly confined optical fields consider a quantum dot in the vicinity of a laser-illuminated metal tip. The strongest light confinement is achieved when the metal tip is irradiated with light polarized along the tip's axis. For this situation.

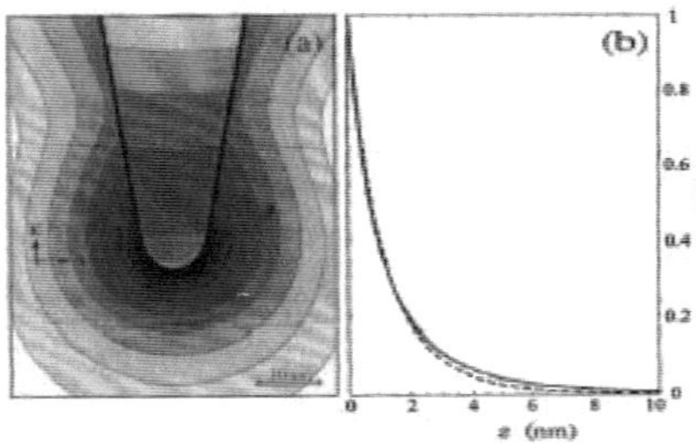

Figure 3. illustrates the comparison of Thx DUO2 in magnetic dipole and magnetic quadrupole (Courtesy of Betha Group LHC-CERN nuclear reactor, Lyon, France, 2016)

2.3. *Discussion of the ThxDUO$_2$ Nano Material Structure with Quantum Quadrupole*

We analyze absorption rates for quantum dots with the two different radii, $a = 5$ nm and $a = 10$ nm. For $a = 5$ nm the magnetic quadrupole transition is excited at a wavelength of $\lambda \approx 500$ nm; the magnetic dipole transition, at $\lambda \approx 550$nm. The quadrupole transition for a quantum gyro-magnetic of radius $a = 10$ nm occurs at $\lambda \approx 615$ nm, and the magnetic quadrupole transition at $\lambda \approx 630$ nm.

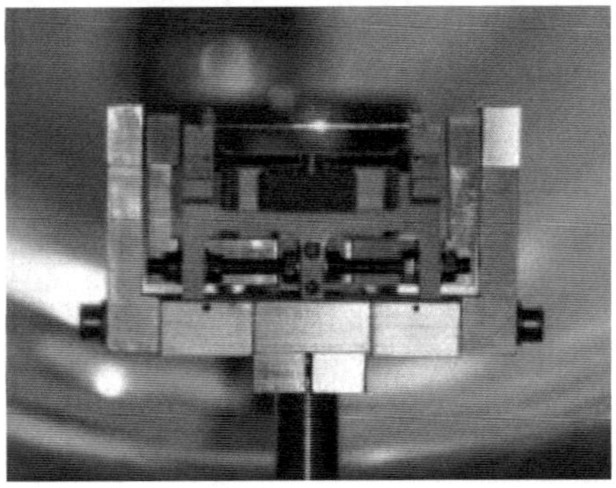

Figure 4. The Th_x DUO_2 nano structure at 489.3teslasgyro-magnetic field after blasting by muon hadron absorption in 1.8 x 10^3currie/mm(Courtesy of Betha Group LHC-CERN nuclear reactor, Lyon, France, 2016)

According to quantum quadrupole magnetic-spin and magnetic resonance, which are expressed by Equations (3) until (6) that was describing how strongest impact from muon-hadron bombarding to Fermi-Dirac interstellar area wide in Josephson's tunneling. However, using by ABR formalism for Th_xDUO_2nano material structure in 1.8 x 10^3currie/mm the wide is 0.001127 x 10^{-10} meters having magnetic field deflection on the resonance of spin-rotate around 6772.55 cm per each Fermi's cloud active reaction. This matter in 510 MHz up to 545 MHz magnetic-spin frequency with Na_2SO_3 liquid moderator's water-cooling at reactor chamber. For showing up the impact and existing of Fermi-Dirac's surface effect caused by Anderson's tunnel for floating of thermal neutron electrical charge in quantum quadrupole magnetic-spin also resonance, it has figured out from Gell-Mann spectroscopy.

The Volkov's detector is specified to have 10 nm pass bands centered at either 740 nm or 760 nm. The pass band center of these it can be tuned in wavelength by slightly rotating them such that the incident wave is no longer normal to the surface.

Table 2. The properties of quantum magnetic field states for Th_xDUO_2nano material.

Materials	Quantum magnetic-spin value	Quantum states
	7.2104/eV	$C^F_{kij\uparrow};E_{(FD,J)}$
Th_x	7.8384/eV	$C^F_{kij\uparrow};E_{(FD,P,J)}$
	7.8551/eV	$C^F_{kij\uparrow};E_{(FD,J)}$

Source: courtesy of Betha Group LHC-CERN nuclear reactor, Lyon, France, 2016

The fitting parameter $a_1 = 0.027$ counts/pulse is the total photon detection efficiency, which agrees within the margin of error with the independently measured value of 4 $\pm$ 1%. The fitting parameter $b_1 = 8.7x10^{-9}$ counts/muon-hadron is the leakage through the detection filter and/or the contribution of spontaneous Raman scattering. In this experiment, FPS, Josephson's scattering, and linear loss are all of the same order of magnitude. Based on the formulation for Fermi-Dirac's active cloud in coherent Compton's wavelength, will be adjoined of critical modulation phonon's vibration and Cooper's pair cloud such as.

Figur 5. Th_xDUO_2nano materials structure was scattering material where one counter is used to trigger the acquisitionand the order is taken to be the PCMs signal at 561 MHz quantum gyro-magnetic (Courtesy of Betha Group LHC-CERN nuclear reactor, Lyon, France, 2016)

For efficient detection of the signal-idler pairs, pump muon-hadron suppression by greater than 90 dB is required. We obtain this suppression by use of three diffraction gratings (Richardson Grating Laboratories, 600 lines/nm, nominally blazed for 800 nm) arranged so that two different spectral regions can be detected on separate photon-counting modules (PCMs). The PCMs are available devices (Perkin-Elmer, model SPCM-AQR-16), which count individual photons at 750 nm wavelength with a quantum efficiency of $\cong$ 75% for new fuel energy in nuclear ordered chain reaction especially on Th_2DUO_2nano material with Josephson's tunnelling and ABR formalism according for new method in nuclear reactor.

3. Conclusions

We have analyzed higher-order multipolar muon-hadron absorption interactions between Th_x DUO_2nano structure with quantum quadrupole and a strongly confined gyro-magnetic optical field. Expressions have been derived for the quadrupole interaction Hamiltonian, the associated absorption rate, and selection rules. The magnetic quadrupole absorption strength depends on the bulk properties of the material (ABR formalism) as well as on the envelope functions (confinement functions).

When the quantum quadrupole with radius a interacts with the confined optical field produced by a sharply pointed tip, the ratio between the magnetic quadrupole absorption rate and the electric dipole absorption rate can be as high as 561 MHz frequency for a = 10 nm. Magnetic quadrupole transitions could be ignored in the extreme near field. This was done to separate between tip and quantum dot smaller than muon-hadron flux around 1.8 x 10^3currie/mm and 485-tesla gyro-magnetic field for Josephson's tunneling. Moreover, the values were used to count individual photons at 750 nm wavelength with a quantum efficiency of $\cong$ 75% for new fuel energy in nuclear ordered chain reaction especially on Th_2DUO_2nano material with Josephson's tunneling and ABR formalism according for new methods in nuclear reactor.

Acknowledgments
We thanks to Thomas Kuntz, Bernard Cambou and Helena Duprix for valuable input to this study. This research was supported by the Betha Group LHC-CERN Deuterium-Uranium Nuclear Reactor, Lyon, France through project DE-FG45-01ER18245 and work done at LHC-CERN, Lyon, France also Max Planck Institute, Deyna, Damstadt, Germany.

References

[1] Abrikosov, Elsaesser T, Kuntz T and Russel 2016 *Ultra Microscopy Inter. J.* **166** 405–412
[2] Knorr A, Koch S W and Helena Duprix 2015 *Opt. Commun.* **213** 167–178
[3] von der Heydt A, Hanewinkel B and Duprix H 2015 *J. Am. Phys.* **162** 9831–9838
[4] Hanewinkel B, Knorr A and Koch S W 2016 *Phys. Rev.* **55** 715–725
[5] Chavez and Duprix H 2015 *Appl. Phys. Lett.* **74** 1507–1509